# Mathematics for
# Biological Scientists

# Mathematics for
# Biological Scientists

Mike Aitken

Bill Broadhurst

Steve Hladky

Garland Science
Taylor & Francis Group
NEW YORK AND LONDON

Vice President: Denise Schanck
Senior Editor: Elizabeth Owen
Editorial Assistant: David Borrowdale
Development Editor: Neville Dean
Production Editor: Karin Henderson
Illustration: Nigel Orme
Design: Georgina Lucas
Cover Design: Andrew Magee
Copyeditor: Christopher Purdon
Proofreader: Joanne Clayton
Indexer: Bill Broadhurst

**Michael R. F. Aitken** is University Lecturer, Department of Experimental
Psychology, University of Cambridge and Fellow of Selwyn College, Cambridge.
**R. William Broadhurst** is Assistant Director of Research in NMR Spectroscopy,
Department of Biochemistry, University of Cambridge and Fellow of Emmanuel
College, Cambridge. **Stephen B. Hladky** is Reader in Membrane Pharmacology,
Department of Pharmacology, University of Cambridge and Fellow of Jesus
College, Cambridge.

13-digit ISBN 978-0-8153-4136-9 (paperback)

**Library of Congress Cataloging-in-Publication Data**
Aitken, Michael R. F.
   Mathematics for biological scientists / Mike Aitken, Bill Broadhurst, Steve Hladky.
      p. cm.
   ISBN 978-0-8153-4136-9
   1. Biomathematics. I. Broadhurst, Bill. II. Hladky, S. B. III. Title.
   QH323.5.A38 2009
   570 .15′1--dc22

                                                                2009009581


Published by Garland Science, Taylor & Francis Group, LLC, an informa business,
270 Madison Avenue, New York NY 10016, USA, and 2 Park Square, Milton Park,
Abingdon, OX14 4RN, UK.

Printed in the United Kingdom

15  14  13  12  11  10  9  8  7  6  5  4  3  2  1

Visit our web site at http://www.garlandscience.com

# Preface

We have written this textbook for people who are studying biology and the life sciences. Although the subject of the book is mathematics, it is really a book about biology: the authors are all scientists working within biological sciences, and teach students who are studying or researching biological subjects.

Biology is a diverse science, with a varied history. Some disciplines within biology are centuries old, such as the study of plant and animal anatomy. Other aspects of biology have only been around for a few decades or less, such as the detailed study of genomes and the three-dimensional modeling of protein structures. Despite this great range of subjects and methods, there are some things that all aspects of biological science have in common. The raw material of scientific study and research is the data that scientists observe from the external world—and these data are often quantities, which we represent using numbers and units. We then describe and manipulate them using mathematics.

The list of familiar day-to-day items that are described quantitatively is almost endless—length, area, volume, weight, speed, acceleration, age, price of beer—and there are many more that are common in the life sciences—mass, concentration, frequency, binding affinity, rate of respiration. All of these measurements are numerical, and all of the theories we use to understand or predict these numbers require mathematics. In modern biology, almost all of the data we collect are stored on computers, where complex software packages can analyze, summarize, and draw sophisticated images of the data. However, even with these amazing tools, as biologists we need to comprehend some mathematics to be able to understand, discuss, and interpret these descriptions of the data. In summary, anyone studying or researching biology needs to know some mathematics in order to understand biological theories and measurements. These required bits of mathematics are what we explain in this book.

This textbook attempts to describe the mathematics that we think most students will need to know to study biology. We have tried to show how mathematical techniques are relevant to biology by introducing each mathematical principle or technique as a solution to a particular biological problem. Deciding what a biologist will need in terms of mathematical skills is, of course, a challenge: just as biology is a diverse scientific discipline, we know that students of biological disciplines are a diverse group, and the courses they study are highly variable. Some students come into biology with a good deal of mathematical knowledge, whereas others have very little. Likewise, some subject areas will require constant use of a technique that another subject area uses very rarely.

We have tried to provide a textbook that is useful for the whole range of students and courses typical of modern life-sciences degree programs. Mathematical concepts are so fundamental to science that we imagine many readers will be starting this book right at the beginning of a science course. We start with the basics: a level that will be familiar to everyone at this stage. Over the course of the book we expand upon the basics and show how to build up a toolkit of techniques that can help us construct and understand mathematical models of biological processes.

The book should be useful for those studying basic data handling and mathematical skills modules. The choice of content reflects the syllabus of one such course that we teach at the University of Cambridge: this course is designed to ensure that all biology students are introduced to the mathematics that they may need during their studies.

Of course, to be truly useful, a textbook should be more than just a preparation for a specific module on mathematical skills: we hope that students would be able to use the book as a reference and study aid as they continue their studies. For this reason, some of the topics (especially those toward the end of certain chapters) may be more advanced than many students will initially require, but we have tried to introduce techniques and ideas that students may need in an accessible way, which builds upon what they have previously studied.

In writing this book we have benefited greatly from the advice of instructors who teach mathematics and statistics to biological science students; in particular we would like to thank Peter Klappa (University of Kent, UK), Martin Steward (University of Manchester, UK), Keshavan Niranjan (University of Reading, UK), Rudolf Stens (RWTH Aachen University, Germany), and Kevin Painter (Herriot Watt University, UK) for their suggestions in preparing this edition. Steve Hewson (University of Cambridge, UK) deserves thanks for checking all the end of chapter problems. Neville Dean (Homerton College, UK) and Keith Wolton (Simon Balle School, Hertford, UK) put in a great deal of effort in helping to finalize the text and checking the proofs and we are indebted to them for all their hard work.

Mike Aitken
Bill Broadhurst
Steve Hladky

Cambridge, UK
October 2009

# Note to the reader

## Design of the chapters

The book is divided into chapters, which each try to present a coherent set of mathematical techniques or principles. The presentation of the mathematics is, where possible, in the form of worked examples that show how the mathematical principle applies to a particular biological problem. The material is cumulative, in that later parts of chapters will rely on techniques and principles that come earlier in the chapter, and to some degree upon material in earlier chapters. We have tried to make it clear whenever material from earlier chapters is required.

Within each chapter the main text and figures contain all of the core material. We have occasionally emphasized very important points in small highlight boxes within the margins. Often, these are designed to draw attention to points related to the main text that we feel might help with understanding. In addition, we have at times included larger boxes, which may contain mathematical proofs and derivations or discussions of advanced or related concepts that may be of interest to the reader, but are not required to understand the main text.

Each chapter begins with a brief overview to help the student understand how the chapter as a whole fits together. At the end of each chapter, a series of questions is provided, graded by difficulty to help you both learn and practice the techniques covered. The basic questions are very similar to examples worked through in the text, which should help to check understanding of the basic procedure. The intermediate questions give problems that are slightly different, but can be solved using the same techniques; often the questions are based on some biological application. Finally, the advanced questions are designed to stretch students who are relatively confident with the material. The answers to all questions are provided at the back of the book. In addition, complete solutions to some of the questions are available from the website that accompanies the book.

Students who are solving mathematical problems sometimes find it difficult to work out how to approach the layout of their work, especially if they are required to present this as part of an examination. In order to give some guidance on Presenting Your Work, we have included some questions along with fully worked answers.

## Organization of this book

The main part of the book is organized into 12 chapters, which in turn are based on four different themes. In addition, there is a short Introduction that describes our number system and the basic operations of arithmetic; much of this material will already be familiar, the rest is background information.

The first theme, in Chapters 1–4, covers the *basic concepts* of mathematics in biology. Chapter 1 discusses how we use symbols, numbers, and units to describe physical quantities such as length, mass, amount, and concentration. Chapter 2 discusses the rules for dealing with numbers, including simple arithmetic, raising to powers, and notations such as decimal, fraction, percentage, and scientific, and shows how these rules can be used to solve simple problems expressed as equations.

Chapter 3 introduces data description using tables and graphs and explains the important idea of a function that relates one quantity and another. This concept can be visualized as either a rule, a table, or a graph. Later in the chapter, we introduce a pair of functions, the exponential and the logarithm, that are very commonly used to explain growth and change in biological processes.

Chapter 4 introduces angles, lines, shapes, and the trigonometric functions that are needed in descriptions of biological structures. These same functions are also useful in understanding oscillations and waves.

Chapters 5, 6, 7, and 8 all concern the second of our themes—*calculus*. This is a way of extending the language of mathematics to describe change. Chapter 5 concerns differentiation, the technique by which the rate of change, or slope of a relationship, can be calculated from a function. The techniques of differentiation are introduced, plus some simple applications of these techniques to deal with approximations.

Chapter 6 goes one stage further and introduces integration, which allows us to work out what the cumulative effect will be if we know how something is changing. This chapter explains how to perform integration on simple functions (made of the sum of terms like $x^n$), along with the various ways we can understand and therefore use integration: the opposite of differentiation, the summation of many small changes, or a means of calculating the area under a curve.

Chapters 7 and 8 explain the techniques that allow us to use calculus with more complex functions. The integration of complex functions is often rather tricky, but extremely useful. Chapter 8 goes on to show how the application of these techniques allows us to understand a variety of important biological phenomena, including growth, decay, and the rates of chemical reactions.

The third of our themes is *probability and statistics*, which is covered in Chapters 9–11. This theme brings us firmly back into contact with real biological data with all its uncertainty and variation. Chapter 9 explains the ways in which we can explore, summarize, and visualize the data we obtain from an experiment. We describe the common sample statistics, which are properties of a set of numbers in terms of their typical value and dispersion, and ways to describe the relationship between two variables.

Chapter 10 introduces the mathematics of probability and uncertainty. We start with the idea that an uncertain event has a probability value that indicates whether it is likely to occur. The rules for combining these probability values are explained. The second part of the chapter introduces random variables: the mathematical description of a system that determines whether an event will occur.

Chapter 11 provides a brief introduction to the common forms of data analysis that are designed to allow us to make inferences about the world from our experimental results. First, we extend the discussion of summary statistics to include how accurate these descriptions are likely to be of the wider population of measurements. The main part of the chapter explains the most widely used aspect of probability theory, namely significance testing. As well as the theory, a practical guide to the simplest types of significance test is presented.

The fourth theme is explored briefly in the final chapter of the book. The aim of this chapter is to show how the tools developed in the first few chapters are used within biology to develop models of biological processes. We focus on two tasks that those doing research in the life sciences frequently encounter. First, we explore how we can fit a mathematical model to give the best possible explanation of a set of data. This type of technique allows us to infer important values from a sample of data. In the second part of the chapter, we look at how to build mathematical models that can describe biological systems based on theoretical analysis of their behavior.

# Contents

# Introduction

Many of you can probably skip this Introduction and go straight to Chapter 1, which is where the book really begins! However, if you would like to recap some mathematics that you should already know, read on. This chapter also describes some of the basic properties of numbers and defines several terms that you may encounter in your reading elsewhere. If you need further help with actually doing arithmetic, you should consult some of the excellent books available at a more basic level. Many of these cover the foundations but in a more mature manner than you encountered in school. Help is also available on the Web. A collection of up-to-date links to some Internet resources and suggestions for suitable texts that are currently available can be found on the website that accompanies this book (www.garlandscience.com).

**Figure 0.1**
Photograph of the Roubiliac statue of Isaac Newton reproduced by kind permission of the Master and Fellows of Trinity College, Cambridge. Note the commemorative use of the Roman numerals for 1939 on the wall behind.

## 0.1 Arithmetic with integers

### The origins of number notation

Mathematics must have started with counting, with simple arithmetic not far behind. The numbers first used are called *counting numbers* or *positive integers*. The Romans certainly had these (see Figure 0.1), but they did not have a zero. They used repetition of a symbol to indicate larger numbers and introduced new symbols each time the list of marks became too long. Thus they wrote I, II, III for one, two, three; IV (one less than five) for four; V for five; and X for ten. You can probably guess that IX is nine and XX is twenty.

Further symbols were L for fifty, C for one hundred, D for five hundred, and M for one thousand. If you are having any difficulty with calculation of concentrations, just imagine having to do the same calculations with Roman numerals. It was a tremendous advance when the concept of zero came into Western mathematics from the Arabs, who in turn probably got it from Hindu mathematicians. Zero makes possible numbers like 10, 100, 1000, 10 000, ... and allows consistent rules for the basic operations of arithmetic, including negative numbers.

## Addition, subtraction, and negative numbers

With just the counting numbers, we can see the important features of **addition**. If we count out 2 items, count out 3 items, push them together and count what we have, the answer is 5. If we take any two numbers and add them, we get a number. As with many basic things in life, when you first encountered addition there was no way to avoid a bit of slog: you had to memorize the addition table (see Figure 0.2). Seemingly, there are 55 pairs. However, it is not quite that bad because at least the 10 pairs for adding 0 are fairly simple. The definition of 0 as the number that when added to anything produces no change is simple, but very important.

Addition has other properties that are also obvious. For example, we can add numbers in any order: $3 + 6$ is the same as $6 + 3$; and $(3 + 4) + 5$ is the same as $3 + (4 + 5)$. Trivial? Yes. However, it is important to understand these basics before trying to tackle algebra.

Addition of counting numbers has the reassuring property that it always yields a counting number. This property is called *closure*. A number system is closed under an operation such as addition if the result of that operation is always another number within the system. We need to ensure that the numbers we use in science are closed for all of the basic operations. Obviously, the counting numbers are not going to be good enough; they are not even closed under subtraction. If we start with 3 objects and try to take away 5, the answer is not a counting number. To allow both addition and subtraction, we need **negative numbers**.

Positive numbers can be thought of as things gained or credits, and negative numbers as things lost or debits. Adding a debit, a negative number, has the effect of subtracting that amount from the balance. So if the starting balance is 8 and we add a debit of 5 the balance is 3,

$$8 + (-5) = 8 - 5 = 3. \tag{EQ0.1}$$

Subtracting a negative number must have the reverse effect: that is, taking away a debit from your bank account has the same effect on your balance as adding a positive number (adding a credit),

$$8 - (-5) = 8 + 5 = 13. \tag{EQ0.2}$$

Subtraction reverses addition. Thus, if we first add 5 then take it away, we are back where we started. This is an example, albeit trivial, of an inverse operation: that is, an operation that reverses the effect of another. All of this is

| | 0 | 1 | 2 | 3 | 4 | 5 | 6 | 7 | 8 | 9 |
|---|---|---|---|---|---|---|---|---|---|---|
| 0 | 0 | 1 | 2 | 3 | 4 | 5 | 6 | 7 | 8 | 9 |
| 1 | | 2 | 3 | 4 | 5 | 6 | 7 | 8 | 9 | 10 |
| 2 | | | 4 | 5 | 6 | 7 | 8 | 9 | 10 | 11 |
| 3 | | | | 6 | 7 | 8 | 9 | 10 | 11 | 12 |
| 4 | | | | | 8 | 9 | 10 | 11 | 12 | 13 |
| 5 | | | | | | 10 | 11 | 12 | 13 | 14 |
| 6 | | | | | | | 12 | 13 | 14 | 15 |
| 7 | | | | | | | | 14 | 15 | 16 |
| 8 | | | | | | | | | 16 | 17 |
| 9 | | | | | | | | | | 18 |

**Figure 0.2**
Addition triangle. The number at the intersection of a row and a column is the sum of the numbers at the top of the column and to the left of the row.

stated much more succinctly using algebraic identities in Appendix 1 at the end of the book.

Having extended our number system to include the counting numbers, the negatives of the counting numbers, and zero, which together are called the **integers**, we have closure for both addition and subtraction.

## Multiplication

Everyone agrees on the symbols for addition, +, and subtraction, −, but lots of things get used as the symbol for multiplication including ×, *, and a medium-size raised dot ·. Frequently the multiplication sign is left out altogether, as in $x = vt$. From the context we are supposed to realize that the combination $vt$ represents the product of $v$ and $t$. The whole spirit of writing algebra is to keep it compact − so this type of omission is widely accepted provided everyone knows what is meant. It would be unusual to write $5 \times b$ as almost everyone would shorten this to $5b$ (but not $b5$). However, beware! Computer programs will not let you get away with it. For instance, in a spreadsheet program like Microsoft Excel®, the expression $(a + b)(c + d)$ must be written as (a + b)*(c + d).

Multiplication simply means that we add up multiple copies of something (see Figure 0.3). Thus $5 \times 2$ means add on 2 things 5 times, which gives the same answer as adding on 5 things 2 times. You can, at least in principle, work out the answer by counting blocks. In practice when you come to actually performing multiplication, either you must know the multiplication tables (see Figure 0.4) or you need to 'know someone who does' like the programmer who wrote the code for your calculator.

It takes a bit more block pushing than in Figure 0.3 to illustrate the point, but if we repeat the whole operation of adding 2 things 5 times 3 times over, $3 \times (5 \times 2)$, we get the same answer as if we repeated the operation of adding 5 things 3 times twice, $2 \times (3 \times 5)$. It is the same for any other tongue-twisting combinations of repeated multiplication you like. What this means is that we can multiply in any order we like.

The number 0 is very special in the process of addition because it does nothing. In multiplication it does a great deal; it gets rid of anything! The number that does nothing in multiplication is 1; we can multiply anything by the number 1 without changing its value. The identities that say all of this algebraically are in Appendix 1 at the end of the book.

It is also worth spending a little time on how we can combine addition and multiplication. Suppose we add five 2s then five 3s, which means adding 10 and then 15 to get 25. We can see an important point by writing this out in full $2 + 2 + 2 + 2 + 2 + 3 + 3 + 3 + 3 + 3$, reordering the terms $2 + 3 + 2 + 3 + 2 + 3 + 2 + 3 + 2 + 3$, adding adjacent pairs of 2 and 3, and noting that $5 + 5 + 5 + 5 + 5$ is just $5 \times 5$. Thus adding five 2s then five 3s produces the same result as adding five 5s (see Figure 0.4). That statement can be written more compactly as $5 \times 2 + 5 \times 3 = 5 \times (2 + 3) = 5 \times 5$. Another example of the same sort of thing, which you can check by doing the arithmetic, is that $27 + 18 = 3 \times 9 + 2 \times 9 = (3 + 2) \times 9 = 5 \times 9 = 45$. Breaking 27 into two factors (things that multiply each other), here 3 and 9, is called **factorization**.

**Figure 0.3**
$5 \times 2$ means add on 2 things five times. This can be imagined as arranging blocks in a rectangle 2 blocks across and 5 blocks down. To determine the result of the multiplication you just count the blocks giving 10. Adding on 5 objects 2 times means taking the blocks 5 at a time twice, i.e. 2 rows of 5 blocks. The answer obtained by counting is still 10 so $5 \times 2 = 2 \times 5$.

| | 0 | 1 | 2 | 3 | 4 | 5 | 6 | 7 | 8 | 9 |
|---|---|---|---|---|---|---|---|---|---|---|
| 0 | 0 | 0 | 0 | 0 | 0 | 0 | 0 | 0 | 0 | 0 |
| 1 | | 1 | 2 | 3 | 4 | 5 | 6 | 7 | 8 | 9 |
| 2 | | | 4 | 6 | 8 | 10 | 12 | 14 | 16 | 18 |
| 3 | | | | 9 | 12 | 15 | 18 | 21 | 24 | 27 |
| 4 | | | | | 16 | 20 | 24 | 28 | 32 | 36 |
| 5 | | | | | | 25 | 30 | 35 | 40 | 45 |
| 6 | | | | | | | 36 | 42 | 48 | 54 |
| 7 | | | | | | | | 49 | 56 | 63 |
| 8 | | | | | | | | | 64 | 72 |
| 9 | | | | | | | | | | 81 |

**Figure 0.4**
Multiplication triangle. The number at the intersection of a row and a column is the product of the numbers at the top of the column and to the left of the row.

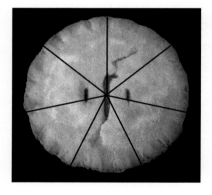

**Figure 0.5**
A pie cut into seven equal slices. Each slice is 1/7th of the whole. Two slices are 2/7ths, and so on.

Applied here it is just a game, but applied to symbols it will be an important tool that will allow us to rearrange our statements to provide new insights. We will return to this in later sections.

The rules for multiplication with negative numbers are consequences of the rules for addition and subtraction. $(-5) \times 2$ means take away 2 objects 5 times, that is, remove 10 objects. How about $(-5) \times (-2)$? We can think of that as taking away debits. Think about your bank balance. Removing debits has the same effect on the final balance as adding credits — so the answer must be $+10$. We are saying the same thing about negative numbers as before: taking away negative numbers is the same as adding positive numbers.

# 0.2 Division, reciprocals, fractions, and rational numbers

The integers are closed for addition, subtraction, and multiplication but they are not closed for division.

Division, $a \div b$, asks 'how many $b$s are there in $a$'. Thus we require that if we take the answer and multiply that by $b$ we must end up back where we started. Division and multiplication must reverse each other. For example because $5 \times 2 = 10$, we know that there are five 2s or two 5s in 10; that is, $10 \div 2 = 5$ and $10 \div 5 = 2$. That is fine. However, what happens if we try to divide 10 by 3 or 2 by 9? Clearly, there are more than three but fewer than four 3s in 10, but the answer when we divide 10 by 3 is not any integer. Similarly, $2 \div 9$ is less than 1 but greater than 0. To handle these calculations the number system must be extended to fill in the gaps between the integers. We need to learn how to subdivide 1; the important new concept we need is the **reciprocal** of an integer. Once we have reciprocals, fractions and division are 'easy'.

To consider a specific example, what is 1/7th? The essential point is that, if we have 7 of them, we must be back to a whole. Graphically (see Figure 0.5), if we take a pie (I'll have apple please) and divide it into 7 equal slices, then each slice is 1/7th of the pie. Thus the reciprocal of 7, written as 1/7, is defined as 1/7th of a whole, that is 1/7th of 1. The essential point is that $7 \times (1/7) = 1$. That defines the reciprocal, 1/7th. A **fraction** is any integer multiple of a reciprocal. Thus $2 \times (1/7)\text{th} = (2/7)\text{ths}$ means take two of the seven pieces, 3/7ths means take three, etc. The number above the line in the fraction is called the **numerator**; the number below it, the **denominator**. Note that $2 \times (1/7)\text{th}$ of a whole pie and $(1/7)\text{th}$ of 2 whole pies are exactly the same amount of pie. We can multiply integers and reciprocals in either order.

With this definition of a fraction we can now return to division. As a simple example, what is $3 \div 2$? Whatever it is, when we take 2 of them (multiply by 2) we must get back to 3. Take 3 pies and cut each into two equal slices, so 3 pies is the same as 6 slices. Division by 2 means we must separate these slices into two equal piles. Each of these will contain 3 slices which is the same as $3 \times (1/2)$ pie which is 3/2 of a pie. Thus (3 pies) $\div$ 2 is the same thing as 3 half pies: $3 \div 2 = 3 \times (1/2) = 3/2$. This is obvious in terms of pies (or pieces of paper) and there is nothing special about our choices of the numbers 3 and 2 used in the example.

The only exception to the preceding discussion is that we cannot divide by zero. The number 0 does not have a reciprocal value and division by 0 is impossible.

Note that unlike $+$ and $\times$, the $\div$ operation requires us to be careful about order: $2 \div 3$ means that we take the reciprocal of 3, not of 2. Once we have taken the reciprocal we can multiply in either order. For example, $2 \div 3 = 2 \times (1/3) = (1/3) \times 2 = (2/3)$ is less than one, whereas $3 \div 2 = 3 \times (1/2) = (1/2) \times 3 = (3/2)$ is greater than one.

In Chapter 2 we shall define a ratio as something like $3 : 4$ which can be said as 'three to four'. However, mathematicians regard a ratio of numbers simply as being one number divided by another, so $3 : 4 = 3 \div 4$. Thus a fraction, that is an integer times the reciprocal of an integer, can be called a *ratio* of integers. Numbers that can be expressed exactly as a fraction are called **rational numbers**.

## 0.3 Decimal numbers

Most arithmetic is done with decimal numbers. Decimals are just a compact way of writing fractions where the denominator is some power of 10 such as 10, 100, or 1000. For instance 173.8 is the same as 1738/10, whereas 17.38 is 1738/100. Once we can do arithmetic with fractions, there are no new rules for decimal numbers.

Any rational number can be written either as an exact decimal number or as a decimal approximation, which when extended to greater and greater accuracy, is seen to have a repeating pattern, such as:

$(1/2) = 0.5$ exactly;
$(1/8) = 0.125$ exactly;
$(1/125) = 0.008$ exactly;
$(1/3) = 0.333333\ldots$ (3 repeating forever);
$(1/7) = 0.142857142857142857\ldots$ (142857 repeating forever);
$(9/14) = 0.642857142857142857\ldots$
                    (142857 repeating for ever after an initial portion).

## 0.4 Real numbers

Numbers with decimal approximations with sequences that repeat forever are approximations to rational numbers. They can be written exactly as a fraction using some integer in the denominator. Other numbers that can only be written approximately as decimal numbers do not have repeating patterns. These numbers are not rational; for example we will see in Chapter 2 that it is impossible to write the square root of 2, $\sqrt{2}$, as a rational number. We can use rational numbers, usually written in decimal notation, to get as close to the value of $\sqrt{2}$ as we like, but no rational number can ever be exactly equal to $\sqrt{2}$. Numbers like $\sqrt{2}$ that, as it were, fall into the holes between the rational numbers are called **irrational numbers**. Other examples of irrational numbers discussed later in this book are $\sqrt[3]{2}$ (the cube root of 2), $\pi$ (the ratio of the circumference of a circle to its diameter), and $e$ (the base of the natural logarithms). Taken together, rational and irrational numbers are said to comprise **real numbers**. The results of measurements in science are assumed to represent approximations to an underlying 'correct' theoretical value that is described by its units and a real number (see Chapter 1).

### The importance of rational numbers

Although real numbers are the numbers we actually need for most of science, the rational numbers still have a special importance. All the arithmetic we can actually do by writing marks on paper, pushing buttons on calculators, or typing characters into computers is the arithmetic of rational numbers. For real numbers that are not rational we can define a name for the number, such as $\sqrt{2}$ or $\pi$, and we can find rational numbers that are within any specified tolerance, but we cannot actually write down the irrational number itself using any combination of the digits 0 to 9, even allowing a decimal point or the use of fractions. Whenever we want to use an irrational number in arithmetic we have to make do with using an approximation. You may have been taught that $\pi$ is $(22/7) = 3.142857142857\ldots$ (see Chapter 4). That is close, but not exact. The value 3.1416 is closer, but still not actually equal to $\pi$. Nevertheless we will frequently write something like $\pi = 22/7$. What a scientist means by a statement like $\pi = 3.14$ is that *to sufficient accuracy* $\pi$ is 3.14. If we wanted to emphasize that this is an approximation we could write $\pi \approx 3.14$. You can think of $\approx$ as meaning 'as close to equal as we need'. We will be more careful about writing approximate values of numbers when scientific notation is considered in Section 2.6.

# 0.5 Forbidden operations using real numbers: complex numbers

There are two arithmetical operations on numbers that cannot be performed using real numbers. The first of these is division by zero. Recall that $a \div b$ is defined by the rule: if $a = b \times c$ then $a \div b = c$. However, what happens if $b$ is zero? Because 0 times anything is 0, if $b = 0$ then $a$ must also equal 0, but $c$ can take on any value. We can say nothing about its value from just knowing that $a$ and $b$ are zero. Often there is nothing we can do about it: division by zero just has to be left undefined. There is, however, one very important and very common exception. If all we know is that $a$ and $b$ are 0, we are stuck: the answer cannot be defined. However, usually when 0/0 arises in real life we know more: we know the value of $a/b$ when both $a$ and $b$ are very small. Interpreting $a/b$ when $a$ and $b$ get smaller together is the fundamental problem that was solved by Leibniz and Newton when they independently invented differential calculus (see Chapter 5).

The second type of operation we cannot perform using just real numbers is illustrated by trying to calculate $\sqrt{-1}$, the square root of $-1$. The answer is neither 1 nor $-1$ because the square of either of these is equal to 1. In fact there is no rational or irrational number equal to $\sqrt{-1}$. To be able to find the square root of any number we need to extend the number system one last time by defining the complex numbers.

A complex number has two components, called the real and imaginary parts. Mathematicians like complex numbers because *all* of the arithmetical operations (except that pesky division by zero) applied to complex numbers produce complex numbers – at last the number system really is closed. Physicists come to like complex numbers because they make calculations much easier in optics, electrical circuit theory, and quantum mechanics. A few biologists use complex numbers because they are helpful in the analysis of responses to light or sound signals and in the theory behind spectroscopic techniques like nuclear magnetic resonance (NMR). However, most people likely to read this book will never use them, so in this book we shall stick to the real numbers.

## 0.6 Summary

We can talk of several different kinds of number. The earliest numbers that any of us met were the counting numbers (one, two, three, ...); these are often referred to as positive integers. Later we met zero and negative integers. We talk of integers as including all these numbers: positive integers, zero, and negative integers. We also met fractions such as 1/4 or 3/2. Any number of the form $p/q$, where $p$ and $q$ are integers, is said to be a rational number. Note that because we can always write an integer $p$ as $p/1$, rational numbers include all the integers, not just what we normally think of as fractions. We next met irrational numbers, which cannot be written as $p/q$. The real numbers are all the rational and irrational numbers taken together. These are the numbers that can be used together with units to describe physical quantities. Describing physical quantities in the next chapter marks the real start of this book.

# CHAPTER 1

# Quantities and Units

**Figure 1.1**
The authors, Mike, Steve, and Bill, holding a protein sample destined for the 500 MHz NMR spectrometer in the background. Image courtesy of Mike Aitken, Bill Broadhurst, and Steve Hladky.

As scientists, we need to make quantitative statements about the physical quantities measured in our experiments. Algebra provides the language and grammar to make these statements. In this language the sentences are equations or inequalities whereas the words are *symbols*. A symbol may stand for a physical quantity or a number; for an operation such as addition or multiplication; or for a relationship such as 'is equal to' or 'is greater than'. Often we use letters such as $x$, $t$, $m$, or $A$ to stand for physical quantities such as distance, time, mass, or area. Symbols can also be special characters such as + for addition, or a combination of letters such as 'sin' for the sine function introduced in Chapter 4.

A physical quantity is a combination of a numerical value and a unit, for example a length of 1 m, a time of 2 s, or a mass of 70 kg, where the 'm' stands for meter, 's' for second, and 'kg' for kilogram. Both are needed; if we change the unit the number changes accordingly. Many of the laws of science are expressed as simple equations relating physical quantities. A familiar example is $F = ma$ where $F$, $m$, and $a$ stand for force, mass, and acceleration. There are various systems for choosing units and conventions for how physical quantities are to be described. In this book we use the *Système Internationale* (SI) system of units, which has become standard for scientists and engineers throughout the world.

# 1.1 Symbols, operations, relations, and the basic language of mathematics

In the language of mathematics, the words are symbols like $x$, $t$, $m$, $+$, $\times$, $\div$, $=$, $>$. Symbols can stand for numbers or for physical quantities; they can indicate operations or they can state relationships like 'is equal to' or 'is greater than'. You first started using many of these symbols back in primary school where you learned what $+$ and $=$ mean. Even then you also used symbols to stand for unknown numbers in exercises like that shown in Figure 1.2.

**Figure 1.2**
In elementary arithmetic, we may have thought of □ as just a space holder to tell us where to write the answer, but it can also be regarded as a symbol that stands for a number whose value is not already known.

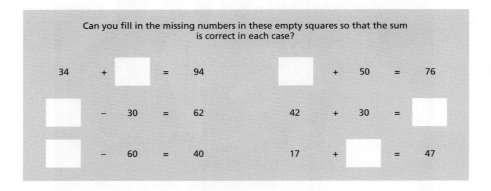

Can you fill in the missing numbers in these empty squares so that the sum is correct in each case?

| 34 | + | ☐ | = | 94 |  ☐ | + | 50 | = | 76 |
| ☐ | − | 30 | = | 62 |  42 | + | 30 | = | ☐ |
| ☐ | − | 60 | = | 40 |  17 | + | ☐ | = | 47 |

You may have thought of □ as just a box to tell you where to write the answer, but it can also be regarded as a symbol called 'box' that stands for a number whose value is not already known. The equation $\square + 3 = 8$ tells us a relation between □ and the numbers 3 and 8, and this relation allows us to solve for the value, 5, to be assigned to the variable 'box'. That really is the crux of using algebra; it allows us to state relations *before* we know the actual values. Of course the relations between our symbols are going to be a bit more complicated – but the principle behind the use of algebra is still the same.

Note that whenever algebraic expressions are typeset, the letters used in a symbol are written in italics if the symbol represents either a number or a physical quantity. By contrast plain roman type is used for symbols that represent units or labels. Typographical conventions like these are fiddly but they can be very important. For example, in the equation for the gravitational force on an object at the earth's surface,

$$F = mg = m \times 9.8 \text{ m s}^{-2}, \tag{EQ1.1}$$

$m$ and m are completely different. The italic type tells us that $m$ stands for a physical quantity, mass, which might be expressed in kilograms; the plain roman type for the m after the 9.8 tells us it stands for the unit, meter.

Symbols and algebra can be used to express very profound notions. For instance $E$ can represent the total energy of a chunk of matter, $m$ its mass, and $c$ the speed of light. Combining these with the symbol for 'is equal to' and the notation for raising to a power Einstein wrote

$$E = mc^2. \tag{EQ1.2}$$

That bit of shorthand is a lot more compact and a lot more famous than its equivalent in English, 'The total energy of an object is equal to its mass multiplied by the square of the speed of light.' However, and this is the important point for now, the algebra and the English are being used to say exactly the same thing.

Now consider a very simple example. We can say in English:

'John is thirty centimeters taller than Robert.' (EQ1.3)

How can that be converted to an algebraic equation? First rewrite the sentence to emphasize that it is telling us something about heights:

'The height of John is equal to the height of Robert plus thirty centimeters.' (EQ1.4)

This sentence is a bit stilted but is still correctly constructed in English. Now introduce some symbols: $J$ to stand for 'the height of John', $R$ for 'the height of Robert', $=$ for 'is equal to', $+$ for 'plus', and the abbreviation 'cm' for 'centimeter'. This yields the equation

$$J = R + 30 \text{ cm}. \tag{EQ1.5}$$

This statement is called an **equation** because it says that one expression is **equal** to another. The expressions are symbols combined according to the grammar of algebra. Even though the algebraic grammar is rigid and restricted, the equation is still a sentence that can, with a bit of effort, be read in English.

A more sophisticated approach is to use subscripts. Let $H$ stand for 'height' whereas J and R stand for 'John' and 'Robert', respectively. The subscripted symbols $H_J$ and $H_R$ then stand for the height of John and the height of Robert and the sentence becomes the equation

$$H_J = H_R + 30 \text{ cm}. \tag{EQ1.6}$$

Note that the subscripts in this case are not italic because they are labels attached to known individuals.

When reading an equation the equals sign is the verb. There are several other verbs used in elementary algebra. For instance $H_R < H_J$ is a relation that can be read as the height of Robert is less than the height of John. The algebraic verbs you will encounter in this book are: $=$, is equal to; $\neq$, is not equal to; $\equiv$, is identical to (is always equal to for any values of the variables); $<$, is less than; $\leqslant$, is less than or equal to; $>$, is greater than; $\geqslant$, is greater than or equal to; $\approx$, is approximately equal to (as close as we need); and $\sim$, is similar to. There are others that are very common in mathematics, for example $\in$ which means 'is an element of', but they are rarely seen in science and so will not be covered here.

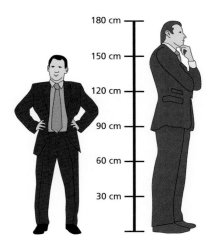

**Figure 1.3**
The fact that John is 30 cm taller than Robert can be represented algebraically as: $H_J = H_R + 30$ cm.

Using algebraic notation as in EQ1.6 is more compact than saying the same thing in plain English, and it leads directly to a solution for the height of either John or Robert in terms of the other. However, we have paid a high price for this compactness. We have introduced five symbols, an abbreviation, and strange-looking combinations of symbols and labels like $H_J$ and $H_R$. We run the risk that these symbols and expressions will be seen as just jargon, and like all jargon will get in the way of understanding. For the simple example given here this price is too high to pay – it would be much better to tell someone about the heights using plain English. However, that is to miss the main point: when the relations get a bit more complex, the English versions become so clumsy that we would just give up. We must pay the high price to use symbols and algebra because the gains are much greater. Furthermore, obeying the rules of algebra, we can safely manipulate equations to gain insights which would be beyond us by other means.

The rules of algebra will be explained more thoroughly in Chapter 2. There are not many, they are very simple, you have already met them, and they are easy to remember.

We can add zero to any expression without changing its value.

We can multiply any expression by 1 without changing its value.

We can swap the two sides of an equation.

We can add or subtract the same thing to both sides of an equation.

We can multiply both sides of an equation by the same thing.

We can divide both sides of an equation by the same thing (other than zero).

We can replace any expression by another equal expression.

Mathematics as it is taught in school deals with equations and symbols that refer to numbers. In the same spirit, Chapter 2 will review arithmetic, algebra, and solving equations. You will already know much of the information but do not skip it because the story is told in a different way using algebra and its grammar. Thus you can both refresh your understanding of numbers and arithmetic and learn how to use algebra.

The emphasis on numbers and structure in Chapter 2 is fine if you like that sort of thing. However, if that were the only use of algebra most of us as scientists simply would not bother. Why do we bother? Is not arithmetic with a calculator enough? In a sense the answer to that is given throughout the rest of this book, but something a little less sweeping should be said here at the beginning.

## 1.2 Physical quantities and physical value equations

Trying to be a scientist without being able to use basic algebra is a bit like trying to live in France without being able to speak French. Algebra is the basic language for stating many of the fundamental relationships between the real physical quantities that we describe. In fact, if you are reading this book, it is very likely that you already know some famous equations, for instance:

$d = vt$   distance traveled at constant velocity equals velocity multiplied by time,                    (EQ1.7)

$F = ma$   force equals mass multiplied by acceleration,          (EQ1.8)

and

$E = mc^2$   energy equals mass multiplied by the speed of light squared.

(EQ1.2)

Here the symbols, $d$, $v$, $t$, $F$, etc. are not used to stand for numbers: they stand for **physical quantities**, for example $a$ is an acceleration, $m$ is a mass, and $F$ is a force. The equations that relate them are **physical value equations**, not just equations relating numbers. A slightly more complicated example is the equation describing how the number of occupied (or bound) receptors, $N_b$, varies with the concentration of a drug, $c$, the total number of receptors, $N_T$, and the affinity constant for binding, $K_a$

$$N_b = \frac{K_a c N_T}{1 + K_a c}.$$
(EQ1.9)

This equation, which reappears several times in the book, will be introduced fully in the next chapter. Note that $c$ has very different meanings in the last two equations: physicists use $c$ to stand for the speed of light but biologists almost always use it for concentration. That raises a very important point: it is absolutely crucial that you always make sure that the definitions of your symbols are clear to those who will read them.

## 1.3 Physical quantities, numerical values, and units

The symbols like $a$, $t$, and $K_a$ in EQ1.7–EQ1.9 stand for the combination of a **numerical value** and a **unit**. These symbols for physical quantities are combined in physical value equations to make statements about the real world. This use of physical value equations is almost universal among scientists so it is vitally important to understand the distinction between a number and a physical quantity.

---

Consider the length of the line just above this text. If we measure its length in centimeters, we get 7.62 cm. Choosing different units of measurement the answers might be 0.0762 m, 76.2 mm, or 76 200 μm; whereas in US or British units the answers might be 3 inches, or 0.25 feet. Thus we can write

7.62 cm = 0.0762 m = 76.2 mm = 76200 μm = 3 in = 0.25 ft.   (EQ1.10)

If you were asked the length of the line and replied just 7.62 or 0.25, that would be useless. The numerical value 7.62 is meaningful as the length of the line only when it is combined with the statement 'when the length is measured in centimeters'. You can say 'the length is 7.62 cm' or you can say 'the length expressed in centimeters is 7.62', but one way or the other you *must* specify the unit. The number and unit should be thought of as one whole, the quantity. For instance, in the equation for the distance traveled by a car accelerating at a constant rate from a standing start,

$$d = (1/2)at^2, \tag{EQ1.11}$$

when the time $t$ is squared both the number and the unit must be squared. If $t = 3$ s, then $t^2 = 9$ s$^2$.

## 1.4 Conversion of units

There are several different ways to remember how to change units. One of the safest (i.e. the easiest to get right every time) is based on a simple rule of algebra; we can always multiply anything by the number 1 without changing its value. The secret is to write the number 1 in a clever way. For example, because 1 kg = 1000 g, 1000 g/1 kg = 1, and a conversion from kilograms to grams can be written as

$$3.7 \text{ kg} = 3.7 \text{ kg} \times \frac{1000 \text{ g}}{1 \text{ kg}} = 3700 \text{ g}. \tag{EQ1.12}$$

Note that we have also used kg/kg = 1 (as anything except 0 divided by itself equals 1).

Any conversions we ever need to do are just applications of this principle. Suppose we have watched a leafcutter ant proceeding in a very businesslike manner carrying its leaf segment homeward. Because the pace was steady, the relation between the distance it moved $d$, the velocity $v$, and the time we watched, $t$, is

$$d = v \times t. \tag{EQ1.13}$$

**Figure 1.4**
A leafcutter ant bearing its leaf back to the nest. To find out how fast it is moving, we can record its position at two times and divide the distance traveled by the time taken. Image courtesy of Bristol Zoo Gardens.

The equation above relates the actual physical quantities: how fast it was traveling, the time, and the distance. If the elapsed time was 3 s and the distance traveled 36 mm then the velocity can be calculated as

$$v = \frac{d}{t} = \frac{36 \text{ mm}}{3 \text{ s}} = 12 \text{ mm s}^{-1}. \qquad \text{(EQ1.14)}$$

The answer, which can be written as either 12 mm s$^{-1}$ or 12 mm/s, is read as 12 millimeters per second. The unit s$^{-1}$ or 1/s is called a reciprocal second.

The validity of the relation between distance, velocity, and time does not depend on our choice of units. For example, we might, somewhat perversely, have said that the distance traveled was 0.036 m and the time taken was 0.05 min and thus that its velocity was 0.036 m / 0.05 min = 0.72 m min$^{-1}$. We can check that the two versions of the velocity are indeed the same by converting units.

To convert from meters to millimeters note that 1 meter is the same thing as 1000 millimeters, that is, 1 m = 1000 mm and

## Box 1.1

The bookkeeping required for conversion of units can be written down using the rule that anything divided by itself is just the number 1. For instance using 1000 g/1 kg = 1,

$$3.7 \text{ kg} = 3.7 \text{ kg} \times \frac{1000 \text{ g}}{1 \text{ kg}} = 3700 \text{ g}.$$

Alternatively the same thing can be written down using the rule that any physical quantity in an algebraic expression or equation can be replaced by another of equal value, e.g.

$$3.7 \text{ kg} = 3.7 \times 1 \text{ kg} = 3.7 \times 1000 \text{ g} = 3700 \text{ g}$$

in which 1 kg is replaced by 1000 g. As we normally wouldn't bother to write down all the steps, this version is faster to write and most people would do it this way. However, this example was so easy we could do it in our heads. The extra work of using multiplication by 1 becomes increasingly worthwhile as the conversions become more complicated. For example the conversion from 120 km h$^{-1}$ to 33.3 m s$^{-1}$ can be written as either

$$120 \text{ km h}^{-1} = 120 \frac{\text{km}}{\text{h}} \times \frac{1000 \text{ m}}{1 \text{ km}} \times \frac{1 \text{ h}}{60 \text{ min}}$$

$$\times \frac{1 \text{ min}}{60 \text{ s}} \approx 33.3 \text{ m s}^{-1}$$

which lays out each step carefully or

$$120 \text{ km h}^{-1} = 120 \frac{1 \text{ km}}{1 \text{ h}} = 120 \frac{1000 \text{ m}}{60 \text{ min}}$$

$$= \frac{120 \times 1000}{60} \times \frac{1 \text{ m}}{1 \text{ min}}$$

$$= \frac{120 \times 1000}{60} \times \frac{1 \text{ m}}{60 \text{ s}}$$

$$= \frac{120 \times 1000}{60 \times 60} \text{ m s}^{-1}$$

$$\approx 33.3 \text{ m s}^{-1}$$

which, unless you were feeling very nervous, would be shortened to

$$120 \text{ km h}^{-1} = \frac{120 \text{ km}}{1 \text{ h}} = \frac{120 \times 1000 \text{ m}}{3600 \text{ s}}$$

$$\approx 33.3 \text{ m s}^{-1}.$$

The first method provides the security of specifying each step, the latter allows you to save time writing. Whichever way you choose to do it, the conversion does emphasize that a fast-moving car travels a long way in just a single second.

$$\frac{1000 \text{ mm}}{1 \text{ m}} = 1. \qquad\qquad\qquad \text{(EQ1.15)}$$

Similarly to convert from minutes to seconds note that $60 \text{ s} = 1 \text{ min}$ which is the same as

$$\frac{1 \text{ min}}{60 \text{ s}} = 1. \qquad\qquad\qquad \text{(EQ1.16)}$$

Combining all of these and canceling out units that appear in both the numerator and the denominator (again anything except 0 divided by itself equals 1)

$$
\begin{aligned}
0.72 \ \frac{\text{m}}{\text{min}} &= 0.72 \ \frac{\text{m}}{\text{min}} \times \frac{1000 \text{ mm}}{1 \text{ m}} \times \frac{1 \text{ min}}{60 \text{ s}} \\
&= \frac{0.72 \times 1000}{60} \times \frac{\text{m}}{\text{min}} \times \frac{\text{mm}}{\text{m}} \times \frac{\text{min}}{\text{s}} \qquad \text{(EQ1.17)} \\
&= 12 \text{ mm s}^{-1}
\end{aligned}
$$

which is the same as before.

Conversion of reciprocal units sometimes causes confusion. Thus it is worthwhile looking specifically at these to show that application of the standard procedures produces sensible answers. First, look at a time expressed in milliseconds and change the units to seconds

$$500 \text{ ms} = 500 \text{ ms} \times \frac{1 \text{ s}}{1000 \text{ ms}} = 0.5 \text{ s}. \qquad\qquad \text{(EQ1.18)}$$

That makes sense. A large number, 500, of a small unit of time like a millisecond will be equal to a part, $0.5 = 1/2$, of a much longer unit of time like a second.

Now look at a rate constant, something that says how much happens per unit of time. Suppose this is 500 events per millisecond which is written as $500 \text{ ms}^{-1}$. How do we convert this to events per second?

$$500 \text{ ms}^{-1} = \frac{500}{\text{ms}} \times \frac{1000 \text{ ms}}{\text{s}} = \frac{500000}{\text{s}} = 500000 \text{ s}^{-1}. \qquad \text{(EQ1.19)}$$

Again this makes sense: if 500 events occur in a millisecond, then 1000-fold more must occur in the 1000-fold longer time of a second.

When we try to add terms in an equation, each of those terms must have the same unit. For example, in the US and British systems of units, the length of one foot, written as 1 ft, is defined as being equal to twelve inches, written as 12 in, and one inch is defined as being equal to 0.0254 m. How do we add 1 m and 1 ft? In this case we can convert 1 ft into the equivalent length in meters using an exact conversion before adding the two lengths:

$$1 \text{ m} + 1 \text{ ft} = 1 \text{ m} + 1 \text{ ft} \times \frac{12 \text{ in}}{1 \text{ ft}} \times \frac{0.0254 \text{ m}}{1 \text{ in}} = 1 \text{ m} + 0.3048 \text{ m} = 1.3048 \text{ m}.$$

(EQ1.20)

Units can sometimes protect us from silly mistakes. It does not make any sense to try to add a force to a distance – and reassuringly we cannot convert a unit for force into a unit for distance. To consider a slightly more complicated example, look at the equation given earlier for the number of occupied binding sites when a drug binds to a receptor,

$$N_b = \frac{K_a c N_T}{1 + K_a c}.$$

(EQ1.21)

Because in the denominator the unit of the combination $K_a c$ is added to the number 1, it must have the same unit as a number, that is none. Thus when calculating the value of the product $K_a c$ the unit of $K_a$ must be the reciprocal of the unit of the concentration, $c$. For instance if we have chosen the unit of $c$ to be nmol $l^{-1}$ (nanomoles per liter) then, when we calculate the value of the product $K_a c$, the unit of $K_a$ must be $l$ nmol$^{-1}$.

Converting between all of these different units is a major nuisance. Thus wherever possible it makes excellent sense for everyone to agree to use the same ones. Furthermore it helps a lot if everyone agrees on a standard set of abbreviations, for example m, cm, mm, μm, km, etc. However, even with such an agreed list, you still have to convert units – it is just that the numbers in most of the conversion factors are powers of ten.

## 1.5 SI units

The only serious candidate for an agreed list of units is the international system, SI. Because the French got there first, the definitive documents for the international treaties are all in French.

| Table 1.1 SI base units | | |
|---|---|---|
| base quantity | name | symbol |
| length | meter | m |
| mass | kilogram | kg |
| time | second | s |
| electric current | ampere | A |
| thermodynamic temperature | kelvin | K |
| amount of substance | mole | mol |
| luminous intensity | candela | cd |

The SI system is based on a few base units such as the meter for distance and the second for time. The complete list is given in Table 1.1. You will immediately recognize most of the units because they are taught in school. All physical quantities we can measure can be expressed using combinations of these seven base units. For example, within the SI system, volumes can be measured in cubic meters, $m \times m \times m$, written $m^3$; velocity, distance divided by time, in $m / s = m \, s^{-1}$; concentrations, amount / volume, in mol / m$^3$ = mol m$^{-3}$; and the association constant between a drug and a receptor, which has the units of 1 / concentration, in m$^3$ mol$^{-1}$. Some examples of derived units are given in Table 1.2.

| Table 1.2  Examples of SI derived units | | |
|---|---|---|
| **derived quantity** | **derived unit** | **symbol** |
| area | square meter | m$^2$ |
| volume | cubic meter | m$^3$ |
| speed, velocity | meter per second | m s$^{-1}$ |
| acceleration | meter per second squared | m s$^{-2}$ |
| mass density | kilogram per cubic meter | kg m$^{-3}$ |
| amount of substance concentration | mole per cubic meter | mol m$^{-3}$ |
| mass fraction | kilogram per kilogram, which may be represented by the number 1 | kg kg$^{-1}$ = 1 |
| mole fraction | amount of substance per amount of all substances present | mol mol$^{-1}$ = 1 |
| per cent | per cent | % |

Some of these derived quantities are assigned special names, because it is useful to think of these combinations of units together. For instance, force is sufficiently important that it is measured in its own unit, the newton, with symbol N. However, because 1 N is defined as that force which would accelerate a 1 kg mass at 1 m s$^{-2}$,

$$1\ N = 1\ kg \times 1\ m\ s^{-2} = 1\ kg\ m\ s^{-2}, \tag{EQ1.22}$$

the symbol N is just a synonym for kg m s$^{-2}$. Note that the first letter of the symbol used for an official SI unit is written in upper case if the symbol is named for a person; for example, we write 'N' because the newton was named after Sir Isaac Newton. By contrast, the first letter of the name of the SI unit is written in lower case, which allows us to distinguish whether we mean the name of the unit, newton, or the name of the person, Newton. Note that 'Celsius' by itself is not an SI unit, so this convention does not apply in 'degree Celsius'.

Some other examples of special names for derived units are given in Table 1.3. Energy is even more important than force in chemistry and biology, so it should come as no surprise that it too has its own derived unit called the joule, with symbol J. The definition of the joule is the work done if a force of 1 N moves an object through 1 m,

$$1\ J = 1\ N \times 1\ m = 1\ N\ m = 1\ kg\ m^2\ s^{-2}. \tag{EQ1.23}$$

Both joules and newtons illustrate part of the reason that combinations of units are named. The equation 1 J = 1 N m tells us the relation between force and work very nicely: work equals force times distance. The message is a lot less obvious if we write 1 J = 1 kg m$^2$ s$^{-2}$.

## Table 1.3 Examples of SI approved derived units with special names and symbols

| derived quantity | name | symbol | expression in terms of other SI units |
|---|---|---|---|
| frequency | hertz | Hz | $s^{-1}$ |
| force | newton | N | $m\,kg\,s^{-2}$ |
| pressure | pascal | Pa | $N\,m^{-2}$ |
| energy | joule | J | $N\,m$ |
| power | watt | W | $J\,s^{-1}$ |
| charge | coulomb | C | $A\,s$ |
| Celsius temperature* | degree Celsius | °C | K |

*Conversion between temperature scales is not just a change of units, it also involves a shift in the zero point.

Note that in the SI convention the symbol '%' stands for 'multiply by the number 0.01'. For instance saying that a mass fraction is 10 % means that the actual mass fraction is $10 \times 0.01 = 0.1$.

The quantities we deal with are not always similar in size to the base units, so one important way in which new units are derived is by considering multiples and fractions. It is always possible to create a unit which is 1000 times bigger or 1/1000th the size of any named unit by using the standard prefixes (see Table 1.4). Note these are never combined. Thus you do not use things like 'millimicro': instead you use nano.

SI units are correct and consistent. In many cases they are also convenient, but not always. There is a long and noble tradition of powerful groups forcing exceptions to be made. One of these exceptions is the liter. If you have ever seen a cubic meter of water you will know that it is not a practical unit for work in the laboratory; after all, its mass is 1000 kg! The liter on the other hand is very convenient, as every shopper will know. The SI committees wanted to dispose of the liter as a unit of volume because originally it was defined in terms of the volume occupied by a kilogram of water at a certain temperature rather than in terms of the meter. There was a good practical reason for this choice; it was a lot easier to load 1 kg of water into a

## Table 1.4 SI prefixes

| symbol | prefix | factor | symbol | prefix | factor |
|---|---|---|---|---|---|
| Y | yotta | $10^{24} = (10^3)^8$ | d | deci | $10^{-1}$ |
| Z | zetta | $10^{21} = (10^3)^7$ | c | centi | $10^{-2}$ |
| E | exa | $10^{18} = (10^3)^6$ | m | milli | $10^{-3} = (10^3)^{-1}$ |
| P | peta | $10^{15} = (10^3)^5$ | μ | micro | $10^{-6} = (10^3)^{-2}$ |
| T | tera | $10^{12} = (10^3)^4$ | n | nano | $10^{-9} = (10^3)^{-3}$ |
| G | giga | $10^{9} = (10^3)^3$ | p | pico | $10^{-12} = (10^3)^{-4}$ |
| M | mega | $10^{6} = (10^3)^2$ | f | femto | $10^{-15} = (10^3)^{-5}$ |
| k | kilo | $10^{3} = (10^3)^1$ | a | atto | $10^{-18} = (10^3)^{-6}$ |
| h | hecto | $10^{2}$ | z | zepto | $10^{-21} = (10^3)^{-7}$ |
| da | deka | $10^{1}$ | y | yocto | $10^{-24} = (10^3)^{-8}$ |

Examples: 1 nm is 1 nanometer $= 10^{-9}$ m; 1 mV is 1 millivolt $= 10^{-3}$ V.
The only exception to the rules for using prefixes occurs for mass, where the base unit is kg, 1 gram is 1 g, and the base for all other multiples is g – for example, mg and μg.

volumetric flask and put a mark on the neck than it was to make a box exactly 1 dm (one decimeter) on each side. However, this sort of definition wreaks havoc with the logical structure of units. In fact, as long as you stick with this sort of definition, 1 l is almost certain not to be exactly 1 dm$^3$ because we do not know the exact density of water. This bothered some physicists but the difference was much too small for biologists to worry about and they stuck by the liter. The SI committees finally gave in and allowed the liter as a unit – but only as a proper unit of volume redefined to be exactly 1 dm$^3$:

$$1 \text{ l} = 1 \text{ dm}^3 = 10^{-3} \text{ m}^3. \hspace{3cm} \text{(EQ1.24)}$$

Table 1.5 lists other derived units still in common use.

| Table 1.5 Other derived units for volume and pressure still in common use | | | |
|---|---|---|---|
| derived quantity | name | symbol for the units | expression in terms of SI units |
| volume | liter | l | 1 dm$^3$ or 10$^{-3}$ m$^3$ |
| pressure | atmosphere | atm | 1.013 × 10$^5$ Pa |
| pressure | millimeters of mercury | mmHg | 133 Pa |
| pressure | centimeters of water | cmH$_2$O | 98 Pa |

The meaning of 'one liter' is unambiguous and clear but the same cannot be said for the symbol for the unit. In most of the world the symbol is a lowercase l leading to ml, µl, nl, etc. In the USA the symbols are L, mL, µL, nL, etc. The uppercase L has the advantage that it doesn't look like the number one. By contrast in many fonts, the shapes of the number 1 and the lowercase letter l are similar, sometimes even identical. So one liter becomes 1 l which can look a bit odd. However, the general principle in SI is clear: liter is not the name of a person so the symbol for liter should be lower case. In this book we follow most of the world in using the standard convention, lower case.

Most conversions of units follow the general pattern illustrated above. (A huge compilation of conversion factors can be found in Appendix B of SP811 available from the US National Institute of Standards and Technology (NIST) in either html or pdf format at http://physics.nist.gov/Pubs/ .) There are, however, a few conversions that cannot be done by the standard methods. One of these is between different temperature scales.

Temperature is special because the different scales in common use differ not only in the size of the basic unit but also in the point defined as zero. Originally, the Fahrenheit scale set the temperature of the freezing point of water to 32 °F and the boiling point of water to 212 °F whereas the Celsius temperature scale set these values to 0 °C and 100 °C. Although it may seem a good thing to have a nice round number of degrees between these two important temperatures, to say that the temperature is zero at the freezing point of water is still completely arbitrary. A much more natural choice exists. The random movements of atoms and molecules increase with temperature. That also works in reverse and the random movements come to a stop at approximately −273.15 °C. If we shift the zero point of our temperature scale to this point, then, and only then, we get the very convenient result that the energy of the random movements, the thing we are actually measuring, is proportional to the temperature. The kelvin scale does just that. With zero now at its correct position, called absolute zero, and the size of the degree as defined by Celsius, water freezes at 273.15 K and boils at 373.15 K. Notice that there is no degree sign '°' before the K for temperatures given in the kelvin scale.

When used in a physical value equation, temperature should *always* be expressed using kelvins. The physical quantity is absolute temperature and only the kelvin scale puts zero at the right place. Note that the Celsius scale is now a derived temperature scale within SI, and is defined in such a way that 0 °C is set to be *exactly* 273.15 K. The precise definitions of the kelvin and Celsius scales can be found on the internet at sites such as http://physics.nist.gov/cuu/Units/kelvin.html .

There are simple formulae to convert between temperatures measured on the Celsius, Fahrenheit, and kelvin scales. Suppose that we indicate the numerical value on the Celsius scale of a temperature $T$ as $\{T\}_C$, with similar notation for the Fahrenheit and kelvin scales, then:

$$\{T\}_C = \{T\}_K - 273.15, \qquad \{T\}_K = \{T\}_C + 273.15, \qquad \text{(EQ1.25)}$$
$$\{T\}_F = 9\{T\}_C/5 + 32, \qquad \{T\}_C = 5(\{T\}_F - 32)/9. \qquad \text{(EQ1.26)}$$

## Amount versus mass

There are two sensible ways to measure how much of something we have: count up the **number** or determine the **mass**. In SI, mass is properly called exactly that: mass. However, because we almost always determine the mass of an object by weighing it, mass is often a little carelessly called weight. Strictly, the weight of an object is the force exerted on that object by gravity so the proper units for weight are the units of force. At the surface of the earth a 1 kg mass weighs 9.8 N. That is completely straightforward and logical, but you (and I) would always say that the mass weighs 1 kg. Whoever said life was logical? The use of 'weight' to mean either mass or weight, depending on the context, is so well entrenched that we are stuck with it.

In SI units 'amount of a substance' essentially means a count. The ultimate, intuitive measurement of amount is the count of the individuals present. When the number is relatively small, as in the number of frogs in a pond, this is exactly what is used. However, in a scientific measurement the number of entities present is often very, very large. Thus one entity, as in one individual ion or atom or molecule, is not a very convenient unit for specifying amount. The unit which has been chosen is the **mole**, with the symbol for the unit written as mol. One mole is the amount of substance that contains as many elementary entities as there are atoms in 12 g of carbon-12 ($^{12}C$).

The number of elementary entities per mole has to be determined experimentally. It is known as Avogadro's constant in English and the Loschmidt constant in German. Its officially recognized value in March 2007 was $(6.022\ 141\ 79 \pm 0.000\ 000\ 30) \times 10^{23}$ mol$^{-1}$ (scientific notation for numbers, for example $6 \times 10^{23}$, is explained in Section 2.6). Avogadro's constant allows us to convert from an amount in moles to a pure number equal to the number of individuals. Choosing to use a weird quantity that must be determined experimentally as the basis for amount may seem fairly peculiar, but in fact it is very convenient because it makes it easy to calculate the mass of 1 mol of any chemical substance. All we need to know is the chemical formula and thus the relative molecular mass.

The **relative molecular mass** of a substance, $M_r$, is the average mass per molecule divided by 1/12th of the mass of an atom of $^{12}$C. As such $M_r$ is a pure number. The mass of 1 mol of substance, properly called the **molar mass** but often called the molecular mass or molecular weight, is then $M_r \times 1\ \mathrm{g\ mol}^{-1}$. Whatever you call it, the mass you start from in practical calculations is the mass of a mole of the substance, not that of a single molecule. In general, if $n$ is the amount of the substance and $m$ is the mass,

$$m = M_r \times 1\ \mathrm{g\ mol}^{-1} \times n. \qquad \text{(EQ1.27)}$$

> The **relative molecular mass** of a substance, $M_r$, is the average mass per molecule divided by 1/12th of the mass of an atom of $^{12}$C.

Consider the example of glucose, $C_6H_{12}O_6$. The relative molecular mass could be calculated simply from this formula as $(6 \times 12) + (12 \times 1) + (6 \times 16) = 180$ if we take 12, 1, and 16 as the relative atomic masses of carbon, hydrogen, and oxygen. More accurately, the natural mixture of carbon isotopes has an average relative atomic mass of 12.011; likewise, we can find more accurate values for the average relative atomic masses of hydrogen and oxygen. With these more accurate values, the answer to five significant figures is 180.16. So 180.16 g of glucose contains 1 mol of glucose. The mole is the obvious choice for specifying amount in chemistry and the molecular aspects of biology.

> The **molar mass** of a substance is the mass of 1 mol, namely $M_r \times 1\ \mathrm{g\ mol}^{-1}$. This is often imprecisely called the molecular weight.

If you never read American medical or physiology textbooks, skip this paragraph. If you do, there is one additional unit for amount, the equivalent, with symbol Eq, that you are very likely to encounter. Equivalents can be used to state the amount of ions present. In effect this unit tells you that instead of stating how many moles of ions are present, you are stating the number of moles of unit charges. Thus 1 mmol of $Na^+$ ions, one charge per ion, is 1 mEq of $Na^+$ whereas 1 mmol of $Ca^{2+}$ ions, two charges per ion, is 2 mEq of $Ca^{2+}$. (Equivalents were introduced because they made charge balance calculations simpler. Thus in 1 mmol of $CaCl_2$ the 1 mmol of $Ca^{2+}$ balances not 1 mmol but 2 mmol of $Cl^-$. By contrast, using milliequivalents, this becomes 2 mEq $Ca^{2+}$ balancing 2 mEq of $Cl^-$, which at least to some eyes looks a bit tidier.) The unit, equivalent, was rooted in the physical chemistry of the 1930s but has largely been abandoned by chemists. It has been excluded from SI.

## 1.6 Concentration

Amount tells us how much or how many. Concentration tells us how crowded things are. Applied to a solution in a basic, primitive sense,

$$\text{concentration} = \frac{\text{amount of solute}}{\text{volume of solution}}. \qquad \text{(EQ1.28)}$$

Note that if we are dealing with gases, then we talk about space rather than solution. By far the most common unit for concentrations in biology, molar with symbol M, is defined as the concentration of a solution containing 1 mol of solute for each liter of solution volume. Concentrations based on solution volume are popular partly because they correspond to the most intuitive definition but also because they are very convenient when preparing dilutions (see Section 1.7 Dilutions and doses).

For many purposes, it is far more important to know the concentration rather than the amount. The basis for this is a large topic in physical chemistry. For example: consider two chambers each containing potassium chloride and separated by a membrane as shown in Figure 1.5. We want to know how often potassium ions will be striking the membrane from each side. Now imagine that we double the amount present in each chamber by doubling the size of the chambers but keep the same composition of the fluid and the same membrane. If we keep the concentration the same, then the number of potassium ions close to the membrane will be the same, so the number hitting the membrane will be the same. In other words, if we keep the concentration constant, things like rates of reaction at the surface, rates of diffusion across the membrane from one chamber to the other, and numbers of molecules evaporating each second from each unit of surface area, will all be the same. They do not depend on the amount of potassium chloride present in the chamber: they depend on the concentration. Also, very importantly, as rates of reaction within solutions depend on collisions between the reactants, these rates depend on how many particles are close at hand, i.e. again it is the concentrations not the total amounts in the container that determine how many reactions occur per unit volume of solution (see example in the End of Chapter Questions).

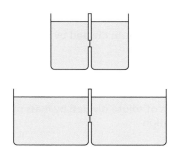

**Figure 1.5**
The importance of concentration rather than amount. Doubling the size of the containers and the amounts of everything within them does not change the rates of collisions with a membrane separating them. Nor does it change the rates of reactions within the solutions.

## Units for concentrations

The basic idea of concentration is clearly indicated in EQ1.28. However, you will encounter many variations on the theme. More generally we can measure how crowded the solute is in a solution by comparing some indication of how much solute we have relative to how much there is of everything else, namely

$$\text{concentration} = \frac{\text{some measure of solute}}{\text{some measure of the solvent, solution, or space}}.$$

(EQ1.29)

The different units for concentrations of solutions correspond to different ways of specifying how much of the solute is dispersed in how much solvent or solution. There are three measures of the solute, mass, amount, or volume, that we can choose and the same three choices for measuring the extent of either the total solution or just the solvent. Fortunately, of the 18 ($3 \times 3 \times 2$) possible combinations, *only* the five combinations defined and described in Table 1.6 are commonly used. Each of these has circumstances where it is particularly convenient. However, there is a major downside of having more than one: we have to learn how to convert between them.

## Molarity

In biology the most commonly used of all units for concentration is molar. This is not an officially sanctioned SI unit but its meaning is precise in terms of units that are:

## Table 1.6 Concentrations in SI

| basis of definition | units | notes |
|---|---|---|
| amount of solute divided by volume of solution | mol m$^{-3}$, mol dm$^{-3}$, or mol l$^{-1}$ | In SI, known as *amount of substance concentration*. Biologists usually refer to these concentrations using the non-SI unit M and refer to them as molarities. 1 M = 1 mol l$^{-1}$ |
| mass of solute divided by volume of solution | 1 kg m$^{-3}$ = 1 g l$^{-1}$ = 1 mg ml$^{-1}$ or 1 µg ml$^{-1}$ = 10$^{-3}$ g l$^{-1}$ = 10$^{-3}$ kg m$^{-3}$ | |
| mass of solute divided by mass of solution | kg kg$^{-1}$ | In SI, known as *mass fraction of the solute*. Mass fraction solutions are easy to prepare without volumetric glassware; all that is needed is a balance. |
| amount of solute divided by mass of solvent | mol kg$^{-1}$ | These concentrations are often called molalities. |
| volume of solute divided by volume of solution | m$^3$ m$^{-3}$ | |

$$1 \text{ M} = 1 \text{ mol l}^{-1} = 10^3 \text{ mol m}^{-3},$$

$$1 \text{ mM} = 1 \text{ mmol l}^{-1} = 1 \text{ mol m}^{-3}.$$

Such concentrations expressed in mol l$^{-1}$ are often referred to as molarities but more often just as 'concentrations'. Using molarity is convenient and relatively harmless, and all efforts by the 'standardizers' to stamp it out have so far failed. Perhaps someday this unit will follow the liter and be allowed in from the cold.

Concentration in terms of the amount of solute is what we want whenever we are considering reactions or even just diffusion. Concentration in terms of (amount of solute)/(volume of solution), namely molarity, is just what we want whenever we want to use a pipette to deliver a known amount of solute.

## Molality

In most laboratories, mass can be measured much more accurately than volume. Thus in the most careful work, it is very useful to refer concentration not to the volume of the solution but rather to the mass of solvent. The reason for choosing mass of solvent rather than mass of solution is considered in the next section. The SI unit for concentration, defined as amount of solute divided by the mass of solvent, is mol kg$^{-1}$. The obsolete name for this unit was molal, with symbol m. This unit and its symbol have largely disappeared from use, which is just as well because m is already overworked as meter and $m$ as mass. However, concentrations measured in mol kg$^{-1}$ are still often referred to as molalities. A non-SI unit, mol (liter of cell water)$^{-1}$, has been used in cell physiology to mean essentially the same as mol kg$^{-1}$ because to sufficient accuracy the mass of 1 l of cell water is 1 kg. In this case the SI unit does exactly what is wanted and is easier to write so the non-SI unit should be abandoned.

## Molarity versus molality

In a relatively dilute solution such as 0.15 M NaCl, the concentration can be expressed as either 0.15 M or 0.15 mmol kg$^{-1}$ because almost all of the solution is solvent and 1 l of solution will contain very close to 1 kg of solvent. If these numerical values are so similar, why bother with the distinction between molarity and molality? The answer is, of course, that they are not always close – molarity and molality differ if *either* some of the solutes in the solution are sufficiently large *or* the solution is sufficiently concentrated that the solutes occupy a significant proportion of the volume. Normally, the solutions we deal with outside of cells are relatively dilute and it really does not make any difference which type of concentration we use. However, things are very different when we come to consider the solution inside a cell.

Cells contain lots of large solutes, like proteins and nucleic acids, that are in solution but nevertheless occupy a lot of space that is not available for the solvent or for the smaller solutes (see Figure 1.6). Perhaps the simplest example is the red blood cell where 30 % of the volume is occupied by hemoglobin. The rest of the cell interior, 70 % of the volume, is occupied by a solution with fairly normal properties. For instance sugars, alcohols, and small ions like K$^+$ and Cl$^-$ dissolve in the cell water of a red cell much as they do in a simple solution. Thus at equilibrium the concentrations of these solutes inside and outside are the same if we express them as molalities. However, because the volume of the red cell exceeds that of the cell water by a factor of 100/70, when we divide the amount of solute by the cell volume instead of the water volume we will get a concentration (a molarity) that is only 70 % as large as expected inside the cell. That might lead us to conclude,

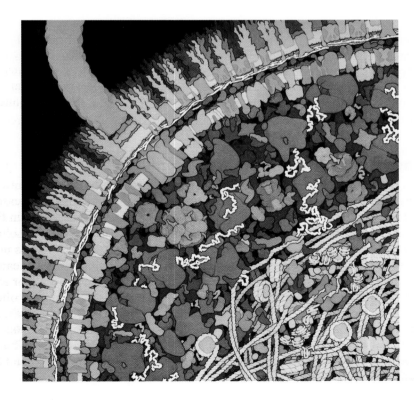

**Figure 1.6**
The crowded cell. A cartoon drawn to scale of a tiny region of the interior of an *Escherichia coli* bacterium. The volumes occupied by individual ribosomes, proteins, and RNA molecules are shown. Water molecules which are much smaller occupy the space remaining between the solutes – about 70 % of the total volume of the cell. Image courtesy of David S. Goodsell, Scripps Research Institute.

incorrectly, that somehow the solutes interact very differently with water inside the cell than outside it.

Molality and molarity also differ for very concentrated solutions because in these concentrated solutions even small solutes like glucose, $Na^+$, and $Cl^-$ can occupy a large proportion of the volume. Very concentrated solutions are rarely encountered with living cells but they are commonplace in any laboratory that uses a technique known as density gradient centrifugation.

## Mass concentration

Stating concentrations in terms of the mass of solute can be very convenient when we want to make up a solution. To prepare a $1\ mg\ ml^{-1}$ solution, you might weigh out 10 mg of the solute, and make it up to 10 ml by adding the solvent. In practice, when making up a solution with much less solute than solvent, it is accurate enough to just add 10 ml of solvent using a pipette. Now compare this with what we have to do to make up a 1 mM solution. We need to convert from amount, how many millimoles, to mass, how many milligrams, so that we can weigh out the solute using our balance. If the relative molecular mass of glucose is 180.16, the molar mass is $180.16\ g\,mol^{-1}$ and the concentration in terms of mass of a 1 mM solution will be

$$1\ mmol\ l^{-1} \times 180.16\ g/mol \times 1\ mol/1000\ mmol = 180.16\ mg\ l^{-1}.$$

To prepare 10 ml of this we thus need $180.16\ mg\,l^{-1} \times 0.01\ l = 1.8016\ mg$. So, we need to know the relative molecular mass, we have a lot more calculation to do, *and* we end up with a number that is difficult to remember. Nevertheless, because we usually want to end up with a known molarity, these are conversions that must be mastered.

In general, all you do to convert from concentration expressed as $mol\,l^{-1}$ to concentration expressed as $g\,l^{-1}$ is multiply by the relative molecular mass. It is very easy – but it does still take time and when you are in a hurry it is one more step you can get wrong. These are the main reasons why recipe books for making up solutions in the laboratory often use mass concentrations measured in $mg\,ml^{-1}$ or $g\,l^{-1}$ rather than amount concentrations measured in M or mM.

## Concentrations expressed as percentages

Many find that it is more convenient to express concentrations in terms of percentages than in terms of any of the basic units we have considered so far. However, carelessness in the use of percentages has caused a great deal of confusion because it is often not made clear what is a percentage of what. In SI, the symbol '%' stands for multiply by 0.01. Thus the only quantities, usually ratios, that can be expressed as a percentage are those with no units. Thus the ratio of mass of solute to mass of solution and the ratio of volume of solute to volume of solution can properly be expressed as percentages, often written by biologists as % (w/w) and % (v/v) respectively. Unfortunately, in older literature, a 1 % solution often means a solution containing 1 g solute per 100 ml of solution. Use of % (w/v) is *not* permitted in SI because a mass cannot be a fraction or a percentage of a volume. It is indefensible to use 1 % to mean $1\ g\ (100\ ml)^{-1}$. It would be best even to avoid 1 % (w/v); use

10 g l$^{-1}$ or 10 mg ml$^{-1}$ instead. Nevertheless such terminology continues to be used. Table 1.7 summarizes the various ways in which concentrations are expressed using percentages.

Having defined the various units for concentrations, the challenge now is to understand how to convert between them.

## Conversion between molarity and g l$^{-1}$

How much NaCl must we weigh out to prepare a solution with concentration $c = 0.15$ M? This is probably the simplest conversion and, fortunately, it is the only one you need to do frequently. To calculate the mass/volume concentration we need to know that the relative molecular mass is $M_r = 58.45$. The mass required for a volume of solution $V$ is then (see EQ1.27)

$$m = M_r \times 1 \text{ g mol}^{-1} \times c \times V \qquad \text{(EQ1.30)}$$

and

$$m/V = 0.15 \text{ mol l}^{-1} \times 58.45 \text{ g mol}^{-1} = 8.77 \text{ g l}^{-1}. \qquad \text{(EQ1.31)}$$

In practice, it is often more convenient to write this as 8.77 mg ml$^{-1}$. Whichever, you now choose a convenient volume and mass so that the mass you need to weigh is large enough for your balance while the volume of solution is small enough for your bottle.

### Table 1.7 Concentrations expressed as percentages

| designation | how written | explanation | notes |
|---|---|---|---|
| % by weight | % or %(w/w) | 1 %(w/w) = 1 g (100 g)$^{-1}$ = 10$^{-2}$ mass of solute divided by total mass of the solution × 100 % | Very useful when accuracy is important as mass can be determined more accurately than volume in most laboratories. |
| % by volume | % or %(v/v) | 1 %(v/v) = 1 ml (100 ml)$^{-1}$ = 10$^{-2}$ | Useful only when the solute is a liquid as in 10 %(v/v) ethanol in water or medium made up with 10 % bovine serum. The latter should be 10 %(v/v) rather than just 10 % but because serum is a liquid and the medium was almost certainly prepared by pipetting the (v/v) is often omitted because it is assumed to be 'obvious'. |
| %(w/v) | %(w/v) Not just % | mass of solute divided by total volume of the solution × 100 % 1 %(w/v) = 1 g (100 ml)$^{-1}$ | Because a mass cannot be a fraction of a volume it also cannot be a percentage of a volume and thus this unit is prohibited in SI. Nevertheless %(w/v) is easily the most common version of percentage units encountered in biology. Think of it as an abbreviation for g (100 ml)$^{-1}$. You must recognize it when you see it and be able to use it to calculate a mass of solute or volume of solution when you know the other. |

## Conversion between molarity and molality

This is perhaps the most complicated conversion you will encounter because you need to convert between volume of solution and mass of solvent. Thus, as an intermediate step, you need to know both the mass of the solute and the mass of the solvent in a liter of the solution. To find this you need to know the density of the solution, i.e. its total mass per unit volume.

Call the molarity of the solution $c$, its volume $V$, the relative molecular mass of the solute $M_r$, the density of the solution $\rho$, and the molality $b$. The amount of solute present is then $c \times V$. The mass of the solution is $m_{solution} = \rho \times V$, the mass of solute present is $m_{solute} = c \times V \times M_r \times 1 \text{ g mol}^{-1}$, and by difference the mass of the solvent is

$$m_{solvent} = m_{solution} - m_{solute} = V \times (\rho - c \times V \times M_r \times 1 \text{ g mol}^{-1}).$$

(EQ1.32)

The molality can now be calculated directly as

$$b = \frac{(\text{amount of solute})}{(\text{mass of solvent})} = \frac{c \times V}{(\rho - c \times M_r \times 1 \text{ g mol}^{-1})V}$$
$$= \frac{c}{\rho - c \times M_r \times 1 \text{ g mol}^{-1}}.$$

(EQ1.33)

Thus, for 0.1500 M NaCl, which has a density of 1.0064 kg l$^{-1}$, the molality is

$$b = \frac{0.1500 \text{ mol l}^{-1}}{1.0064 \text{ kg l}^{-1} - 0.15 \text{ mol l}^{-1} \times 58.45 \times 1 \text{ g mol}^{-1}}$$
$$= \frac{0.1500 \text{ mol l}^{-1}}{(1.0064 - 0.00877) \text{ kg l}^{-1}} = 0.1504 \text{ mol kg}^{-1}.$$

(EQ1.34)

That confirms what we said before. For relatively dilute solutions the numerical values of the molarity in mol l$^{-1}$ and the molality in mol kg$^{-1}$ are almost the same. Accurate conversion between them requires you to know the density of the solution – which requires delving into esoteric sources like the *Handbook of Physics and Chemistry* or the International Critical Tables.

# 1.7 Dilutions and doses

In virtually every branch of biology you need to add chemicals to solutions to produce effects. In biochemistry this might be an inhibitor to change the rate of a reaction. In pharmacology the substance added is called a drug and the amount added the dose. If the volume to which the doses are added can be defined, the doses are chosen to produce known concentrations.

Much of pharmacology is based on experimental determination of dose–response curves using *in vitro* assay systems that produce some measurable response to the drug of interest. It is an intriguing fact that the muscle of the ileum, the longest part of the small intestine, has receptors for a huge variety

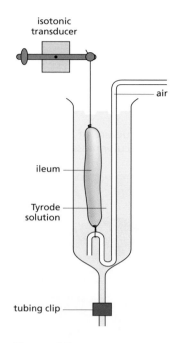

of interesting chemicals. Some say if there is a drug receptor in the brain you will also find it somewhere in the wall of the gut. However true that may be, a lot has been learned from applying drugs to lengths of ileum and noting how much they contract. As indicated in Figure 1.7, the length of gut is suspended in a bath of known volume, for example 10 ml, with one end tethered and the other attached to a transducer which detects changes in length. The experiments measure the shortening produced by various concentrations of the drug, for instance acetylcholine, and how these responses are altered by other drugs, for instance atropine. Typically a concentration–response curve (see Figure 1.8) for acetylcholine might be determined using a range of concentrations from 10 nM to 50 μM acetylcholine in the absence of atropine and compared with those measured in the presence of a low concentration, for example 1 nM, of atropine. How the bath concentrations of acetylcholine are achieved is considered in the exercises at the end of this chapter. Here we will consider how to add 1 nM atropine. 'Adding a concentration of a drug to a solution' means you add the amount of the drug needed to produce the stated concentration.

Suppose we have cells or a tissue suspended in 10 ml of a buffered solution (often called 'the buffer') and we want to add atropine (see Figure 1.9), relative molecular mass 289, to achieve a concentration of 1 nM. What mass of drug must be added? The answer is (10 ml = $10^{-2}$ l):

$$1\ \text{nM} = 10^{-9}\frac{\text{mol}}{\text{l}} \times 10^{-2}\ \text{l} \times \frac{289\ \text{g}}{\text{mol}} = 2.89 \times 10^{-9}\ \text{g} = 2.89\ \text{ng}. \quad (\text{EQ1.35})$$

(Scientific notation for numbers in which $10^{-9}$ is used as a shorthand for 1/1 000 000 000 is discussed in Section 2.6.) Routinely, we can easily weigh out 10 mg to 1 % accuracy, but 2.89 ng is not something that is exactly easy to handle; in fact you probably could not see it on the end of a spatula. So we will need to add the required amount of drug by pipetting a small quantity of a concentrated stock solution into our final solution. How do we prepare this stock solution and what volume of the stock must we add? To consider the volume first, the smallest amount we can add is limited by the accuracy of our pipettes. Modern pipettes (like the one in Figure 1.10) can deliver as little as 0.5 μl but to achieve 1 % accuracy it is best to handle volumes of 10 μl or greater. The largest volume of stock we can add is limited by the amount of its solvent we can tolerate in the final solution. If the drug is dissolved in a

**Figure 1.7**
Organ bath apparatus for measuring contractions of a segment of ileum. Drugs in concentrated stock solutions can be added to the bathing solution by simple pipetting. The dilution factor is just the volume added divided by the final volume present.

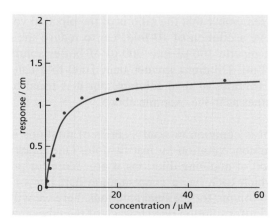

**Figure 1.8**
Shortening of a 4 cm length of guinea-pig ileum in response to a range of concentrations of acetylcholine. Note that in order to display the data for the high concentrations, 10, 20, and 50 μM, the data points for the low concentrations are plotted very close to each other.

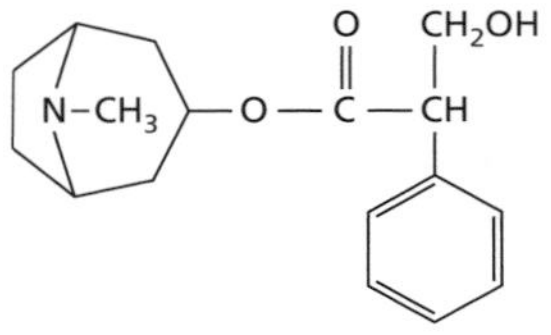

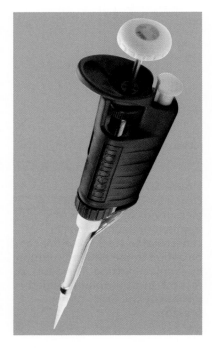

**Figure 1.9**
Molecular structure of atropine, relative molecular mass 289.

**Figure 1.10**
Gilson pipette. Image courtesy of Anachem Ltd.

solvent like ethanol we would normally ensure that the volume pipetted was less than 0.1 % of the final volume. This limit can be increased to 1 % if the solvent is water but if we must add an even larger volume then the stock we add must have been prepared by adding drug to a solution with the same concentrations of everything else, namely ions, sugars, etc., as the final solution. The solution with all of these 'ordinary' components is often called the buffer solution or even just 'the buffer'. Strictly, of course, the 'buffer' is the pH buffer which is almost always included in the solution. In effect the buffer solution can then be thought of as the solvent for the drug.

Pipetting 10 µl of our stock solution to a final volume of 10 ml will produce a 1000-fold dilution of the stock, thus the concentration of atropine in the stock solution must be 1 µM so that the final concentration can be 1 nM. Preparing a 1 µM stock solution is still problematic. To make up 10 ml we would need to weigh out 2.89 µg (1000 times more than before), which is still too small for most laboratory balances. So the procedure has to be a little more complicated: we must prepare an initial stock at a higher concentration, say 1 mM; use that to prepare a second, working stock 1000 times more dilute; and finally use the second stock to add the atropine to the experimental solution.

The principle behind making dilutions is simple; we add a small volume of a concentrated solution of drug to a volume of solvent or buffer solution. The key point is that the amount of drug in the pipetted volume equals the amount of drug in the total volume of the final solution. If $V_p$ is the volume pipetted, $V_f$ is the final volume, $c_p$ is the concentration in the stock solution, and $c_f$ is the concentration in the final solution (same units), then the amount of drug is just

$$\text{amount} = c_p \times V_p = c_f \times V_f. \tag{EQ1.36}$$

The concentrations can be either amount/volume or mass/volume, but they must both be expressed the same way. The ratio of the concentrations is just the inverse of the ratio of the volumes (you can think of inverting a ratio as turning it upside down)

$$\frac{c_f}{c_p} = \frac{V_p}{V_f}. \tag{EQ1.37}$$

It is important to remember that the ratio is of the pipetted volume to the *final* volume. To achieve a dilution of 10-fold, i.e. to reduce the concentration to 1/10th, we might pipette 100 µl into 900 µl of buffer solution to achieve a final volume of 1 ml. Dilutions greater than 1000-fold can be prepared by making up large volumes of final solution, but it is usual to do it in stages with smaller volumes as in the example above.

In microbiology it is common to want a series of dilutions where successive solutions differ in concentration by just twofold. One quick way to produce these is the method of doubling dilutions, a procedure for producing a series of dilutions in which each has a concentration half that of its predecessor. Pick a convenient volume for the final solutions, here we will use 0.5 ml. We add 0.5 ml of the buffer solution *once* to each of the tubes in the series except the first. Then we add 0.5 ml of the original stock in buffer to the first tube *twice*. Now we take 0.5 ml from the first tube, add it to the second and mix. The concentration in the second tube is now half that in the first. Next we take 0.5 ml from the second tube and mix it into the third. The concentration in the third tube is now $1/2 \times 1/2 = 1/4$th. If then we take 0.5 ml of the third solution and mix it into the fourth, the concentration in the fourth tube is $1/2 \times 1/4 = 1/8$th and so on. This procedure is simple and can be repeated as often as needed. It has several advantages. Only one pipette and tip are required, the volume pipetted is always the same, and the accuracy of the dilutions is limited only by the reproducibility, sometimes called the precision, of the pipette and not by its absolute accuracy. Equally importantly, the simplicity of the procedure helps us to avoid mistakes. For instance the tube from which we must take the 0.5 ml is always the one that has 1 ml in it. All we need to remember while pipetting is which tube is next for the length of time it takes to move the pipette tip from one to the other. This sounds silly but when doing really tedious tasks any trick that helps to prevent errors is very welcome. (Why doubling? Probably because we keep doubling the volume of successive tubes.)

## Box 1.2

Acetylcholine is to be added to the bathing solution for a length of ileum at concentrations 1, 2, 5, 10, 20, 100, 200, 500, 1000, 2000, or $5000 \times 10^{-8}$ M. The drug is available as a 10 mM solution in water. What other stocks in water should be prepared?

The intermediate stock solutions will also be in water – so it will be acceptable for the volume of stock pipetted into the final solution to be as high as $1\% \times 10$ ml = 100 μl. The smallest volume we can pipette accurately is 10 μl. Thus for the solution with the least drug we will want a stock such that the smallest volume we can pipette accurately, 10 μl, into 10 ml will give the correct concentration. The ratio of these volumes is 1 : 1000 so the ratio of the

concentration in the most dilute stock to the final concentration must be 1000 : 1; that is, the concentration must be $1 \times 10^{-8}$ M $\times 10^3 = 10$ μM. Using additions between 10 and 100 μl this stock can be used to produce 1, 2, 5, and $10 \times 10^{-8}$M. 10 μM is a 1000-fold dilution of the 10 mM stock. A 100-fold dilution of the 10 mM stock will suffice for 20, 50, and $100 \times 10^{-8}$ M, a 10-fold dilution for 200, 500, and $1000 \times 10^{-8}$ M and the full-strength stock can be used for 2000 and $5000 \times 10^{-8}$ M. So, to cover the entire range of concentrations, we need to make three dilutions, 10-fold, 100-fold, and 1000-fold. Each of these could be prepared in a single step from the original stock or they could be prepared using three 10-fold serial dilutions.

## 1.8 Numerical value equations

So far in this chapter almost all of the equations have been physical value equations that relate real physical quantities. These are often just what we want, especially when writing down equations that are to be applied to many situations. However, you will have noticed that when using these physical value equations it is necessary to write out the units whenever specific values are included. That can get really tedious. There is another way to write equations that can greatly reduce the number of times we need to write out units. These are numerical value equations and you will see a lot of them in Chapters 5 to 8.

You will have met the equation $F = mg$ where $m$ is the mass of an object, $F$ is the force due to the gravity of the earth, and $g$ is the acceleration due to gravity. The constant $g$ is not a fundamental constant of physics. It depends on exactly where we are: the further we are from the center of the earth the smaller it is. If our interest were calculating satellite orbits we would need to worry about how $g$ varies with position. However, to two significant figures it is the same everywhere on the earth's surface and can be written as 9.8 m s$^{-2}$. So we can write

$$F = m \times (9.8 \text{ m s}^{-2}). \qquad \text{(EQ1.38)}$$

This equation is the subject of one of the more interesting legends in physics, that Galileo dropped different masses from the Leaning Tower of Pisa (see Figure 1.11) and discovered that they all fell with the same acceleration, which according to this equation is just 9.8 m s$^{-2}$. EQ1.38 is a **physical value equation**; $F$ and $m$ represent the actual force and the actual mass. Thus if the mass is 70 kg

$$F = 70 \text{ kg} \times 9.8 \text{ m s}^{-2}$$
$$= 686 \text{ N}. \qquad \text{(EQ1.39)}$$

(The standard 70 kg man weighs 686 N at the surface of the earth.)

We can convey exactly the same information using a **numerical value equation** by writing the following:

$$F = m \times 9.8 \qquad \text{(EQ1.40)}$$

where $F$ is the force measured in newtons and $m$ is the mass measured in kilograms. Here, in contrast to the equations before, $F$ and $m$ are pure numbers. This numerical value equation correctly relates the numerical values of the force and mass only when the physical quantities have been expressed using a particular choice of units. If we change the units we have to change the numerical constant.

Consider again our standard man with mass 70 kg. We can define a symbol, $m$, to represent the mass of an individual and for this man $m = 70$ kg. However, we could also define a different symbol, let us call it $\{m\}$, by saying the mass of an individual is $\{m\}$ kg. So defined, $\{m\}$ is just a number that gives the

A **physical value equation** states a relation between physical quantities. When using the equation for calculations each variable standing for a quantity must take on a numerical value and its associated unit.

A **numerical value equation** states a relation between numerical values of physical quantities that depend on a specific choice of units. The units chosen *must* be stated somewhere close-at-hand and the equation must be rewritten if they are changed.

**Figure 1.11**
Leaning Tower of Pisa. According to legend, Galileo dropped objects of different masses while hanging over the edge and determined that they fell with the same acceleration. This is the prediction from Newton's second law, $F = ma$, and the law of gravitation near the earth's surface, $F = mg$, because these imply $a = g$. Image courtesy of M4rvin under Creative Commons Attribution–Share Alike 2.0 Generic.

numerical value of the mass when it is expressed in kilograms; $\{m\}$ itself has no units. Now what happens if we want to express the mass in grams? The physical quantity is not affected by our choice of units, thus we write

$$m = 70 \text{ kg} = 70\ 000 \text{ g}. \tag{EQ1.41}$$

However, the numerical value most definitely does change. If the mass is expressed in grams rather than kilograms, $\{m\}$ becomes 70 000! Hopefully, this is clear. Now to add confusion, most of the time we do not bother to write the curly brackets! You have got to work out from context whether a symbol like $m$ is referring to the physical quantity, $70 \text{ kg} = 70\ 000 \text{ g}$, or to the numerical value which would be 70 if mass is being expressed in kilograms or 70 000 if it is being expressed in grams.

How do you know whether a symbol represents a physical quantity or a numerical value? First, look for the definition of the symbol. In the paragraph just above, the indication that $\{m\}$ is a numerical value (in addition to the dead giveaway of the curly brackets) is that it is immediately followed by the units! The combination of the numerical value and the units describes the physical quantity. By contrast, as $m$ is defined at the start of this section as 'the mass of an object' then it must be a physical quantity, a mass. What do you do if the definition is not immediately to hand? You look for clues in the context. Look at the equation and its surrounding text. EQ1.38 must be a physical value equation because the constant is written with its units, $9.8 \text{ m s}^{-2}$. By contrast, with EQ1.40 you are given two clues that it is a numerical value equation: the 9.8 is written without units and the units needed to interpret $F$ and $m$ are given immediately after the equation.

Why do we use numerical value equations? Why put up with the complications of having two sorts of equation to deal with? This may be made clearer by the following example.

Suppose we have measured the rate of a reaction as a function of the concentration of the substrate. The measurements can be reported in a table, as in Table 1.8, or in a graph, as in Figure 1.12. In the SI convention, entries listed in a table or plotted on a graph should be pure numbers. We get the number by dividing the physical quantity by a standard reference value. Here the concentrations have been divided by 1 mM and the rates by $1 \text{ mol s}^{-1}$. Ignoring the units, the numbers in this table, the numerical values, can be fitted with a smooth curve (see Figure 1.12) with equation

| **Table 1.8 Rate of reaction versus concentration** | |
|:---:|:---:|
| *C* / mM | *R* / (mol s$^{-1}$) |
| 0 | 0.000 |
| 1 | 0.017 |
| 5 | 0.079 |
| 10 | 0.147 |
| 15 | 0.202 |
| 20 | 0.246 |
| 25 | 0.277 |
| 30 | 0.296 |
| 35 | 0.303 |
| 40 | 0.298 |
| 45 | 0.281 |
| 50 | 0.253 |

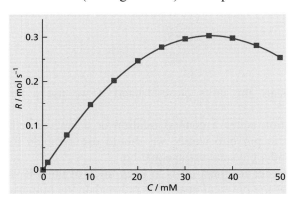

**Figure 1.12**
Rate of reaction versus concentration. The points are taken from Table 1.8. The smooth curve is a plot of EQ1.42.

$$R = 0.0171C - 0.000241C^2. \tag{EQ1.42}$$

$C$ and $R$ in Table 1.8 and Figure 1.12 represent the physical quantities concentration and rate. In the numerical value equation EQ1.42, these symbols have different meanings. In this equation they represent the numerical values obtained when the concentration and rate are divided by particular reference values, 1 mM for concentration and 1 mol s$^{-1}$ for the rate of reaction. Any other choice of reference values would require different values of the constants in the numerical value equation. For instance, if we wanted to express the concentrations with unit M instead of mM (using reference value 1 M instead of 1 mM), the equation would have to be rewritten as

$$R = 17.1C - 241C^2. \tag{EQ1.43}$$

As a check, if the physical value of the concentration is 0.02 M = 20 mM, either EQ1.42 with $C = 20$ or EQ1.43 with $C = 0.02$ tells us that $R \approx 0.25$, but if we use the wrong numerical value for $C$ in either equation we get the wrong answer!

It is of course possible to describe the data in Table 1.8 using a physical value equation where now $R$ and $C$ have the same meanings as in Table 1.8 and Figure 1.12.

$$R = \left(0.0171 \text{ mol s}^{-1} \text{ mM}^{-1}\right) \times C - \left(0.000241 \text{ mol s}^{-1} \text{ mM}^{-2}\right) \times C^2. \tag{EQ1.44}$$

This equation has the advantage that it remains valid for any choice of units; or, to put it another way, before we can use it to calculate actual values the equation forces us to change units if $C$ and $R$ are initially expressed in units other than mM and mol s$^{-1}$. Thus in a sense this equation is safe; it forces us to sort out the units. However, because EQ1.44 is so much messier, most people would just make sure they got the reference values right before deciding on the values of $C$ and $R$. Numerical value equations often save a lot of writing (compare EQ1.43 and EQ1.44), but they leave sorting out the units entirely to you. It is a little like removing the crash barriers on a twisting mountain road. The bad news is you will frequently encounter both physical value and numerical value equations (and you will probably have to work out just what somebody meant when they have not given clear clues). The good news is both are very useful and a wise choice of which to use can make life a lot easier.

Advice:

1.  We **must** state the units close-at-hand if we use a numerical value equation.
2.  If the equation states a general relation or a law, something we might want to try to put into words, we should use symbols for the physical quantities and a physical value equation.
3.  It is much easier for most of us to keep track of units if we use symbols that stand for the physical quantities rather than the numerical values.

4.  If the equation states an empirical relation, where the constants are represented by their values rather than by symbols, we are probably better off using numerical value equations.

Although we may sometimes curse units because they can take so much time to write, it is often worth the effort just to make sure that you have all the factors in the right places. If the units come out right at the end of the calculation, then there is a good chance that we have done things correctly. If they come out wrong, then we know there must be an error, so we have to go back and find the mistake.

---

### Box 1.3

In official documentation an unambiguous convention is needed to distinguish between physical quantities and numerical values of those quantities. Sometimes this extra notation actually aids understanding and an example crept into the discussion of temperature (see EQ1.25 and EQ1.26) because we thought it helped. Using the same curly-bracket notation for numerical values the physical value equation,

$$F = m \times (9.8 \text{ m s}^{-2})$$

becomes

$$\{F\}_N = \{m\}_{kg} \times 9.8$$

where $\{F\}_N = F / 1 \text{ N}$ and $\{m\}_{kg} = m / 1 \text{ kg}$.

Similarly we could rewrite EQ1.42 as

$$\{R\}_{\text{mol s}^{-1}} = 0.0171\{C\}_{mM} - 0.000241\{C\}_{mM}^2.$$

The main reason to use numerical value equations is to save the tedium of writing units. Most people, including us, decide that the tedium of writing all the brackets is still too great. You will rarely see the curly brackets, even in this book.

---

## 1.9 Scalar and vector physical quantities

So far, we have discussed each physical quantity in terms of its unit and numerical value. A physical quantity that is completely specified by the unit and a single numerical value is called a scalar. These are what we will be considering in the rest of this book. If we do not say otherwise, physical quantity means scalar physical quantity. However, if you think for a minute about physical quantities like force, velocity, or acceleration, you will realize that we need to specify not only their magnitudes but also their directions. To specify a force completely takes not one magnitude but three, which we would often choose to be values along three perpendicular axes, $x$, $y$, and $z$. A physical quantity that requires a direction as well as a magnitude is called a vector. (The word vector also has other meanings in pathology and molecular biology; it can mean an agent that spreads a disease or an agent used to insert genetic material into a cell.)

Physicists need to use vector quantities like force, acceleration, and electric field in almost everything they do, but biologists need vector quantities only rarely so we will say very little more about them.

## Presenting Your Work

### QUESTION

Stock solutions for preparation of a 'physiological' saline.

It is common practice to make up and store concentrated stock solutions so that working buffer solutions can be prepared quickly by simple dilutions. Complete the table below to indicate (i) how much of each substance should be weighed out to prepare the stocks and (ii) the volumes to be added to prepare 250 ml of final solution in a volumetric flask. Note that 'how much' in this context means how much mass.

| substance | $M_r$ | $C_{stock}$ / mM | $V_{stock}$ / ml | mass of substance to prepare stock $m_{stock}$ / g | $C_{final}$ / mM | volume to pipette to prepare 250 ml of buffer $V_p$ / ml |
|---|---|---|---|---|---|---|
| NaCl | 58.45 | 1000 | 1000 | | 140 | |
| KCl | 74.56 | 500 | 50 | | 4 | |
| $MgCl_2 \cdot 6H_2O$ | 203.33 | 500 | 10 | | 0.5 | |
| $CaCl_2$ (anhydrous) | 110.99 | 500 | 20 | | 1.5 | |
| Hepes (pK 7.49) | 238.3 | 100 | 500 | | 10 | |
| NaOH | 40.01 | 1000 | 100 | | ~5 [†] | |
| water | – | – | – | – | – | approximately [‡]: |
| glucose | 180.16 | – | – | – | 10 | mass to add [§]: $m_{glu} =$ |

Notes:

[†]The NaOH would in fact be added dropwise to titrate the final solution to pH 7.4 (or whatever is desired). Because the pK of Hepes is 7.49, slightly less than half as much NaOH must be added as Hepes.

[‡]You can assume for this relatively dilute solution that all of the volumes must add up to 250 ml.

[§]Glucose solutions grow bacteria unless sterilized. It is thus common practice not to use a glucose stock solution but rather to add glucose as solid when making up the final solution.

For NaCl  amount of NaCl in stock $= C_{stock} \times V_{stock} = 1$ mol
mass of NaCl in stock $=$ amount $\times M_r = 58.45$ g
volume to pipette $=$ (amount in buffer) $/ C_{stock}$
$= 35$ mmol $/ 1000$ mmol$^{-1} = 35$ ml

| substance | $M_r$ | $C_{stock}$ / mM | $V_{stock}$ / ml | mass of substance to prepare stock $m_{stock}$ / g | $C_{final}$ / mM | volume to pipette to prepare 250 ml of buffer $V_p$ / ml |
|---|---|---|---|---|---|---|
| NaCl | 58.45 | 1000 | 1000 | 58.45 | 140 | 35 |
| KCl | 74.56 | 500 | 50 | 1.864 | 4 | 2 |
| MgCl$_2$·6H$_2$O | 203.33 | 500 | 10 | 1.017 | 0.5 | 0.25 |
| CaCl$_2$ (anhydrous) | 110.99 | 500 | 20 | 1.110 | 1.5 | 0.75 |
| Hepes (pK 7.49) | 238.3 | 100 | 500 | 11.91 | 10 | 25 |
| NaOH | 40.01 | 1000 | 100 | 4.001 | ~5 | 1.25 |
| water | – | – | – | – | – | 185.75 |
| glucose | 180.16 | – | – | – | 10 | 0.4504 g |

## End of Chapter Questions

(Answers to questions can be found at the end of the book)

### Basic

**1.** In the British system of units, length is measured in *feet* and *inches*. One foot is defined as twelve inches (1 ft = 12 in), whereas one inch is defined as 1 in = 25.4 mm. Suppose that a leafcutter ant travels at 2.4 ft min$^{-1}$; what is its speed in mm s$^{-1}$ to three significant figures?

**2.** Convert the following percentages into decimal numbers using the SI convention that % stands for 'multiply by the number 0.01':
(a) 45 %
(b) 2 %
(c) 0.00374 %
(d) 237 %

**3.** Convert the following into percentages:
(a) 0.67
(b) 0.00000024
(c) 1/5
(d) 5.671

**4.**
(a) Express 20 °C in Fahrenheit and kelvin.
(b) Express 98.6 °F in Celsius and kelvin.
(c) The temperature in outer space is said to be about 4 K. What is this temperature in Celsius and Fahrenheit?

**5.** If 180.16 g of glucose contains 1 mol of glucose calculate how many glucose molecules there are in 1 pg. Take Avogadro's constant (Loschmidt constant) to be $6.022142 \times 10^{23}$ mol$^{-1}$. Give your answer to the nearest million.

**6.** Find the molality of a 1 % (w/w) solution of glucose. (180.16 g of glucose contains 1 mol.)

**7.** Find the molarity of a 0.05 % solution of NaCl given that the relative molecular mass of NaCl (one Na$^+$ ion and one Cl$^-$ ion) is 58.45. What assumption do you need to make?

**8.** The molar mass of sucrose is 342.3 g mol$^{-1}$ and the density of a 2.555 M solution of sucrose is 1.2887 kg l$^{-1}$. What is the concentration of sucrose expressed as % (w/w)? What is the concentration expressed as molality (mol kg$^{-1}$)?

### Intermediate

**9.** The flux of bicarbonate ions across the membrane of a cultured brain endothelial cell has been reported as 1.25 pmol cm$^{-2}$ s$^{-1}$. Express this in terms of the derived SI unit for flux, mol m$^{-2}$ s$^{-1}$.

**10.** Write out a protocol for producing a series of 100-fold dilutions with final volumes close to 1 ml.

**11.** The initial rate of a bimolecular reaction (i.e. a reaction involving 2 reactants) depends on the number of collisions between the reactants. How much larger will the total amount of product produced per second be if we:
(a) double the volume of the solution keeping its composition the same;
(b) double the volume of the solution keeping the amounts of the reactants present constant;
(c) double the concentration of one reactant, keeping the concentration of the other and the volume constant;
or
(d) double the concentrations of both reactants while keeping the volume constant?

**12.** In a simple bimolecular reaction

$$A + B \underset{k'}{\overset{k}{\rightleftharpoons}} C$$

the initial rate of formation of C per unit volume, $R_{initial}$, is given by

$$R_{initial} = k[A]_{initial}[B]_{initial}$$

where $[A]_{initial}$ means initial concentration of A. For a 100 ml reaction flask, 1 µM initial concentrations of A and B, and $k = 2 \times 10^3$ M$^{-1}$ s$^{-1}$, calculate the total initial rate of formation of C.

**13.** What is the mass of a hydrogen molecule expressed in kilograms?

**14.** If the charge on a proton is $1.6 \times 10^{-19}$ C what is the sum of the charges on the positive ions in one liter of a 1 M NaCl solution?

**Advanced**

**15.** You are provided with a 60 % (w/w) solution of cesium chloride.

The relative molecular mass of CsCl is 168.37 and the specific gravity of a 3 M solution is 1.385.

How would you prepare:
(a)  15 g of a 20 % (w/w) solution;
(b)  15 g of a solution whose molality is 5 mol kg$^{-1}$;
(c)  10 ml of a 3 M solution? (You do not have access to a 10 ml volumetric flask.)

Hint: In general, you should first calculate how much CsCl you need in the final solution – then determine how to obtain this from the stock. With a 60 % (w/w) stock is it more appropriate for 'how much' to mean amount or mass?

**16.** What considerations are taken into account when deciding on the concentrations and volumes of stock solutions? Devise a choice of stock concentrations for NaCl, KCl, $MgCl_2$, $CaCl_2$, and Hepes such that 10 ml of each stock solution must be added to make up 250 ml of the final buffer solution described in Presenting Your Work.

# CHAPTER 2

# Numbers and Equations

In Chapter 1 we saw how we can give values for physical quantities in terms of numbers and units and then use algebra to state relations between these quantities. In this chapter we develop the use of algebra further while reviewing some aspects of arithmetic: fractions; powers and roots; order of calculations; and ratios and percentages. We shall see how the summation sign can be used to denote the addition of many values. Scientific notation for numbers is introduced to let us write very small and very large numbers and even ordinary-sized numbers when we know them to only a certain accuracy.

Algebra is the natural language to use to describe the basic operations we can do with numbers. Such operations include the simple things like adding, subtracting, multiplying, and dividing, and more advanced operations like raising to a power and extracting a root. Algebra is also a useful tool when we consider how numbers can be combined to make fractions, ratios, and percentages. Finally, we return to algebra itself to consider what it means to solve equations. A very important example of solving equations is provided by the prediction of the amount of a small molecule, which could be a drug or hormone, that will be bound to its receptor if we know the amount of the receptors, the concentration of the small molecule, and the affinity constant for the binding.

This chapter has three main objectives. The first is to help you understand what you are doing when you perform calculations, regardless of whether you use pencil and paper, a calculator, or a computer. The second is to show that algebra greatly aids us in describing the relations between numbers and, by extension, physical quantities. The third is to show you how to tackle algebraic problems.

## 2.1 Arithmetic with fractions

Any number we can write down, punch into a calculator, or type into a spreadsheet on a computer is a rational number; that is, it can be written as a fraction. Any integer is a fraction, $n = n/1$, and furthermore any decimal number that can be displayed is a fraction, for example a number like 3.14 can be regarded as a shorthand for 314/100. Arithmetic has been completely mastered when we can perform all the basic operations with fractions. And once we can do arithmetic with fractions, we can handle algebraic fractions. Practice with both the arithmetic and the algebra of fractions is provided in the End of Chapter Questions.

## The fundamental fraction is the reciprocal of an integer

The key to understanding fractions is first to understand what is meant by the reciprocal of an integer. The reciprocal, $1/n$, of any integer, $n$, is defined by the fact that if we add up $n$ of them, or equivalently if we multiply $1/n$ by $n$, we are back to 1 whole. It is the mathematical equivalent of pie slices (hence pie charts as in Chapter 9). If we slice a pie into 21 pieces, each is 1/21st of a pie. Once we have the reciprocal, fractions follow immediately, they are just integers multiplied by reciprocals; that is, if $a$ and $b$ are integers the general fraction can be written as

$$\frac{a}{b} \equiv a \times \frac{1}{b} \equiv \frac{1}{b} \times a. \qquad \text{(EQ2.1)}$$

## The reciprocal of a product is the product of the reciprocals

To take an example suppose we want 21 slices. We can get to our 21 slices in at least three ways. We could just set about slicing 21 pieces. We could first cut the pie into 7 equal slices and then divide each of these into 3. Finally, we could first cut 3 slices and then divide each of these into 7. In other words if we define multiplication of reciprocals as successively subdividing into pieces then $1/21 = (1/3) \times (1/7) = (1/7) \times (1/3)$; that is, $1/(3 \times 7) = (1/3) \times (1/7)$ and we have the reciprocal of a product is the product of the reciprocals and furthermore we can multiply reciprocals in any order. Showing this formally using symbols and algebra is good practice in algebraic techniques – see End of Chapter Question 18. If we let $a$ and $b$ be any integers other than zero, then

$$\frac{1}{a \times b} \equiv \frac{1}{b} \times \frac{1}{a} \equiv \frac{1}{a} \times \frac{1}{b}. \qquad \text{(EQ2.2)}$$

## Factorization and the reduced form of a fraction

Now suppose we have divided the pie into some multiple of 7 slices, for example 14 or 21 or, more generally, $7n$ where $n$ is an integer. If now we collect $n$ pieces we are back to 1/7th of the pie, if we collect $2n$ we are back to 2/7ths of the pie, and so on. Thus 2/7, 4/14, 6/21, and $2n/7n$ where $n$ is any integer are all the same fraction of the whole. If we can write the numerator and denominator as products of integers, an example of factorization, any factors in common can be struck out or canceled without changing the value of the fraction. A fraction is said to be in its unique, reduced form (also called its simplest form) if the numerator and denominator do not share any common factors other than 1. It is normal to state fractions in reduced form, but we are at liberty to use any other form when it is convenient, for example in the next section when adding fractions.

## Addition and subtraction

How do we add 2/7ths and 1/3rd? The general rule is to find an integer, preferably the smallest, that is exactly divisible by both denominators – for 3 and 7 this is 21. 2/7 is the same as 6/21 and 1/3 is 7/21 so the addition becomes $(6/21) + (7/21) = 13/21$. This process can be summarized algebraically as

$$\frac{a}{b} + \frac{c}{d} \equiv \frac{ad}{bd} + \frac{cb}{db} \equiv \frac{ad + cb}{bd}. \tag{EQ2.3}$$

where $a$, $b$, $c$, and $d$ are integers with $b \neq 0$ and $d \neq 0$. The rule for addition of fractions is an example of just how compact and time saving using symbols and algebra can be. The same statement in words would go something like this (take a deep breath): 'To calculate the sum of two fractions the numerator of the result is the first numerator times the second denominator plus the second numerator times the first denominator and the denominator of the result is the first denominator times the second denominator'. (Breathe again.)

We do not need separate rules for subtraction, it is just addition of a negative number, so

$$\frac{2}{7} - \frac{1}{3} = \frac{6}{21} - \frac{7}{21} = \frac{6 - 7}{21} = -\frac{1}{21}. \tag{EQ2.4}$$

In algebraic terms we can write

$$\frac{a}{b} - \frac{c}{d} \equiv \frac{a}{b} + \frac{-c}{d} \equiv \frac{ad - cb}{bd}. \tag{EQ2.5}$$

## Multiplication

What is $(2/7) \times (2/3)$, which can be read as either 2/7ths times 2/3rds or as 2/7ths of 2/3rds? Now 2/3 is the same as 14/21 and if our pie has been sliced into 21 pieces we have 14. To find 2/7ths of these 14, for each 7 slices we need to keep 2 (2/7ths or 2 out of 7), so we keep 4 and the answer is 4/21. To summarize

$$\frac{2}{7} \times \frac{2}{3} = \frac{2}{7} \times \frac{14}{21} = 2 \times \frac{1}{7} \times 14 \times \frac{1}{21} = 2 \times \frac{14}{7} \times \frac{1}{21} = 2 \times 2 \times \frac{1}{21} = \frac{4}{21} \tag{EQ2.6}$$

or just

$$\frac{2}{7} \times \frac{2}{3} = \frac{2 \times 2}{7 \times 3} = \frac{4}{21}. \tag{EQ2.7}$$

In general,

$$\frac{a}{b} \times \frac{c}{d} \equiv \frac{a \times c}{b \times d} \equiv \frac{ac}{bd}. \tag{EQ2.8}$$

To put it into words: in the product of two fractions, the numerator of the result is the product of the numerators and the denominator of the result is the product of the denominators.

## Reciprocal of a fraction

From the definition of reciprocal, the reciprocal of a fraction times that fraction must be 1. Because $(2/7) \times (7/2) = 1$ the reciprocal of 2/7 is 7/2. To find the reciprocal of a fraction all you need to do is swap the numerator and denominator, i.e. you turn the fraction upside-down.

$$\frac{1}{\left(\dfrac{a}{b}\right)} \equiv \frac{b}{a}.$$

(EQ2.9)

## Division

Division by any number (except zero) is the same as multiplication by the reciprocal of the number. So

$$\frac{a}{b} \div \frac{c}{d} \equiv \frac{a}{b} \times \frac{d}{c} \equiv \frac{ad}{bc}.$$

(EQ2.10)

For instance,

$$\frac{2}{7} \div \frac{2}{3} = \frac{2}{7} \times \frac{3}{2} = \frac{3}{7}.$$

(EQ2.11)

What this says is that 2/7ths is a larger fraction of 2/3rds of a pie than it is of a whole pie, in fact it is 3/7ths of 2/3rds of a pie but only 2/7ths of a whole pie. This calculation is indicated in terms of pie slices in Figure 2.1.

**Figure 2.1**
A 21-segment pie showing 2/7ths of the pie (red), 2/3rds of the pie (red plus blue), and thus also the fraction of 2/3rds represented by 2/7ths, namely (2/7)/(2/3). 2/7ths is 6 slices, 2/3rds is 14 slices, and 2/7ths as a fraction of 2/3rds is 6 out of 14 slices or 3/7ths.

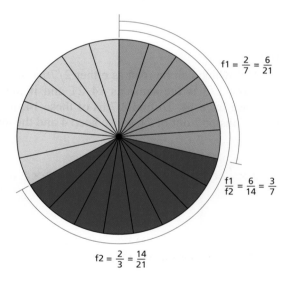

$$f1 = \frac{2}{7} = \frac{6}{21}$$

$$\frac{f1}{f2} = \frac{6}{14} = \frac{3}{7}$$

$$f2 = \frac{2}{3} = \frac{14}{21}$$

# 2.2 Addition of many terms

Frequently we have to add up not just two numbers, but lots of them. Doing it with actual numbers and a calculator is not hard; just a 'bit' tedious. However, writing 999 plus signs to indicate that 1000 values need to be added is just going too far. As we frequently do need to say 'add up a list of variables' (see Chapter 9) we need a notation for addition of many terms that is both compact and convenient. For example, suppose we want to add up a list of numbers that begins:

1.2,  2.3,  1.7,  3.4,  2.0,  ...

The secret is to use subscripted variables, which in this context are called indexed variables: in this case we could let $x_i$ stand for the $i$th number in the list:

$$x_1 = 1.2, \ x_2 = 2.3, \ x_3 = 1.7, \ x_4 = 3.4, \ x_5 = 2.0 \ldots$$

We also define a symbol $\sum$ (read sigma) to mean 'add them up'.

Thus

$$\sum_{i=1}^{n} x_i = x_1 + x_2 + x_3 + \ldots + x_{n-2} + x_{n-1} + x_n \qquad \text{(EQ2.12)}$$

means add up all $n$ of the values of the variables $x_i$ that correspond to values of the index $i$ from 1 to $n$ in steps of 1. In other words we add up all the values from $x_1$ to $x_n$. For the values given above the sum of the first five values is given by

$$\sum_{i=1}^{5} x_i = x_1 + x_2 + x_3 + x_4 + x_5 = 1.2 + 2.3 + 1.7 + 3.4 + 2.0 = 10.6.$$

$$\text{(EQ2.13)}$$

If we wanted to add up only the 5th to 10th values we would write $\sum_{i=5}^{10} x_i$.

The convenience of the sigma notation for summation means we must learn how to use subscripts. We also need to know a few basic properties of indexed summation.

- Multiplying every term by the same factor $a$ and adding up gives the same answer as first adding up the terms and then multiplying the sum by $a$:

$$\sum_{i=1}^{n} a x_i = a \sum_{i=1}^{n} x_i. \qquad \text{(EQ2.14)}$$

- The order of addition still does not matter:

$$\sum_{i=1}^{n} (x_i + y_i) = \sum_{i=1}^{n} x_i + \sum_{i=1}^{n} y_i. \qquad \text{(EQ2.15)}$$

- Finally, if all the variables have the same value, call it $b$, the sum is just $n$ times $b$:

$$\sum_{i=1}^{n} x_i = \sum_{i=1}^{n} b = nb. \qquad \text{(EQ2.16)}$$

## 2.3 Powers and roots

Repeated multiplication of a number by itself is known as raising to a power. Multiplication was invented so that we could add on the same thing many times. Similarly 'raising to a power' was invented to indicate repeated multiplication. Repeated multiplication is encountered a lot more frequently than might at first be thought. For compound interest on a savings account, each month the amount saved increases by a constant factor, say 1.004, so that an initial investment of £1000 would be worth £1000 × 1.004 after one month, £1000 × 1.004 × 1.004 after two, and so on. To take a more biological example, bacterial cells might divide once every twenty minutes. If there were $n$ at the start, then after 20 min there would be $n \times 2$, after 40 min $n \times 2 \times 2$, after 60 min $n \times 2 \times 2 \times 2$, and so on.

Raising a number to the power of two is often referred to as squaring that number. For example the square of five is the same as the result of raising five to the power of two: $5^2 = 5 \times 5 = 25$. Similarly cubing a number is the same as raising that number to the power three: five cubed is $5^3 = 5 \times 5 \times 5 = 125$. Beyond that we just state the power to which a number is being raised. For example, we refer to $5^4 = 5 \times 5 \times 5 \times 5 = 625$ as 'five raised to the fourth power' or just 'five to the fourth'.

The number used in repeated multiplication is called the **base**, and number of items multiplied is the **exponent**. For instance, in $2^5 = 2 \times 2 \times 2 \times 2 \times 2 = 32$, the base is 2 and the exponent is 5. This is fine for exponents that are positive integers. But what do we mean if we use an exponent that is a negative integer, zero, or a fractional number?

Negative powers are used to indicate the reciprocals of numbers raised to powers:

$$2^{-1} = \frac{1}{2} \text{ and } 2^{-3} = \frac{1}{2^3} = \frac{1}{2 \times 2 \times 2} = \frac{1}{8} = 0.125. \tag{EQ2.17}$$

In algebraic terms we can say that

$$a^{-n} \equiv \frac{1}{a^n}, \quad \text{provided } a \neq 0. \tag{EQ2.18}$$

Zero raised to a negative power is not defined.

To understand what is meant by raising to the power zero, we first need to make use of some simple rules for combining multiplication and raising to powers. For example both

$$2^5 \times 2^3 = (2 \times 2 \times 2 \times 2 \times 2) \times (2 \times 2 \times 2)$$
$$= 2 \times 2 \times 2 \times 2 \times 2 \times 2 \times 2 \times 2 = 2^8 \tag{EQ2.19}$$

and

$$2^4 \times 2^{-3} = \frac{2 \times 2 \times 2 \times 2}{2 \times 2 \times 2} = 2 \qquad \text{(EQ2.20)}$$

are examples of the general rule that the product of a number $a$ raised to two powers equals $a$ raised to the sum of the powers:

$$a^x \times a^y \equiv a^{x+y}. \qquad \text{(EQ2.21)}$$

EQ2.20 is an instance of the general rule that the quotient of a number $a$ raised to two powers equals $a$ raised to the difference of the powers:

$$\frac{a^x}{a^y} \equiv a^{x-y}. \qquad \text{(EQ2.22)}$$

A very important special case of this rule is when one power is the negative of the other. From EQ2.22

$$a^x \times a^{-x} \equiv a^0. \qquad \text{(EQ2.23)}$$

However, because $a^{-x}$ is the reciprocal of $a^x$ and by the very definition of reciprocal anything multiplied by its own reciprocal equals 1:

$$a^x \times a^{-x} \equiv \frac{a^x}{a^x} = 1. \qquad \text{(EQ2.24)}$$

Therefore

$$a^x \times a^{-x} \equiv a^0 = 1. \qquad \text{(EQ2.25)}$$

Anything raised to the 0th power must be 1. For instance, $2^0 = 1$. However, as with division, there is an important exception: zero raised to the power zero, $0^0$, is not defined.

Another example,

$$2^3 \times 5^3 = 2 \times 2 \times 2 \times 5 \times 5 \times 5 = 2 \times 5 \times 2 \times 5 \times 2 \times 5 = (2 \times 5)^3$$

$$= 10^3 = 1000,$$

illustrates the use of a second general rule, the product of two numbers each raised to the same power is the same as the product of the numbers raised to that power.

$$a^x \times b^x \equiv (ab)^x. \qquad \text{(EQ2.26)}$$

An interesting case is raising to a power a number that was itself obtained by raising another number to a power. $(3^2)^3 = 9^3 = 729$ can also be written as $(3^2)^3$, $3^6$ or $(3^3)^2$, which is an example of another general rule. Raising a number first to one power then another is the same as raising the number to the product of the powers:

$$(a^x)^y \equiv (a^y)^x \equiv a^{x \times y}. \qquad \text{(EQ2.27)}$$

All of the preceding relations look fairly obvious when $x$ and $y$ are integers, but what does $a^x$ mean when $x$ is a fraction? As an example consider $2^{7/3}$ and $2^{8/3}$. Because $2 < 7/3 < 8/3 < 3$ we expect $2^{7/3}$ and $2^{8/3}$ to both fall between $2^2 = 4$ and $2^3 = 8$, and in addition that $2^{7/3} < 2^{8/3}$, but how do we calculate actual values? Recall that in Section 2.1 we constructed rational numbers by first considering reciprocals of integers and then their multiples, that is all the numbers that can be expressed as the ratio of two integers. By analogy with defining the reciprocal, we need to define what it means to raise a number to the $1/n$th power. To do that, we first need to define 'taking the $n$th root' of a number.

## Finding the *n*th root of a number

In much the same way as subtraction is the reverse of addition, and division the reverse of multiplication, *taking a root* is the reverse of *raising to a power*. For example, because we know that five to the power four is 625, we can say that a fourth root of 625 is five. We can write this using a special symbol as $\sqrt[4]{625} = 5$. Likewise we can write down that a square root of 25 is five: $\sqrt[2]{25} = 5$. Because we frequently need to write down the *square root* of a number, we usually omit the little '2' before the square root symbol: for example $\sqrt{25}$ for the square root of 25. Note that there can be more than one square root or fourth root of a number. For instance if we square $-5$ we also get 25 and if we raise $-5$ to the fourth power we get 625, thus $-5$ is a square root of 25 and a fourth root of 625. We often write this by saying that the square roots of 25 are $\pm\sqrt{25} = \pm 5$ and the real fourth roots of 625 are $\pm 5$. (If we allow complex numbers, there are two more fourth roots of 625. These are $\pm 5i$ where $i = \sqrt{-1}$ but we promised in the Introduction to ignore these.)

Expressed generally if $a^n = b$, then $a$ is an $n$th root of $b$. Whenever we need to specify only one $n$th root we require that the root be positive and call it the principal $n$th root, $\sqrt[n]{b}$. Provided the original number, $b$, is positive there is always a principal $n$th root.

Sometimes the principal root is reasonably obvious. For instance, 2 and $-2$ are square roots of 4 because either of these when squared gives 4. Similarly 3.1 is a square root of 9.61 (another is $-3.1$) because $3.1^2 = 9.61$ and 3.1 is a cube root of 29.791 because $3.1^3 = 29.791$. In each case the principal root is the positive value.

However, sometimes the root is not obvious at all. Indeed sometimes it is impossible to write down an exact value of a principal root although for positive numbers we can always get as close as we like. Take a simple example: what is $\sqrt{2}$? The principal square root of 1.9881, $\sqrt{1.9881}$, is 1.41 and $\sqrt{2.0164}$ is 1.42 (you can check these by simple multiplication) but neither of these is $\sqrt{2}$. We say that $\sqrt{2}$ is bracketed by the values 1.41 and 1.42. We can name the value, we just have, it is $\sqrt{2}$, and we can always improve on the estimate by trial and error (see Box 2.1) but that only gives us a better approximation. In fact, we can prove that $\sqrt{2}$ is not a rational number (see Box 2.2). Any answer on a display or printout can only be an approximation. For instance a Casio FX-115W calculator gives $\sqrt{2} = 1.414213562$, but the square of this number is in fact closer to 1.999999999 than to 2, so the actual square root must be a little larger. The answer was the best the calculator could do. A more precise answer, 1.4142135623731, which can be coaxed out of Microsoft Excel®, is closer, but it is still an approximation as its square is 2.000000000000014 (to 16 significant figures). Thus the actual square root must be a little smaller. No matter how closely we bracket the answer, it will still be in between the two bracketing values.

## Box 2.1

It is possible to improve upon the estimate of a square root by trial and error, eventually reaching any accuracy we want. For instance, because $1.4^2 = 1.96$ and $1.5^2 = 2.25$, $\sqrt{2}$ must lie between 1.4 and 1.5 and it would appear to be closer to 1.4. So try values a little larger, namely 1.41 and 1.42. As shown in the second line below, these also bracket the correct value. So try values in between, for example 1.414 and 1.415 and so on. To summarize

$$1.4^2 = 1.96 < 2 < 2.25 = 1.5^2$$
$$1.41^2 = 1.9881 < 2 < 2.0164 = 1.42^2$$
$$1.414^2 \approx 1.9994 < 2 < 2.0022 \approx 1.415^2$$
$$1.4142^2 \approx 1.99996 < 2 < 2.00024 \approx 1.4143^2.$$

This can be continued forever or until you lose patience with the game. However, no matter how closely we bracket the answer, it will still be in between two bracketing values. Fortunately, being able to get as close as we like to a number (or a physical quantity) is good enough for every calculation that we can actually do.

We never need to do trial-and-error approximations like this one in real life. To calculate a square root, we just use a calculator. Nevertheless the simple, tedious trial-and-error method is important because it shows that there is no magic. Using only very simple operations it really is possible to calculate roots to however many decimal places we want.

## Box 2.2 The Theorem of Theaetetus

The square root of 2 is not a rational number. A simple demonstration that $\sqrt{2}$ cannot be a rational number was presented in Plato's Theaetetus dialog and is now called the theorem of Theaetetus. If $\sqrt{2}$ is rational then we must be able to write

$$\sqrt{2} = \frac{a}{b} \qquad \text{(EQB2.1)}$$

where $a$ and $b$ are integers that have no common factors other than 1 (the fraction is in reduced form). If we can show that in order for EQB2.1 to be true, $a$ and $b$ must have a common factor then we have shown that EQB2.1 cannot be true. If EQB2.1 is true then

$$2 = \left(\frac{a}{b}\right)^2 = \frac{a^2}{b^2}$$
$$2b^2 = a^2 \qquad \text{(EQB2.2)}$$
$$a^2 = 2b^2$$

must also be true. That means that $a^2$ must be an even integer, i.e. divisible by 2. Because the squares of even integers are even and the squares of odd integers are odd (these statements are easy to show: do it yourself) this also means that $a$ must be even and $c = a/2$ is an integer. Substituting $a = 2c$ into the last line of EQB2.2

$$(2c)^2 = 2b^2$$
$$4c^2 = 2b^2 \qquad \text{(EQB2.3)}$$
$$2c^2 = b^2$$

and both $b^2$ and $b$, like $a^2$ and $a$, must be even numbers. Thus both $a$ and $b$ are exactly divisible by 2 and the premise that $a$ and $b$ share no common factor is false. That in turn means that the starting premise that $\sqrt{2}$ is rational, namely that $\sqrt{2}$ can be written as the ratio of two integers with no common factors, must be false.

Fortunately, being able to get as close as we like to a number (or a physical quantity) is good enough for every calculation that we can actually do, so the approximation produced by a calculator is all we need.

### Raising a number to any fractional power

Having found out how to calculate the principal $n$th root of a positive number to any desired accuracy, we can proceed to interpret $a^{1/n}$ for $a \geqslant 0$.

Suppose we raise $a^{1/n}$, whatever it is, to the $n$th power,

$$\left(a^{1/n}\right)^n \equiv a^{n \times (1/n)} \equiv a^1 = a. \tag{EQ2.28}$$

For instance, $(625^{1/4})^4 = 625^{(1/4) \times 4} = 625^1 = 625$. However, $(a^{1/n})^n \equiv a$ means, by the definition of taking roots, that $a^{1/n}$ is an $n$th root of $a$. In fact we define $a^{1/n}$ to be the *principal* (positive if $a > 0$) $n$th root of $a$,

$$a^{1/n} \equiv \sqrt[n]{a}. \tag{EQ2.29}$$

Again, there is an important proviso: $a^{1/n}$ is defined only if $a \geqslant 0$. This is because it would not be possible, for example, to write down a real square root of $-4$.

Discovering that $a^{1/n}$ is just the $n$th root of $a$, which we can calculate, is an incredibly useful result, because we can now write

$$a^{m/n} \equiv \left(a^{1/n}\right)^m. \tag{EQ2.30}$$

This means that for positive bases we have defined raising to any fractional power we like. *All* we have to do is write the power as $m/n$ where $m$ and $n$ are integers, find the $n$th root, and then raise that result to the $m$th power (or raise to the $m$th power and then find the $n$th root). To take a very simple example $27^{2/3} = (27^{1/3})^2 = (3)^2 = 9$. A more difficult example is $7^{2/3}$. This can be tackled by first raising 7 to the power of 2, $7^{2/3} = (7^2)^{1/3} = 49^{1/3}$ and then finding the cube root of 49 by trial and error. You will be relieved to know that we would never do this in practice, we would use a calculator. For example, on a Casio FX-115 just enter $7x^y(2 \div 3) =$ and out pops an approximate answer, 3.65930571 to more accuracy than we will ever need.

## 2.4 Order of precedence or sequence of operations

It is easy to see that we need some rules for the order in which to perform the various arithmetical operations when evaluating expressions. For instance, what is the value of $3 + 4 \times 2 + 5$? Do we just work from left to right and first add the 3 and the 4, then multiply by 2 then add 5 to get a final answer of 19; or do we do the multiplication before the additions to obtain 16? Either could make sense. If readers are to have any chance of cracking the algebraic code, everyone must agree to the same rules or conventions for how we perform the calculations. We would also expect that there must be some means for the person writing down an expression to say 'do it my way'. And of course there is. We can impose any order of evaluation using brackets. Everything inside a pair of brackets must be evaluated before it is combined with anything outside.

The agreed order of precedence is

1.  **B**rackets (which include the brackets in the symbols for functions).
2.  **E**xponentiation, that is raising to a power or taking a root.
3.  **D**ivision or **M**ultiplication.
4.  **A**ddition or **S**ubtraction.
5.  If there is still any ambiguity, proceed from left to right.

Some people find it helps to remember the acronym **BEDMAS** to get the operations in points 1 to 4 in the right order.

What this means is best seen with examples:

- $9 \div 3^2 = 9 \div 9 = 1$ because we raise 3 to the 2nd power before we divide;
  - but $(9 \div 3)^2 = 3^2 = 9$ because we first evaluate the expression inside the brackets.

- $1 + 3^2 \div 9 = 1 + 9 \div 9 = 1 + 1 = 2$ because we raise to a power first, then divide, then add.

- $(3 + 4) \times (2 + 5) = 7 \times 7 = 49$ because we first evaluate the expressions in the brackets;
  - but $(3 + 4) \times 2 + 5 = 7 \times 2 + 5 = 14 + 5 = 19$ because we first evaluate the expression inside the brackets, then multiply, then add;
  - and $3 + 4 \times 2 + 5 = 3 + 8 + 5 = 16$ because, in the absence of brackets, we first multiply then add.

- $24 \div (4 \div 2) = 24 \div 2 = 12$ because we work inside the brackets first;
  - but $24 \div 4 \div 2 = 6 \div 2 = 3$ because we work left to right.

- $6 \div (3 \times 2) = 6 \div 6 = 1$ because we work inside the brackets first;
  - but $6 \div 3 \times 2 = 2 \times 2 = 4$ because we work left to right (or in any order after replacing division by 3 with multiplication by the reciprocal, $1/3$);
  - and $6 \times 2 \div 3 = 12/3 = 4$ because we work left to right (or in any order after replacing division by 3 with multiplication by the reciprocal $1/3$). Note that this gives the same answer as the previous case.

The rules above have been devised so that expressions can be made as tidy as possible with brackets only where they are needed. For instance by universal convention $(3 \times 2) + 5$ would be written $3 \times 2 + 5$. One of the few instances where you might want to insert unnecessary brackets is in an expression like $24 \div 4 \div 2$. Enough people get this wrong that you might just want to write it as $(24 \div 4) \div 2$ to make sure your meaning is completely clear. Notice that we are less likely to make a mistake if we replace each division with a multiplication by a reciprocal, because we can do the multiplications in any order. Thus 24/4/2 is regarded as very poor style mainly because it can cause a lot of confusion – it is best not to use more than one solidus (the '/' symbol) in a term. However, if it is used then it must mean the same as $24 \div 4 \div 2$. There is no agreed convention allowing the use of '/' characters of different lengths.

There is also ambiguity when we have a power of a power, such as $2^{3^4}$; some people argue that this should mean $2^{3^4} = 2^{(3^4)} = 2^{81} = 2.42 \times 10^{24}$ rather than $2^{3^4} = (2^3)^4 = 8^4 = 4096$ as required by the rule that we work from left to right. To be on the safe side, it is best to use brackets to make the meaning clear.

## 2.5 Ratios and percentages

A **ratio** is a comparison between two quantities. It is written as

amount of first sort : amount of second sort.

For instance in 1962 for every 7 men admitted to Dartbridge University there were 2 women. We say that the ratio of men to women admitted was 7 : 2. A ratio remains unchanged if we multiply both sides by the same thing. If 30 women were admitted, how many men were admitted? 30 is $15 \times 2$ women, so the number of men must have been $15 \times 7 = 105$.

Nothing requires the numbers in a ratio to be integers. For instance in 1992 the ratio of men to women admitted to Dartbridge was 2.1 : 2 (a distinct improvement in many people's view). That is the same as saying 21 : 20 or 1.05 : 1.

Suppose we measure two variables in an experiment. If the ratio of the value of one variable to that of another is always the same, then the variables are said to be **proportional** to each other (or directly proportional to be precise). If you double one, you double the other. The reason we sometimes say 'directly proportional' is to emphasize that it is the opposite of 'inversely proportional'. When two variables are inversely proportional, doubling one halves the other.

Ratios and fractions are closely related. The following are all equivalent statements: the ratio of the number of men admitted to the number of women admitted was 7 : 2; the number of men admitted was seven halves of the number of women admitted; or the fraction of those admitted who were men was 7/9ths. This illustrates a useful, economical convention. We use the word 'ratio' for the comparison of the numbers in the separate categories, such as men and women, while we use the word 'fraction' for the comparison of each of these numbers to the total. Thus we say that the ratio of men to women was 7 : 2 while the fractions of men and women admitted were 7/9ths and 2/9ths, respectively. In general, to get from a ratio to the fractions of the total (provided the total makes sense) just divide each of the numbers in the ratio by their sum.

An equation saying that two ratios are equal to each other can be written using fractions. To a mathematician a fraction must be a number that can be expressed as a ratio of two integers, but to us as scientists fractions can be ratios of any physical quantities. It is just that we must be careful when writing the fractions to include the units. For instance, if the ratio of mass to volume for a liquid is 0.7 g : 2 ml and we want to calculate the mass of 8 ml we proceed knowing that 0.7 g : 2 ml and $x$ : 8 ml are equal, which we can write as 0.7 g / 2 ml = $x$ / 8 ml. Thus $x = 8$ ml $\times$ (0.7 g / 2 ml) = 2.8 g. Stated formally, if $a : b$ is the same as $c : d$ then $a / b = c / d$. This is sometimes described by saying that you can cross-multiply two ratios that are equal. To cross-multiply, you multiply the right-hand part of each ratio by the left-hand part of the other and set the two expressions equal to each other, i.e. $ad = cb$.

Ratios and fractions are two ways of expressing relative numbers. A third, percentage, is even more common in everyday usage. Thus I might well say that by this stage of reading a discussion on ratios and percentages 10 % of students have fallen asleep but I would be less likely to say that 1/10th of the students are asleep, and it would seem peculiar if I chose to express the same observation as the ratio of asleep to awake students is 1 : 9. The word 'percentage' means number out of 100. Thus 10 % of the students means 10 out of 100, that is the fraction of students asleep would be $10/100 = 1/10 = 0.1$. Similarly 0.3 % means $0.3/100 = 0.003$ whereas 200 % means $200/100 = 2$.

**A number expressed as a percentage is just a fraction multiplied by 100.** Another way of saying the same thing is to define % as a symbol standing for $\times 0.01$, and that is the SI convention:

$$17.5 \% = 17.5 \times 0.01 = 0.175. \qquad\qquad \text{(EQ2.31)}$$

One place where all Europeans encounter percentages is in the calculation of value added tax (VAT), which is usually, but not always, included in the marked prices of goods and services (in the USA sales tax is similar but is added on at the time of purchase). If the price of an item excluding VAT is $x$ and VAT is charged at 17.5 %, then the VAT is

$$x \times 17.5 \% = x \times 17.5 \times 0.01 = x \times 0.175 \qquad\qquad \text{(EQ2.32)}$$

and the total marked price including VAT, $y$, is

$$y = x + x \times 0.175 = x(1 + 0.175) = 1.175x. \qquad\qquad \text{(EQ2.33)}$$

Alternatively, if you know that $y$ is the total price with VAT then the price excluding VAT is just $y \div 1.175$.

Really, there is no more to percentages than that, but still people often have difficulty with them. The trouble usually occurs when the percentages are used to describe large changes, especially if the changes are to smaller values. It is usually much more meaningful to talk about changes of 'so many fold' rather than percentage changes when the changes are large, say 50 % or more. Thus to talk of a 200 % increase is less clear than to talk of a 3-fold increase. The relationship between the two is given by

$$\text{after/before} = (\text{before} + 200 \% \times \text{before})/\text{before} = (1 + 2)/1 = 3.$$

$$(EQ2.34)$$

Likewise, a 75 % decrease is a 4-fold decrease:

$$\text{before/after} = \text{before}/(\text{before} \times (1 - 75 \%)) = 1/25 \% = 1/0.25 = 4.$$

$$(EQ2.35)$$

## 2.6 Scientific notation and significant figures

Science deals with numbers that can be very large or very small and are almost always imperfectly known. For example, the number of carbon atoms in 0.012 kg of solid $^{12}$C is approximately 602 000 000 000 000 000 000 000. Not only is this awkward to write, but if we could compare it with the exact value, we would find that most of those zeros are not really zeros at all. For instance, how would we indicate whether the digit after the 2 really was a zero in the exact value? In fact, a more accurate value is 602 214 200 000 000 000 000 000, so we can see that the digit after the first 2 is not actually zero. There ought to be a better way to write such a number down. The better way should be tidier, easier to write, and should allow us to indicate only the digits we know accurately. The better way is called scientific notation. Thus we could write $6.02 \times 10^{23}$ where $\times 10^{23}$ means multiply by 10 twenty-three times. If we know the number more accurately, we simply use more digits, hence $6.022 \times 10^{23}$ or, the best there is at the moment, $6.022142 \times 10^{23}$. Writing the numbers this way means we write only the digits we really mean.

Just as some numbers in science are very large, others are very small. For instance the mass of a hydrogen molecule is close to 0.000 000 000 000 000 000 000 000 003 32 kg, which is pretty horrible to write. So it is very convenient to be able to write instead $3.32 \times 10^{-27}$ kg. Recall that $10^{-1} = 1/10$ so $10^{-27}$ means divide by 10, or equivalently multiply by 1/10th, 27 times. The rules for scientific notation are that we multiply by 10 or 1/10th enough times that the result, called the **coefficient**, is greater than or equal to 1 and less than 10. We use as many digits to the right of the decimal point as we have real information (or often we just use two). Then we multiply the coefficient by the power of 10 that takes us back to the actual number. The power required is called the **exponent**. This convention allows us to use as many **significant figures**, that is as many digits in the coefficient, as are needed to express the accuracy of the data. Thus $6.022 \times 10^{23}$ expresses Avogadro's number to four significant figures, whereas $6.0 \times 10^{23}$ uses two and $6 \times 10^{23}$ uses just one. The last two say different things. Using one significant figure makes no statement about the next digit, using two significant figures says that the second digit is in fact zero.

To make scientific notation easy to enter when typing, programs like Microsoft Excel® use a letter 'E' to separate the coefficient and the exponent, thus $6.022 \times 10^{23}$ becomes 6.022E23 and $3.32 \times 10^{-27}$ becomes 3.32E$-$27. On many calculators we would enter something like 6.022 <EXP> 23 where <EXP> means press the EXP key.

## Truncation and rounding

We *truncate* a number by just omitting all figures to the right of those needed. Thus both 3.5432 and 3.5487 become 3.54 when they are truncated to three significant figures. This is fine if that is what we want, but more often we would like the shortened number to be as close as possible to the original number. To three significant figures 3.54 is as close as we can get to 3.5432, but 3.5487 should go up to 3.55. Taking the number to the closest shortened number is called *rounding*. The ambiguous case is of course 3.545; should it go up or down? There are several possible methods for handling this situation. The simplest convention is that the value goes up, so that 3.545 rounds to 3.55 and 3.555 rounds to 3.56. You should always do it this way when dealing with something like exam marks as all candidates must be treated the same. However when you are dealing with large amounts of data and are not concerned with the individual values, many scientists and mathematicians argue that always rounding the middle value upward introduces a systematic error into calculations and that it is therefore better to round to the nearest even digit: thus 3.545 now rounds to 3.54, although 3.555 rounds to 3.56 as before. With this convention as many values are rounded down as are rounded up on average which removes the rounding bias.

How about 3.5548? If we are rounding to three significant figures this would go down to 3.55. But if first we round to four significant figures, 3.555, then round to three significant figures we will get 3.56. We should always try to avoid multiple rounding operations and so usually carry more figures than we think are needed through all the calculations then round only once at the end. It is considered very poor style to retain more significant figures in the final answer than are needed to convey what is actually known, but rounding only once is even more important.

## Arithmetic in scientific notation

Addition, subtraction, multiplication, and division for numbers written in scientific notation are straightforward using the rules we already know. To add two numbers first write the numbers with the same exponent, then add or subtract the coefficients and multiply the answer by 10 raised to the exponent. For example we could write either

$$7.31 \times 10^3 + 7.82 \times 10^2 = 73.1 \times 10^2 + 7.82 \times 10^2 = (73.1 + 7.82) \times 10^2$$

$$= 80.92 \times 10^2 \approx 8.09 \times 10^3$$

or

$$7.31 \times 10^3 + 7.82 \times 10^2 = 7.31 \times 10^3 + 0.782 \times 10^3 = (7.31 + 0.782) \times 10^3$$

$$= 8.092 \times 10^3 \approx 8.09 \times 10^3.$$

Needless to say, whichever way we choose to do it, we would not normally write out all the steps.

To multiply or divide numbers in scientific notation we multiply or divide the coefficients and add or subtract the exponents, for example

$$3.45 \times 10^3 \times 2.71 \times 10^{-1} = (3.45 \times 10 \times 10 \times 10) \times (2.71 \div 10)$$

$$= 3.45 \times 2.71 \times 10 \times 10 \times 10 \div 10$$

$$= 9.35 \times 10 \times 10 = 9.35 \times 10^2$$

or, more concisely

$$3.45 \times 10^3 \times 2.71 \times 10^{-1} = 3.45 \times 2.71 \times 10^{(3-1)} = 9.35 \times 10^2.$$

The 9.35 is just the product of the coefficients; the 2 in the exponent of $10^2$ counts up the number of times we need to multiply by 10.

There is a very similar procedure for division, for example

$$3.45 \times 10^{-3} \div 2.3 \times 10^1 = (3.45/2.3) \times 10^{-3-1} = 1.5 \times 10^{-4}.$$

If, after multiplying or dividing the coefficients and rounding the answer to the desired number of significant figures (three, say), the portion of the number to the left of the decimal point is not in the range from 1 to 10 (excluding 10 itself), then we multiply or divide by 10 so that it is, and adjust the exponent accordingly. For example:

| initial answer | intermediate step | final form |
|---|---|---|
| $13.3 \times 10^3$ | $(13.3/10) \times (10 \times 10^3)$ | $1.33 \times 10^4$ |
| $0.678 \times 10^{-5}$ | $(0.678 \times 10) \times (10^{-5}/10)$ | $6.78 \times 10^{-6}.$ |

## 2.7 Solving equations

Scientists frequently need to solve equations. In this section we shall see some methods that can be used, but first we need to define some of the language and look at the basic manipulations of expressions and equations that we need to be able to do.

### Some basics

An **equation** is a statement that two expressions are equal to each other; for example, $2 + 3 = 5$ or $2x - 3az = 2y$. An **expression** is a combination of symbols; for example, $2 + 3$ or $2x - 3az$. A part of an expression is called a sub-expression of that expression; for example, 2 is a sub-expression of $2 + 3 = 5$ and $3a$ is a sub-expression of $2x - 3az$. If an expression can be written as a *sum* of sub-expressions, then each of these sub-expressions is called a **term**; for example, in $2x - 3az$, the sub-expressions $2x$ and $-3az$ are terms. If a term, which might be the whole expression, can be written as the *product* of sub-expressions then each of these is called a **factor**; for example, in $-3az$, the sub-expressions $-3$, $a$, and $z$ are factors. Thus in $y = (x + 2)x + 3x + 5$ the right-hand side of the equation has three terms: $(x + 2)x$, $3x$, and $5$. The first of these terms $(x + 2)x$ has two factors, $(x + 2)$ and $x$, the second term $3x$ has two factors, $x$ and $3$, whereas the third term $5$ has only one factor, namely $5$ itself. Similarly, in $y = (x + 2)(x + 5)$ the right-hand side has two factors each of which has two terms. A polynomial in $x$ is a sum of terms each of which is either a constant or the product of a constant and a positive integer power of $x$. The degree of a polynomial is the highest power of $x$ it contains; that is $ax^3 + bx^2 + cx + d$ where $a$, $b$, $c$, and $d$ are constants is a third-degree polynomial in $x$.

With an expression like $(x + 2)x$ we can multiply each term in the first factor by the second factor: $(x + 2)x = xx + 2x = x^2 + 2x$. The resulting expression now has two terms, namely $x^2$ and $2x$, compared with the single term of the original expression; we say that $(x + 2)x$ has been **expanded** to give $x^2 + 2x$. If necessary, the process can be repeated. For example $(x + 2)(x + 5)$ can be expanded to give $x(x + 5) + 2(x + 5)$, but each term of this new expression can itself be expanded: $x^2 + 5x + 2x + 2 \times 5 = x^2 + 5x + 2x + 10$. We now have four terms. Note that the middle two terms share a common factor, namely $x$. We can take this factor outside a pair of brackets: $x^2 + 5x + 2x + 10 = x^2 + (5 + 2)x + 10$. However, we can always add two numbers together; in this case we get $x^2 + (5 + 2)x + 10 = x^2 + 7x + 10$ and this is the expanded form we would normally use. If two terms differ only by a constant factor they will usually be added together in writing the final expanded answer. For example consider $7y^2 + 9y^2 + 3x^2 + x^2 + 4x - 2x$; we can simplify this expression by adding the first two terms together; then the third and fourth terms; and finally the last two terms:

$$7y^2 + 9y^2 + 3x^2 + x^2 + 4x - 2x = (7 + 9)y^2 + (3 + 1)x^2 + (4 - 2)x$$

$$= 16y^2 + 4x^2 + 2x. \qquad \text{(EQ2.36)}$$

Sometimes it may be necessary to rearrange the order in which the terms are added in order to simplify. For example

$$5ab + 2b - 3ab + 7a + b - a = 5ab - 3ab + 7a - a + b + 2b$$

$$= 2ab + 6a + 3b. \qquad \text{(EQ2.37)}$$

Note that each term always carries its sign with it. Lots of examples that give you practice in doing algebra ask you to 'simplify' or 'evaluate' something, usually involving products and ratios of polynomials. Just what does 'simplify' mean? When applied to a fraction or product of polynomials simplify means multiplying out all factors, dividing where possible, and reducing the original expression to a form in which there is one polynomial in its simplest form divided by another polynomial in its simplest form. In the simplest form of a polynomial each power of the variable, or combination of powers of the variables, occurs only once with all possible arithmetic operations completed and no remaining brackets.

That is fine, but you may not be done. If you have a ratio of polynomials you need to look to see if they can both be written as products of factors (this factorization is considered below) and if so whether any of the factors in the numerator are the same as any of those in the denominator. If so, provided the denominator isn't zero the ratio can be written as

$$\frac{a \times b}{a \times c} = \frac{b}{c} \qquad \text{(EQ2.38)}$$

where $a$, $b$, and $c$ can represent any expressions. Removing ratios like $a/a$ that equal 1 is often called canceling. The second example below illustrates this cancellation.

Example 1: to simplify $1/(x+1) - 1/(x-1)$ we can proceed as follows:

$$\frac{1}{x+1} - \frac{1}{x-1} = \frac{x-1}{(x+1)(x-1)} - \frac{x+1}{(x+1)(x-1)} = \frac{(x-1)-(x+1)}{(x+1)(x-1)}$$

$$= \frac{x-1-x-1}{(x+1)(x-1)} = \frac{-2}{(x+1)(x-1)}.$$

$$\text{(EQ2.39)}$$

Example 2: to simplify $(x^2 + 5x + 6)/(x^3 + 3x^2 + 2x)$ we first note that the numerator can be written as the product of two factors $x^2 + 5x + 6 = (x+2)(x+3)$ while the denominator can be written as the product of three factors $x^3 + 3x^2 + 2x = x(x+2)(x+1)$ and that these have a common factor $(x+2)$. Thus

$$\frac{x^2 + 5x + 6}{x^3 + 3x^2 + 2x} = \frac{(x+2)(x+3)}{x(x+2)(x+1)} = \frac{x+3}{x^2+x}, \ (x \neq -2). \qquad \text{(EQ2.40)}$$

The restriction that $x$ is not equal to $-2$ is necessary. The initial expression cannot be evaluated when $x = -2$ because division by 0 is not defined.

'Evaluate' is used when it is possible to get a number as the result of the simplification. If this number is a rational number, that is it can be written as a ratio of integers, the final result should be written in reduced form or sometimes as a single decimal number.

## Basic manipulations on equations

Solving an equation means rewriting it so that the quantity we want to know is alone on the left and the things we do know are on the right. The basic manipulations we are allowed to perform on equations were summarized in simple language in Chapter 1. To remind you with examples:

(1) We can add zero to any expression without changing its value:

$$5 + 0 = 5;$$
$$9.8(m + 0) = 9.8m.$$

(2) We can multiply any expression by 1 without changing its value:

$3 \times 1 = 3$;
$C_0 \times 1 = C_0$.

(3) We can swap the two sides of an equation:

since $3 \times 2 = 5 + 1$ then $5 + 1 = 3 \times 2$;
if $F = ma$ then $ma = F$.

(4) We can add or subtract the same thing to both sides of an equation:

since $3 \times 2 = 6$ then $3 \times 2 + 7 = 6 + 7$;
if $F = ma$ then $F - mg = ma - mg$.

(5) We can replace any expression by another equal expression:

since $3 \times 6 = 18$ and $6 = 5 + 1$ then $3 \times (5 + 1) = 18$;
if $F = ma$ and $a = \alpha + \beta t$ then $F = m(\alpha + \beta t)$.

(6) We can multiply or divide both sides of an equation by the same thing:

since $3 \times 2 = 6$ then $3 \times 2 \times 4 = 6 \times 4$;
since $5 + 1 = 6$ then $(5 + 1) \times 4 = 6 \times 4$ and $(5 + 1) \div 2 = 6 \div 2$;
if $F = ma$ then $F/a = m$.

Notice, that it may be necessary to introduce brackets around the whole of one or both sides of an equation before we multiply or divide by a value; this is to ensure that the *whole* expression on each side is multiplied or divided (see Section 2.4 on BEDMAS).

There is one very important exception to this last rule: we must not try to divide by zero. For example, $5 \times 0 = 6 \times 0$ but 5 is not equal to 6, i.e. $5 \neq 6$. Thus, if we have a non-zero quantity $m$ (as would be the case if $m$ represents the mass of a real object) we can divide both sides of $F = ma$ by $m$ to give $F/a = m$. By contrast, suppose we want to estimate the acceleration of a car, $a$, from measurements of the initial velocity, $u$, and the velocity measured after a time, $t$,

$$v = u + at \tag{EQ2.41}$$

which we can solve for the acceleration

$$a = (v - u)/t, \quad (t \neq 0). \tag{EQ2.42}$$

You can read the comma following this equation as the words 'provided that'. The extra statement after the comma is needed to indicate that the main equation before it is not valid when $t = 0$ s. When $t = 0$, it makes no sense either physically or mathematically.

We can combine these rules to produce more complex algebraic manipulations. For example suppose we have $z = 1/(t-3)$ and we want to get an equation of the form $t = ?$; this is sometimes known as 'making $t$ the subject of the equation' or 'solving the equation for $t$'. We can do this in several stages:

Multiply both sides of the equation by $(t-3)$ to give $z(t-3) = 1$.

Next divide both sides by $z$, or equivalently multiply by $1/z$, to give $t - 3 = 1/z$. (The brackets around $t-3$ are no longer needed.) Note this is allowed only if $z \neq 0$, but from the starting equation, it never can be for any value of $t$.

Finally we can add three to both sides to give $t = (1/z) + 3$. Furthermore, the rules for the order of precedence (BEDMAS) enable us to leave out the brackets and so we can write $t = 1/z + 3$.

Before proceeding it is recommended that you do End of Chapter Questions 9–11.

## Solving first-order equations

The very simple example of the algebraic expression for heights used in Chapter 1

$$H_J = H_R + 30 \text{ cm} \tag{EQ2.43}$$

was written to tell us John's height if we know Robert's. Notice the use of subscripts to distinguish between the two heights: subscripts are frequently used in scientific work, so it is best to get familiar with them even for simple problems. We can 'solve' this equation to tell us Robert's height if we know John's. Formally, add $-30$ cm to both sides

$$H_J - 30 \text{ cm} = H_R \tag{EQ2.44}$$

then reverse the two sides of the equation

$$H_R = H_J - 30 \text{ cm}. \tag{EQ2.45}$$

This is an example of what is called a first-order equation: the variables, $H_R$ and $H_J$, appear only as themselves, i.e. raised to the first power, and not raised to any other power or as part of more complicated expressions that involve other variables.

An equation is said to be first-order in $x$ if it can be written without brackets such that at least one term contains $x$, but all other terms do not contain $x$ at all. No other powers, roots, or functions of $x$ are allowed. For example: $2x + 3 = 5$ and $xy^2 + z = 5$ are both first-order in $x$; $xy^2 + z = 5$ is also first-order in $z$; but neither $2x + 3 = 5$ nor $xy^2 + z = 5$ is first-order in $y$. A first-order equation is one that is first-order in *all* of its variables and contains no products of variables. For example: $2x + 3 = 5$ and $2x + 3y = 5z$ are both first-order, but $2x^2 + 3 = 5$ and $xy + 3 = 5$ are not. Thus the most general first-order equation in $x$ can be written as

$$ax + b = 0, \quad (a \neq 0). \tag{EQ2.46}$$

As indicated $a$ must not be 0, because if it were this wouldn't be a first-order equation. The general first-order equation has the solution $x = -b/a$ (because $a$ cannot be zero, the division is legal). There's nothing more to say about first-order equations. Practice in solving them really means practice in algebraic juggling of expressions.

## Quadratic equations

An equation is called second-order, or quadratic, in $x$ when it can be written without brackets such that $x$ appears as $x^2$, $x$, or not at all in each term and at least one term contains $x^2$. No other powers, roots, or functions of $x$ are allowed. An equation in more than one variable is called second-order if it can be written without brackets so that the sum of the exponents of the variables in every term is 0, 1, or 2 and the sum of exponents is equal to 2 in at least 1 term. Thus $x^2 + 3x + 2 = 0$, $x^2 + y^2 = 1$, and $x + xy = 2$ are all quadratic equations. The first is quadratic in $x$, the second is quadratic in $x$ and $y$, but the third is first-order in both $x$ and $y$.

Quadratic equations are very common in science. A simple example from your school lessons is the relation for the distance traveled, $d$, after a time $t$ if a car is initially traveling at a speed $v_0$ and then changes speed with a constant acceleration $a$

$$d = v_0 t + \tfrac{1}{2} a t^2. \tag{EQ2.47}$$

Another example is the equilibrium obtained in a bimolecular reaction where species A at initial concentration $a$ and species B at initial concentration $b$ combine to form C. If the concentration of C formed at equilibrium is $x$, then from the principle of mass action

$$K_D = \frac{(a - x)(b - x)}{x} \tag{EQ2.48}$$

where $K_D$ is the dissociation constant for the reaction. Other examples include: the relation between the surface area of a cell and its diameter; and the probability that two alleles of a gene will have particular values (the Hardy–Weinberg principle). It is clearly a good idea to know how to solve quadratic equations.

How do we solve a quadratic equation? Suppose we have the equation

$$x^2 + 3x + 2 = 0. \tag{EQ2.49}$$

If we can see how to rewrite the equation as

$$(x - a)(x - b) = 0 \tag{EQ2.50}$$

for some values of $a$ and $b$ then the solutions are easy to find, because the product of two factors can equal zero only if at least one of the factors is equal to zero. Thus either $x - a = 0$ or $x - b = 0$, so the solutions are just $x = a$ and $x = b$. In this case we can write EQ2.49 as

$$(x + 1)(x + 2) = 0 \tag{EQ2.51}$$

that is, as

$$(x - (-1))(x - (-2)) = 0 \tag{EQ2.52}$$

from which we can see that the solutions are $x = -1$ and $x = -2$.

More generally, when we want to rewrite a quadratic equation in the form of EQ2.50, how do we find $a$ and $b$? The most common method used is trial and error, but we do not have to guess pairs of values at random. We can first recast the problem so that for each guess of $a$ or $b$ we only have to consider one value of the other. To do this we use the general rule for multiplication of factors in brackets:

$$\begin{aligned} (x - a)(x - b) &= x(x - b) - a(x - b) \\ &= x^2 - bx - ax + ab \\ &= x^2 - (a + b)x + ab. \end{aligned} \tag{EQ2.53}$$

This is the same as our starting equation, EQ2.49, if

$$a + b = -3 \quad \text{and} \quad ab = 2. \tag{EQ2.54}$$

So we can guess a value for $a$, calculate $b = -3 - a$, and see if our values of $a$ and $b$ satisfy $ab = 2$. Thus if we guess $a = 1$, then $b = -4$, giving $ab = -4$ which certainly is not correct. How about $a = 2$? If $a = 2$ then $b = -5$ and $ab = -10$, which is even further from the right answer. So we go the other way and try smaller values of $a$. Now $a = 0$ is impossible because $ab$ is then 0. But if $a = -1$ then $b = -2$ and $ab = 2$ as required. (There is another possible value for $a$. Can you find it? Hint: it is staring at you from earlier in this paragraph.)

Solving EQ2.49 was easy because we could quickly 'see' or 'guess' the constants. This breaking down of an expression into a product of factors is called **factorization**. It is often the quick way to solve equations. For higher-order equations, if factorization does not work we often need to resort to numerical methods to get approximate solutions).

Equations that are quadratic in the unknown can also be solved *without* trial and error. The most general form of a quadratic equation can be written as

$$ax^2 + bx + c = 0, \quad (a \neq 0). \tag{EQ2.55}$$

Here $a$, $b$, and $c$ stand for three constants and the expression after the comma means that $a$ must not be 0, otherwise the equation is not a quadratic.

We use symbols for the constants rather than numbers because we want an answer which is true for any possible values that $a$, $b$, and $c$ can take.

The solution is

$$x = -\frac{b}{2a} \pm \frac{\sqrt{b^2 - 4ac}}{2a}. \tag{EQ2.56}$$

The symbol $\pm$ means either $+$ or $-$. In general there are two solutions for $x$; one uses the $+$ sign, the other uses the $-$ sign. As an example of the use of this formula, let us consider again the quadratic equation, $x^2 + 3x + 2 = 0$, that we previously solved by the factorization method. This is an example of the general equation with $a = 1$, $b = 3$, and $c = 2$, thus

$$
\begin{aligned}
x &= -\frac{3}{2} \pm \frac{\sqrt{3^2 - 4 \times 2}}{2} \\
&= -\frac{3}{2} \pm \frac{\sqrt{9 - 8}}{2} \\
&= -\frac{3}{2} \pm \frac{1}{2} \\
&= -2 \text{ or } -1.
\end{aligned}
\tag{EQ2.57}
$$

Deriving the general solution to the quadratic equation (see Box 2.3) has given us a powerful result. We have taken one equation, which previously we had to solve by inspection or by trial and error, and have changed it into another equation that we can use to find the answer by simply plugging in the numbers. In addition to producing a convenient formula for calculation, the algebra has also revealed something that was not at all obvious from the starting equation: it tells us which combinations of values of $a$, $b$, and $c$ lead to real-value solutions of the equation and which ones do not. If $b^2 - 4ac > 0$ there are two real-value solutions to the equation:

$$x = -\frac{b}{2a} + \frac{\sqrt{b^2 - 4ac}}{2a} \quad \text{and} \quad x = -\frac{b}{2a} - \frac{\sqrt{b^2 - 4ac}}{2a}.$$

If $b^2 - 4ac = 0$, there is only one solution: $x = -b/2a$. If, however, $b^2 - 4ac < 0$ then we are trying to take the square root of a negative number, the solutions involve complex numbers, and there are no real-value solutions.

## Box 2.3 Solution of a quadratic equation

In the general quadratic equation

$$ax^2 + bx + c = 0$$

$a$, $b$, and $c$ stand for three known constants; $x$ represents the unknown: the variable we need to find given our known constants.

First divide both sides of the general equation by $a$

$$x^2 + \frac{b}{a}x + \frac{c}{a} = 0.$$

Then add $(b/2a)^2 - c/a$ to both sides of the equation

$$x^2 + \frac{b}{a}x + \left(\left(\frac{b}{2a}\right)^2 - \frac{c}{a}\right) + \frac{c}{a} = \left(\frac{b}{2a}\right)^2 - \frac{c}{a}$$

$$x^2 + \frac{b}{a}x + \left(\frac{b}{2a}\right)^2 = \left(\frac{b}{2a}\right)^2 - \frac{c}{a}.$$

This may look much messier, but if you look carefully you can now see why we chose to add $(b/2a)^2 - c/a$. The left side of this equation is a perfect square.

An expression is a perfect square if it is identical to the square of something else. Here

$$x^2 + \frac{b}{a}x + \left(\frac{b}{2a}\right)^2 \equiv \left(x + \frac{b}{2a}\right)^2.$$

We can make the quadratic equation look a lot simpler if we define a new unknown

$$y = x + \frac{b}{2a}.$$

Whenever we know $x$ we know $y$ and vice versa, so solving for $y$ will be just as good as solving for $x$. In terms of this new unknown or variable our general quadratic equation now becomes

$$y^2 = \left(\frac{b}{2a}\right)^2 - \frac{c}{a} = \frac{b^2}{4a^2} - \frac{c}{a}$$

$$= \frac{b^2}{4a^2} - \frac{c}{a} \times \frac{4a}{4a} = \frac{b^2}{4a^2} - \frac{4ac}{4a^2}$$

$$= \frac{b^2 - 4ac}{4a^2}.$$

We now take the square root of both sides remembering there are two solutions

$$y = \pm\sqrt{\frac{b^2 - 4ac}{4a^2}} = \pm\frac{\sqrt{b^2 - 4ac}}{2a}.$$

To get back to an equation for $x$ add $-b/2a$ to both sides of the equation used to define $y$ to obtain

$$x = y - \frac{b}{2a}.$$

Substituting in the expression for $y$ we have the final answer

$$x = -\frac{b}{2a} \pm \frac{\sqrt{b^2 - 4ac}}{2a}.$$

In most of this chapter, algebra has been used as a shorthand: a compact, very careful way of writing down statements. Producing the general solution to a quadratic equation is different: this is our first real example of obtaining a new result we almost certainly could not have seen without *using* algebra. The next example follows in Section 2.8.

## 2.8 Simultaneous equations

If we have just one unknown and one linear equation, the solution is very simple. However, if our problem is at all interesting we will have two or more unknowns and various relations between them. If all of these relations can be stated as linear equations, and there are enough of them, there are procedures for finding the solution without guesswork.

## Simultaneous linear equations

The most general form for two simultaneous linear equations with two unknowns, $x$ and $y$, is

$$a_1 x + b_1 y = c_1, \tag{EQ2.58}$$

$$a_2 x + b_2 y = c_2. \tag{EQ2.59}$$

In these equations $x$ and $y$ represent the unknowns whereas $a_1, a_2, b_1, b_2, c_1$, and $c_2$ are constants. Why use three letters and two subscripts rather than using six letters to represent the constants? We do not have to – it is a choice. However, with a little practice we will see that it makes it easier to read the equations. The three different letters are used to indicate that these constants appear in the equations in different ways: each $a$ multiplies an $x$; each $b$ multiplies a $y$; and each $c$ appears on its own. The subscripts tell us whether the constant comes from the first or second equation. Using a different letter for each of the six constants would obscure these points. Subscripts may look complex, but in fact they make the code of the symbol jargon much easier to crack.

We can solve these two equations for $x$ and $y$ by first solving one of them for $x$ in terms of $y$, then substituting that expression into the other and solving for $y$. That would be completely correct and for some values of the constants it can be much the fastest way. However, there is a systematic way to write down the solution which leads on to the methods needed for more complicated expressions. For an example of this method, suppose we have the following two equations:

$$3x - 5y = 4,$$
$$2x + 3y = -1. \tag{EQ2.60}$$

To solve these equations we multiply the first equation through by 2/3 so that the number multiplying $x$ becomes the same as in the second equation:

$$\frac{2}{3}(3x - 5y) = \frac{2}{3} \times 4,$$
$$2x - \frac{10}{3}y = \frac{8}{3}. \tag{EQ2.61}$$

We now have two equations where the terms in $x$ are identical.

$$2x - \frac{10}{3}y = \frac{8}{3},$$
$$2x + 3y = -1. \tag{EQ2.62}$$

We can thus eliminate $x$ from the equation by subtracting one equation from the other:

$$-\frac{10}{3}y - 3y = \frac{8}{3} - (-1)$$
$$-\frac{10}{3}y - \frac{9}{3}y = \frac{8}{3} + \frac{3}{3} \tag{EQ2.63}$$
$$-19y = 11.$$

Thus we find

$$y = -\frac{11}{19}.$$  (EQ2.64)

You can now solve for $x$ by substituting this value of $y$ into either of the original equations. Alternatively, you could repeat the procedure used to solve for $y$ only this time multiplying the first equation by $3/5$ before adding the two equations. Completing the solution is left for you to do as an exercise.

In general, the systematic or mechanical way to solve for $y$ is to multiply the first equation EQ2.58 by the ratio of the constant which multiplies $x$ in the second equation, $a_2$, to that which multiplies $x$ in the first, $a_1$:

$$\frac{a_2}{a_1}(a_1 x + b_1 y) = \frac{a_2}{a_1}c_1.$$  (EQ2.65)

This multiplies out to

$$a_2 x + \frac{a_2 b_1}{a_1}y = \frac{a_2 c_1}{a_1}.$$  (EQ2.66)

We now have two equations, EQ2.59 and EQ2.66, where the terms in $x$ are identical. We can thus eliminate $x$ from the equation by subtracting one equation from the other

$$a_2 x + b_2 y - \left(a_2 x + \frac{a_2 b_1}{a_1}y\right) = c_2 - \frac{a_2 c_1}{a_1}$$

$$\left(b_2 - \frac{a_2 b_1}{a_1}\right)y = c_2 - \frac{a_2 c_1}{a_1}.$$  (EQ2.67)

All that remains is to multiply both sides by $a_1$

$$(a_1 b_2 - a_2 b_1)y = a_1 c_2 - a_2 c_1$$  (EQ2.68)

and then divide by $(a_1 b_2 - a_2 b_1)$

$$y = \frac{a_1 c_2 - a_2 c_1}{a_1 b_2 - a_2 b_1}.$$  (EQ2.69)

The fastest way to obtain the general solution for $x$ is to repeat the same procedure but this time eliminating $y$ from the initial equations. That is left for an exercise.

Solutions for three equations in three unknowns can be tackled in just the same way; none of the steps in the algebra is any more difficult, it is just that there are more of them.

Simultaneous linear equations can be solved by straightforward but tedious algebra. Now anything that can be done by repetitive rules or procedures can be done faster on a computer. In practice if you have to solve more than three simultaneous equations or you need to do this often, search the Web for a description of Cramer's Rule, which allows the equations to be solved using 'determinants'. An important advantage of this method is that all the arithmetic can be done by a program like Microsoft Excel® (for more information search Excel® help for 'determinant' or 'MDETERM').

## Simultaneous nonlinear equations

Sometimes these can be solved by elementary methods, sometimes not. If we have one linear equation and one nonlinear equation, we use the linear equation to eliminate one of the variables from the other and then try to solve the resulting equation. As an alternative to solving the equations algebraically it may be easier to find an approximate solution by plotting the equations (see Section 3.2), then refine the estimate by trial and error.

## Binding of small molecules to large ones

The need to describe how small molecules bind to large ones recurs frequently in science. In physical chemistry it arises when gases or solutes adsorb onto surfaces which may be catalysts for reactions. In biochemistry it is at the core of explaining how enzymes work. However, describing the binding of drugs to specific receptors in biological tissues is simpler than either of these. Drug–receptor binding will be our second example of using symbols and algebra to reach conclusions that would have been very difficult using only ordinary language.

We can think of a 'receptor' as a component of the cell, probably a protein, that has several groups, at least three, of just the right types and in just the right positions to bind a small molecule with high affinity. We shall call the small molecule the drug, but it could be a hormone, or a neurotransmitter, or a pheromone. Usually there will not be very many of the high-affinity sites. In many studies, densities of receptors would be expressed in picomoles per milligram of protein in the tissue.

A biological tissue has been isolated and is bathed in an appropriate buffer solution. A drug is added. How do we expect the amount of the drug bound in the tissue to vary with the concentration of the drug, the density of the receptors in the tissue, and the affinity constant for the drug–receptor binding reaction? We need a theoretical relation to compare with the experimental data. The theoretical relation can be derived using algebra from reasonable assumptions about how the binding occurs. These assumptions are:

1.  The tissue is immersed in a large volume of solution such that there are always many more drug molecules remaining in the solution than are bound to the tissue. The concentration, $c$, of the drug in the solution is then regarded as constant, independent of the amount bound.
2.  The tissue contains a small number of receptors with a concentration, $N_T$, which might be expressed in picomoles per milligram of tissue. A receptor can either be empty, when it is said to be 'free'; or it can have an attached drug molecule, when it is said to be 'occupied' or 'bound'. The total concentration of the receptors, $N_T$, the concentration of free receptors, $N_f$, and the concentration of complexes, $N_b$, are related very simply by

$$N_T = N_f + N_b \text{ or } 1 \times N_f + 1 \times N_b = N_T. \tag{EQ2.70}$$

The equation is written first as we would normally write it and then in the same form as the most general linear equation for the two variables, here $N_f$ and $N_b$. This equation is sometimes called the 'master equation' because it ties together the relations between all the different states or forms of the receptor (here just bound or free – see Presenting Your Work, Question A for an example with three forms).

3. The drug, the free receptors, and the complexes are related by a simple chemical equilibrium:

$$\text{bound receptor} \rightleftharpoons \text{free receptor} + \text{drug} \tag{EQ2.71}$$

which can be described by the usual sort of binding equation (sometimes called the law of mass action)

$$N_b = K_a c N_f \text{ or } K_a c \times N_f - N_b = 0. \tag{EQ2.72}$$

$K_a$ is the affinity constant for the reaction, which has units that are the reciprocal of the units for the drug concentration. Thus if the concentration of the drug is expressed in molarity M, the units of the affinity constant are $M^{-1}$.

What we want to know is how the concentration of complexes will vary with the concentration of the drug but it is actually easier to solve first for the concentration of free receptors. We can do that immediately by using the chemical equilibrium equation, EQ2.72, to eliminate the concentration of bound receptors from the master equation, EQ2.70,

$$N_T = N_f + K_a c N_f = N_f(1 + K_a c), \tag{EQ2.73}$$

which can be solved for $N_f$ by dividing both sides by $(1 + K_a c)$:

$$N_f = \frac{N_T}{1 + K_a c}. \tag{EQ2.74}$$

However, now we can use the equation for chemical equilibrium, EQ2.72, to get the expression we want for the concentration of complexes

$$N_b = K_a c N_f = K_a c \frac{N_T}{1 + K_a c} \tag{EQ2.75}$$

$$N_b = \frac{K_a c N_T}{1 + K_a c}. \tag{EQ2.76}$$

Note that the fraction of receptors occupied, $N_b/N_T$, will be one half when the concentration equals $1/K_a$ because then $K_a c = 1$. The equation for the concentration of complexes is often written as

$$N_b = \frac{c N_T}{K_D + c} \tag{EQ2.77}$$

where $K_D = 1/K_a$ is called the dissociation constant. Both forms have their advantages. The affinity constant has the nice feature that larger constants

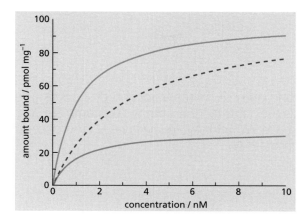

**Figure 2.2**
The theoretical relation for the amount of bound drug to receptors per milligram of tissue as a function of the concentration of the drug free in solution, $N_b = N_T K_a c / (1 + K_a c)$, where $N_b$ is the amount bound, $N_T$ is the amount of receptors, $K_a$ is the affinity constant for the binding, and $c$ is the free concentration. The curves are plotted for the following values of $N_T$ and $K_a$: 100 pmol mg$^{-1}$ and 1 nM$^{-1}$ (solid red curve); 100 pmol mg$^{-1}$ and 0.333 nM$^{-1}$ (dashed blue curve); 33.3 pmol mg$^{-1}$ and 1 nM$^{-1}$ (solid green curve). Note that much higher concentrations would be needed for $N_b$ to exceed $0.99 \times N_T$: 100 nM, 300 nM, and 100 nM, respectively.

mean stronger binding, and it makes the algebra easier. The dissociation constant has the nice feature that it is equal to the concentration of the drug at which the occupancy of the receptor reaches half the maximum.

Armed with this algebraic solution we can now plot a theoretical curve for comparison with data. Figure 2.2 shows curves for three different combinations of the amount of receptors and the affinity constant. Note that when the amount of receptors is reduced, the binding is reduced by the same fraction at all drug concentrations; the red curve can be made into the green curve by squashing it downward. By contrast, when the affinity constant is reduced, the red curve cannot be squashed downward to match the dashed blue curve. Instead the red curve must be stretched toward the right; higher concentrations are required to achieve the same level of binding. To compare the theoretical expression with data, $N_T$ is adjusted so that the equation gives the appropriate maximum response, and $K_a$ (or equivalently $K_D$) is adjusted so that the theoretical curve reaches half the maximal response at an appropriate concentration. An 'eyeball' test of theory is then to look to see if the whole curve is a good description of the data. For instance: does the theoretical curve have the same steepness as the data at low concentrations? Statistical criteria to help decide if the fit really is any good are considered in Chapter 12. We shall return to binding curves and better ways to display these data in Section 3.8 when we can use logarithms.

We started with a plausible but untested description of the binding process: one drug to one receptor, the law of mass action, and a limited supply of receptors. Deriving EQ2.76 allows this description to be recast into a prediction that can actually be tested for how the amount bound varies with the concentration of the drug. This recasting has been used repeatedly in science starting with the description of Michaelis–Menten kinetics for enzymes (1913), the theory of the adsorption of gases onto solid surfaces (by Langmuir in 1916), and the underpinnings of dose–response relations in pharmacology (by Clark in 1926).

# Presenting Your Work

## QUESTION A
### Competitive binding to a receptor

Derive the relation that describes the binding of one drug, A, when a second drug, B, is present that also binds to the same sites. The drug concentrations are $c_A$ and $c_B$, the affinity constants are $K_{aA}$ and $K_{aB}$, the concentrations of receptors with A and B bound are $N_A$ and $N_B$, respectively, the concentration of free receptors is $N_f$, and the total concentration of receptors is $N_T$.

Hint: the master equation is now

$$N_T = N_f + N_A + N_B$$

and

the equations for the chemical equilibria are now

$$N_A = K_{aA} c_A N_f \text{ and } N_B = K_{aB} c_B N_f.$$

---

Eliminate $N_A$ and $N_B$ from the master equation

$$N_T = N_f + K_{aA} c_A N_f + K_{aB} c_B N_f$$

and solve for $N_f$

$$N_f = \frac{N_T}{1 + K_{aA} c_A + K_{aB} c_B}$$

and thus

$$N_A = K_{aA} c_A N_f = \frac{K_{aA} c_A N_T}{1 + K_{aA} c_A + K_{aB} c_B}.$$

---

## QUESTION B

Measuring the turbidity or absorbance, *abs*, of a bacterial suspension is a quick and easy way to estimate the number of bacteria present. The absorbance is a dimensionless number that is proportional to the number of bacteria. We have measured the absorbance at three different times and would like to fit these data with a smooth curve to allow estimation of the number of bacteria present at other times.

| *t* / h | *abs* |
|---------|--------|
| 1 | 0.1369 |
| 2 | 0.3730 |
| 3 | 2.1171 |

We can fit a quadratic function

$$abs = a_0 + a_1 t + a_2 t^2$$

where $t$ is the time measured in hours. (i) Calculate the values of $a_0$, $a_1$, and $a_2$. (ii) What happens if you extrapolate the theoretical relation to $t = 0$? Can you see why we might want to find a better theoretical relation than a quadratic to fit these data?

Hint: when we come to use this equation to predict absorbances at new times, $abs$ and $t$ will be the variables and the values of $a_i$ will be the constants. However, right now, the three values of $a_i$ are the unknowns and the specific values of the absorbances and times are known constants. Thus you can write down three equations in three unknowns.

---

The three equations in the three unknowns are:

$$a_0 + a_1 + a_2 = 0.1369$$
$$a_0 + 2a_1 + 4a_2 = 0.3730$$
$$a_0 + 3a_1 + 9a_2 = 2.1171.$$

Subtracting the first from the second

$$(2 - 1)a_1 + (4 - 1)a_2 = 0.3730 - 0.1369$$
$$a_1 + 3a_2 = 0.2361 \qquad \text{(Equation 1)}$$

and similarly also subtracting the first from the third

$$(3 - 1)a_1 + (9 - 1)a_2 = 2.1171 - 0.1369$$
$$2a_1 + 8a_2 = 1.9802. \qquad \text{(Equation 2)}$$

Multiply both sides of Equation 1 by 2 and subtract that from Equation 2

$$(8 - 6)a_2 = 1.9802 - 2 \times 0.2361$$
$$2a_2 = 1.5080$$
$$a_2 = 0.7540.$$

We can now calculate $a_1$ using Equation 1

$$a_1 = 0.2361 - 0.7540 \times 3$$
$$= -2.0259.$$

---

From the first equation

$$0.1369 = a_0 - 2.0259 \times 1 + 0.7540 \times 1$$
$$a_0 = 0.1369 + 2.0259 - 0.7540$$
$$= 1.4088.$$

Check:     $a_0 + 2a_1 + 4a_2 = 0.3730$     $a_0 + 3a_1 + 9a_2 = 2.1171.$

The fitted equation is

$$abs = 1.4088 - 2.0259\,t + 0.7540\,t^2$$

where $t$ is measured in hours.

Extrapolating the equation to $t = 0$ gives $abs = 1.4088$. This is much higher than the value at, say, $t = 1$, which is $0.1369$. This does not look right as we expect the bacteria to be growing during this period.

## End of Chapter Questions
(Answers to questions can be found at the end of the book)

### Basic

**1.** Evaluate:
(a) $6 + 5 \times 3^2$
(b) $3^3 + 5(3 + 1)$
(c) $7(4(3^2 - 9) + 1)$
(d) $7(4(3^2 - 9)) + 1$
(e) $7(4 \times 3^2 - 9) + 1$

**2.** Evaluate:
(a) $\frac{1}{2} + \frac{1}{7}$
(b) $\frac{1}{2} - \frac{1}{7}$
(c) $\frac{1}{2} \times \frac{1}{7}$
(d) $\frac{1}{2} \div \frac{1}{7}$
(e) $\frac{7}{4} \div \frac{1}{3} \div \frac{2}{3}$

**3.** Change:
(a) 11/8 to a decimal
(b) 13 % to a decimal
(c) 1/16 to a percentage

**4.** Round off the following numbers to three significant figures and express the answers in scientific notation. Also round off to two significant figures and again express the answers in scientific notation.
(a) 4 384 556
(b) 2348
(c) 0.000033678

**5.** What percentage change is a 4-fold increase or a 4-fold decrease?
(With the SI definition of percentage, the ordinary language phrase 'percentage change' means the fractional change expressed as a percentage.)

**6.** What fold change is a 500 % increase or an 87.5 % decrease?

**7.** Working by hand (use your calculator only to perform simple multiplications or to check your answer) find the *exact* square roots of
(a) 0.0004
(b) 0.04
(c) 0.0036

**8.** Working by hand (use your calculator only to perform simple multiplications or to check your answer) estimate the square root of 5 to three decimal places.

### Intermediate

**9.** Simplify the following expressions:
(a) $x + 5 + 3x - 6 + 9x + 3 - 4$
(b) $(x + 3)(x + 2) + 3x^2 - 2$
(c) $(x + 2)^4$
(d) $3(xy^3)^4$
(e) $(x + a)^4$

**10.** Simplify the following expressions:
(a) $3/x - 1/(x - 2)$
(b) $1/(1 + 1/x)$
(c) $1/(1 + 1/(1 + 1/x))$
(d) $(x + 2)/(x^2 - x - 6) + (x + 1)/(x - 3)$

**11.** Write $(a + b)^n$ as a polynomial for $n = 2, 3, 4,$ and 5.

**12.** Rearrange each of the following equations.
(a) $W = 1000 + 2x$ to give an equation of the form $x = ?$ with no $x$ on the right-hand side.
(b) $N = 1/(2 - 3t)$ to give an equation of the form $t = ?$ with no $t$ on the right-hand side.
(c) $T = az/(b + cz)$ to give an equation of the form $z = ?$ with no $z$ on the right-hand side.

**13.** Solve $3.7x - 8.2 = 1.5$, giving your answer to three significant figures.

**14.** Solve $x^2 + 8x + 12 = 0$ by factorization.

**15.** Estimate by a trial-and-error method to three significant figures the roots (solutions) of the equation $x^2 + 8x + 11 = 0$.

**16.** Estimate by a trial-and-error method to three significant figures a real root (solution) of equation $x^3 + 8x + 11 = 0$.

**17.** Solve for $x$, $y$, and $z$ in the simultaneous equations:

$$3x + 2y + 2z = 1$$
$$x + y - z = 0$$
$$2x + 2y + 3z = -2.$$

**Advanced**

**18.** If $a$ and $b$ are integers show that the reciprocal of the product is the product of the reciprocals,

$$\frac{1}{a \times b} \equiv \frac{1}{b} \times \frac{1}{a} \equiv \frac{1}{a} \times \frac{1}{b}.$$

Hint: start with the definition of the reciprocal of the product

$$\frac{1}{a \times b} \times (a \times b) \equiv 1$$

and multiply both sides of the equation successively by the reciprocals of $b$ and $a$.

**19.** It sometimes happens that we want to find the reciprocal of the sum of reciprocals, for example, $1/(1/a + 1/b)$. Write this as a simple fraction.

**20.** We have followed the usual convention of writing the general quadratic equation

$$ax^2 + bx + c = 0$$

with three constants. However, both the general equation and the solution can be expressed in terms of just two constants simply by dividing all terms by $a$. Write the general solution in terms of the two new constants.

**21.** Show that if $b^2 - 4ac = 0$, the quadratic equation $ax^2 + bx + c = 0$ can be rewritten as $(x + b/(2a))^2 = 0$ and thus there are still two roots to the quadratic equation. It is just that they are both the same.

**22.** With $y$ as the unknown, what are $a$, $b$, and $c$ in the standard form of the quadratic equation when $x^2 + y^2 = 1$? Use the values of $a$, $b$, and $c$ to obtain the solutions from the general solution. For what range of values of $x$ does this equation have real-value solutions for $y$? Verify the solutions by solving the starting equation for $y$.

**23.** Find the values of $x$ and $y$ such that $x^2 + y^2 = 1$ and $x + y = 1$.

**24.** Find the values to three significant figures of $x$ and $y$ such that $x^2 + y^2 = 16$ and $2x + y = 3$.

# CHAPTER 3

# Tables, Graphs, and Functions

The relations between quantities, how one depends on another, are at the core of science. The main use of elementary mathematics is to describe these relations. If we are to discuss them, we must be able to display them, and it helps if we can refer to them by name. The relations can be displayed in several ways: we can list values in a table, we can draw a graph, we might be able to use a formula, or we might be content just to give a verbal description. Functions provide a sort of packaging that allows us to encapsulate certain relations and refer to them by name. Functions will be central to the material presented in most of the rest of the book.

After illustrating how functions can be used with simple examples, two new functions will be introduced. Exponentials and logarithms arise frequently whenever we need to consider change, for example in descriptions of the growth of bacterial populations or the decline of drug concentrations between doses. Logarithms are part of the very definitions of pH, used to indicate $H^+$ ion concentrations, and of decibels, used to measure sound intensities. Logarithms are also employed to aid the presentation of any data that are spread over a large range of values, for example the more than 1000-fold range of concentrations often employed when measuring drug or hormone binding.

## 3.1 Tables

Experimental results are usually reported in tables or graphs. Tables are particularly useful when the data are qualitative (for an example see Section 3.3) or when the precise values listed will be wanted for subsequent use. In addition, they are often the starting point for preparing a graph. The entries in a table are organized into rows and columns. Each column refers to a particular property stated in a column heading. Everything below the heading should be a value of that property. A row represents a group of the different properties for a particular experiment, individual, or object. For instance, in Tables 3.1 or 3.2 one column lists the times of the measurements, the other the recorded distances. Each row reports a distance at a particular time. Whenever, as here, one value is chosen and the other measured, values of the chosen (independent) variable are conventionally listed toward the left and the measured (dependent) variable to the right.

Table 3.1 presents the measured distances at various times in the most literal manner possible, a list of times in the left column and a list of distances in the right column. For such a simple example, this is fine, but a moment's thought

### Table 3.1 Distance traveled at various times

| time | distance |
|------|----------|
| 0 s  | 0 m      |
| 1 s  | 0.5 m    |
| 2 s  | 2 m      |
| 3 s  | 4.5 m    |
| 4 s  | 8 m      |
| 5 s  | 12.5 m   |
| 6 s  | 18 m     |

The entries in the columns are the values of the physical quantities, time and distance.

### Table 3.2 Distance traveled at various times

| time / s | distance / m |
|----------|--------------|
| 0        | 0            |
| 1        | 0.5          |
| 2        | 2            |
| 3        | 4.5          |
| 4        | 8            |
| 5        | 12.5         |
| 6        | 18           |

Each entry is a number. The reference value or unit is indicated as part of the heading of each column.

should make it clear that this is not a good general template for tables. Imagine what it would look like if instead of listing times and distances, we were listing concentrations in the first column and observed fluxes across a cell membrane in the second. This could entail repeatedly writing units like mol $l^{-1}$ and mol $\mu m^{-2}$ $s^{-1}$, respectively. With the repetition of the units, the table would look terribly cluttered and for no good purpose. It would be so much tidier if we could agree a convention that allows us to write just the numbers, leaving out the units for the individual values. SI has just such a convention.

In SI, the numbers to be entered in each column are obtained by dividing the physical values by a reference value. The column heading lists the name or symbol for the quantity divided by the reference value. In Table 3.2 the reference time value is 1 s, the numbers in the time column are obtained by dividing the times by 1 s, and the column heading is time / s. Similarly, the reference value for the distances is 1 m and the heading for the distance column is distance / m. The column heading or label tells us exactly what has been done to obtain the numbers below it in the column. Because in practice the reference value is always chosen to be 1 of a unit, we do not normally bother to write the 1. Thus you use distance / mm but not distance / 1 mm and certainly not distance / $10^{-3}$ m, even though these are logically the same.

The SI convention for labeling tables (and, as we shall see, graphs) is so simple and logical that it is hard to understand why anyone does it differently. However, old habits die hard. Perhaps the most common alternative is to put the units in brackets after the name of the quantity, such as 'time (s)'. That loses the direct connection with division by a reference value but is acceptable provided we read it as 'time when it is expressed in seconds'.

To keep the table itself as uncluttered as possible, it is normal to have a brief legend. Sometimes this follows the title above the table, but more often it will be just below the table body, often separated from it by a line. The legend must provide all the additional information needed to define or interpret the table entries. For instance, a sentence like 'Data values are shown as mean $\pm$ s.e.m.' would occur in many legends. The abbreviation s.e.m. stands for standard error of the mean, which you will encounter in Chapter 9.

## Box 3.1 Interpolation

Even in this computer age, you still need to be able to look things up in tables. Most commonly, this occurs when you have experimental data at intervals and want to estimate what you would have found had you looked for values in between those that were measured. This estimation process is called interpolation.

The simplest interpolation procedure, usually all you will need, assumes that the variation between the data points is linear. For instance, you might want to ask for the distance traveled after 4.4 s for the experiment reported in Table 3.2. The interval between 4 s and 4.4 s is 40 % of the interval between 4 s and 5 s. The distance traveled between 4 s and 5 s is 12.5 m − 8 m = 4.5 m, so the estimate of the distance traveled between 4 s and 4.4 s is 40 % × 4.5 m = 1.8 m, and the total distance traveled at 4.4 s becomes

$$d(4.4 \text{ s}) = d(4 \text{ s}) + 1.8 \text{ m} = 8 \text{ m} + 1.8 \text{ m} = 9.8 \text{ m}.$$

Exactly how you lay out a table is a matter of taste. Some people like to have lines between rows, some between columns, some both, some put double lines under the heading or just above the legend, and so on. There is one overriding rule: the table must be clear. It must state exactly what is being listed and it should be easy to see which row and column each item is in. There really is no more to it than that. However, there is no accounting for taste. When preparing tables for journals, you do as the editors tell you.

## 3.2 Graphs

Even the best tables can be dry. Graphs are the application of the principle 'one picture is worth a thousand words' to the reporting of quantitative data. There are many sorts: bar charts; pie charts; box-and-whisker plots; $x,y$ plots and so on. The first three of these are dealt with in Chapter 9, which considers how you describe data using statistics and report the description using graphs. What we will describe here is the simplest way of displaying the relation between two variables, $x$ and $y$.

To prepare an $x$, $y$ graph, sometimes called a Cartesian graph, we draw horizontal and vertical lines called the axes. Along each axis we put equally spaced marks, called tick marks, with tick labels such as numbers for some of them. How many we use in a particular case is largely a matter of style. There should be enough to be informative, which means having at least three labels and usually at least four ticks so that the reader can see that the progression of values is linear with distance along the axis. It is best to avoid having so many that the graph looks cluttered.

It is permissible to use units in the tick labels, but it is usually much neater if the tick labels are just numbers. To achieve this you must convert the values of the physical quantities you want to display to numbers, just as when you want to use a numerical value equation or tabulate values in a table. In SI, the convention for how you do this is exactly the same for graphs and tables. You produce the number by dividing the value of the physical quantity by a reference value. Thus in SI the axis label is the name or symbol for the physical quantity divided by the reference value. The labels on a graph and the column headers in a table do exactly the same job.

When a graph is included in a paper or book, it is usually called a figure and it must be accompanied by a legend. The figure legend is just as important as the graph because it tells us what quantities are being plotted. All of this is much easier to see with some examples.

The data from Table 3.2 are plotted in Figure 3.1. The points could be connected with straight-line segments and this is often done when you want to emphasize that you are not claiming anything about a theoretical relation. However, in this figure we have chosen to put a smooth curve through the points because we do have good reason to believe that a smooth curve is correct. In fact, Figure 3.1 is a plot of $d = (1/2)at^2$, where $d$ is the distance measured in meters (m), $t$ is the time measured in seconds (s), and the

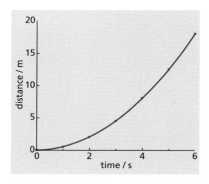

**Figure 3.1**
Distance traveled versus time since start according to $d = 1/2at^2$. Data are taken from Table 3.2. Instead of 'distance / m' and 'time / s' the axis labels could be just '$d$' and '$t$' because these symbols represent the numerical values.

numerical value of the acceleration is $a = 1$. (Strictly we should say that $d$ and $t$ are the numerical values when the distance and time are measured in the stated units, but you will rarely see anyone being so careful.) The tick marks on each axis are equally spaced and each interval represents the same amount of either time or distance, so this is a plot on normal linear axes.

Any spreadsheet program will draw the graphs for you if you first type the data table. The graph in Figure 3.1 was drawn using Microsoft Excel®, where it is called an 'XY' or 'Scatter' chart. The steps needed to produce this graph are the following:

1.  Enter the labels and numbers from Table 3.2 on Sheet 1 and select the entire table; that is, a region of 8 rows and 2 columns.
2.  Select Insert Chart and choose 'XY Scatter' with subtype showing symbols at the points connected by a curve rather than by a sequence of straight lines, then click Next. With this selection Microsoft Excel® calculates and plots what are called splines so that the curve goes through every point without any sharp corners. When your curve has a simple shape, as here, it will come very close to getting it right.
    For reasons best known to Microsoft, the plot of $y$ versus $x$ is called a scatter plot even though we will usually select an option to draw a curve or line. Do not be tempted by what they call a Line Chart. Each line in a Line Chart is a plot of the values in a column versus either the row number or labels entered in the leftmost column. Sometimes you can get away with entering the $x$ values as labels. However, no matter what intervals you have between successive values of $x$, they will all end up equally spaced which is not what you usually want! Furthermore, Microsoft Excel® cannot fit a spline curve to data in a Line Chart.
3.  On the next screen just click Next – we have already selected our data. The next screen after that offers us the chance to set several properties of the chart. Look at each in turn. For this chart, there is no chart title, the axis titles are time / s and distance / m, respectively, gridlines were turned off, there is no legend, and there are no data labels. Once you are happy, click Next.
4.  Choose where to put the chart, either as an 'object' on the worksheet or on a new chart sheet of its own. Click Finish.

You can spend 'many happy hours' fiddling with the formatting of Microsoft Excel® charts. You select the entire chart by clicking near the edges, or the plot region by clicking a blank bit in the middle, an axis and its labels by clicking on a label, or the data series with its symbols and curve by clicking on the curve or a symbol. Once you have selected something you can use the formatting commands to determine how it looks. For Figure 3.1 the box around the chart was removed and a background color was chosen. The box around the plot area was removed, the shading was made the same as the background, the color and size of the symbols, the color and thickness of the curve, and the scale and thickness of the horizontal axis were adjusted. All of this is much the same, no matter which version of Microsoft Excel® you use (Apple Macintosh or PC), though where you find some of the commands can vary. In modern versions of Microsoft Excel® you can either save or print the chart to a high-resolution pdf file suitable for publication or for importing into graphics software for further adjustments.

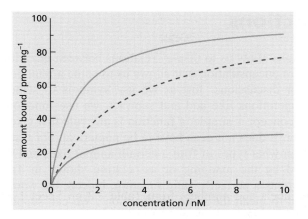

**Figure 3.2**
Plots of the binding equation,
$N_b = N_T K_a c/(1 + K_a c)$, for the
following values of $N_T$ and $K_a$:
100 pmol mg$^{-1}$ and 1 nM$^{-1}$ (red
curve); 100 pmol mg$^{-1}$ and
0.333 nM$^{-1}$ (blue dashed curve);
33.3 pmol mg$^{-1}$ and 1 nM$^{-1}$
(green curve).

A slightly more complicated graph is the plot of the binding equation in Chapter 2, repeated here in Figure 3.2. In this graph, the reference value for the concentration is 1 nM whereas the reference value for the amount bound per milligram of tissue is 1 pmol mg$^{-1}$. Because this figure shows three curves or sets of points, not just one, some way must be found to indicate which curve is which. If there are data points plotted, the usual method is to use different symbols for each curve, such as filled circles, open circles, or filled squares. If there are no symbols, the next best option is to use easily distinguished lines, such as different colors or, if all must be black, solid, dashed, and dotted. Sometimes you distinguish both symbols and lines. In any case, the legend must explain what has been done. Another possibility is to put labels near the curves on the plot itself. This is frowned upon because it can make figures look messy. Nevertheless, sometimes it can be very convenient for both author and reader. Judge for yourself in Figures 3.5 and 3.6, where direct labeling allows easy comparison of the graphs without repeated reference to the legends. Regardless of what you decide to do, the legend must make it obvious to a reader who cannot read your mind.

Figure 3.3 illustrates yet another variation on the use of graphs. Here the intention is to find values of $x$ and $y$ that are solutions of both

$$y = x^2 - x + 1 \text{ and } y = 10 - x. \tag{EQ3.1}$$

To do this, each of these has been plotted. The solutions are the intersections of the two curves. In this case there are two, at $(-3, 13)$ and $(3, 7)$.

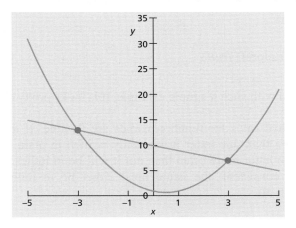

**Figure 3.3**
Plots of $f(x) = x^2 - x + 1$ (green)
and $g(x) = 10 - x$ (red). The
points of intersection give the
solutions of both $y = x^2 - x + 1$
and $y = 10 - x$.

## 3.3 Functions

Tables and graphs allow us to display relations between quantities. Functions provide a sort of encapsulation that allows us to name a relation and handle it in equations or discussions. Just as we use symbols to stand for physical quantities or operations like add and subtract, we use functions to describe relations. The concept is simple: a function is any rule that takes us from one thing to another. It does not even have to be numerical. For instance, suppose we have data on what types of fruit a certain animal likes and that these data are represented by the mapping or association shown in Table 3.3. This mapping represents a perfectly valid function; indeed, the word 'mapping' is frequently used to mean the same as 'function'. We can supply the value of the independent variable, the fruit, and the function supplies the value of the dependent variable, the preference. The possible values of the independent variable, here the list of fruits, taken together are called the domain of the function. The possible values of the dependent variable, here just like or dislike, taken together are called the range of the function. Stated a little more carefully, a function is any rule that assigns one and only one value in its range to every value in its domain. You are always supposed to specify the domains and ranges of functions and occasionally you must but biologists often do not even think about them.

Of course, most of the functions we will deal with in this book will relate numbers or physical quantities. You have already met many such functions like $x^2$, $x^3$, $\sqrt{x}$, and possibly $\log(x)$, $\exp(x)$, and $\sin(x)$ that are readily available in scientific calculators. A great many more can be found in spreadsheet programs like Microsoft Excel®. For each of these functions you supply a value of the independent variable, often called $x$, and the function returns a *unique* answer, the value of the dependent variable, often called $y$.

If $y = x^2 + 2x + 1$ we can say that $y$ is a function of $x$, written $y = f(x)$, where the function or rule is $f(x) = x^2 + 2x + 1$. In the expression $f(x)$, the '$f$' is the name of the function. The independent variable, the one we can choose, goes inside the brackets and is sometimes called the *argument* of the function. The dependent variable, here $y$, is supplied by the function dependent on our choice of $x$. It is sometimes called the *result* of the function. There is nothing special about these choices of symbols. Instead of the letter '$f$', we could use '$g$' or '$h$' or combinations of letters like 'log', 'sin', 'cos', or even 'fruitpref', so the function defined in Table 3.3 could be used in an equation such as

$$preference = \text{fruitpref}(fruit) \tag{EQ3.2}$$

though, as usual with such a simple example, this looks pretty silly.

The typographical rules for function names are flexible. If the name of the function is more than one letter it is usually written in plain type, but if the name is a single letter more often than not it will be in italics. The important point is to be consistent: the name must either be in italics whenever it appears or not at all.

**Table 3.3 An animal's fruit preferences**

| fruit | preference |
|---|---|
| grape | like |
| apple | like |
| strawberry | like |
| plum | like |
| peach | dislike |
| pear | dislike |
| apricot | dislike |

The real use of functions only becomes apparent when we consider rules or mappings that are hard or impossible to express in terms of the simple operations, constants, and variables. To illustrate this we now need to consider a more complicated example. You have already met a good one, the process of finding the $n$th power or $n$th root of a number $x$; that is, of finding $y = x^n$. When $n$ is an integer this is just a simple operation of arithmetic, but when $n$ is not an integer the calculation becomes complicated. In principle, we have seen in Section 2.3 that it can be performed by successive approximations (trial-and-error) but we certainly do not want to write out a description of that procedure each time we need to specify the relation. What we do instead is define a function which can be called either power $(x,n)$ or $x^n$. The latter does not obey the usual rules for naming functions, but that's worth tolerating because it reminds us that the function power $(x,n)$ is equal to $x$ raised to the $n$th power whenever $n$ is a positive integer. Notice that power $(x,n)$ is a function with two arguments; there is no reason why a function need be restricted to just one argument.

The function, power $(x,n)$, is defined by the procedure in Section 2.3 together with rules for the domain and range. The procedure alone is not enough because it does not satisfy the condition that it must produce one and only one answer. For instance there are two real values of $y$ such that $y = 4^{1/2}$, either $+2$ or $-2$, and there is no real value of $y$ that satisfies $y = (-1)^{1/2}$. However, if we restrict $x$ so that it must be greater than or equal to 0, written $x \geqslant 0$, and we insist that the answer is also positive, $y \geqslant 0$, then we do have a function that produces a unique answer to any accuracy we desire. The power function implemented in calculators and computers is defined for $x$ and $n$ as numbers. It sometimes makes sense to allow $x$ to represent a physical quantity, but $n$ will always be just a number.

There are three main methods for displaying a rule for calculating the result of a function: state an expression or a procedure for calculation; list values in a table; or draw a graph. We have already looked at examples of expressions and procedures that define functions. Some of these are simple enough to be evaluated just using the basic operations of arithmetic. These include polynomial functions that are defined by rules like $f(x) = x^2 + 2x + 1$. In the general case, a polynomial function involves only a sum of terms in integer powers of $x$. Rational functions are a more general class of functions that are ratios of two polynomials. With enough terms, you can make a graph of one of those look like almost anything. It is rare for a polynomial or rational function to be exactly what we want in a biological application. Nevertheless, they are very important as they are easy-to-use approximations to more awkard functions. We have also seen a more

## Box 3.2

You can think of power $(x, n)$ or $x^n$ as a way of referring to the answer given by your calculator or spreadsheet program when you supply the values of $x$ and $n$. Being able to refer to the answer, before we do the calculation, lets us include the function in other expressions. As we shall see in later chapters, using the function notation also allows us to talk about things that are true for many functions without specifying exactly which one we mean.

complicated procedure that defines a function: finding $x^n$ for $x \geqslant 0$ when $n$ is not an integer. Of course, we never actually do that by hand, we use a calculator or a computer. However, regardless of how we get the values, encapsulating the procedure as a function allows us to refer to it and use it whenever we want, without first having to calculate the values.

The only restriction on using tables and graphs to define functions is that there must be a clear description that always allows you to find a unique answer. This means that when any properly defined function $y = f(x)$ is plotted versus $x$, a vertical line can intersect the curve produced only once. If it were to intersect twice we would not be able to choose which value of $y$ to use and we would not have a properly defined function.

## 3.4 Inverse functions

Our objective over the next few sections is to set the stage and introduce a pair of functions, exponentials and logarithms, which are encountered repeatedly in biology. In this pair, one function is the inverse of the other. Therefore, we first need to consider what 'inverse function' means.

A function that undoes or reverses the effect of another is called an inverse function. Inverse functions and inverse operations are closely related. For instance we can define a 'multiply by 5 function' $y = f(x) = 5x$. The inverse operation that undoes the effect of multiplication by 5 is division by 5, and thus the inverse function that undoes the effect of $f(x)$ is $z = f^{-1}(y) = y/5$. Similarly we could define an 'add 7 function', $y = g(x) = x + 7$ and the inverse function that undoes the effect of $g(x)$ is then subtraction of 7, $z = g^{-1}(y) = y - 7$. The superscript '$-1$' applied to a function in $f^{-1}(y)$ and $g^{-1}(y)$ means inverse, it does not mean reciprocal. In the second example if we start at $x$, calculate $y = g(x) = x + 7$ and then $z = g^{-1}(y) = y - 7 = x + 7 - 7 = x$, we are back where we started. That is what we mean when we say $g(x)$ and $g^{-1}(y)$ are inverse functions. Generally that can be written as $g^{-1}(g(x)) = x$. If you find the formality of this expression a bit daunting, do not despair. The important point has already been said in words – the inverse function reverses the effects of the function itself.

It may help to look at this graphically. Take as an example $y = f(x) = x^2 + 2x + 1$ for $0 \leqslant x \leqslant 5$ which is plotted in Figure 3.4A. We can find the inverse function by solving this equation for $x$ in terms of $y$ (see Section 2.7), $x = f^{-1}(y) = \sqrt{y} - 1$ where $\sqrt{y}$ is the principal square root of $y$. This inverse function is plotted in Figure 3.4B. Look at these two graphs. Take either one and flip it across the 45° diagonal through the origin and you get the other! That is a general result: take a graph of a function, flip it across the 45° diagonal and, if the resulting plot is also a plot of a function, you have a plot of the inverse. Do not be afraid of inverse functions, you can always get approximate values from a graph of the original function.

Does a function always have an inverse? No. Why not? In Figures 3.4A and 3.4B the function $f(x)$ was only defined for positive values of $x$. What happens if we change the definition of the function so that it is also defined for negative values of $x$; that is, $g(x) = x^2 + 2x + 1$ for $-5 \leqslant x \leqslant 5$. There is

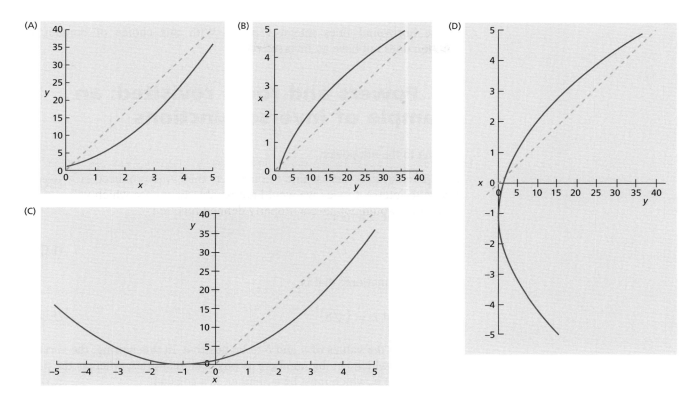

**Figure 3.4**
Plots of functions and inverse functions. (A) Plot of the function $y = f(x) = x^2 + 2x + 1$ for $0 \leqslant x \leqslant 5$. (B) Plot of the inverse function $x = f^{-1}(y) = \sqrt{y} - 1$. Note that the plot in (B) can be obtained from that in (A) by flipping or reflecting the curve across the 45° diagonal running through the origin. (C) Plot of the function $y = f(x) = x^2 + 2x + 1$ for $-5 \leqslant x \leqslant 5$. (D) Plot obtained by flipping that in (C) across the 45° diagonal. Note that this is not the plot of a function because values of the independent variable, here $y$, correspond or map to more than one value of the dependent variable, here $x$.

nothing wrong with this as a function, but it does not have an inverse. If all we have is a result, say 4, we do not know if we started with a positive or negative value of $x$; that is, 1 or $-3$, so we cannot guarantee to get back where we started. This is shown graphically in Figure 3.4C and in Figure 3.4D, which is the same graph flipped across the 45° diagonal through the origin. In Figure 3.4D, a vertical line can intersect the curve twice. The relation plotted in Figure 3.4D does not yield a unique value so it is not a function.

Inverse functions have a sort of symmetry. If $f^{-1}(y)$ is the inverse function for $f(x)$ then $f(x)$ is the inverse function for $f^{-1}(y)$. You can see this graphically. We do not even need to draw the graph again. A graph of $y$ versus $x$ is a graph of a function if every vertical line from the lowest to the highest value of $x$ in the domain intersects the curve exactly once. This function has an inverse if every horizontal line from the lowest to the highest value of $y$ within the range $f(x)$ intersects the curve exactly once. For example, in Figure 3.4A the domain is set to be $0 \leqslant x \leqslant 5$ and the range of the function is $0 \leqslant y \leqslant 36$. For these values every vertical and horizontal line intersects the blue curve once and the function has an inverse. By contrast for Figure 3.4C the domain is set to be $-5 \leqslant x \leqslant 5$ while the range is still $0 \leqslant y \leqslant 36$. With these values every vertical line intersects the blue curve once, but now some of the horizontal lines intersect twice. With this choice of domain, the function does not have an inverse.

## 3.5 Powers and roots revisited: an example of inverse functions

Raising to the $n$th power,

$$b = a^n \tag{EQ3.3}$$

and taking the $n$th root

$$a = \sqrt[n]{b} \tag{EQ3.4}$$

are inverse functions; that is,

$$a = \sqrt[n]{a^n} \text{ and } b = \left(\sqrt[n]{b}\right)^n \tag{EQ3.5}$$

if we require the values of $a$ and $b$ to be positive. In this pairing, the variables are $a$ and $b$, which can be numbers or physical quantities, whereas the power (or exponent) $n$, which must be a number, is a constant.

The inverse relationship between raising to a power and taking roots is shown graphically in Figures 3.5 and 3.6. The first of these shows $x$ raised to several different powers for positive values of $x$, that is, $y = x^n$ for $x \geqslant 0$. There is nothing unusual here. The second is the same figure flipped across the 45° diagonal passing through the origin. Plotting the power laws and flipping the graph solves the problem of plotting and calculating roots at least in principle. The only difficulty is that we cannot read the graph accurately enough to be useful in most calculations. All your calculator or spreadsheet does is provide the extra accuracy you need to make this useful. There is no magic here. It does not matter whether you use a magnifying glass on a flipped chart, struggle through the tedium of the trial-and-error procedure, or simply trust Casio and press the buttons on your calculator: all of these procedures allow you to calculate the value of an $n$th root. Regardless of which you do, naming it as a function – 'taking the $n$th root' – allows us to use this relation in describing our science even *before* we put in specific values. That is what the use of algebra in science is all about.

## 3.6 Exponentials and logs

Raising to a power is a function of the base when the exponent is constant. However, raising to a power can be looked at differently. When the base $a$, which should be positive, is held constant and the exponent is varied, raising $a$ to the power $x$, $y = a^x$, is called the exponential function with base $a$. Both $a$ and $x$ must be numbers, not physical quantities. So far, there is nothing new. The procedures we have already described allow us to calculate exponentials just as they allow us to raise a base to a power. However, there is a change in emphasis focusing now on the exponent rather than the base. It is now very

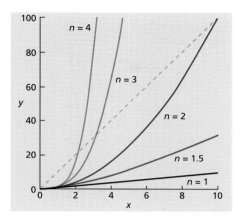

**Figure 3.5**
Plots of $x^n$ for the indicated values of $n$ and $x \leqslant 10$.

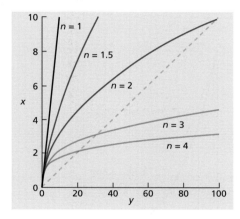

**Figure 3.6**
Plots of the $n$th root of $y$ for the indicated values of $n$ and $y \leqslant 100$.

important to remember that the exponent can be any real number. To reflect this change in emphasis, but for no other reason, it is normal to call the exponent in the exponential function $x$ rather than $n$. Graphs of the exponential functions $y = a^x$ with $x \geqslant 0$ are shown in Figure 3.7 for three special values of the base, $a = 2$, $a = e \approx 2.718$, and $a = 10$. Exponential functions have the very important property that each time $x$ is increased by the same amount, that is, the same amount is *added*, $y$ is *multiplied* by the same factor. This behavior is called **exponential growth**. In Figure 3.7 for base 2, each time $x$ increases by 1, $y$ is multiplied by 2; for base 10, each time $x$ increases by 1, $y$ is multiplied by 10; for base $e$, each time the multiplication is by $e$. An important example of exponential growth is the early stage of bacterial growth when the bacteria are not crowded and the number present doubles at constant intervals called, not surprisingly, the doubling time.

Figure 3.7 shows the exponential growth when the exponent is positive. Equally important is the **exponential decay** when the exponent is negative, $y = a^{-x}$ with $x \geqslant 0$ as in Figure 3.8. Now if $x$ is increased by 1, for base 2, $y$ is halved; for base 10, it is reduced 10-fold, that is, to 1/10th the value; and for base $e$ it is reduced $e$-fold. Exponential decay is seen whenever the rate of loss of something is proportional to the amount remaining. Good examples are the radioactive decay of isotopes used as tracers and the decrease in concentrations of many drugs within the body sufficiently long after drug administration has stopped.

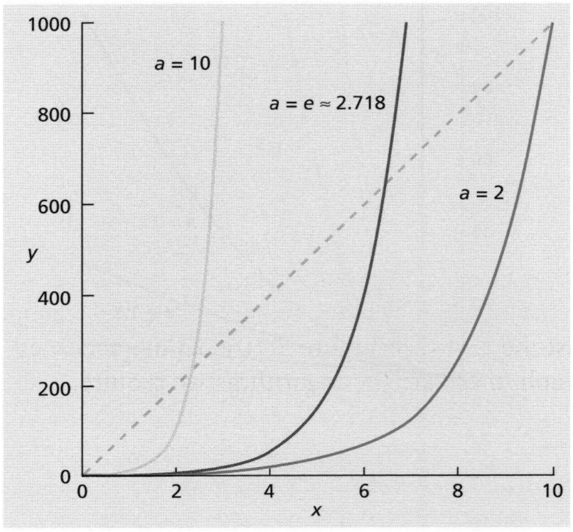

**Figure 3.7**
Plots of exponential functions for positive exponents for three different bases. These three curves intersect at the point $x = 0$, $y = 1$.

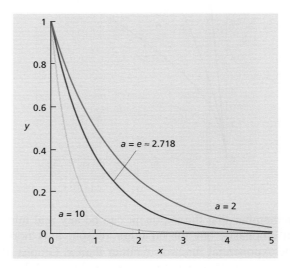

**Figure 3.8**
Plots of exponential functions for negative exponents for three different bases.

The three values of the base, 2, 10, and $e$, chosen for Figures 3.7 and 3.8 are special for different reasons. The number 2 is special because it corresponds nicely to concepts like the doubling time of a bacterial population or the half-time for the decrease in drug concentration in the body. It is also the base for counting in computers. The number 10 is special because it is the base for the decimal system that ordinary mortals use for counting. The value we have called '$e$' is special because it arises 'naturally' in calculus; this will be explained properly in Section 8.1. Because $e$ is the most convenient base to use in calculus, it is preferred by mathematicians and the name 'exponential function', with no mention of which base is being used, almost always means that the base is $e$ and another name for the function $e^x$ is $\exp(x)$. The origin of using the symbol $e$ to stand for 2.718... is unknown. The first person to make use of the value in a consistent manner was Napier, but the first person to use the symbol $e$ to stand for this constant was probably Euler, though apparently no one knows why. He may have chosen it as the first letter of the word exponential or as the first letter in the alphabet that he had not already used as a symbol at the time. In any case '$e$' is fitting as Euler did much to establish its properties.

The inverse function for an exponential – the function that reverses the effect of the exponential function – is called a logarithm.

If $y = a^x$ for $a > 0$ and $x$ is equal to a real number, then $x = \log_a(y)$.

(EQ3.6)

For example, we know that $8 = 2^3$ so we can also write $3 = \log_2(8)$. Graphically, to get a plot of a logarithm, all we need to do is plot the exponential for the same base, flip it across the 45° diagonal through the origin, and adjust the text (see Figure 3.10). To the accuracy we can read the values from a graph, these are the logarithms we require.

Just as a logarithm is the inverse function for an exponential, an exponential is the inverse function for a logarithm:

if $x = \log_a(y)$ for $a > 0$ and $y$ is equal to a positive real number,
then $y = a^x$.

(EQ3.7)

Because the exponential is the inverse function for a logarithm, it is sometimes called the antilogarithm.

Some exact values of logarithms are easy to see, because $\log_a$ of any number that can be written as $a^x$ is just $x$. Thus $\log_2(2) = 1$; $\log_2(8) = 3$; $\log_2(1/2) = -1$; $\log_2(1/8) = -3$, and so on. Similarly $\log_{10}(10) = 1$; $\log_{10}(1000) = 3$; $\log_{10}(0.1) = -1$; $\log_{10}(0.0001) = -4$, and so on. A negative value of the logarithm means that the original value was less than 1; a positive value that it was greater than 1. Note also that $\log_a(1) = 0$ for any base $a$.

The base 10 is used so frequently that $\log_{10}(x)$ is called the 'common logarithm' and in scientific work $\log(x)$, written without any base, usually means $\log_{10}(x)$. The other logarithmic base that occurs frequently is $e$, so the **n**atural **l**ogarithm is also given a special, easy-to-write label: $\ln(x) = \log_e(x)$. Unfortunately for us, mathematicians often use $\log(x)$ with no indication of

**Figure 3.9**
Leonhard Euler.

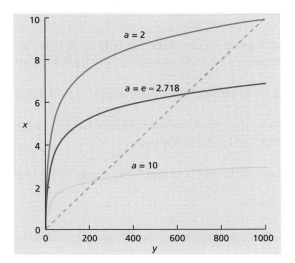

**Figure 3.10**
Plots of log functions for the three bases indicated. These curves are the same as in Figure 3.7 but reflected across the 45° diagonal line passing through the origin. They intersect at the point $y = 1$, $x = 0$.

the base to mean $\log_e(x)$. In the remainder of this book $\log(x)$ will always mean $\log_{10}(x)$ and $\ln(x)$ will mean $\log_e(x)$. The *natural logarithm* function $\ln(x)$ is considered in more detail in Chapter 8. In the rest of this chapter, we shall concentrate on base 10.

A simple table for $\log_{10}(x)$ for $x$ from 1 to 10 is given in the appendix at the end of the book. More extensive tables can be found in the *Handbook of Chemistry and Physics* published by CRC Press, and in the *Handbook of Mathematical Functions* compiled by Abramowitz and Segun (see References at the end of the book). Specific values can be obtained using any good scientific calculator.

Logarithms have some very useful properties. For $x > 0$ and $y > 0$

$$\log(xy) = \log(x) + \log(y), \tag{EQ3.8}$$

for $x > 0$

$$\log(x^y) = y \log(x), \tag{EQ3.9}$$

and, as a special case of the preceding

$$\log\left(\frac{1}{x}\right) = -\log(x). \tag{EQ3.10}$$

> There is *no* simple rule for relating $\log(x + y)$ to $\log(x)$ and $\log(y)$.

We have all the knowledge needed to prove EQ3.8 and EQ3.9 from the properties of the exponential function and the definition of the logarithm as its inverse; the proofs are given in the solution to Question A in Presenting Your Work at the end of this chapter. For now, we can check very quickly that these properties are true for powers of 10. For example if $x = 10$ and $y = 100$, then $\log(x) = 1$, $\log(y) = 2$ and

$$\log(xy) = \log(1000) = 3 = 1 + 2 = \log(x) + \log(y).$$

Similarly

$$\log(10^6) = 6 \log(10) = 6 \times 1 = 6.$$

We can also check other values with the help of a calculator or a table of logarithms. For example, we look up $\log(2.73)$ to be 0.4362 so

$$\log(273) = \log(2.73 \times 10^2) = \log(2.73) + \log(10^2) = 0.4362 + 2 = 2.4362.$$

We can check this is the correct answer directly using a calculator. Similarly, we can look up $\log(7.95)$ to be 0.9004 so

$$\log(0.0795) = \log(7.95 \times 10^{-2}) = \log(7.95) + \log(10^{-2}) = 0.9004 - 2$$

$$= -1.0996.$$

Provided we can find logarithms to base 10, we can also calculate logarithms to any other base.

For example, suppose we want to find $\log_2 (100)$. First we write $y = \log_2 (100)$ so that $100 = 2^y$. Next, we take logarithms to base 10 on both sides of this equation to give

$$\log_{10} (100) = y \log_{10} (2).$$

From this equation we find that $y$ is given by

$$y = \log_{10} (100) / \log_{10} (2) = 2/\log_{10}(2) = 6.644.$$

In general, if we know $y = \log_{10} (x)$ and want to know $z = \log_a (x)$ then invert the second equation, giving $x = a^z$, and insert this expression for $x$ into the first equation:

$$y = \log_{10} (x) = \log_{10} (a^z) = z \log_{10} (a). \qquad (\text{EQ3.11})$$

Solving for $z$ gives

$$z = y / \log_{10} (a), \qquad (\text{EQ3.12})$$

or writing it out in full

$$\log_a (x) = \frac{\log_{10} (x)}{\log_{10} (a)}. \qquad (\text{EQ3.13})$$

As there are only three bases in common use, the only examples you are ever likely to see are

$$\log_2 (x) = \log_{10} (x) / \log_{10} (2) \approx 3.322 \times \log_{10} (x) \qquad (\text{EQ3.14})$$

and

$$\ln (x) = \log_{10} (x) / \log_{10} (e) \approx 2.303 \log_{10} (x). \qquad (\text{EQ3.15})$$

Finally, a note of terminology: the word 'logarithm' is frequently shortened to just 'log'. We shall frequently make use of this abbreviation in the remainder of this book, particularly when, as in the next few paragraphs, we need to make frequent use of the word.

## 3.7 Log plots and log paper

You need to understand a log plot when you see one. It is also worthwhile learning how to plot data with a logarithmic scale. In a semi-log graph or plot, one of the axes is logarithmic whereas the other is linear. What is a logarithmic axis? It has a special nonlinear scale that spaces the numbers themselves so that the logarithms of the numbers increase linearly with distance along the axis. Suppose we have data that decrease exponentially with time. When plotted on normal linear scales the results would look like Figure 3.11.

**Figure 3.11**
Exponential decrease of
concentration with time.

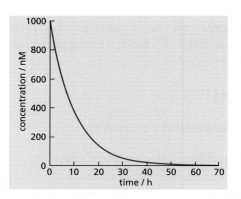

**Figure 3.12**
Data from Figure 3.11 plotted
using a logarithmic scale for the
concentrations. The linear scale
on the right is not normally
shown. It displays the values of
logarithms of the concentrations
that are plotted on the
logarithmic scale to the left.

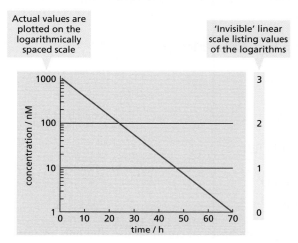

The same data are plotted, without calculating logs, on the special three-cycle log scale in Figure 3.12. With three log cycles, the log varies by 3, thus concentrations spanning a 1000-fold range can be plotted.

The key to a log axis is that each time the number increases by the same factor, the position for plotting that number must move an equal distance along the axis. Thus 1, 2, 4, 8, 16, 32, 64, 128, and 256 must all be equally spaced, separated by $\log(2)$, on the invisible log axis. Similarly 1, 3, 9, 27, 81, and 243 must be separated by $\log(3)$ and 1, 10, 100, and 1000 must be separated by $\log(10)$. The lowest point can be any power of 10 we choose. In Figure 3.12 it is (concentration / 1 nM) = 1.

The shape of the curve that results from using a plot with a logarithmic axis is just the same as if the logarithms of the values were calculated and plotted with a linear axis. Obtaining a straight line when you use a log plot reveals that the data follow an exponential pattern.

## 3.8 What use are logarithms?

Now that calculators have replaced the use of log tables in practical calculations, why are logarithms still important tools for a scientist? Perhaps the best way to approach this is to discuss several examples, each illustrating an important use.

## Plotting the logarithm of the number of cells displays whether cells are growing in an 'exponential phase'

When bacteria are plated onto nutrient agar in a Petri dish the number of cells present initially grows exponentially, doubling at regular intervals. However, eventually the number approaches a limit imposed by the size of the plate, the availability of nutrients, and the products they produce. A plot of number versus time might look like that shown in Figure 3.13.

At the foot of this curve the number of bacteria doubles each time $t$ increases by one doubling-time; that is, the increase is exponential as the number increases proportional to $2^{t/\text{doubling-time}}$ (which is the same as $e^{t/\tau}$ where $\tau = \text{doubling-time}/\ln(2) \approx \text{doubling-time}/0.69$). However, it is very difficult to see from the plot how long the exponential phase lasts or even if it is exponential at all. It can be made much more obvious by replotting the data using logarithms. Whenever a variable, $y$, increases exponentially with time, the logarithm of that variable to any base increases linearly with time. Thus although the increase is exponential each time the number of cells increases by a factor of 2, the common logarithm of the number increases by $\log(2) = 0.301$ and a plot of $\log_{10}$ (number of bacteria) versus time should give a straight line. The variation is replotted in Figure 3.14 and it is now obvious that the exponential phase of growth lasts for approximately 3 h. (Because $\log_2$ (number) has the tidy feature that it increases by 1 each time the number doubles, you will occasionally see $\log_2$ (number) instead of $\log_{10}$ (number).)

## Plotting the logarithm of the plasma concentration of a drug shows whether the concentration is decreasing exponentially with time

After a drug is injected intravenously, the concentration of the drug in the blood initially decreases by a relatively large factor in each small unit of time as a result of both drug diffusion out of the blood into the tissues and delivery of the drug to sites like the kidneys and the liver where it is eliminated from the body. Later, in each small unit of time, the concentration decreases by a relatively smaller factor because, although elimination continues, drug now diffuses out of the tissues back into the blood, partly replacing that which is being lost. In this later phase the rate of drug elimination, the amount left in the body, and the plasma concentration are expected to be proportional to each other and at these times the concentration should fall exponentially. A typical variation is plotted in Figure 3.15 as plasma concentration versus time. It is exceedingly difficult to see which portions of this curve display an exponential variation and which do not. The exponential nature can be seen much more clearly by replotting the data using logarithms.

Whenever a variable decreases exponentially with time, for example as $2^{-t/\text{half-life}}$ or $e^{-kt}$ where $k$ is a rate constant, the logarithm of that variable decreases linearly with time. Each factor of 2 decrease in the concentration corresponds to a decrease in $\log(\text{concentration})$ of $\log(2) = 0.301$ (or a decrease in $\ln(\text{concentration})$ of $\ln(2) = 0.69$). The variation from Figure 3.15 is replotted as $\log(\text{concentration})$ versus time in Figure 3.16. It is now clear that the decrease in concentration is indeed exponential for times longer than about 2 h.

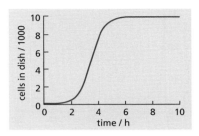

**Figure 3.13**
Number of bacteria in a Petri dish as a function of time after plating.

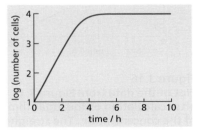

**Figure 3.14**
Plot of the data from Figure 3.13 after calculation of the logarithm of the number of cells. The initial straight line indicates that during this period the number of cells was increasing exponentially with time.

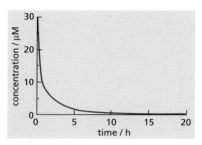

**Figure 3.15**
Plasma concentration of a drug after an intravenous injection.

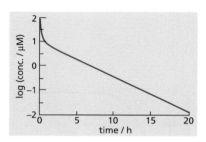

**Figure 3.16**
Plot of the data from Figure 3.15 after calculation of the logarithm of the concentration. The straight line indicates that after an initial period the plasma concentration was decreasing exponentially with time.

## Logarithms are used to display data that are spread over a large range of values

Drug, hormone, or neurotransmitter interaction with receptors can be investigated directly by measuring the binding of the small molecules to the receptors. An example of this type of data for a drug with a dissociation constant $K_D = 0.316$ μM is shown in Figure 3.17. The theory developed in Chapter 2 predicts that the amount bound will increase nearly proportionally with drug concentration, $c$, at low concentrations but approach a limiting or saturating value for high concentrations. The binding reaches half the maximum value when the concentration equals the dissociation constant, but reaches 95 % only for concentrations 19 times larger. Thus to display the amount bound when it is near maximum, allowing calculation of this maximum, it has been necessary to extend the concentration scale to 10 μM. This does allow us to display the data for high concentrations that demonstrate the approach to a limiting value. However, for most drugs the data for low concentrations are much more interesting because these are the concentrations used to produce specific effects. The amount bound for a low concentration is almost impossible to read from the plot because all of the data are squashed together against the vertical axis. There really should be a better way to display data that extend over such a large range of values.

A better way to display these data uses logarithms. The same data are replotted in Figure 3.18 as the amount bound versus the $\log_{10}$ of the concentration. By this simple device it is possible in a single graph to display data for even higher concentrations than before, here up to 100 μM, together with the data for low concentrations. For instance, at 0.1 μM, the value $-1$ on the log plot, the percentage bound can now be read from the plot as being close to 25 % of the maximum.

**Figure 3.17**
Percentage of receptors with drug bound as a function of the concentration of the free (i.e. unbound) drug in the solution.

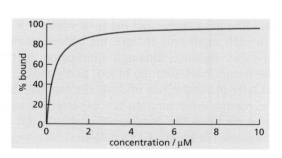

**Figure 3.18**
Data from Figure 3.17 and for higher concentrations after taking the logarithm of the concentration. Note that data are shown for a much greater range of concentrations and that the percentage bound for low concentrations is much more clearly displayed.

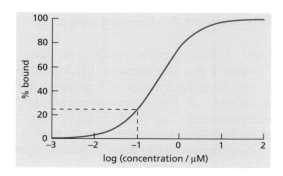

## pH

Because the concentration of $H^+$ ions in aqueous solution, often written as $[H^+]$, can be anywhere between $10$ M and $10^{-15}$ M, and we commonly encounter solutions in the range from $0.1$ M to $10^{-9}$ M, it is customary not to state $[H^+]$ directly but instead to refer to the pH defined as

$$pH = -\log\left([H^+]/1\ M\right). \tag{EQ3.16}$$

To say it in words: pH is the negative of the base 10 logarithm of the numerical value of the hydrogen ion concentration when that concentration is expressed in molarity. (Physical chemists are a bit more careful and define pH in terms of activities and ion-selective electrode potentials. Along with almost all biologists, we shall use the 'rough and ready' definition.) The minus sign is introduced in EQ3.16 so that most pH values we ever see will be positive numbers. For instance, pH 7.4 means that $[H^+] = 10^{-7.4}\ M \approx 4 \times 10^{-8}\ M$.

EQ3.16 has been written carefully; unfortunately, it is usually written without reference to the 1 M. This really is lazy because the logarithmic function is only defined for positive **numbers**, not for physical quantities. Whenever we see a log in an equation we need to ask ourselves whether the argument is a number and if not what did the author mean to write? Specifically, we need to find the reference value they used to obtain the numerical value given to the log function. The reason people can be allowed to be lazy when writing this equation for pH is that the reference value is always 1 M.

The equilibrium equation relating the concentrations for the protonation–deprotonation of a weak acid, HA, to yield its conjugate weak base $A^-$ and a hydrogen ion, $H^+$, can be written as

$$[HA] \times K_a = [H^+][A^-]. \tag{EQ3.17}$$

The subscript a is used here because it is an acid dissociation constant: it is $H^+$ rather than $OH^-$ that dissociates. Square brackets are used to indicate the concentration of the species enclosed. Weak acids or bases present in solution can take up or release $H^+$ when other acids or bases are added, a process called buffering. Thus these are often called buffers.

How can this equilibrium equation be rewritten in terms of pH?

Dividing and taking logs of both sides

$$\frac{K_a}{[H^+]} = \frac{[A^-]}{[HA]} \tag{EQ3.18}$$

$$\log\left(\frac{K_a}{[H^+]}\right) = \log\left(\frac{[A^-]}{[HA]}\right). \tag{EQ3.19}$$

Just as we want to talk about pH instead of $[H^+]$, it is also very convenient to replace $K_a$ by

$$pK_a = -\log(K_a/1\ M). \tag{EQ3.20}$$

The value of $pK_a$ is the negative of the logarithm of the numerical value of the acid dissociation constant for the weak acid when that is expressed in molarity. Sorry, that last sentence is so complicated to read. It really is better to learn to read the algebra!

These definitions allow us to rewrite the left side of EQ3.19 as

$$\log\left(\frac{K_a}{[H^+]}\right) = \log\left(\frac{K_a/(1\ M)}{[H^+]/(1\ M)}\right) = pH - pK_a \tag{EQ3.21}$$

which leads directly to the Henderson–Hasselbalch equation for a weak acid,

$$pH - pK_a = \log\left(\frac{[A^-]}{[HA]}\right), \tag{EQ3.22}$$

$$pH = pK_a + \log\left(\frac{[A^-]}{[HA]}\right). \tag{EQ3.23}$$

The use of these equations to describe the changes in pH when acids and bases are added to a solution is illustrated in Presenting Your Work.

## Box 3.3

For a weak base, B, that can associate with a hydrogen ion, $H^+$, to form the conjugate acid

$$[BH^+]K_a = [B][H^+]$$

and

$$pH = pK_a + \log\left(\frac{[B]}{[BH^+]}\right).$$

More generally for the acidic and basic forms of a buffer that are related by

$$base + protein \rightleftharpoons acid$$

$$pH = pK_a + \log\left(\frac{[base]}{[acid]}\right)$$

$$pH = pK_a + \log\left(\frac{[proton\ acceptor]}{[proton\ donor]}\right).$$

Note that in these equations [base] and [acid] are the actual concentrations of the basic and acidic forms present at equilibrium not the concentrations of the free acid and free base that may have been added to the solution.

## Bels and decibels

Sound is propagated as oscillations of pressure. We can perceive sounds over a very wide range of pressures, more than 100 000-fold. Because the intensity of a sound, which indicates the ability to transmit energy or do work, is proportional to the square of the pressure, we can hear sounds over an intensity range of $10^{10}$-fold, from the faintest sound we can detect to an intensity so loud that damage occurs. Over this huge range it is approximately true that equal increases in perceived loudness occur for each doubling of intensity. Another way of saying the same thing is that our perception of loudness varies approximately with the logarithm of the intensity rather than with the intensity itself.

The straightforward log scale for intensities relative to any chosen standard level, $I_0$, leads to Bels (named in honor of Alexander Graham Bell the inventor of the telephone). The Bel relative to a standard is defined as

$$\text{Bel}_I = \log_{10}\left(\frac{I}{I_0}\right). \tag{EQ3.24}$$

However, because the smallest difference in loudness we can reliably detect is about 1/10th this size, it is more convenient to use the decibel defined by

$$\text{dB}_I = 10\log_{10}\left(\frac{I}{I_0}\right) \tag{EQ3.25}$$

which makes the just-perceptible change 1 dB.

Decibels can be used to measure the log of the intensity of just about anything, such as the power in an electrical circuit, the intensity of sound, or the brightness of light, but by far the most common application is with sound and acoustics. In acoustics, the reference intensity is usually taken to be that associated with a standard sound pressure $\text{SPL} = 20\ \mu\text{Pa} = 2 \times 10^{-5}\ \text{N m}^{-2}$, which corresponds to a just-audible sound under optimal conditions. Because the intensity of a sound wave is proportional to the square of its amplitude, the pressure ($P$), we have

$$\frac{I}{I_0} = \left(\frac{P}{\text{SPL}}\right)^2 \tag{EQ3.26}$$

and

$$\text{dB SPL} = 10 \times \log_{10}\left(\frac{I}{I_0}\right) = 10 \times \log_{10}\left(\left(\frac{P}{\text{SPL}}\right)^2\right)$$

$$= 20 \times \log_{10}\left(\frac{P}{\text{SPL}}\right). \tag{EQ3.27}$$

For example, if the pressure $P$ is 0.4 Pa, then we can refer to the sound intensity as $20\log_{10}[0.4/(2 \times 10^{-5})]\ \text{dB} = 86\ \text{dB SPL}$. In common use we do not bother to write the SPL.

Very roughly, we have 100 perceptible levels of loudness, spaced 1 dB apart (with perhaps another 20 at the loud end of the range, which damage our ears). Table 3.4 lists various levels of sound and the corresponding intensities and decibel levels.

**Table 3.4  Various levels of sound and the corresponding intensities and decibel levels (measured at around 2 kHz)**

| dB | $I/I_0$ | $P$/SPL | notes |
|---|---|---|---|
| 0 | 1 | 1 | faintest audible sound under optimal conditions |
| 3 | 2 | $\sqrt{2}$ | |
| 6 | 4 | 2 | |
| 10 | 10 | $\approx 3$ | |
| 20 | 100 | 10 | |
| 30 | 1000 | $\approx 30$ | hushed audience |
| 40 | 10 000 | 100 | |
| 80 | $10^8$ | 10 000 | busy street |
| 90 | $10^9$ | $\approx 30\,000$ | orchestra, loud |
| 100 | $10^{10}$ | $10^5$ | rock bands, hearing damage after long-term exposure |
| 120 | $10^{12}$ | $10^6$ | damage after short-term exposure |
| 130 | $10^{13}$ | $\approx 3 \times 10^6$ | pain |
| 194 | $\approx 2.5 \times 10^{19}$ | $\approx 5 \times 10^9$ | one atmosphere, $10^5$ Pa, the divide between sound waves and shock waves |

# Presenting Your Work

## QUESTION A

Starting from the rules for powers

$$10^{\alpha} \times 10^{\beta} = 10^{\alpha+\beta}$$

and

$$(10^{\alpha})^{\beta} = 10^{\alpha \times \beta}$$

show for $x$ and $y$ both $> 0$

(i)   $\log(xy) = \log(x) + \log(y)$

and for $x > 0$

(ii)  $\log(x^{y}) = y \log(x)$

and, as a special case of the preceding,

(iii) $\log(1/x) = -\log(x)$.

---

Note that

$x = 10^{\log(x)}$ and $y = 10^{\log(y)}$.

Thus

i) $xy = 10^{\log(x)} \times 10^{\log(y)}$,

applying the rule for a product of powers,

$xy = 10^{\log(x)+\log(y)}$.

Taking logs of both sides

$\log(xy) = \log(10^{\log(x)+\log(y)})$.

Let $z = \log(x) + \log(y)$

$\log(xy) = \log(10^{\log(x)+\log(y)}) = \log(10^{z}) = z$
$\qquad\quad = \log(x) + \log(y)$

which proves the first relation.

ii)  $\log(x^{y}) = \log((10^{\log(x)})^{y}) = \log(10^{y\log(x)}) = y\log(x)$.

iii)  Because $1/x = x^{-1}$

$\log(1/x) = \log(x^{-1}) = -1 \times \log(x) = -\log(x)$.

## QUESTION B

(i) The $pK_a$ value for the buffer Hepes is 7.5. Calculate the pH that will be obtained if

(a) 10 mM Hepes acid and 10 mM NaHepes

or

(b) 4 mM Hepes acid and 16 mM NaHepes
   are added to distilled water.

(ii) The $pK_a$ values for the buffers Hepes and Tris are 7.5 and 8.08. Find the pH and the concentrations of $Tris^0$ (the neutral, base form), $TrisH^+$, HHepes (the neutral, acid form), and $Hepes^-$ if 50 mM free Hepes acid and 50 mM free Tris base are added to distilled water.

Hint: unless the pH takes on extreme values the concentrations of $H^+$ and $OH^-$ are both much less than the concentrations of any of the forms of the buffers. Thus in part (i) the net association or dissociation of $H^+$ with the buffer must be very small compared with the initial concentrations, and we can assume that the final concentrations of HHepes and $Hepes^-$ are equal to the concentrations added. Similarly in part (ii) we can assume that each $H^+$ that dissociates from HHepes will associate with $Tris^0$ to form $TrisH^+$.

(i) Using

$$pH = pK_a + \log\left(\frac{[base]}{[acid]}\right) = pK_a + \log\left(\frac{[Hepes^-]}{[HHepes]}\right)$$

for (a)

$$pH = 7.5 + \log\left(\frac{10 \text{ mM}}{10 \text{ mM}}\right) = 7.5 + \log(1) = 7.5.$$

(This illustrates the general rule that $pH = pK_a$ when the final acid and base forms are at the same concentration.)

For (b)

$$pH = 7.5 + \log\left(\frac{16 \text{ mM}}{4 \text{ mM}}\right) = 7.5 + \log(4) = 8.1.$$

(ii) Let the final concentrations of $Hepes^-$ and $TrisH^+$, which will be equal, be called x. The concentrations of Tris base and of Hepes free acid will thus both be 50 mM − x.

Using

$$pH = pK_a + \log\left(\frac{[base]}{[acid]}\right)$$

for Hepes

$$pH = 7.5 + \log\left(\frac{[Hepes^-]}{[HHepes]}\right) = 7.5 + \log\left(\frac{x}{50 \text{ mM} - x}\right)$$

and for Tris,

$$pH = 8.08 + \log\left(\frac{[Tris^0]}{[TrisH^+]}\right) = 8.08 + \log\left(\frac{50 \text{ mM} - x}{x}\right).$$

These two expressions for the pH must be equal. (There is only one value of pH in the solution, the same for the two buffers.) Thus

$$7.5 + \log\left(\frac{x}{50 \text{ mM} - x}\right) = 8.08 + \log\left(\frac{50 \text{ mM} - x}{x}\right)$$

$$\log\left(\frac{x}{50 \text{ mM} - x}\right) - \log\left(\frac{50 \text{ mM} - x}{x}\right) = 8.08 - 7.5$$

$$2 \log\left(\frac{x}{50 \text{ mM} - x}\right) = 0.58$$

$$\log\left(\frac{x}{50 \text{ mM} - x}\right) = 0.29.$$

Taking the antilogs

$$\frac{x}{50 \text{ mM} - x} = 10^{0.29} = 1.950$$

$$x = (50 \text{ mM} - x)1.950$$

rearranging

$$2.950x = 1.950 \times 50 \text{ mM}$$

$$x = 33.05 \text{ mM}.$$

Thus

$$[Hepes^-] = [TrisH^+] = 33.05 \text{ mM},$$

$$[HHepes] = [Tris^0] = 50 \text{ mM} - 33.05 \text{ mM} = 16.95 \text{ mM}.$$

The pH can be calculated from the Henderson–Hasselbalch equation for Hepes

$$pH = 7.5 + \log\left(\frac{33.05}{16.95}\right) = 7.79.$$

Check using equation for Tris

$$pH = 8.08 + \log\left(\frac{16.95}{33.05}\right) = 7.79.$$

## End of Chapter Questions

(Answers to questions can be found at the end of the book)

### Basic

**1.** Find the values of $x$ and $y$ such that $x^2 + y^2 = 1$ and $x + y = 1$ by plotting graphs for each of these relations on the same axes.

**2.** Given that $\log_{10}(4.182) = 0.6214$ find
(a) $\log_{10}(418.2)$
(b) $\log_{10}(0.00004182)$
(c) $\log_{10}(4.182^2)$
(d) $\log_{10}(1/4182.0)$
(e) $\log_{10}(\sqrt{0.4182})$

**3.** The inverse function of the logarithm is known as the antilogarithm. Given that the antilogarithm to base 10 of 0.0347 is 1.0832, write down the antilogarithm of 5.0347.

**4.** What is the 1.7th root of 5 accurate to four significant figures?

**5.** Calculate the following, giving your answers to four significant figures.
(a) $\exp(2.3)$
(b) $\exp(-4)$
(c) the value of $x$ for which $\exp(x) = 231$
(d) the value of $t$ for which $23.9 \exp(-t/112.4) = 15$

### Intermediate

**6.** Use the formula

$$\text{distance} = \sqrt{(\text{change in } x)^2 + (\text{change in } y)^2}$$

to find the distance between the points $(-2, 3.3)$ and $(4, -4.7)$.

**7.** Calculate the pH of human skin if $[H^+]$ is equal to $3 \times 10^{-6}$ M.

**8.** Express the sound intensity of 5 mPa in decibels; take SPL, the reference intensity, to be 20 μPa.

**9.** If $y = 300 \times 2^{t/t_{\text{double}}}$ with $t_{\text{double}} = 25$ min, what are $B$ and $\tau$ in $y = B \exp(t/\tau)$?

**10.**
(a) If $N = N_0 \exp(-t/\tau)$ make $t$ the subject of the equation to get $t$ as a function of $N$.
(b) If $P = 18 \exp(-t/2.375)$ calculate the value of $t$ at which $P = 5$.

### Advanced

**11.** A landowner who is particularly fond of a certain species of animal, introduced 200 of them into a wood. Initially, this population grew exponentially with a doubling time of 2 years. Assuming exponential growth, plot the population growth as a function of time for the first six years both as an ordinary plot and as a semi-logarithmic plot. Which plot gives a straight line? Write down an equation in terms of the exponential function that describes this population growth and use this to estimate the population after 20 years. This estimate in fact is substantially larger than the actual population. Suggest reasons why this might be the case.

**12.** The log function is special in a number of ways, one of which is that it is completely defined by just two conditions. Show that if we can find a function such that

(1) $f(x^y) = yf(x)$ for every $x$ and $y$, $(x > 0)$

and

(2) $f(a) = 1$,

then this function is just the logarithm to the base a, $\log_a(z)$, i.e. $f(z) = \log_a(z)$.

Hint: consider $f(u)$ when $u = a^v$ which implies $v = \log_a(u)$.

# CHAPTER 4

# Shapes, Waves, and Trigonometry

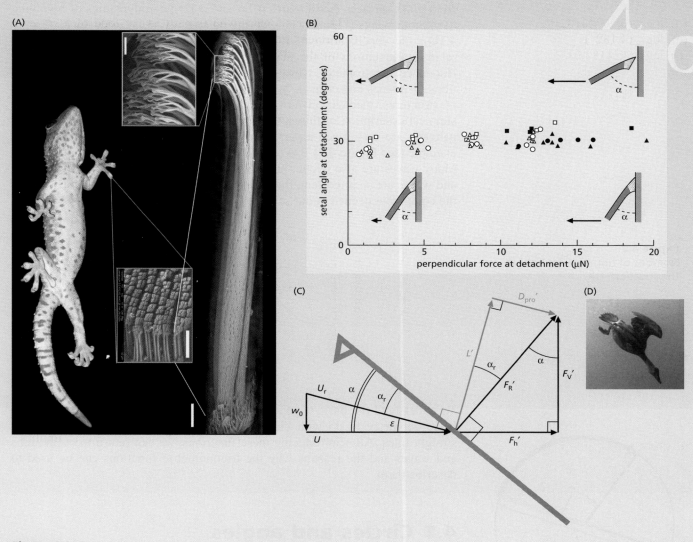

**Figure 4.1**
Angles in the description of mechanical movements. (A) Gecko on a vertical sheet of glass showing the fibrillar structure of the toe pads (modified from Autumn, K. *et al.* (2006) *Journal of Experimental Biology* 209: 3569) and (B) the definition of the angle between an individual foot-hair and the surface, which is important for the adhesion or detachment that allow it to stick to such a smooth surface (Autumn, K. *et al.* (2000) *Nature* 405: 681). (C) Flows and forces for an element of hawkmoth wing moving through air (Usherwood, J.R. and Ellington, C.P. (2002) *Journal of Experimental Biology* 205: 1547). (D) Dive angle for an arctic eider duck under water (Heath, J.P. *et al.* (2006) *Journal of Experimental Biology* 209: 3974).

Trigonometry started life as the study of triangles, but it is now much broader: the study of just about anything to do with either quantitative aspects of angles or processes that repeat at regular intervals. Trigonometry is clearly important for surveyors and navigators, but why should a biologist tackle the subject? Angles themselves are important in any study of structure or the forces involved in mechanical movements. Examples are: the variation of the force of biting with the gape angle of the jaws of a bat; the forces needed in muscles that contract body segments in the salamander (similar considerations apply to our breathing muscles); the detachment of adhesive gecko toes from glass surfaces (see Figures 4.1A and 4.1B); aerodynamics of revolving hawkmoth wings (Figure 4.1C); diving patterns of arctic eider ducks (Figure 4.1D); the specification of the positions of atoms in molecules; and the relative orientations of the magnetic fields and atomic spins in a magnetic resonance spectrometer. However, trigonometry and the trigonometric functions, like sines and cosines, are important to biologists even outside the area of structures, mechanical movements, and forces because they provide the simplest description of things that change in cycles. Cycles, regular repeated patterns, are very much a part of life. We cannot escape them; for a blatant example (see Figure 4.2) you only need to look up, is it light or dark outside. Many waves are also cyclic, notably sound waves, and light. Because cycles and waves are so important, the ability to describe them is an essential part of the equipment needed in our science.

**Figure 4.2**
Number of hours of daylight in the southern UK.

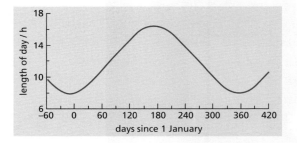

To understand the basic trigonometric functions we first need to know some basic geometry to do with circles, angles, and simple shapes made up of straight lines. After introducing these functions, we can return to oscillations and waves and the reasons why the trigonometric functions can be used to describe them.

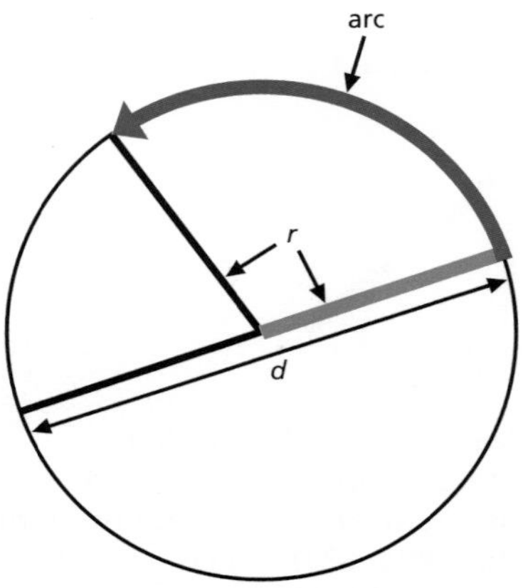

## 4.1 Circles and angles

In geometry, a **circle** is very special: it is the only shape for which every point on the perimeter is the same distance from a single point called the **center**. A straight line from the center to the circle is called a **radius**. A straight line from the circle to the center extended until it hits the circle again is called a **diameter**. An **arc** is any portion of the circle and the **arc length** is the distance from one of its ends to the other: the distance you would walk along the circle if you started at one end of the arc and proceeded to the other. The distance all the way around is called the **circumference**.

**Figure 4.3**
A circle indicating radii of length, r, from the center to the circle, an arc, and a diameter of length d = 2r.

From antiquity it has been known that the ratio of circumference to diameter for all circles is always the same. This constant value is denoted as π:

$$\pi = \text{circumference/diameter}. \qquad (EQ4.1)$$

The value of π has now been estimated to more than one billion significant figures. Thankfully, we can usually make do with the simple approximation, $\pi \approx 22/7$.

What is an angle? An angle is a difference between two directions from a point. Imagine that you are standing at the center of a circle facing in one of the directions. Now turn to face in the new direction. Can you think of any more natural way to describe how far you have turned, the angle, than as a fraction of a complete rotation? How should this be measured? As indicated in Figure 4.4, the fraction of a rotation is just the arc length divided by the circumference,

$$\text{fraction of a rotation} = \text{arc length/circumference of the circle}. \qquad (EQ4.2)$$

Note that it does not matter how big you draw the circle, as the arc length and circumference are both proportional to the radius.

Expressing an angle as a fraction may seem a bit strange. Ask most people about a direction or an angle and if we get any answer more precise than 'turn left at the next cross-roads', it might be in terms of points of a compass, in terms of an imaginary clockface as in 'aeroplane at 2 o'clock' or, if they are trying to be quantitative, in degrees. There is nothing wrong with these. Each has its proper place, but none of them is ideally suited for use in mathematics.

Ever since degrees were invented by the Babylonians they have been the units used in navigation and surveying. The Babylonians did not have our modern number system but they did want to specify angles to a quite surprising accuracy. So what they did was divide the circle into 360 equal parts each called a 'degree of arc'. So 1 degree is 1/360th of a circle – it is a fraction really. Each degree was in turn divided into 60 minutes of arc and each minute into 60 seconds (see Box 4.1). In this system, as written on the dial of a compass, a quarter turn to the right is +90°, a quarter turn to the left is −90°. Note that a turn of −90° points us in the same direction as a three-quarters turn to the right, that is +270°, whereas an about-face is represented by either +180° or −180° (which we often shorten to ±180°). A similar system of degrees, minutes, and seconds is still very much in use for

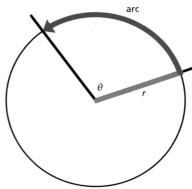

**Figure 4.4**
An angle measures the distance between two directions. The arc length, the length of the arc between the two directions, is the length measured along the curved arrow. The angle, $\theta$, is proportional to the fraction of a complete rotation that separates the two directions. By convention the constant of proportionality is chosen as 2π so that
$\theta$ = arc length / radius.

---

**Box 4.1**

Why 60 min, 60 s, and 360°? The 60 is fairly easy. The Babylonian number system used a base of 60. It seems strange to us but it has some very nice features. The number 60 is the smallest number exactly divisible by 2, 3, 4, 5, and 6; this is very convenient when we need to do lots of division. Dividing 10 into 3 equal lots can be inconvenient, but 60/3 is just 20. So 60 is fine, but why 360? One can only speculate. They compared circles with hexagons. A hexagon can be drawn so that its six corners just touch the circle dividing it into six arcs. It might be that they decided to divide each of these arcs rather than the entire circle into 60 bits – and 6 times 60 is indeed 360.

specifying positions on the surface of the earth. Latitudes are specified as north (N) or south (S) of the equator and longitudes as east (E) or west (W) of the Greenwich meridian. The typographical convention is that there is no space between the number and the degree symbol or the symbols for the subdivisions; for instance 43° 49′ 23″ N and 69° 40′ 51″ W, which are the latitude and longitude, courtesy of Google Earth, of the viewpoint in Maine for the picture in Figure 4.5.

**Figure 4.5**
View of a Maine coastal inlet taken from 43° 49′ 23″ N and 69° 40′ 51″ W.

Although degrees are very handy for everyday use, there is a better way to define angles when they are used in mathematics. It is now universally agreed to define the angle between two directions as the length of the arc between them divided by the radius of the circle,

$$\text{angle} = \frac{\text{arc length}}{\text{radius}}. \tag{EQ4.3}$$

How does this definition of angle compare with the fraction of a rotation? Remember that

$$\text{circumference} = 2\pi \times \text{radius} \tag{EQ4.4}$$

so

$$\text{radius} = \frac{\text{circumference}}{2\pi} \tag{EQ4.5}$$

and

$$\text{angle} = \frac{\text{arc length}}{\text{radius}} = \frac{\text{arc length}}{\left(\dfrac{\text{circumference}}{2\pi}\right)} \tag{EQ4.6}$$

$$= 2\pi \times \frac{\text{arc length}}{\text{circumference}} = 2\pi \times \text{fraction of a rotation}.$$

To say the same thing another way, one revolution has an arc length of $2\pi r$, so the angle represented by a complete revolution is $2\pi$, which is approximately $2 \times 3.14159 = 6.28318$. The reason for introducing the $2\pi$ and making a complete revolution such an awkward angle will be made clear later after we have defined the trigonometric functions (see EQ4.16).

In SI, it is optional whether an angle is written as a pure number, as here, or with the unit radians. We can say that a complete revolution is an angle of $2\pi$ or that it is an angle of $2\pi$ radians. We do whichever is more convenient and switch between them whenever we want. The word radians is called a unit; but it is really more a way of indicating that a number represents an angle. More often than not people do say 'radians' if only to emphasize that they are not using degrees.

Because degrees and radians are both used to express angles, it must be possible to convert from one to the other. The conversion is simple; the degree symbol, '°', is defined to mean 'multiply by the number $2\pi/360$'. (This is the same sort of thing as defining the % sign to mean multiply by 0.01.) Because there are $2\pi$ radians and $360°$ in a complete revolution, this gets it exactly right. Thus $90° = \pi/2 \approx 1.57$ and $180° = \pi \approx 3.14$. With this definition for the degree symbol, angles are *always* just numbers regardless of whether we choose to state them in degrees or radians. Note that for simple fractions of a complete rotation, such as $1/2$ or $1/6$, we often use a $\pi$ when expressing the angle in radians. For example, $30°$ is $1/12$th of a complete revolution so we write the angle as $2\pi/12 = \pi/6$. However, for more general angles we often give a numerical value in which we have substituted the value of $\pi$. Thus an angle of $23°$ is converted to radians as follows:

$$23° = 23 \times \frac{2\pi \text{ radians}}{360} = 0.4014 \text{ radians}$$

to four significant figure accuracy.

There is one further twist. Which way is positive? Is it turning to the right (clockwise) or left (anticlockwise)? If you look at a compass the degrees increase as you go around in a clockwise direction. However, in mathematics the normal convention is that angles increase as you go round in an anti-clockwise direction.

## 4.2 Straight lines, angles, and a parallelogram

We shall treat several properties of lines and angles as obvious or as matters of definition that do not need any further explanation. Thus, without any more ado, here is the list:

1.  Two infinitely long straight lines in a **plane** (a flat surface) intersect once or not at all. If no matter how much they are extended in length they never intersect they are **parallel**. If they do cross each other, there are two pairs of equal angles at the intersection as shown in Figure 4.6.
2.  If all four angles at the intersection of two lines are the same, the lines are said to be **perpendicular** and the angles must be $90° = \pi/2$. A $90°$ angle is often called a '**right angle**'. A shape bounded by two sets of parallel lines is called a **parallelogram**.

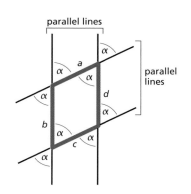

**Figure 4.6**
A parallelogram. All of the angles marked $\alpha$ are equal to each other. Opposite sides of a parallelogram have equal lengths, i.e. $a = c$ and $b = d$.

3. If one line intersects two parallel lines, then each angle at one intersection is equal to the corresponding angle at the other (see Figure 4.6).
4. Two parallel lines are always the same distance apart. More generally if two parallel lines intersect two other parallel lines, then the opposite sides of the parallelogram that is formed, shown bold, are equal; that is, in Figure 4.6, $a = c$ and $b = d$.
5. A **rectangle** has four equal angles of 90° (see End of Chapter Questions) with opposite sides of equal length. If all four sides are the same length it is a **square**.
6. Triangles are closed shapes having three sides and three internal angles. If one of the angles of a triangle is 90° (or $\pi/2$), then we say that the triangle is a *right angled triangle*. The side opposite the right angle is called the *hypotenuse*. Right angled triangles have special properties, and lead to the definition of trigonometric functions as we shall see in Section 4.5.

## 4.3 Area and volume

The **area** of a rectangle is defined to be

$$\text{area} = (\text{length of base}) \times (\text{length of height}) \qquad \text{(EQ4.7)}$$

which is almost always shortened to

$$\text{area} = \text{base} \times \text{height}. \qquad \text{(EQ4.8)}$$

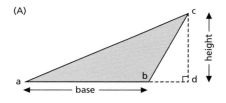

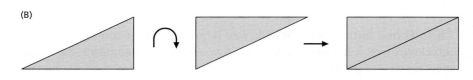

**Figure 4.7**
Area of a triangle. (A) The area of a triangle can be calculated as base × height. (B) Two identical right angle triangles can be stacked to form a rectangle.

What is the area of a triangle? We can draw the triangle with one of the sides horizontal and call that side the base as in Figure 4.7A. We can then drop a line (from point c to point d) from the corner opposite to the base so that it meets the base or an extension of it at right angles. This perpendicular line is the height and the area of the triangle is then $1/2 \times$ length of the base $\times$ length of the height or, as it is usually said

$$\text{area} = 1/2 \times \text{base} \times \text{height}. \qquad \text{(EQ4.9)}$$

If the angle at either end of the base is a right angle this is easy to see. Create an identical triangle, rotate this entire copy through 180°, and align the two triangles along the hypotenuse as in Figure 4.7B. The combined shape is a rectangle. The area of the rectangle is just base $\times$ height and since the triangles are identical they must each have half the area. If none of the angles in the triangle is a right angle as in Figure 4.7A it takes a little more work to show that the formula in EQ4.9 is correct. This is left as an End of Chapter Question.

Areas of more complicated shapes can be estimated to any desired accuracy by superimposing on them rectangles, possibly very many very small ones, calculating the areas of these, and adding up the results. Sometimes this process can be made exact using calculus as we shall see in Chapter 6. Fortunately for us, exact formulae for many frequently occurring shapes have already been derived using calculus. For example, the area of a circle of radius $r$ is equal to $\pi r^2$. It is also possible to find formulae for the surface area of a three-dimensional object; for example, the area of the surface of a sphere of radius $r$ is equal to $4\pi r^2$.

Suppose that we need to calculate the cross-sectional area of a nerve fiber that has a diameter of 6 μm. The first step in using the formula area $= \pi r^2$ is to write down the value of the radius: $r = 6$ μm$/2 = 3$ μm. We now substitute this value into the equation to get:

$$\text{area} = \pi r^2 = \pi(3 \text{ μm})^2 = \pi \times 9 \text{ μm}^2 \approx 28.3 \text{ μm}^2.$$

Note that we may sometimes want the area expressed in units of square meters using scientific notation. One way of doing this is to express the radius in meters *before* substituting into the equation for area:

$$r = 3 \text{ μm} = 3 \text{ μm} \times \frac{1 \text{ m}}{10^6 \text{ μm}} = 3 \times 10^{-6} \text{ m}.$$

The calculation of area now becomes:

$$\text{area} = \pi r^2 = \pi(3 \times 10^{-6} \text{ m})^2 = \pi \times 9 \times 10^{-12} \text{ m}^2 \approx 28.3 \times 10^{-12} \text{ m}^2$$

$$= 2.83 \times 10^{-11} \text{ m}^2.$$

Another way to the same answer is to convert units after calculating the area,

$$\text{area} = 28.3 \text{ μm}^2 = 28.3 \text{ μm}^2 \times \left(\frac{1 \text{ m}}{10^6 \text{ μm}}\right)^2 = 28.3 \times 10^{-12} \text{ m}^2$$

$$= 2.83 \times 10^{11}\text{m}^2.$$

Similar arguments apply to **volume**. The volume of a rectangular box (often referred to as a **cuboid** or rectangular prism) is defined to be

volume = width × height × depth.                    (EQ4.10)

A cuboid whose faces are all squares is a **cube**.

Another way of looking at the volume of a rectangular box is as the area of one of the rectangular faces multiplied by the distance between this face and its parallel face. This latter method is an example of a more general result that the volume of a right **cylinder** with cross-section of any shape is the area of the cross-section multiplied by the depth of the cylinder; thus a right circular cylinder with radius $r$ and length $l$ has volume equal to $\pi r^2 l$.

The volumes of more complicated solids can be estimated to any desired accuracy by building up the solids from rectangular boxes, possibly very many very small ones, calculating the volumes of these, and adding up the results. Again, this process can be made exact using calculus, although this is not covered in this book. For example, the volume of a sphere of radius $r$ is equal to $(4/3)\pi r^3$.

Suppose we want to find the volume of an unfertilized sea-urchin egg whose radius is 80 μm. The volume is

$$4\pi(80 \text{ μm})^3/3 = \frac{4\pi \times 512000}{3} \text{ μm}^3 \approx 2144660.6 \text{ μm}^3 \approx 2140000 \text{ μm}^3,$$

where the final answer is given to three significant figures. This answer is correct, but it's not in a very convenient form; it would be better to use scientific notation: $2140000 \text{ μm}^3 = 2.14 \times 10^6 \text{ μm}^3$. Notice how in scientific notation we can easily see that the answer has been given to three significant figures so we do not need to state this separately. Another alternative would be to calculate the value in terms of the base unit, cubic meters, directly, again using scientific notation:

$$4\pi(8.0 \times 10^{-5} \text{ m})^3/3 = \frac{4\pi \times 512}{3} \times 10^{-15} \text{ m}^3 \approx 2144.66 \times 10^{-15} \text{ m}^3$$
$$\approx 2.14 \times 10^{-12} \text{ m}^3.$$

Often a volume is calculated in order to calculate a concentration. For example, suppose we observe 6 cells in a volume measuring 2 mm × 2 mm × 100 μm. What is the cell concentration? We shall begin the calculation in scientific notation. The volume is

$$2 \times 10^{-3} \times 2 \times 10^{-3} \times 100 \times 10^{-6} \text{ m}^3 = 4 \times 10^{(-3-3+2-6)} \text{ m}^3$$
$$= 4 \times 10^{-10} \text{ m}^3.$$

We can now complete the calculation to give concentration in cells per cubic meter as

$$6 \text{ cells}/(4 \times 10^{-10} \text{ m}^3) = (6/4) \times 10^{10} \text{ cells m}^{-3} = 1.5 \times 10^{10} \text{ cells m}^{-3}.$$

Because we never deal with a cubic meter of a cell suspension, it is more convenient if this can be expressed as cells per milliliter. Because $1 \text{ m}^3 = 10^3 \text{ l} = 10^3 \times 10^3 \text{ ml} = 10^6 \text{ ml}$, this is

$$1.5 \times 10^{10} \text{ cells m}^{-3} = \frac{1.5 \times 10^{10} \text{ cells}}{1 \text{ m}^3} \times \left( \frac{1 \text{ m}^3}{10^6 \text{ ml}} \right)$$
$$= 1.5 \times 10^{(10-6)} \text{ cells ml}^{-1}$$
$$= 1.5 \times 10^4 \text{ cells ml}^{-1}.$$

# 4.4 Pythagoras' theorem

## Box 4.2 Pythagoras' theorem

Consider the right angled triangle with base of length $a$, height $b$, and hypotenuse $c$ as shown in bold in the figure. Now make three identical copies of this triangle and arrange them so that they enclose a square with side $c$. The entire figure is a square with side $a + b$.

The area of the large square is $(a + b)^2 = a^2 + 2ab + b^2$, the area of the central square is $c^2$, and the area of the four triangles together is $4 \times (1/2)ab = 2ab$. Adding up the areas of the central square and the four triangles and setting this equal to the area of the large square

$$c^2 + 2ab = a^2 + 2ab + b^2$$

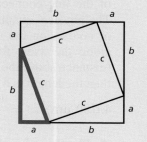

**Figure 1**
Pythagoras' theorem: $c^2 = a^2 + b^2$. The side of the triangle with length $c$ is the hypotenuse.

which, after subtracting $2ab$ from both sides, becomes

$$c^2 = a^2 + b^2.$$

This astonishingly simple result is Pythagoras' theorem.

One very useful property of a right angled triangle is known as Pythagoras' theorem. For any right angled triangle

$$c^2 = a^2 + b^2 \qquad\qquad (EQ4.11)$$

where $c$ is the length of the hypotenuse and $a$ and $b$ are the lengths of the other two sides, as shown in Figure 4.8. A simple proof of this relation is shown in Box 4.2.

**Figure 4.8**
A right angled triangle. The side of the triangle with length $c$ is the hypotenuse.

One very important use of Pythagoras' theorem is to determine the distance between two points if we know their positions on an $x,y$ grid as in Figure 4.9.

$$\text{distance} = \sqrt{(\text{change in } x)^2 + (\text{change in } y)^2}. \qquad\qquad (EQ4.12)$$

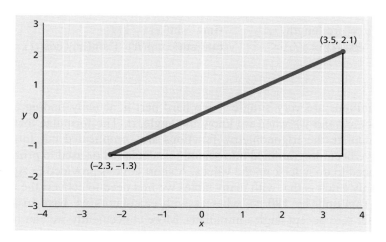

**Figure 4.9**
The distance between two points is given by the formula:

$$\text{distance} = \sqrt{(\text{change in } x)^2 + (\text{change in } y)^2}.$$

For instance, to find the distance between the point $x = -2.3$, $y = -1.3$, often written as $(-2.3, -1.3)$, and the point $(3.5, 2.1)$ we note that the horizontal and vertical separations are just $3.5 - (-2.3) = 5.8$ and $2.1 - (-1.3) = 3.4$, respectively. Thus from Pythagoras' theorem the length of the line connecting the points is just $\sqrt{(5.8)^2 + (3.4)^2} = \sqrt{45.2} \approx 6.72$. The physical quantity, if any, corresponding to this numerical value depends, of course, on the unit (which must be the same) for the quantities plotted along the axes; for example if the coordinates of the points plotted are (distance above the origin) / mm and (distance to the right of the origin) / mm, then the length is $6.72 \times 1$ mm $= 6.72$ mm.

# 4.5 Basic trigonometric functions: sine and cosine

The basic functions of trigonometry are the sine, cosine, and tangent functions of an angle. In science, these same functions describe at least approximately the oscillations of objects such as atoms in a molecule, the propagation of sound and light waves, and the hours of daylight throughout the year. Indeed, even those cycles that are not sinusoidal are often analyzed as sums of sine and cosine functions, an advanced topic called *frequency analysis*, which will not be considered in this book.

The properties of the trigonometric functions are easiest to understand when they are considered as functions of an angle $\alpha$. Note it is conventional to use Greek letters like $\alpha$, $\beta$, $\gamma$, $\theta$, and $\phi$ to stand for angles. We could use any symbols we like but it is easier to communicate if we use the symbols that others expect. The definitions of $\sin(\alpha)$ and $\cos(\alpha)$ when $\alpha$ is an angle are shown in Figure 4.10. This shows a circle with radius $r$ with a line drawn from the center to the circle at angle $\alpha$ relative to the $x$ axis. We can draw a right angle triangle with base $x$, height $y$, and hypotenuse $r$, as shown. The trigonometric functions, angle, sine, cosine, and tangent, are defined as

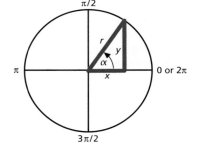

**Figure 4.10**
Definitions of the sine, cosine, and tangent functions when the argument, $\alpha$, is an angle. $\sin(\alpha) = y/r$, $\cos(\alpha) = x/r$, and $\tan(\alpha) = y/x$. The line, whose length is $y$, is perpendicular to the *x* axis.

**angle**: $\alpha = $ arc length$/r$,          **sine**: $\sin(\alpha) = y/r$,
**cosine**: $\cos(\alpha) = x/r$,          **tangent**: $\tan(\alpha) = \sin(\alpha)/\cos(\alpha) = y/x$.

$$(EQ4.13)$$

The definitions are very simple – the hard part is actually finding accurate values for $x/r$ and $y/r$ when we know the angle, $\alpha$. The formulae that enable us to calculate the trigonometric functions are complicated but fortunately these have been programmed into scientific calculators or used to print tables. All we need to do is look up the values. A short table is provided in Appendix 3. The only catch is that with many tables we first need to convert the angle to degrees (multiply by $180/\pi$), whereas with calculators we have to find out how to set it to accept angles in radians. More conveniently we can just type something like '$= \sin(2.73)$' into a spreadsheet program like Microsoft Excel®.

One note of caution: the word tangent has two meanings in geometry and trigonometry. One is the name of a trigonometric function, $\tan(\alpha)$, the other refers to a line that just grazes a curve with the same slope or gradient as the curve at that point. These two meanings are closely related because the slope of a curve at a point is $\tan(\alpha)$ where $\alpha$ is the angle between the tangent to the curve and the $x$ axis.

## Box 4.3

Key values of $\sin(\alpha)$ and $\cos(\alpha)$ can be calculated from the definitions in terms of angles. The notation is the same as that used in Figure 4.10 together with the letter $n$, which represents any integer.

When $\alpha = 0$ or any multiple of $2\pi$, $x = 1$, $y = 0$, $r = 1$

$\cos(0) = \cos(2n\pi) = x/r = 1$ and
$\sin(0) = \sin(2n\pi) = y/r = 0$.

Similarly when $\alpha = \pi/2 + 2n\pi$, $x = 0$, $y = 1$, $r = 1$

$\cos(\pi/2 + 2n\pi) = x/r = 0$ and $\sin(\pi/2 + 2n\pi) = y/r = 1$,

and continuing the pattern

$\cos(\pi + 2n\pi) = -1$, $\sin(\pi + 2n\pi) = 0$,
$\cos(3\pi/2 + 2n\pi) = 0$, and $\sin(3\pi/2 + 2n\pi) = -1$.

That locates the peaks and troughs and the points where the curves in Figure 4.11 must cross the $\alpha$ axis.

We can also find values for $\alpha = \pi/6$, $\pi/4$, or $\pi/3$. For $\pi/4$ (45°), the defining triangle is an isosceles triangle with $r = 1$ and $x = y$. But then, by Pythagoras' theorem

$r^2 = x^2 + y^2 = 2x^2$
$x = y = r/\sqrt{2}$

and

$\sin(\pi/4) = \cos(\pi/4) = 1/\sqrt{2}$.

For $\pi/6$ we have to do a little more work as shown in Figure 1. The defining triangle plus an identical triangle together make up an equilateral triangle; that is, the length of the vertical line is the same as the radius $r$. But that means that $y$ is just half the radius, and

**Figure 1**
Calculation of the trigonometric functions for $360° = \pi/6$. A right angled triangle as shown by the heavy lines is reflected across the $x$ axis to create an equilateral triangle with all three angles equal to $60° = \pi/3$.

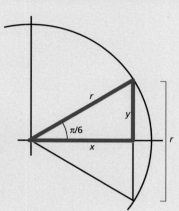

$\sin(\pi/6) = y/r = 1/2$.

Again, using Pythagoras' theorem

$r^2 = x^2 + y^2 = x^2 + r^2/4$
$x^2 = (3/4)r^2$
$x = \sqrt{3} \times r/2$

and

$\cos(\pi/6) = x/r = \sqrt{3}/2$.

You should show that the same argument leads to

$\cos(\pi/3) = 1/2$    and    $\sin(\pi/3) = \sqrt{3}/2$.

These and similar values for $2\pi/3$, $5\pi/6$, $7\pi/6$, $4\pi/3$, and so on account for all those plotted in Figure 4.11. That is enough to let us draw the smooth curves for the functions shown in the figure.

A very important property that relates sine and cosine functions is a direct consequence of Pythagoras' theorem. Referring again to Figure 4.10, we know that $r^2 = x^2 + y^2$, so dividing this equation through by $r^2$ and using the definitions of $\sin(\alpha)$ and $\cos(\alpha)$ we have

$$[\cos(\alpha)]^2 + [\sin(\alpha)]^2 = 1. \tag{EQ4.14}$$

In fact, we have a special notation for the squares of trigonometric functions: $[\sin(\alpha)]^2$, for example, is written simply as $\sin^2(\alpha)$. Thus this last equation is usually written

$$\sin^2(\alpha) + \cos^2(\alpha) \equiv 1 \tag{EQ4.15}$$

where, because EQ4.15 is true for every angle $\alpha$ (that is, it is an identity), we have written it with the special version of the equals sign with three lines instead of the usual two.

Although there is no need for us to do the hard work of finding accurate values for the trigonometric functions, we do need to understand how $\sin(\alpha)$ and $\cos(\alpha)$ vary with $\alpha$. One way to see this is to plot the functions, as has been done in Figure 4.11.

**Figure 4.11**
Plots of $\sin(\alpha)$ (solid line, square symbols) and $\cos(\alpha)$ (dashed line, circular symbols). Note that the cosine function looks just like the sine function with the only difference being that it has been shifted to the left by $\pi/2$.

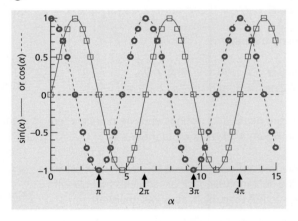

Note that in Figure 4.11 the curve for $\sin(\alpha)$ is almost linear near the origin so for very small values of $\alpha$

$$\sin(\alpha) \approx \alpha. \tag{EQ4.16}$$

This very simple result is the reason why radians are preferred compared to degrees, fractions of a revolution, or any other measure of angles. If, for instance, angles had been defined as fractions of a revolution, then the result would have been $\sin(\alpha) \approx \alpha/2\pi$. Keeping EQ4.16 simple is very important in the calculus of trigonometric functions (see Chapter 7).

The plots of sine and cosine in Figure 4.11 and the continuations for negative values illustrate several important properties and identities.

1.  The sine and cosine functions are periodic, they repeat at intervals of $2\pi$: for any number $\alpha$ and any integer $n$,

$$\sin(\alpha + 2n\pi) \equiv \sin(\alpha) \tag{EQ4.17}$$

and

$$\cos(\alpha + 2n\pi) \equiv \cos(\alpha). \qquad \text{(EQ4.18)}$$

In terms of angles all these say is that each time the angle increases by $2\pi$ or any multiple of $2\pi$ we are back to the same point on the circle.

2.  The sine function for negative values of $\alpha$ is just the negative of the sine function for the corresponding positive values

$$\sin(-\alpha) \equiv -\sin(\alpha) \qquad \text{(EQ4.19)}$$

and the cosine function for negative values of $\alpha$ is just the same as for positive values

$$\cos(-\alpha) \equiv \cos(\alpha). \qquad \text{(EQ4.20)}$$

Properties such as these create a problem when we use the inverse functions such as arcsin (or $\sin^{-1}$), arccos (or $\cos^{-1}$), and arctan (or $\tan^{-1}$). For instance if we know that $\sin(\alpha)$ is 0.5, then we might try to find the value of $\alpha$ as $\sin^{-1}(0.5) = 30° = \pi/6$. However, other values of $\alpha$ also have the same sine values, such as $-210°$, $150°$, and $390°$. Fortunately, we are often able to decide from the context which value of the angle we want.

## Box 4.4 A problem with inverse trigonometric functions

Beware, there is a real inconsistency in the standard notation of trigonometry. Thus $\sin^n(\theta)$ is usually taken as shorthand for $(\sin(\theta))^n$: for instance,

$$\sin^2(\theta) = \sin(\theta) \times \sin(\theta).$$

However, there is a very important exception. The notation $\sin^{-1}(\theta)$ does *not* mean $(\sin(\theta))^{-1} = 1/\sin(\theta)$. Why not? By general agreement if $y = f(x)$ then the inverse function that takes us back from $y$ to $x$ is written as $f^{-1}(y)$. We cannot use the superscript $-1$ to have this meaning of inverse function and at the same time to mean reciprocal. Because it is easy to write the reciprocal of sine as $(\sin(\theta))^{-1}$, the reciprocal is always written out in full whereas $\sin^{-1}(\theta)$ is used to mean inverse sine.

3.  Both the sine and cosine functions give values that vary between $-1$ and $+1$.

4.  Both functions look just the same except that one is shifted along the $\alpha$ axis relative to the other

$$\cos(\alpha) \equiv \sin(\alpha + \pi/2) \qquad \text{(EQ4.21)}$$

or to put it the other way round

$$\sin(\alpha) \equiv \cos(\alpha - \pi/2). \qquad \text{(EQ4.22)}$$

To find the value of the cosine of any number, we could just look at the sine whose argument is that number plus $1.57029\ldots$. Tables, calculators, and computers have both functions solely to save us time and effort.

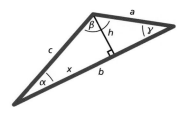

**Figure 4.12**
The sine and cosine rules for a triangle. If the lengths of the sides are a, b, and c and the opposite angles are α, β, and γ then

$$\frac{\sin(\alpha)}{a} = \frac{\sin(\beta)}{b} = \frac{\sin(\gamma)}{c}$$

and $a^2 = b^2 + c^2 - 2bc\cos(\alpha)$. h and x are useful when deriving these relations, see End of Chapter Question 14.

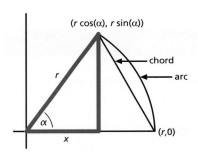

**Figure 4.13**
The length of the chord of a circle of radius r is given by length of chord = $r\sqrt{2(1-\cos(\alpha))}$ where α is the angle made by the chord at the center of the circle.

The tangent function is also cyclical but with period π: half that of sine and cosine. Furthermore, although it is smooth *almost* everywhere, it is very dramatically *not* smooth at $\alpha = \pi/2 + n\pi$. At these values of α, tan(α) is not defined because cos(α) = 0.

### Sine and cosine rules for triangles

Suppose we have made some measurements of angles and lengths in an animal, and that we now want to know other lengths and angles that we have not measured. We can often find these other values by using two rules known as the *sine rule* and the *cosine rule*. Intuitively, we can see that there must be some sort of relation between the size of an angle in a triangle and the length of the opposite side (see Figure 4.12). Thus we fully expect the side opposite the largest angle, β in Figure 4.12, to be longer than the sides opposite α and γ. The precise relation is given by the sine rule:

$$\frac{\sin(\alpha)}{a} = \frac{\sin(\beta)}{b} = \frac{\sin(\gamma)}{c}. \tag{EQ4.23}$$

For example, suppose that we know that $a = 2$ cm, $b = 4.5$ cm, and $\alpha = 20°$. We can calculate the angle β from

$$\sin(\beta) = b\sin(\alpha)/a = (4.5 \text{ cm})\sin(20°)/(2 \text{ cm}) = 2.25 \times 0.342 \approx 0.77.$$

So by using the inverse sine function we find

$$\beta = 0.878 \text{ radians} = 0.878 \times 180°/\pi \approx 50°.$$

But note that we expect the value of the angle to be between 90° and 180° so the inverse sine has not given us the angle we require. In the range required the angle with the correct sine is $180° - 50° = 130°$.

The cosine rule for a triangle is the generalization of Pythagoras' theorem and is true for any triangle.

$$a^2 = b^2 + c^2 - 2bc\cos(\alpha). \tag{EQ4.24}$$

Deriving EQ4.23 and EQ4.24 is left for you to do as an advanced End of Chapter Question. A special case of the cosine rule is when the two lengths b and c are the same: $b = c = r$. The third length a is then the length of a chord in a circle of radius r. (A chord is a straight line connecting two points on a curve.)

$$\text{length of chord} = r\sqrt{2(1-\cos(\alpha))}. \tag{EQ4.25}$$

You can verify this relation directly from Figure 4.13 by using EQ4.12 to calculate the length of the chord from the coordinates of its ends.

## 4.6 Sinusoidal oscillations

Something is said to oscillate if it varies in a regular repeated fashion. Oscillations can be found everywhere. Examples include such diverse effects as the number of hours of daylight through the year, the lengths of chemical bonds, the swing of a pendulum, or the position of a weight bobbing up and down on the end of a spring. All these examples can be described to a reasonable approximation as sinusoidal variations. (The origin of these oscillations will be discussed in Section 7.8 on simple harmonic motion.)

## Amplitude, frequency, and phase of a sinusoidal oscillation

Imagine a child being pushed on a swing (see Figure 4.14). Once someone has given a good push, the child hovers momentarily at one end of the arc, gains speed as they approach the bottom, then slows down again while rising to the other end. A mathematical pendulum as in Figure 4.15A tries to capture the essence of this motion (but misses all the excitement). In the pendulum the swing seat and squirming child are replaced by a heavy weight while the creaking chains, cables, or lengths of old rope are replaced by a long massless, inextensible string. If such an ideal pendulum is pulled a small distance, $A$, to the left and then released, a plot of its subsequent horizontal position versus time looks like that shown in Figure 4.15B. The clock was started at the instant the pendulum reached the midpoint.

We say that the position varies **sin**usoidally in time because this plot looks like the plot of the **sin**e function in Figure 4.11. In fact, the only difference between the plots, apart from the different orientation, is the scales. The sine function repeats every time its argument increases by $2\pi$, while the repeat time or period of the pendulum is $T$, which might be 3 s. Similarly, the amplitude of the sine function (its maximum value) is 1 while the amplitude of the swing is $A$, which might be 37 cm. How do we fit the two together?

The vertical scale is easy. We just multiply the sine function by 37 cm. How do we stretch it horizontally? For each increase in time equal to $T$, the argument of the sine function (the bit inside the parentheses) must increase by $2\pi$. That is just what happens if we make the argument $2\pi t/T$ which, as required, is just a unitless number. The sinusoidal oscillation is then written as

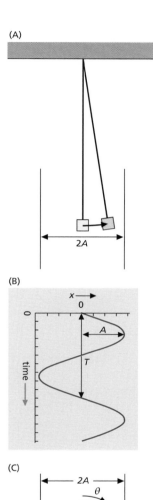

**Figure 4.15**
(A) A mathematical pendulum. A heavy weight or bob is suspended by a massless, inextensible string from a rigid support. The maximum displacement of the pendulum, $A$, is very small compared with the length of the string. (B) Position of the pendulum bob in (A) plotted as a function of time. (C) The horizontal position of a marker moving clockwise around a circle with radius $A$ is the same as that of the pendulum.

**Figure 4.14**
A child's swing. Photograph by Rudolf and Thorn Cardinal.

$$x = A\sin(2\pi t/T)$$
$$= A\sin(2\pi f t)$$

(EQ4.26)

where the frequency is $f = 1/T$. It gets tiresome always writing $2\pi$ in equations, so it is normal to define an angular frequency

$$\omega = 2\pi f$$

(EQ4.27)

so that the equation can be written in the somewhat tidier form

$$x = A\sin(\omega t).$$

(EQ4.28)

Because we have ended up with a sine function, its argument, $\omega t$, can be interpreted as an angle as indicated in Figure 4.15C. If the square marker advances around the circle at a constant rate, $\omega$, so that $\theta = \omega t$ then the horizontal position of the marker will be the same as the position of the pendulum. For this reason $\omega$ is sometimes called the angular velocity rather than the angular frequency. It does not matter which name we use, they both have the same unit, $s^{-1}$. (Remember that a radian is not really a unit, so radians $s^{-1}$ can always be replaced by $s^{-1}$.)

We chose to start the clock just when the pendulum reached the middle of the oscillation. How would we describe the oscillation if someone else controlled the clock and $t = 0$ was at some other time? The easiest way to deal with this is to note the time at which the pendulum passes the midpoint, call it $t_0$, and subtract this value from all the times. We describe the oscillation as

$$x = A\sin(\omega(t - t_0))$$

(EQ4.29)

or, to make things look a little tidier, we can invent a phase angle $\varphi = -\omega t_0$ and write

$$x = A\sin(\omega t + \varphi).$$

(EQ4.30)

This is the most general way of writing a single sinusoid. Of course, we could have chosen to write this equation using a cosine instead of a sine. Because $\sin(\alpha) = \cos(\alpha - \pi/2)$ anything that can be described as a sinusoid can be expressed using either. It is just that the phase angle is different by $\pi/2$. If we allow ourselves both a cosine and a sine term we could also write the general sinusoid as (see End of Chapter Question 13)

$$x = B\sin(\omega t) + C\cos(\omega t).$$

(EQ4.31)

The effects of varying the amplitude, angular frequency, and phase of a general sinusoid are illustrated in Figure 4.16.

## Damped oscillations

The small oscillations of a high quality pendulum really do look sinusoidal. However, if the pendulum were left undisturbed over a long enough period we would see that the amplitude would progressively decrease toward zero. The

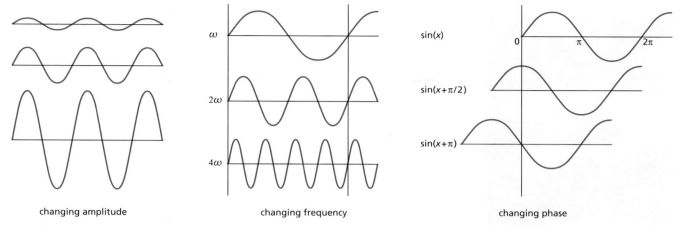

**Figure 4.16**
The effects of varying the amplitude, frequency, and phase of a general sinusoid.

processes that produce the decrease are called damping and the resulting oscillation is called a damped oscillation. An example is shown in Figure 4.17. Mathematically this can be described by

$$x = A \sin(\omega t + \varphi)e^{-t/\tau_{\text{damping}}}. \tag{EQ4.32}$$

Physically, damping results from anything which converts kinetic energy into heat. For the pendulum these losses arise from resistance to movement through the air, the imperfect properties of the string, and movements of the top support. Because the regular swing of a pendulum is one method for keeping time, a great deal of care has gone into devising methods for giving a pendulum regular, gentle shoves without disturbing the period of the oscillation.

Free oscillations are those that can occur in a system while it is isolated from its surroundings, in the absence of a repetitive stimulus from the outside. In many such oscillations no energy input is required to sustain the oscillation, at least for a while. Examples are a swinging pendulum or a vibrating molecular bond. In these, there is a cyclic conversion back and forth between kinetic and potential energy of the objects that are moving. However, many biological oscillations are different in that energy sources are required. For instance, the circadian (daily) rhythms of many organisms are linked to the synthesis of proteins that then suppress transcription of the genes coding for their production. When present the proteins are continually being degraded. They last long enough to turn off transcription, but once no more are being produced their concentrations fall. Transcription then turns on again and protein levels start to go up, which eventually inhibits transcription again and so on. The result is that both the rate of transcription and the amount of protein present go through repeated cycles. Energy sources are required to drive many of the steps in this process.

On a larger scale, oscillations in populations deal with organisms, each of which needs a continual supply of energy to survive.

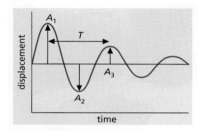

**Figure 4.17**
Damped oscillations. Each maximum in the sinusoidal displacement is less than the previous maximum.

## 4.7 Waves

To say that a wave *propagates* is almost the same as saying that it moves, but there is an additional overlay of meaning. A wave is a disturbance that moves, but the medium through which the disturbance moves ends up in pretty much the same state as before the disturbance arrived. For instance if you pull a washing line out to its full extent, hold the end, and then flick your hand from side to side a solitary wave travels along the cord. Each piece of the line moves from side to side, but only the wave moves along from your hand to the fixed end (where it will be reflected). Another example of a solitary wave is an action potential traveling along a nerve axon. The action potential is associated with ion movements across the axon membrane and currents flowing briefly first one way then the other along the axon, but only the wave, the action potential, progressively propagates along the length of the axon.

Either a wave passing a point on the washing line or an action potential passing along an axon can be described in two different but closely related ways. We can look at a snapshot in time, where the wave will have a particular shape, or we can focus attention on a particular point along the cord or axon and look at how the position or electrical potential varies with time. These two views are so closely related that if you know the velocity of the disturbance you can almost think of them as telling us the same thing. If you have a time course of an action potential at a point along the axon, and the axon's properties remain constant along its length, you can convert this into a snapshot by flipping it across the vertical axis, multiplying the times by the velocity, $v$, and relabeling the horizontal axis. On the relabeled axis the portion of the axis that corresponds to any time interval, $\Delta t$, has a length $\Delta x = v \times \Delta t$ (see Figure 4.19). The prefix '$\Delta$' which is part of '$\Delta t$' and '$\Delta x$' means 'change in'.

Often, when we think of waves, we think of periodic waves; these have a pattern which repeats at regular intervals like sound waves, the electric and magnetic fields on alternating current (AC) power lines, radio waves, light, and to some extent water waves in the open ocean. Such waves can be described using sinusoids, accurately for light of a well-defined color and sound of a pure tone or approximately and locally for water waves under calm to moderate conditions in the open ocean. Just as with solitary waves, we can look at the disturbance either as a snapshot in time, where the sinusoid has a repeat distance or wavelength $\lambda$, or we can look at the time course which will have a period $T$ and frequency $f = 1/T$. The velocity of propagation can be calculated as how far an identifiable feature of the wave moves forward per unit time. Because the velocity is constant, we can calculate it as the distance moved forward in one period; that is, as the wavelength divided by the period

$$v = \lambda/T = f\lambda. \tag{EQ4.33}$$

A sinusoidal traveling wave, or propagating wave, can be described by

$$y = A \sin\left(2\pi\left(\frac{x}{\lambda} - \frac{t}{T}\right) + \varphi\right). \tag{EQ4.34}$$

**Figure 4.18**
Wave on a washing line. View frames from the bottom upward. The first frame was taken just after the near end of the line was flicked from side to side. The wave propagates along the line, but the line itself only moves from side to side.

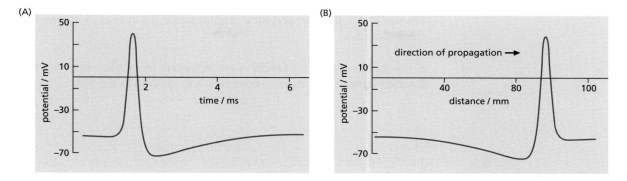

**Figure 4.19**
Action potential in a squid giant axon. (A) Time course as seen at a point along the axon. The potential increases rapidly to a peak at about +40 mV then decreases back toward the resting potential, here about −60 mV. (B) Snapshot of the potential along the axon at an instant in time. The scale has been labeled assuming the velocity is 20 m s$^{-1}$. Adapted from Hodgkin, A.L. and Huxley, A.F., Resting and action potentials in single nerve fibers. *J Physiol*, 1945, Oct 15; 104(2): 176–95 with permission from Wiley-Blackwell.

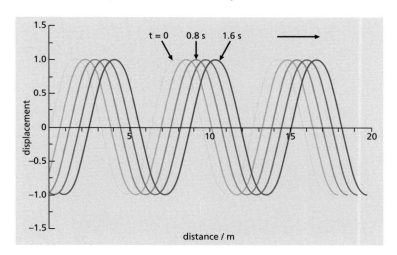

**Figure 4.20**
A propagating sine wave. The waveform is shown at successive times.

To understand what this equation says look at Figure 4.20 which depicts a wave traveling to the right with velocity $v$. The sinusoids are drawn with amplitude $A = 1$, wavelength $\lambda = 2\pi$ m $= 6.28$ m, period $T = 4$ s, velocity $v = \lambda/T = 1.57$ m s$^{-1}$, and phase angle $\varphi = 0$. With these values EQ4.34 says that at $t = 0$ the snapshot of the waveform is just a sinusoid described by

$$y = \sin\left(\frac{2\pi x}{\lambda} + 0\right) = \sin\left(\frac{2\pi x}{2\pi \text{ m}}\right) = \sin(x/1 \text{ m}). \tag{EQ4.35}$$

A short time $t$ later the wave has moved to the right a distance $vt$. The displacement at each point $x$ is now the same as the displacement at $x - vt$ on the first curve, i.e.

$$y = \sin\left(\frac{2\pi}{\lambda}(x - vt)\right) = \sin\left(\frac{2\pi}{\lambda}\left(x - \frac{\lambda t}{T}\right)\right) = \sin\left(\frac{2\pi x}{\lambda} - \frac{2\pi t}{T}\right)$$
$$= \sin\left(\frac{x}{1 \text{ m}} - \frac{\pi t}{2 \text{ s}}\right). \tag{EQ4.36}$$

You can check that this equation does describe the curves in Figure 4.20 using your calculator.

If the minus sign in EQ4.34 were replaced by a plus sign it would still describe a traveling wave, just one that was moving to the left.

# Presenting Your Work

## QUESTION

What is the sum of the internal angles in a triangle, namely $\alpha + \beta + \gamma$ in Figure A?

Hint: the sum of the angles inside any polygon may be found from the sum of the complementary angles outside the shape. These are also shown in Figure A.

(A)

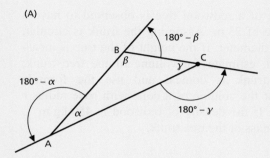

**Figure A**

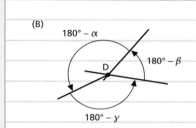

**Figure B**

Lines parallel to AB, AC, and BC are constructed all passing through a single point D as shown in Figure B. Because these new lines are parallel to the original lines in Figure A the angles between them are the same, as marked. The three angles thus must add up to a complete rotation, namely 360°. Thus

$$360° = (180° - \alpha) + (180° - \beta) + (180° - \gamma)$$

$$360° = 3 \times 180° - (\alpha + \beta + \gamma)$$

$$-180° = -(\alpha + \beta + \gamma)$$

$$\alpha + \beta + \gamma = 180°.$$

## Basic

**1.** What is $37.4°$ in radians:
(a) as a fraction of $\pi$;
(b) as a decimal value?

**2.** Find, to four significant figures, the values of:
(a) $\cos(37.4°)$
(b) $\sin(\pi/5)$
(c) $\tan(-239°)$
(d) $0.25\cos^2(37.4°) + 0.25\sin^2(37.4°)$

**3.** In each case find all the values of $x$ in degrees between $0°$ and $360°$ to the nearest degree:
(a) $\sin(x) = 0.79$
(b) $\cos(2x) = 0.03$
(c) $\tan(x) = 3.2$
(d) $\sin(x) = -0.99$
(e) $\sin(x) = 1.01$

**4.** Calculate the area of a rectangular field measuring 203 m by 340 m. Give your answer in square meters and in hectares (1 ha $= 10^4$ m$^2$).

**5.** Calculate the area of a circle with radius 53 μm.

**6.** Calculate the volume of a sphere having radius of 1.3 mm; give your answer in both cubic millimeters and cubic meters (use scientific notation for the latter).

## Intermediate

**7.** Plot graphs of the function $y = 0.5\tan(x)$ and $y = x$ for $0 \leqslant x < \pi/2$. Hence find the solutions for $\theta = 0.5\tan(\theta)$; give your answer in degrees, accurate to the nearest degree.

**8.** A population of foxes fluctuates sinusoidally between 340 and 460 with a period of 2 years. If the population was at a maximum of 460 in the year 2006, write down an expression for the number of foxes $t$ years later.

**9.** Calculate the arc length between two points on a circle of radius 300 mm when there is an angle of $127°$ between the two radii connecting the center of the circle to the points. Give your answer to the nearest millimeter.

**10.** A biologist observes the number of snails in a meter square quadrat at several random locations in a plot and finds that on average there are 31 snails per square meter. If the plot measures 67 m $\times$ 23 m, estimate the total number of snails in the plot.

**11.** The base of a redwood tree is observed to have a circumference of 23 m. Assuming the trunk is circular, calculate the diameter. If the height of the tree is measured as 75 m, estimate the volume of the tree trunk; assume the trunk is conical and use the formula $(1/3)\pi r^2 h$ for the volume of a cone with base radius $r$ and height $h$. If the density of redwood is 450 kg m$^{-3}$, estimate the mass of the tree trunk.

## Advanced

**12.** A triangle has two sides of length 27.3 mm and 15.2 mm, with the angle between them equal to $103°$. Use the cosine and sine rules to find the length of the third side and the other two angles.

**13.** The general rule for the sine of the sum of two angles is

$$\sin(\alpha + \beta) = \sin(\alpha)\cos(\beta) + \cos(\alpha)\sin(\beta).$$

Use this rule to show that the most general sinusoid

$$x = A\sin(\omega t + \varphi)$$

can be rewritten in the form

$$x = B\sin(\omega t) + C\cos(\omega t).$$

Find the relations between $B$ and $C$ on the one hand and $A$ and $\varphi$ on the other.

**14.** Prove the sine and cosine rules stated in the text as EQ4.23, EQ4.24 (see Figure 4.12). Hints: refer to Figure 4.12 and use relations like $\sin(\alpha) = h/c$, $\sin(\gamma) = h/a$, $c^2 = h^2 + x^2$, $a^2 = h^2 + (b-x)^2$, and $\cos(\alpha) = x/c$. The line with length $h$ was drawn perpendicular to the side opposite the angle $\beta$. To draw a similar line for the other angles, you will have to extend the side opposite beyond the triangle.

# CHAPTER 5

# Differentiation

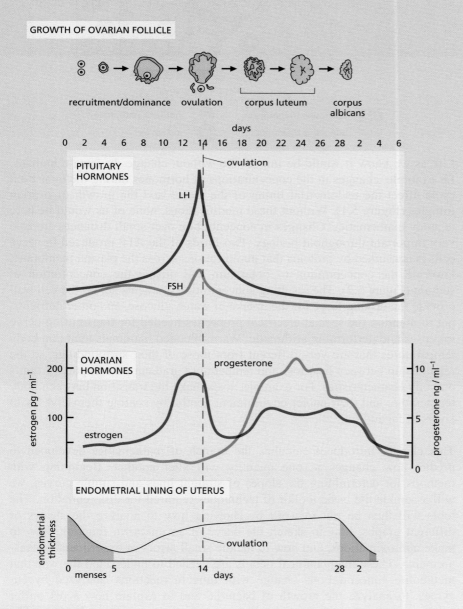

GROWTH OF OVARIAN FOLLICLE

recruitment/dominance    ovulation    corpus luteum    corpus albicans

**Figure 5.1**
Changes in the concentrations of follicle-stimulating hormone (FSH) and
luteinizing hormone (LH) during the menstrual cycle affect the endometrial
lining of the uterus and the growth of ovarian follicles.

**Figure 5.2**
Concentration gradients of
sodium and potassium across the
cell wall provide the driving force
for the uptake of water, glucose,
and other nutrients. Portions
modified from van Wynsberghe,
Noback and Carola, Human
Anatomy and Physiology, 3rd Ed.
p 63, Fig. 3.4. With permission
from the McGraw-Hill Companies
Inc.

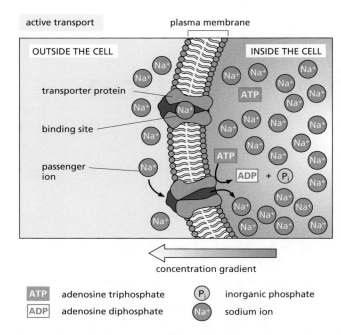

Life as we know it would be impossible without change. In female humans, for example, changes in the concentrations of hormones during the menstrual cycle affect the endometrial lining of the uterus and the growth of ovarian follicles (Figure 5.1). Without these modifications, none of us would be here to study mathematics. Changes in concentration over small distances are also very important throughout biology. Two-thirds of the ATP produced in nerve cells is consumed by proteins that pump cations across the plasma membrane, lowering the concentration of potassium and raising the concentration of sodium (Figure 5.2). The resulting concentration gradients across the cell wall provide the driving force for the uptake of water, glucose, and other nutrients, not to mention the special electrical properties needed for transmitting nerve impulses and performing arithmetic. Warm-blooded mammals maintain body temperatures that are very different from those of their surroundings, so the relationship between heat loss and temperature gradient puts important limits on body characteristics. For example, seals make the transition between body temperature and their cooler environment gentler by coating themselves with layers of fat and fur.

This chapter introduces calculus, the branch of mathematics developed to predict how changes in one quantity will alter another. Beginning with methods for determining the slopes of straight lines and simple curves, we will assemble the basic toolkit of techniques required for differentiation. The tools will then be put to work by showing how to analyze the shapes of different graphs, how to sketch the curve of an unknown function, how to make approximations, and how to handle small errors in experimental measurements. These mathematical details are needed to understand the way that metabolite concentrations change with time in reactions catalyzed by enzymes, to analyze the growth of bacteria, and to explain how a pH buffer solution works, which are all discussed in the next few chapters. The principles developed here also have applications in current research, from magnetic resonance imaging of the human head to generating computational models of the electrical activity of the heart.

# 5.1 The slope of a straight line

Graphs provide an important way of conveying information in biology, as well as many other areas, such as economics. For example, Figure 5.3 shows the cumulative number of cattle infected with the foot and mouth virus each week during an outbreak of the disease. The detection and management of such an outbreak is interesting in terms of the biology of infection and the epidemiology of the spread of the disease, but also has serious financial consequences for farmers, not to mention the impact on the cattle themselves. To interpret the layers of information summarized by a graph of this sort, we need to start by revising some of the basics.

In Section 1.5 we discussed the relationship between temperatures measured on the Fahrenheit (°F) and Celsius (°C) scales: the relationship is given in EQ1.26. Using the function notation introduced in Chapter 3, the Fahrenheit temperature $f$ and the Celsius temperature $c$ are linked by the formula:

$$f(c) = \frac{9}{5}c + 32. \tag{EQ5.1}$$

*Note that this is an example of a numerical value equation.* A plot of this function produces a straight line (Figure 5.4), with an increase in $c$ leading to a proportional increase in $f$. The slope of a straight line is rather like the gradient of a hill: the steeper the incline appears to be, the greater the gradient. On a Himalayan mountain track with a large positive gradient, the altitude of a climber will increase rapidly when only a modest horizontal distance has been traveled (Figure 5.5). By contrast, negative gradients lead downhill, from the summit back to base camp.

The gradient of the straight-line plot of $f$ against $c$ is clearly positive, because $f$ increases when $c$ gets larger. However, to make a more quantitative statement about the slope we need specific information about a pair of points that both lie on the line. Inspection of EQ5.1 reveals that $f = 32$ when $c = 0$

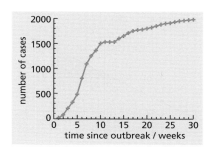

**Figure 5.3**
The cumulative number of cattle infected with the foot and mouth virus each week during an outbreak of the disease.

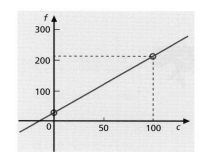

**Figure 5.4**
The functional relationship between temperatures measured on the Fahrenheit (°F) and Celsius (°C) scales.

**Figure 5.5**
On a Himalayan mountain track with a large positive gradient, the altitude of a climber will increase rapidly when only a modest horizontal distance has been traveled. Negative gradients lead downhill, from the summit back to base camp. Image courtesy of Warrenski under Creative Commons Attribution - Share Alike 2.0 Generic.

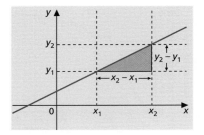

**Figure 5.6**
The gradient (or slope) of a straight line is equal to $(y_2 - y_1)/(x_2 - x_1)$, the increase in $y$ values divided by the increase in $x$ values.

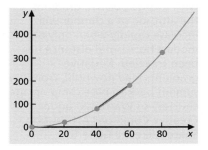

**Figure 5.7**
The position of an apple at 0, 20, 40, and 60 ms after it begins its descent toward the head of Isaac Newton.

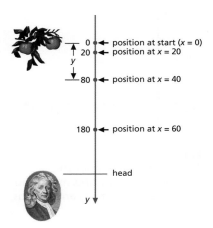

**Figure 5.8**
A plot of $y$ against $x$ for $y(x) = 0.05x^2$. The gradient of the straight line between points at $x = 40$ and $x = 60$ gives a value for the average velocity between 40 and 60 ms.

and that $f = 212$ when $c = 100$. If we define the slope of the line as the increase in the value of the function $f$ for a given increase in the value of the variable $c$, then:

$$\text{Slope} = \frac{f(100) - f(0)}{100 - 0} = \frac{212 - 32}{100} = 1.8 = \frac{9}{5}.$$

Interestingly, the value of the slope determined here is the same as the coefficient of the $c$ term in EQ5.1.

More generally, we can try the same approach using some basic coordinate geometry and the generic formula for a straight line:

$$y(x) = A + Bx. \tag{EQ5.2}$$

Choosing any two pairs of coordinates that lie on this line, $(x_1, y_1)$ and $(x_2, y_2)$, we can determine the slope (Figure 5.6). However, the coordinates of both of these points must satisfy EQ5.2, which means that $y_1 = A + Bx_1$ and $y_2 = A + Bx_2$. Using this information to calculate the slope, namely the rate of increase of $y$ with respect to an increase in $x$, we get:

$$\text{Slope} = \frac{y_2 - y_1}{x_2 - x_1} = \frac{(A + Bx_2) - (A + Bx_1)}{x_2 - x_1} = \frac{B(x_2 - x_1)}{x_2 - x_1} = B.$$

This result proves that when the equation of a straight line is written in the form $y = A + Bx$, the gradient is always represented by the value of the constant $B$.

# 5.2 Average and instantaneous rates of change

Moving on now to consider functions more complex than simple straight lines, imagine that we are watching a slow motion movie of Isaac Newton sitting in the orchard behind his family home in Woolsthorpe, Lincolnshire, UK. If a ripe apple falls off a branch above him, it will approach his head by $y$ millimeters in $x$ milliseconds, given approximately by the formula:

$$y(x) = 0.05x^2. \tag{EQ5.3}$$

*Note that this is another example of a numerical value equation*; that is, $y$ is the distance traveled divided by a reference value of 1 mm and $x$ is the time divided by a reference value of 1 ms. Figure 5.7 uses a coordinate line to show the position of the apple at 0, 20, 40, and 60 ms after it begins its descent toward the cranium of the soon-to-be Lucasian Professor of Mathematics.

The plot of $y$ against $x$ in Figure 5.8 shows that the apple travels ever-larger distances during consecutive 20 ms time intervals. A single gradient value can therefore never provide enough information to describe the complete shape of the curve. Because the rate of change of distance with respect to time is another way of talking about velocity, this is equivalent to saying that a single velocity would not be able to give a full account of the trajectory of the apple as it accelerates downward under the influence of gravity. We can, however, work out the average velocity of the apple over a given time interval. For example, we can consider how far the apple travels between the time points at 40 and 60 ms:

$$\text{Average velocity} = \frac{\text{Distance}}{\text{Time}} = \frac{y(60) - y(40)}{60 - 40}\frac{\text{mm}}{\text{ms}} = \frac{180 - 80}{20}\text{mm ms}^{-1}$$

$$= 5 \text{ mm ms}^{-1}.$$

This value of 5 mm ms$^{-1}$ represents the average rate at which the distance traveled by the apple changes with time in the interval between 40 and 60 ms. The result of such a calculation is equivalent to assuming that a good idea of the shape of a plot of distance against time over this region can be obtained by drawing a straight line between the coordinates (40, 80) and (60, 180), as shown in Figure 5.8.

**Table 5.1 Calculations of the average velocity of the apple for shorter and shorter time intervals ($h$ ms) after 40 ms. As $h$ approaches 0, the average velocity approaches a limit of 4 mm ms$^{-1}$**

| duration of time interval / ms | distance traveled by the end of the time interval / mm | average velocity during the time interval / mm ms$^{-1}$ |
|:---:|:---:|:---:|
| $h$ | $y(40 + h) = 0.05 \times (40 + h)^2$ | $[y(40 + h) - y(40)]/h$ |
| 20 | 180.000000 | 5.0000 |
| 10 | 125.000000 | 4.5000 |
| 5 | 101.250000 | 4.2500 |
| 1 | 84.050000 | 4.0500 |
| 0.1 | 80.400500 | 4.0050 |
| 0.01 | 80.040005 | 4.0005 |

We can use the same approach to work out the average velocity of the apple for shorter and shorter time intervals after 40 ms by setting $x = 40 + h$ and then forcing $h$ to adopt smaller and smaller values, as shown in Table 5.1. As $h$ gets closer and closer to zero, the average velocity approaches a limiting value of 4 mm ms$^{-1}$. Rather than an average velocity, this limit of 4 mm ms$^{-1}$ actually represents an instantaneous velocity: the instantaneous rate of change of distance with respect to time evaluated at 40 ms after the apple began its descent.

Because limits are a significant concept in calculus, special mathematical jargon has been developed to represent them. If the ratio $[y(40 + h) - y(40)]/h$ tends toward a value of 4 as $h$ tends toward zero, we write:

$$\frac{y(40 + h) - y(40)}{h} \rightarrow 4 \qquad \text{as} \qquad h \rightarrow 0.$$

This expression can be compressed even further:

$$\lim_{h \to 0} \left\{ \frac{y(40 + h) - y(40)}{h} \right\} = 4.$$

Before proceeding to the next stage, you should attempt question 1 from the End of Chapter Questions.

## 5.3 The slope of a curve

When we used the ratio $(y_2 - y_1)/(x_2 - x_1)$ to determine the slope of the straight line function $f(c) = (9c/5) + 32$, the result was a constant. We would have obtained the same value by choosing any pair of coordinates $(x_1, y_1)$ and $(x_2, y_2)$ that lie on the line. For a curve, such as the path traced by the plummet of Newton's apple given by $y(x) = 0.05x^2$, the gradient of a line joining two points on the graph will vary depending on which particular pair of points is selected. This is because the instantaneous rate of change is different at each point on a curve. However, we can determine the instantaneous rate of change of the curve at any given point by using the procedure encountered in Section 5.2.

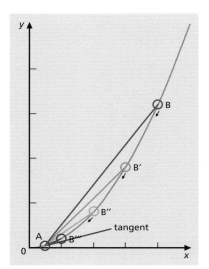

The method is shown graphically for $y(x) = 0.05x^2$ in Figure 5.9. The gradient of the straight line that joins a fixed point A, which has coordinates (1, 0.05), with a moveable point B is evaluated as B approaches A along the curve. When B is distant from A, the line that joins the two points is clearly a poor approximation for the shape of the curve at A. However, as B moves to B′, to B″, and then to B‴ the average rate of change given by the slope of the line AB approaches the value of the instantaneous rate of change of the graph at point A. Throughout this process, the gradient of line AB gets ever closer to the gradient of the tangent to the curve at point A. In fact, the instantaneous rate of change of a function is always equal to the slope of the tangent to the curve at that particular point. The slope of the tangent is clearly different at each point along the graph of $y(x) = 0.05x^2$, becoming more gentle near to the minimum at $x = 0$, but steeper further out along the positive $x$ axis.

**Figure 5.9**
As B moves to B′, B″, and B‴ (points closer to A) the average rate of change given by the slope of the line AB approaches the value of the instantaneous rate of change of the graph at point A; the line AB itself approaches the tangent to the curve at A.

To express this procedure using algebra, we start once again by considering two points that lie on the curve, but this time we will make them more general, giving A the coordinates $(x, y)$ and B the coordinates $(x_1, y_1)$. To save space, we will introduce a new notation, labeling $(x_1 - x)$ as *the increment in x* or $\Delta x$, and $(y_1 - y)$ as *the increment in y* or $\Delta y$. The '$\Delta$' symbol here is the Greek capital letter delta. An increment is similar to an increase, except that the change can be either positive or negative. If we alter the value of the variable from $x$ to $x_1$, then the increment $\Delta x$ will be equal to $(x_1 - x)$. The corresponding increment in $y$ that occurs as a result of the change in $x$, because of the mathematical relationship between them, is given by $\Delta y = y_1 - y$.

If we write down the relationship for point A:

$$y = 0.05x^2, \tag{EQ5.4}$$

then point B, which possesses the coordinates $(x + \Delta x, y + \Delta y)$ must obey the same equation because it lies on the same curve:

$$(y + \Delta y) = 0.05 \times (x + \Delta x)^2$$

$$= 0.05 \times (x^2 + 2x\Delta x + (\Delta x)^2) \tag{EQ5.5}$$

$$= 0.05x^2 + 0.10x\Delta x + 0.05(\Delta x)^2.$$

Subtracting EQ5.4 from EQ5.5 gives an expression for $\Delta y$:

$$(y + \Delta y) - y = 0.05x^2 + 0.10x\Delta x + 0.05(\Delta x)^2 - 0.05x^2$$

that is, $\Delta y = 0.10x\Delta x + 0.05(\Delta x)^2.$ (EQ5.6)

Dividing both sides of EQ5.6 by $\Delta x$ will produce an expression for the ratio of the changes in $x$ and $y$ produced by the increment in $x$:

$$\frac{\Delta y}{\Delta x} = 0.10x + 0.05\Delta x.$$

If $\Delta x$ is small compared with $0.10x$, we can make it obvious that the increment in $x$ is miniscule by replacing the term $\Delta x$ with $\delta x$ (using a lowercase Greek delta), along with a corresponding increment in $y$ of $\delta y$. The equation for the average rate of change for a particularly tiny increment in $x$ is therefore written:

$$\frac{\delta y}{\delta x} = 0.10x + 0.05\delta x. \tag{EQ5.7}$$

Table 5.2 shows what happens to this ratio near $x = 1$ as $\delta x$ gets smaller and smaller. Not surprisingly, as $\delta x$ decreases in size, $\delta y$ gets smaller as well. However, just as we saw for the gradient in Table 5.1, the ratio of the two increments, $\delta y / \delta x$, does not disappear in a puff of smoke, but instead gradually gets ever closer to a limiting value of $0.1$.

**Table 5.2 Calculations of $\delta y / \delta x = 0.10x + 0.05\delta x$ for $x = 1$ as $\delta x$ approaches 0**

| $\delta x$ | $\delta y = 0.10x\delta x + 0.05(\delta x)^2$ evaluated at $x = 1$ | $\delta y / \delta x = 0.10x + 0.05\delta x$ evaluated at $x = 1$ |
|---|---|---|
| 10 | 6 | 0.6 |
| 1 | 0.15 | 0.15 |
| 0.1 | 0.0105 | 0.105 |
| 0.01 | 0.001005 | 0.1005 |
| 0.001 | 0.00010005 | 0.10005 |
| 0.0001 | 0.0000100005 | 0.100005 |

This method has been named 'differentiation from first principles'. It can be summarized in terms of the following steps:

(1) Write down the function $y$ in terms of the variable $x$.
(2) Substitute $(x + \delta x)$ for $x$ and $(y + \delta y)$ for $y$.
(3) Find an expression for $\delta y$.
(4) Divide both sides of the equation by $\delta x$.
(5) Determine the limit of $\delta y / \delta x$ as $\delta x$ tends toward zero.

When the same logic is applied at any point along the curve, we obtain a general expression for the limit of $\delta y/\delta x$ as $\delta x$ approaches zero. The answer can be obtained by setting $\delta x$ to zero everywhere on the right-hand side of EQ5.7:

$$\lim_{\delta x \to 0} \left\{ \frac{\delta y}{\delta x} \right\} = 0.10x. \tag{EQ5.8}$$

Confirming what we suspected from Table 5.1, this equation gives a value of 4 for the slope of the tangent to the curve $y = 0.05x^2$ at the point where $x = 40$.

Expressions like 'the limit of $\delta y/\delta x$ as $\delta x$ approaches zero' or even

$$\lim_{\delta x \to 0} \left\{ \frac{\delta y}{\delta x} \right\}$$

are far too long-winded to say regularly, or even to write down. Instead, it is time to introduce some more jargon by defining a new quantity, $dy/dx$, as the 'derivative of $y$ with respect to $x$', which means exactly the same thing.

$$\frac{dy}{dx} = \lim_{\delta x \to 0} \left\{ \frac{\delta y}{\delta x} \right\}.$$

To summarize the results of this section, we have analyzed the function

$$y = 0.05x^2,$$

and found that the derivative of $y$ with respect to $x$ has the form

$$\frac{dy}{dx} = 0.10x.$$

Another way of expressing this would be to state directly that the derivative of $0.05x^2$ is $0.10x$, which can be done using the following nomenclature:

$$\frac{d}{dx}(0.05x^2) = 0.10x.$$

## 5.4 Differentiating simple expressions

The 'differentiation from first principles method' encountered in Section 5.3 can be used to determine the gradient of almost any kind of function, but sadly it also has the disadvantage of being rather tedious. Fortunately, pioneering mathematicians have worked out rules for differentiating most of the functions we are likely to encounter. As these are simple enough to remember, it is not necessary to repeat the whole process of differentiation from first principles every time we come across a new function. Most importantly, there is a useful formula for differentiating any function that contains a variable raised to a power.

If the multiplier $A$ and the index $n$ are constants, then the derivative of

$$y = Ax^n$$

is

$$\frac{dy}{dx} = Anx^{n-1}.$$

This expression is so helpful that it will be used regularly in the rest of this chapter. It is therefore worth unpacking the mathematics into phrases. On differentiating a power function:

(1) leave any constant multiplier (such as $A$) *unchanged*;
(2) *decrease* the original value of the index ($n$) by 1, giving ($n - 1$);
(3) and finally *multiply* the result by the original value of the index ($n$).

For example, if $y = x^3/5$, the constant $A$ is equal to $1/5$ and the index $n$ is equal to 3. Therefore

$$\frac{dy}{dx} = \frac{1}{5} \times 3 \times x^{3-1} = \frac{3}{5}x^2.$$

Four cases that need to be spelled out in detail here occur when the value of the power $n$ is zero, one, negative, or fractional:

(1) If $y = A$, then the power to which $x$ is raised is zero, because $Ax^0 = A \times 1 = A$. Therefore $dy/dx = A \times 0 \times x^{0-1} = 0$. Hence the gradient of a constant function will always be zero.

(2) If $y = 2x = 2x^1$, then the constant $A$ is equal to 2 and the index $n$ is equal to 1. Therefore $dy/dx = 2 \times 1 \times x^{1-1} = 2 \times x^0 = 2$. Hence the gradient of a linear function will always be a constant.

(3) If $y = 1/x$, then the constant $A$ is equal to 1 and the index $n$ is equal to $-1$, because $1/x = x^{-1}$. Therefore $dy/dx = 1 \times (-1) \times x^{-1-1} = -x^{-2} = -1/x^2$.

(4) If $y = 4\sqrt{x}$, then the constant $A$ is equal to 4 and the index $n$ is equal to $+1/2$, because $\sqrt{x} = x^{1/2}$. Therefore $dy/dx = 4 \times (1/2) \times x^{(1/2)-1} = 2x^{-1/2}$.

As well as power functions, many other types of function are important in biology, each of which can be differentiated using its own special rule, as summarized in Table 5.3. These rules are also well worth remembering; we will analyze the behavior of each function in more detail in Chapters 7 and 8.

Before proceeding to the next stage, you should attempt question 4 from the End of Chapter Questions.

**Table 5.3 Derivatives of some commonly encountered functions**

| function, $f(x)$ | derivative, $\mathrm{d}\,f(x)/\mathrm{d}x$ | notes |
|---|---|---|
| $Ax^n$ | $Anx^{n-1}$ | $A, n$ constants |
| $A\ln(ax)$ | $A/x$ | $A, a$ constants |
| $Ae^{nx}$ | $Ane^{nx}$ | $A, n$ constants |
| $A\sin(ax)$ | $Aa\cos(ax)$ | $A, a$ constants |
| $A\cos(ax)$ | $-Aa\sin(ax)$ | $A, a$ constants |
| $A\tan(ax)$ | $Aa/\cos^2(ax)$ | $A, a$ constants |

# 5.5 Differentiating a sum of two functions

Our toolkit for differentiating basic expressions is nearly complete, but now we have to work out how to deal with more complicated equations. Specifically, we need a rule that will show us how to tackle the sum (or the difference) of two or more simple power functions.

The 'sum rule' states that *the derivative of a sum of functions is equal to the sum of the derivatives of those functions*. To justify using this rule, let us suppose that $y = x^2 + x - 6$. The derivative of $y$ with respect to $x$ can be found from first principles as follows:

$$(y + \delta y) = (x + \delta x)^2 + (x + \delta x) - 6$$
$$= x^2 + 2x\delta x + (\delta x)^2 + x + \delta x - 6.$$

Subtracting $y = x^2 + x - 6$ from this equation gives:

$$(y + \delta y) - y = x^2 + 2x\delta x + (\delta x)^2 + x + \delta x - 6 - (x^2 + x - 6)$$
that is, $\delta y = 2x\delta x + (\delta x)^2 + \delta x.$

Dividing both sides by $\delta x$ gives:

$$\frac{\delta y}{\delta x} = 2x + \delta x + 1.$$

As $\delta x$ tends toward zero, this becomes:

$$\frac{\mathrm{d}y}{\mathrm{d}x} = \lim_{\delta x \to 0}\left\{\frac{\delta y}{\delta x}\right\} = 2x + 1,$$

that is, $\dfrac{\mathrm{d}}{\mathrm{d}x}(x^2 + x - 6) = 2x + 1.$

To suit the present argument, this result can be rewritten in a slightly more helpful form as:

$$\frac{d}{dx}(x^2 + x - 6) = 2x + 1 - 0. \tag{EQ5.9}$$

A careful look at EQ5.9 reveals that each term on the right-hand side is the derivative of the corresponding term on the left-hand side, because:

$$\frac{d}{dx}(x^2) = 2x, \qquad \frac{d}{dx}(x) = 1, \qquad \frac{d}{dx}(6) = 0.$$

Putting all of this together in one step, we can say that:

$$\frac{d}{dx}(x^2 + x - 6) = \frac{d}{dx}(x^2) + \frac{d}{dx}(x) - \frac{d}{dx}(6) = 2x + 1 - 0.$$

We have therefore demonstrated that the sum rule is reasonable and could use a similar argument to show that *the derivative of a difference between two functions is equal to the difference of the derivatives of the two functions.*

This rule has a straightforward application in the treatment of cancer by radiotherapy. Irradiated cells are killed mainly as a result of double strand breaks, a type of damage to DNA that is particularly difficult to repair and can lead to oddly shaped chromosomes. Two days after treating a culture of human white blood cells, $n$, the average number of damaged chromosomes per cell, depends on $r$, the dose of X-ray radiation absorbed, according to the equation:

$$n = Ar + Br^2, \tag{EQ5.10}$$

where $A$ and $B$ are constants (see Figure 5.10). The SI unit for a dose of absorbed radiation is the 'gray', represented by the symbol 'Gy' and corresponding to the absorption of 1 joule of radiation per kilogram of matter.

> **Sum and difference rules**
> If both $u(x)$ and $v(x)$ are functions of the same variable $x$, then
>
> $$\frac{d}{dx}(u \pm v) = \frac{du}{dx} \pm \frac{dv}{dx}.$$

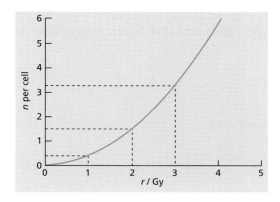

**Figure 5.10**
Two days after treating a culture of human white blood cells, the average number of damaged chromosomes per cell, $n$, has a quadratic dependence on $r$, the dose of X-ray radiation absorbed.

The behavior captured by EQ5.10 is called a *linear–quadratic dose response* because of the two terms on the right-hand side: the first is linear in $r$ (meaning that a plot of the function $n = Ar$ would give a straight line), whereas the other is quadratic in $r$ (that is, a plot of $n = Br^2$ would produce a U-shaped curve). The derivative of $n$ with respect to $r$ can be obtained by adding the derivatives of both individual terms:

$$\frac{\mathrm{d}n}{\mathrm{d}r} = \frac{\mathrm{d}}{\mathrm{d}r}(Ar + Br^2) = \frac{\mathrm{d}}{\mathrm{d}r}(Ar) + \frac{\mathrm{d}}{\mathrm{d}r}(Br^2) = A + 2Br.$$

This shows something a competent radiologist should be aware of: that the rate of change of the response will be different at different dose strengths. To understand more, it can help to put numbers into the equation: if $A = 0.05$ Gy$^{-1}$ and $B = 0.35$ Gy$^{-2}$, then absorbing a 1 Gy dose of X-ray radiation will give rise to an average of 0.4 abnormal chromosomes per cell. Raising the dose by 1 Gy will increase the response to 1.5, whereas a further 1 Gy boost escalates the damage to 3.3 abnormal chromosomes per cell (see Figure 5.10).

## 5.6 Higher derivatives

After a function has been differentiated once, there is no reason why it should not be differentiated a second time, using exactly the same rules. For example, over a period of 20 min after a drug that stimulates reproduction has been introduced into a colony of *Escherichia coli* bacteria, the number $n$ of bacteria after time $t$ min is given by the expression:

$$n = 1000 + 30t^2 - t^3. \tag{EQ5.11}$$

Using the rules from Sections 5.4 and 5.5, we can write down the derivative of $n$ with respect to $t$ as shown.

First derivative:

$$\frac{\mathrm{d}n}{\mathrm{d}t} = 60t - 3t^2. \tag{EQ5.12}$$

EQ5.12 represents what we should more properly call the *first derivative* of $n$. We can differentiate again and then again with respect to $t$ to get *higher* derivatives:

Second derivative:

$$\frac{\mathrm{d}^2 n}{\mathrm{d}t^2} = \frac{\mathrm{d}}{\mathrm{d}t}\left(\frac{\mathrm{d}n}{\mathrm{d}t}\right) = 60 - 6t.$$

Third derivative:

$$\frac{\mathrm{d}^3 n}{\mathrm{d}t^3} = \frac{\mathrm{d}}{\mathrm{d}t}\left[\frac{\mathrm{d}^2 n}{\mathrm{d}t^2}\right] = \frac{\mathrm{d}}{\mathrm{d}t}\left[\frac{\mathrm{d}}{\mathrm{d}t}\left(\frac{\mathrm{d}n}{\mathrm{d}t}\right)\right] = -6.$$

Fourth derivative:

$$\frac{\mathrm{d}^4 n}{\mathrm{d}t^4} = \frac{\mathrm{d}}{\mathrm{d}t}\left\{\frac{\mathrm{d}^3 n}{\mathrm{d}t^3}\right\} = \frac{\mathrm{d}}{\mathrm{d}t}\left\{\frac{\mathrm{d}}{\mathrm{d}t}\left[\frac{\mathrm{d}^2 n}{\mathrm{d}t^2}\right]\right\} = \frac{\mathrm{d}}{\mathrm{d}t}\left\{\frac{\mathrm{d}}{\mathrm{d}t}\left[\frac{\mathrm{d}}{\mathrm{d}t}\left(\frac{\mathrm{d}n}{\mathrm{d}t}\right)\right]\right\} = 0.$$

To avoid confusion between the square of the first derivative, $(dn/dt)^2$, and the second derivative, $d^2n/dt^2$, this is described when speaking aloud as 'dee two $n$ by dee $t$ squared'. Similarly, the third derivative, $d^3n/dt^3$, is referred to as 'dee three $n$ by dee $t$ cubed'.

However, what do these higher derivatives actually *mean*? We have already seen that the *first* derivative $dn/dt$ expresses how rapidly the value of the population $n$ alters when the value of the variable $t$ is changed. In the same vein, the *second* derivative shows how rapidly the gradient $dn/dt$ changes with a change in $t$.

A graphical way of thinking about this is shown in Figure 5.11. Because $dn/dt$ describes the slope of a plot of $n$ against $t$, $d^2n/dt^2$ must express the rate of change of that slope, which is related to the curvature. The slope of the plot of $n$ against $t$ is zero at $t = 0$, becomes positive as $t$ increases from zero, and reaches its steepest at $t = 10$. As $t$ increases further, the tangent becomes less steep until it is horizontal again at $t = 20$ and then starts slanting downward (that is, the gradient is negative for $t > 20$). The slope increases throughout the period $0 < t < 10$: this is a region of *positive* curvature, corresponding to values of $d^2n/dt^2$ that are greater than zero. When $dn/dt$ reaches its maximum value, at $t = 10$, the second derivative is equal to zero. From $t = 10$ to $t = 20$, the slope continues to decrease, corresponding to a region of *negative* curvature in which $d^2n/dt^2$ is less than zero.

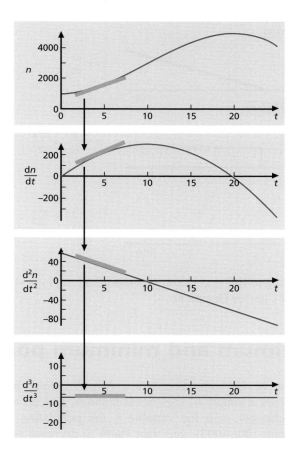

**Figure 5.11**
The first three derivatives of $n(t) = 1000 + 30t^2 - t^3$ shown graphically.

If a graph gives a straight line, then its slope is a constant and its curvature should be zero. We can express this mathematically using the following relationships:

Function: $y = Ax + B$.

Slope: $\dfrac{\mathrm{d}y}{\mathrm{d}x} = A$.

Curvature: $\dfrac{\mathrm{d}^2 y}{\mathrm{d}x^2} = 0$.

Yet another way of understanding higher derivatives is to consider the relationship between *acceleration*, *velocity*, and the *distance* traveled when driving in a car: velocity is the rate of change of distance with respect to time, whereas acceleration describes the change in velocity as a function of time and is the second derivative of distance with respect to time (Figure 5.12). Going further, the derivative of acceleration is called *jerk* – a rather appropriate name. A sudden change in acceleration by a learner driver causes the vehicle to lurch rather abruptly, producing discomfort in passengers, as well as comments of a critical nature from the driving instructor.

**Figure 5.12**
Velocity, acceleration, and jerk are defined as successive derivatives of distance.

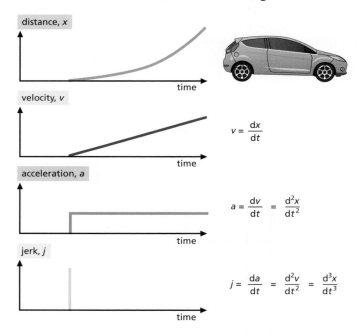

Before proceeding to the next stage, you should attempt questions 5, 6, and 7 from the End of Chapter Questions.

# 5.7 Maximum and minimum points

The time course of the population of our *E. coli* colony, plotted again in Figure 5.13, shows a couple of significant features. Instead of increasing or decreasing indefinitely with the variable $t$, the population $n(t)$ rises to a maximum at point (20, 5000), and then starts to decrease for larger values

## Box 5.1 Forms of notation for differentiation

Throughout this book, derivatives are presented using **Leibniz's notation**. The derivatives with respect to $x$ for the function $y = f(x)$ are written as:

$$\text{first derivative} = \frac{dy}{dx} = \frac{d}{dx}(f(x)),$$

$$n\text{th derivative} = \frac{d^n y}{dx^n}.$$

An advantage of using Leibniz's notation is that the variable used in the differentiation process ($x$) is always mentioned explicitly in the denominator of the derivative.

Other common approaches provide useful abbreviations in the right context, but their meaning is less obvious for newcomers. For example, **Lagrange's notation** uses a 'prime' symbol to distinguish between a function and its derivative, with second and third derivatives marked by additional primes:

for example $f'(x) = \frac{df}{dx}$, $f''(x) = \frac{d^2 f}{dx^2}$, $f'''(x) = \frac{d^3 f}{dx^3}$,

and $f^{(n)}(x) = \frac{d^n f}{dx^n}$.

Higher derivatives can be represented by a superscript containing an Arabic number in brackets. This representation is compact and easy to typeset, so it is often used in mathematics and physics textbooks. However, it has a few disadvantages: the number of prime marks can easily be miscounted, and primes are used to denote many other things. For example, in biology, prime symbols are used to indicate the carbon atoms of the ribose ring of nucleic acids, such as the 3'-end of a strand of RNA.

**Newton's notation** provides a further specialized twist, making use of dots above the function to represent derivatives with respect to time:

for example $\dot{s} = \frac{ds}{dt}$ and $\ddot{s} = \frac{d^2 s}{dt^2}$.

Unfortunately, the dots can be hard to spot, so they are often overlooked.

of $t$. It is worth noting that the slope of the curve is zero at the maximum point, something that is true for any curve that possesses a feature of this sort. However, the converse is *not* always the case. A point with zero slope may also represent a *minimum*, as found at point (0, 1000) in Figure 5.13. Maxima and minima together are both referred to as *turning points*, because the gradient of the curve changes either from positive to negative or from negative to positive as you pass through the point.

A major benefit of learning how to differentiate is that this tool can be used to predict where turning points occur, a vital step toward working out the maximum or minimum values that a particular function can adopt. Because we know that

$$n(t) = 1000 + 30t^2 - t^3,$$

the process of predicting the maximum number of bacteria in the *E. coli* colony begins by differentiating the expression for the population, $n$, with respect to time, $t$, and then identifying the values of $t$ at which the derivative is zero. First, we differentiate the expression:

$$\frac{dn}{dt} = 60t - 3t^2 = 3t(20 - t).$$

Setting the derivative equal to zero gives

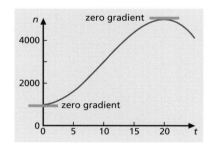

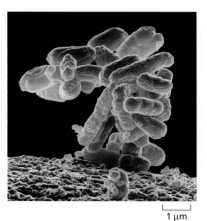

**Figure 5.13**
Variation with time of the population of an *E. coli* colony.

$$3t(20 - t) = 0.$$

However, this can only occur

when $t = 0$    or    when $(20 - t) = 0$, that is,  $t = 20$.

Substituting these values back into the population equation:

when $t = 0$,      $n(0) = 1000 + 30 \times (0)^2 - (0)^3 = 1000$

when $t = 20$,    $n(20) = 1000 + 30 \times (20)^2 - (20)^3 = 1000 + 12\,000 - 8000$
$$= 5000.$$

This leaves us knowing the coordinates of two guaranteed turning points, so our next task is to find out whether these are minima or maxima. In this case, because we have already seen a plot of the function, it is obvious that there is a minimum at (0, 1000) and a maximum at (20, 5000), but this job is not always so straightforward. The ambiguity can be resolved mathematically by examining the sign of the slope on either side of each turning point. As Figure 5.14 shows, on the low $t$ side of the minimum turning point the gradient is negative, while on the high $t$ side the gradient is positive. By contrast, the slope is positive on the low $t$ side of the maximum turning point, but negative on the high $t$ side. Another way of expressing this is to say that at a minimum the gradient increases through zero, whereas at a maximum the gradient decreases through zero.

The bacterial population profile has turning points where the first derivative is zero at $t = 0$ and $t = 20$. If we look at values of $t$ that are a little smaller than 0, say $t = -0.1$, the first derivative is clearly negative:

$$\frac{\mathrm{d}n}{\mathrm{d}t} = 3t(20 - t) = 3 \times (-0.1) \times (20 - (-0.1)) = -6.03.$$

Similarly, when we set $t$ to be a little larger than its value at this turning point, say $t = +0.1$, the first derivative is positive:

$$\frac{\mathrm{d}n}{\mathrm{d}t} = 3 \times (+0.1) \times (20 - 0.1) = +5.97.$$

**Figure 5.14**
A minimum occurs when the gradient goes from negative to positive, whereas a maximum occurs when a gradient goes from positive to negative.

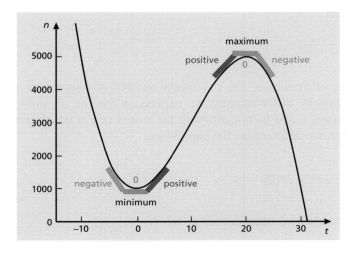

This tells us that near $t = 0$ the slope increases through zero, so the curve must be going through a minimum point. Taking the second turning point, if we set $t$ to be a little smaller than 20, say $t = 19$, the first derivative is now positive:

$$\frac{dn}{dt} = 3 \times (+19) \times (20 - 19) = +57.$$

Similarly, if we make $t$ larger, say $t = 21$, the first derivative is negative:

$$\frac{dn}{dt} = 3 \times (+21) \times (20 - 21) = -63.$$

Because the gradient of the curve decreases through zero near $t = 20$, this must be a maximum point. The same process can be used as a first derivative test to identify maxima and minima in other graphs, as shown in Table 5.4.

**Table 5.4 The first derivative test for recognizing maxima and minima if the first derivative of a function, $dy/dx$, is zero at a particular point, $x = a$**

| at a *minimum* the derivative is negative to the left and positive to the right | | | | at a *maximum* the derivative is positive to the left and negative to the right | | | |
|---|---|---|---|---|---|---|---|
| $x$ | $a - \delta x$ | $a$ | $a + \delta x$ | $x$ | $a - \delta x$ | $a$ | $a + \delta x$ |
| $\dfrac{dy}{dx}$ | $-$ | 0 | $+$ | $\dfrac{dy}{dx}$ | $+$ | 0 | $-$ |
| slope | \ | ___ | / | slope | / | ‾‾ | \ |

Working out values of the first derivative on both sides of a turning point is rather time consuming. As a short cut, the same conclusion can usually be reached by differentiating a second time and considering the curvature of the graph. For the time course of the *E. coli* population, we found in Section 5.6 that:

$$\frac{d^2 n}{dt^2} = 60 - 6t.$$

At $t = 0$, this second derivative is positive ($+60$), indicating that the slope of the curve becomes larger when the value of $t$ increases. This means that the

**Table 5.5 The second derivative test for recognizing maxima and minima for a function $y$ at a particular point, $x = a$**

| | at a *minimum* the second derivative is positive | the test *fails* if the second derivative is zero or undefined | at a *maximum* the second derivative is negative |
|---|---|---|---|
| $\dfrac{dy}{dx}$ | 0 | 0 | 0 |
| $\dfrac{d^2 y}{dx^2}$ | $+$ | 0 | $-$ |
| | ☺ $+ +$ | ☺ $\circ\ \circ$ | ☹ $- -$ |

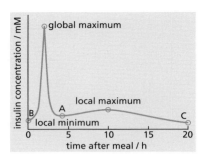

**Figure 5.15**
The blood concentration of the hormone insulin plotted as a function of time (after a meal rich in carbohydrates).

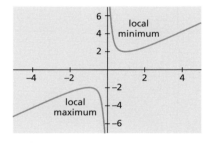

**Figure 5.16**
For $y = x + 1/x$ the local minimum is higher than the local maximum.

turning point must be a minimum, as the gradient increases through zero. At $t = 20$, the second derivative is negative ($-60$) because the gradient decreases through zero, so this must be a maximum point. This **second derivative test** approach is widely used for analyzing the turning points of graphs (see Table 5.5) and is usually more convenient to apply than the first derivative test. However, if the second derivative is zero, the second derivative test fails, so it is then wise to return to the first derivative test to check what is going on.

It is worth emphasizing here that a maximum turning point is not necessarily going to possess the largest value that a function can adopt. In Figure 5.15 the blood concentration of the hormone insulin is plotted as a function of time after a meal rich in carbohydrates. The plot has several turning points, which show the difference between a local maximum and a global maximum. Similarly, the value of the function at the minimum turning point (A) is larger than the values at the limits of the displayed range (B and C). The graph reveals that insulin is secreted in two phases after food is consumed: a rapid release from the cache of hormone stored in the pancreas, followed by a slower phase corresponding to freshly synthesized protein.

In fact, quite simple functions exist for which the local minimum has a higher value than the local maximum, such as the plot in Figure 5.16 of $y = x + 1/x$.

## 5.8 Points of inflexion

The time dependencies of our expressions for $n$, $dn/dt$, and $d^2n/dt^2$ for the *E. coli* colony are summarized in Figure 5.17. At the local minimum of point (0, 1000), the first derivative is zero, whereas the second derivative is positive. This positive curvature means that the slope of the plot of population $n$ against time $t$ increases with increasing values of $t$ throughout the range $-\infty < t < 10$. At (20, 5000), the local maximum, $dn/dt$ is once again zero, but $d^2n/dt^2$ is now negative. Negative curvature causes $dn/dt$ to decrease as $t$ increases throughout the range $10 < t < +\infty$.

However, what happens when the curvature is zero at $t = 10$? As Figure 5.17 shows, this corresponds to a turning point in the graph of the first derivative, indicating that the gradient of the plot of $n$ against $t$ possesses its steepest positive value here. Regions of this sort where the curvature of a graph changes sign, either from negative to positive or from positive to negative, are called points of inflexion.

Points of inflexion are important in biology because they define conditions under which a response (such as the rate of a reaction) is most or least sensitive to a change in conditions (like the concentration of a metabolite). A good example is the ability of a buffer system to resist changes in pH brought about by the addition of alkali, as shown in Figure 5.18. The point of inflexion indicates where the graph of pH as a function of hydroxide concentration moves from a region of negative curvature to a region of positive curvature. A solution that contains a mixture of sodium acetate and acetic acid works best as a pH buffer when it is relatively insensitive to the amount of alkali that is added. This occurs near the point of inflexion in the titration curve, around

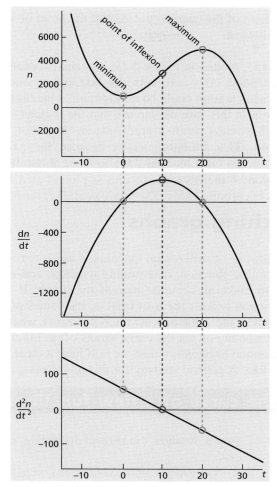

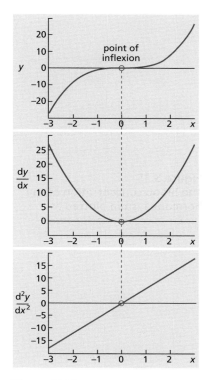

**Figure 5.19**
The graph of $y = x^3$ and its first two derivatives showing that $x = 0$ is a point of inflexion.

**Figure 5.17**
The variations of $n$, $dn/dt$, and $d^2n/dt^2$ with respect to time for the expressions describing the *E. coli* colony. The plot of $n$ has a maximum and a minimum (for which the first derivative is zero) and a point of inflexion (for which the second derivative is zero).

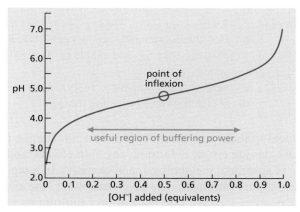

**Figure 5.18**
A plot of pH against concentration of added hydroxide ions for the acetate-acetic acid system. The system works best as a pH buffer when it is relatively insensitive to the amount of alkali that is added: this occurs near the point of inflexion in the titration curve, around pH 4.8.

pH 4.8. The gradient of the curve is not zero at the point of inflexion, but the slope is at its shallowest.

The graph of $y = x^3$, plotted in Figure 5.19, contains an example of a special type of inflexion point. At $x = 0$, both the first and second derivatives are zero. In general, such a point could be a minimum, a maximum, or a point of inflexion. However, in this case we can see that the regions $-\infty < x < 0$ and $0 < x < +\infty$ both possess positive gradients, so this must correspond to a point of inflexion. This situation occurs because the graph of the first derivative only touches the $x$ axis but fails to cross it, so the gradient never adopts values that are less than zero.

## 5.9 Sketching graphs

Another important use of differential calculus is as an aid for sketching the curve of an unfamiliar function, determining its shape and essential features without having to make an accurate plot of the graph. If we can find out where such a function *crosses the axes* (that is, the values of $x$ that give zero values of $y$, as well as the value of $y$ at which $x$ is zero), where it has *maxima and minima*, and whether or not the curve shoots off to infinity (that is, it has a vertical or horizontal *asymptote*), then we will have a clear idea of what the plot should look like. A general strategy for this process is given in Box 5.2.

### Box 5.2 Strategy for sketching the curve of a new function:

**Step 1. Analyze the function, $f(x)$.**

(a) Where does it cut the $x$ and $y$ axes?
(b) If a range of $x$ values is given, what are the boundary values of $y$?
(c) Are there any asymptotes?

**Step 2. Analyze the first derivative, d$f(x)$/d$x$.**

(a) Find any values for which the first derivative is zero.
(b) Construct a sign chart for d$f$/d$x$ to identify local maxima and minima.

**Step 3. Analyze the second derivative, d$^2 f(x)$/d$x^2$.**

(a) Construct a sign chart for the second derivative to confirm local maxima and minima.
(b) Identify any points of inflexion, where d$^2 f$/d$^2 x$ is equal to zero.

**Step 4. Sketch the graph of $f(x)$.**

(a) Collect the coordinates of all points obtained in steps 1–3.
(b) Plot the coordinates of intercepts, boundary points, maxima, minima, and points of inflexion on an axis system.
(c) Note where any asymptotes should appear.
(d) Sketch the curve by joining the dots.

In Phase I clinical trials, a new drug is given to a few human participants to discover the safe dosage range, determine any side effects, and monitor how the medication is absorbed, metabolized, and excreted. In a report on a recent trial of a potential antihistamine hay-fever treatment, the change in body temperature, $T$ degrees Celsius, one hour after $x$ milligrams of the drug are administered in a tablet to a healthy volunteer is described by the expression:

$$T(x) = \frac{x^2}{6}\left(1 - \frac{x}{9}\right),$$

over the range $0 \leqslant x \leqslant 6$. To understand the consequences of this research, it would be helpful to sketch the dependence of $T$ on $x$.

**Step 1.** We can begin by identifying where the curve cuts the $x$ and $T$ axes.

$T$ axis intercept: when $x = 0$, $T(0) = [(0)^2/6] \times (1 - (0)/9) = 0$.

$x$ axis intercepts: when $T = 0$, $[x^2/6] \times (1 - x/9) = 0$ so
either $x^2/6 = 0$,         giving $x = 0$,
or $(1 - x/9) = 0$,         giving $x = 9$.

Intercepts with the axes therefore appear at two points: $(x, T) = (0, 0)$ and $(9, 0)$. Note that the second of these intercepts is outside of the given range, $0 \leqslant x \leqslant 6$. We have already evaluated the change in body temperature for $x = 0$, but when $x = 6$:

$$T(6) = \frac{(6)^2}{6} \left( 1 - \frac{(6)}{9} \right) = 2.$$

The boundary points are therefore at points $(0, 0)$ and $(6, 2)$. The function for $T(x)$ has no $x$ terms in the denominator, so it will not possess an asymptote parallel to the $T$ axis. The function therefore stays at finite values throughout the given range.

**Step 2.** The original form of the function can be simplified slightly:

$$T(x) = \frac{x^2}{6} \left( 1 - \frac{x}{9} \right) = \frac{x^2}{6} - \frac{x^3}{54}.$$

Determining the first derivative,

$$\frac{\mathrm{d}T}{\mathrm{d}x} = \frac{\mathrm{d}}{\mathrm{d}x} \left[ \frac{x^2}{6} - \frac{x^3}{54} \right] = \frac{1}{6} \times 2 \times x^{2-1} - \frac{1}{54} \times 3 \times x^{3-1}$$

$$= \frac{x}{3} - \frac{x^2}{18} = \frac{x}{3} \left( 1 - \frac{x}{6} \right).$$

Turning points therefore appear when $(x/3) \times (1 - x/6) = 0$, so either $x = 0$ or $x = 6$. The turning points are $(0, 0)$ and $(6, 2)$.

We can now construct sign charts for the first derivative at these two points (see Tables 5.6 and 5.7), from which we see that $(0, 0)$ is a minimum and $(6, 2)$ is a maximum.

**Step 3.** The second derivative,

$$\frac{\mathrm{d}^2 T}{\mathrm{d}x^2} = \frac{\mathrm{d}}{\mathrm{d}x} \left[ \frac{x}{3} - \frac{x^2}{18} \right] = \frac{1}{3} \times 1 \times x^{1-1} - \frac{1}{18} \times 2 \times x^{2-1}$$

$$= \frac{1}{3} - \frac{x}{9}.$$

Constructing a sign chart for the second derivative test gives Table 5.8, from which we again see that $(0, 0)$ is a minimum and $(6, 2)$ is a maximum.

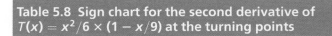

| $x$ | $-1$ | $0$ | $+1$ |
|---|---|---|---|
| $\dfrac{dT}{dx}$ | $\dfrac{(-1)}{3}\left(1 - \dfrac{(-1)}{6}\right) = -\dfrac{7}{18}$ | $0$ | $\dfrac{1}{3}\left(1 - \dfrac{1}{6}\right) = +\dfrac{5}{18}$ |
| | negative | | positive |
| slope | \ | — | / |

**Table 5.6 Sign chart for the first derivative of $T(x) = x^2/6 \times (1 - x/9)$ at the point (0, 0)**

The turning point is a minimum

**Table 5.7 Sign chart for the first derivative of $T(x) = x^2/6 \times (1 - x/9)$ at the point (6, 2)**

| $x$ | $+5$ | $+6$ | $+7$ |
|---|---|---|---|
| $\dfrac{dT}{dx}$ | $\dfrac{5}{3} \times \left(1 - \dfrac{5}{6}\right) = +\dfrac{5}{18}$ | $0$ | $\dfrac{7}{3} \times \left(1 - \dfrac{7}{6}\right) = -\dfrac{7}{18}$ |
| | positive | | negative |
| slope | / | — | \ |

The turning point is a maximum

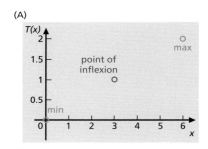

(A)

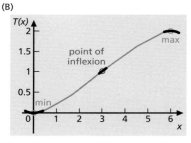

(B)

**Table 5.8 Sign chart for the second derivative of $T(x) = x^2/6 \times (1 - x/9)$ at the turning points**

| $x$ | $0$ | $6$ |
|---|---|---|
| $\dfrac{dT}{dx}$ | $0$ | $6$ |
| $\dfrac{d^2T}{dx^2}$ | $\dfrac{1}{3} - \dfrac{(0)}{9} = +\dfrac{1}{3}$ | $\dfrac{1}{3} - \dfrac{(6)}{9} = -\dfrac{1}{3}$ |
| | positive | negative |
| turning point | minimum | maximum |

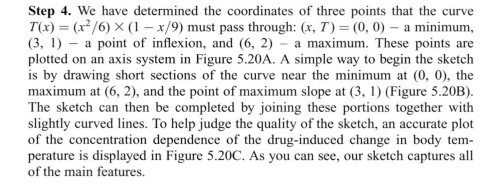

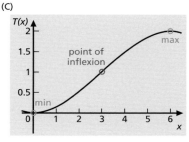

(C)

**Figure 5.20**
Completing the sketch of $T(x) = x^2/6 \times (1 - x/9)$ from the turning points and the point of inflexion.

The second derivative is zero when $(1/3) - (x/9) = 0$, giving $x = 3$. Therefore there is a point of inflexion with coordinates (3, 1).

**Step 4.** We have determined the coordinates of three points that the curve $T(x) = (x^2/6) \times (1 - x/9)$ must pass through: $(x, T) = (0, 0)$ − a minimum, $(3, 1)$ − a point of inflexion, and $(6, 2)$ − a maximum. These points are plotted on an axis system in Figure 5.20A. A simple way to begin the sketch is by drawing short sections of the curve near the minimum at (0, 0), the maximum at (6, 2), and the point of maximum slope at (3, 1) (Figure 5.20B). The sketch can then be completed by joining these portions together with slightly curved lines. To help judge the quality of the sketch, an accurate plot of the concentration dependence of the drug-induced change in body temperature is displayed in Figure 5.20C. As you can see, our sketch captures all of the main features.

# 5.10 Tangents

The fast track toward understanding how much a change in a variable will alter a particular function is to calculate its first derivative. However, as we saw in Section 5.3, the derivative also represents the gradient of the tangent to a graph of the function at a given point. We can use this fact to obtain the equation of a tangent to any curve whose equation we know. As Figure 5.21 illustrates, a tangent is a straight line that touches a curve at only one point, so to determine its equation we will need two pieces of information: the gradient of the curve and the coordinates of the point of contact.

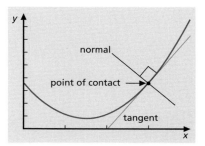

**Figure 5.21**
The tangent to a curve at any point is a straight line through the point and has the same gradient as the curve at that point. The normal is the straight line through the point of contact that is perpendicular to the tangent.

To illustrate how to apply this approach, we will work out the equation of the initial rate tangent for the reaction promoted by catalase, an enzyme that prevents the accumulation of toxic hydrogen peroxide in organisms that live in the presence of oxygen. Overall, the reaction catalyzed by catalase is:

$$H_2O_2 \rightarrow H_2O + \tfrac{1}{2}O_2.$$

The progress of the reaction can be monitored by measuring the volume of oxygen gas that evolves from the sample. The time course of product formation, displayed in Figure 5.22A, initially appears to be linear, but the rate gradually decreases as the reaction progresses. The initial rate is useful for comparing the kinetics of one reaction with another, because it will always be the same for a particular combination of enzyme and substrate at a given pH and temperature in a stated buffer solution.

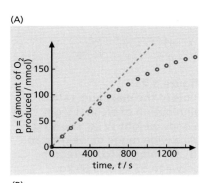

Over the first thousand seconds of the decomposition of hydrogen peroxide by catalase, the dependence of the amount of molecular oxygen produced $p(t)$ at a particular time $t$ is given to a good approximation by $p(t) = 0.19t - 0.00005t^2$. To calculate the initial rate of this reaction, we need the first derivative: $dp/dt = 0.19 - 0.00010t$.

At $t = 0$,

$$\frac{dp}{dt} = 0.19 - 0.00010 \times 0 = 0.19.$$

The tangent to the time course curve at $t = 0$ is a straight line, so its equation must have the form $p = At + B$. The gradient of the tangent, $A$, will be equal to the first derivative evaluated at $t = 0$. Thus $A = 0.19$. The value of the constant $B$ can be found because we know that the tangent has to pass through the point $(0, 0)$, so $0 = A \times (0) + B$ giving $B = 0$.

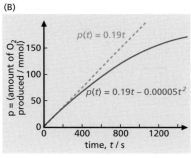

**Figure 5.22**
(A) A plot against time $t$ s of the amount $p$ mmol of oxygen gas that accumulates from the reaction $H_2O_2 \rightarrow H_2O + \tfrac{1}{2}O_2$.
(B) The curve $p(t) = 0.19t - 0.00005t^2$ used to model the volume of oxygen and the tangent to the curve at $t = 0$.

The equation of the initial rate tangent is therefore:

$$p(t) = 0.19t.$$

The product time course curve and the initial rate tangent line are shown in Figure 5.22B.

# 5.11 Linear approximations

It is often useful to have a way of working out the effect that making a small change in a variable has on a particular function. Differentiation provides a simple way of doing this in the form of a linear approximation based on the equation of a tangent to the curve. The following example outlines the method, although in this simple case it would be just as easy to obtain the result using a calculator.

A comparative study of the rate at which ethanol can be cleared from the blood of mammals (including rats, rabbits, sheep, and humans) reveals a strong dependence on body weight (Figure 5.23A). If the maximum elimination rate is $V$ g h$^{-1}$ for a body weight $W$ kg, then

$$V = 0.45 \times W^{0.71}.$$

Suppose we know that alcohol is cleared from the blood of a male student with a body weight of 80 kg at a maximum rate of 10 g h$^{-1}$. A nice idea would be to come up with some sort of approximation that would make use of the information we already have for the man to estimate the rate of elimination for a 70 kg female student. From Section 5.3 we know that the slope of the tangent is equal to the first derivative of the function at the point of contact. In the present case we can make an approximation that the slope of the curve

$$\frac{dV}{dW} \approx \frac{\Delta V}{\Delta W},$$ (EQ5.13)

where $\Delta W$ is a small change in the value of the variable $W$ and

$$\Delta V = V(W + \Delta W) - V(W).$$ (EQ5.14)

Rearranging EQ5.13 gives the expression:

$$\Delta V \approx \frac{dV}{dW} \Delta W.$$ (EQ5.15)

Merging the information in EQ5.14 and EQ5.15, we can show that:

$$V(W + \Delta W) = \Delta V + V(W) \approx \frac{dV}{dW} \Delta W + V(W).$$

The expression

$$V(W + \Delta W) \approx \frac{dV}{dW} \Delta W + V(W),$$ (EQ5.16)

is described as a linear approximation because it has the same form as the generic equation for a straight line:

$$y = Ax + B,$$

except that here $y$ represents $V(W + \Delta W)$, the new value of the function, which in this case is the rate of elimination for the female; $x$ represents $\Delta W$,

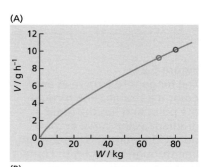

(A)

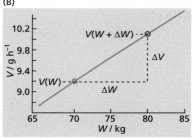

(B)

**Figure 5.23**
Plots of the maximum rate of elimination of alcohol $V$ g h$^{-1}$ for a body weight $W$ kg, showing (A) a wide range of weights and (B) zooming in to the region between 65 kg and 85 kg.

the small change in the value of the variable, corresponding to the weight difference of 10 kg; $A$ represents $dV/dW$, the gradient of the curve at the data point we have information for, whereas $B$ represents $V(W)$, the value of the function at the known point, the rate of elimination for the male.

Because $V(W) = 0.45 \times W^{0.71}$, we can use the formula for differentiating a power function to show that:

$$\frac{dV}{dW} = 0.45 \times 0.71 \times W^{0.71-1} = 0.32 W^{-0.29}.$$

Setting $V(W) = 10$, $W = 80$, and $\Delta W = -10$, we can now estimate that:

$$V(70) \approx 0.32 \times (80)^{-0.29} \times (-10) + 10 = -0.90 + 10 = 9.1.$$

According to our linear approximation, the female student has a maximum elimination rate of $9.10 \, g \, h^{-1}$. The value from the formula $V = 0.45 \times W^{0.71}$ is 9.19, so, interestingly, the approximation yields an error of less than 1 %, even though the weights of the students differ by 12.5 %.

# 5.12 Handling experimental errors

A similar approach can be used to estimate the effect of experimental errors. For example, under an optical microscope the radius, $r$, of an unfertilized sea-urchin egg is measured as 80 μm, to an accuracy of ±5 μm. Assuming that the cell is spherical, it would be useful to get a quick and approximate idea of the likely error in the cellular volume, $V$, calculated using the equation

$$V = \frac{4}{3} \pi r^3.$$

As before, we can approach an answer by using an approximate form of the first derivative:

$$\frac{dV}{dr} \approx \frac{\Delta V}{\Delta r}.$$

If we rewrite this approximation in the form

$$\Delta V \approx \frac{dV}{dr} \Delta r,$$

and set $\Delta r$ to be the error in the measurement of the radius, then $\Delta V$ will be an estimate of the *absolute error* in the volume calculation. The derivative of volume with respect to radius is:

$$\frac{dV}{dr} = \frac{4}{3} \pi \times 3 \, r^{3-1} = 4\pi r^2.$$

If $r = 80$ μm and $\Delta r = 5$ μm, then the absolute error in volume of the egg is:

$$\Delta V \approx 4\pi \times (80 \ \mu\text{m})^2 \times 5 = 4.0 \times 10^5 \ \mu\text{m}^3 \ = \ 4.0 \times 10^{-13} \ \text{m}^3.$$

Rather than dealing with the absolute error in a quantity, it is usually more useful to consider the *relative error*, $|\Delta V / V|$, or the *percentage error*, $100 \times |\Delta V / V| \ \%$. For our sea-urchin egg measurement we find that:

relative error, $\quad \left| \dfrac{\Delta V}{V} \right| = \dfrac{4\pi \ r^2 \Delta r}{(4/3) \ \pi \ r^3} = \dfrac{3 \ \Delta r}{r} = \dfrac{3 \times 5 \ \mu\text{m}}{80 \ \mu\text{m}} = 0.19$ and

percentage error, $\quad 100 \ \% \times \left| \dfrac{\Delta V}{V} \right| = \dfrac{300 \ \% \ \Delta r}{r} = \dfrac{300 \ \% \times 5 \ \mu\text{m}}{80 \ \mu\text{m}} = 19 \ \%.$

Another useful example illustrates how to estimate errors when looking at two inversely proportional quantities. Many biological methods make use of fluorescent tags: chemical markers that react with a strand of DNA or a protein can reveal the location of the target when illuminated with light at a characteristic frequency. The 'fluorescence lifetime' of a particular tag measures the average time that it stays in an electronically excited state. This lifetime is inversely proportional to the rate at which an electronically excited tag molecule decays into the ground state by emitting a photon. If the decay rate is characterized by a rate constant $k$, then the lifetime $\tau$ can be calculated using the equation:

$$\tau = \frac{1}{k}.$$

If the error in measuring the rate constant is $\Delta k$, then is it possible to quantify the corresponding error in the lifetime? We can use the approximation

$$\frac{d\tau}{dk} \approx \frac{\Delta \tau}{\Delta k},$$

to show that

$$\Delta \tau \approx \frac{d\tau}{dk} \Delta k.$$

Because we know how to express $\tau$ in terms of $k$, we can show that

$$\frac{d\tau}{dk} = 1 \times -1 \times k^{-2} = -k^{-2} = -\frac{1}{k^2}.$$

The error in the lifetime can therefore be estimated as:

$$\Delta \tau \approx -\frac{\Delta k}{k^2}.$$

For example, if the decay rate is measured as $(100 \ \pm \ 10) \times 10^6 \ \text{s}^{-1}$, then setting $k = 10^8 \ \text{s}^{-1}$ and $\Delta k = 10^7 \ \text{s}^{-1}$ reveals that:

$$\Delta \tau \approx -\frac{10^7 \ \text{s}^{-1}}{(10^8 \ \text{s}^{-1})^2} = -10^{7-16} \ \text{s} = -10^{-9} \ \text{s}.$$

In other words, when the error in measuring the decay rate is approximately $10^7 \ \text{s}^{-1}$, we should expect errors in the lifetime calculation to be of the order of 1 ns.

## Presenting Your Work

### QUESTION A

Measuring the peak flow rate while exhaling is a simple way to assess lung efficiency and whether respiratory passages are blocked. For women, the peak flow rate $p$ (measured in liters per minute) is related to age $a$ (in years) by the equation:

$$p = 425 + 3a - 0.05a^2.$$

(a) Find an expression for the rate of change of $p$ with respect to $a$ and evaluate it for a 50-year-old woman.

(b) At what age is the peak flow rate a maximum? What is the maximum peak flow rate?

---

(a) If $p = 425 + 3a - 0.05a^2$, then

$$\frac{dp}{da} = \frac{d}{da}(425) + \frac{d}{da}(3a) - \frac{d}{da}(0.05a^2) = 0 + 3 - 2 \times 0.05a = 3 - 0.1a.$$

For a woman with $a = 50$,

$$\frac{dp}{da} = 3 - 0.1 \times 50 = -2 \; \text{l min}^{-1} \text{y}^{-1}.$$

(b) $\frac{dp}{da} = 0$ when $3 - 0.1 \times a = 0$ so age is 30 years. Because $\frac{d^2 p}{da^2} = -0.1 < 0$ there is a maximum peak flow rate at 30 years.

This peak flow rate is $p = 425 + 3 \times 30 - 0.05 \times 30^2 = 470 \; \text{l min}^{-1}$.

**QUESTION B**

For the function $y = x^4 - 18x^2 + 56$ in the range $-4 \leqslant x \leqslant +4$, sketch a graph of $y$ against $x$, indicating all maxima, minima, and points of inflexion.

---

**Step 1**

**Analyze the function, $y(x)$.**

The first step is to evaluate the function at the boundaries of the given range.

When $x = \pm 4$, $y = (16)^2 - 18 \times 16 + 56 = +24$.

The second step is to find where the function cuts the $x$ and $y$ axes.

When $x = 0$, $y = +56$, so the curve does intersect with the $y$ axis.

To determine where the curve cuts the $x$ axis we need to solve the equation

$x^4 - 18x^2 + 56 = 0.$

The expression on the left-hand side of this equation can be expressed as the product of two quadratic functions, since

$x^4 - 18x^2 + 56 = (x^2 - 4) \times (x^2 - 14).$

Both of these quadratic expressions can be rewritten in the form of a difference of two squares, because

$(x + 2) \times (x - 2) = x^2 - 4$

and

$(x + \sqrt{14}) \times (x - \sqrt{14}) = x^2 - 14.$

The curve therefore cuts the $x$ axis at $x = -\sqrt{14}$, $x = -2$, $x = +2$, and $x = +\sqrt{14}$.

There are no asymptotes in the range $-4 \leqslant x \leqslant +4$.

**Step 2**

**Analyze the first derivative, $\dfrac{dy}{dx}$.**

Because $y = x^4 - 18x^2 + 56$, $\dfrac{dy}{dx} = 4 \times x^{4-1} - 18 \times 2 \times x^{2-1} + 0 = 4x^3 - 36x = 4x(x^2 - 9)$.

Local maxima and minima occur when $\dfrac{dy}{dx} = 0$

that is, when $x = 0$ and when $(x^2 - 9) = 0$, namely $x = \pm 3$.

When $x = \pm 3$, $y = (9)^2 - 18 \times 9 + 56 = -25$.

Local maxima and minima might therefore occur at $(-3, -25)$, $(0, +56)$, and $(+3, -25)$.

## Step 3

**Analyze the second derivative, $\dfrac{d^2y}{dx^2}$.**

Because $\dfrac{dy}{dx} = 4x^3 - 36x$,

$$\frac{d^2y}{dx^2} = 4 \times 3 \times x^{3-1} - 36 \times 1 \times x^{1-1} = 12x^2 - 36 = 12(x^2 - 3).$$

So, when $x = 0$, $\dfrac{d^2y}{dx^2} = -36$, which is negative, characteristic of a local maximum.

When $x = \pm 3$, $\dfrac{d^2y}{dx^2} = 12 \times (9 - 3) = +72$; this positive value indicates two local minima.

Points of inflexion occur when $\dfrac{d^2y}{dx^2} = 0$, namely when $(x^2 - 3) = 0$ and when $x = \pm\sqrt{3}$.

Thus there are points of inflexion at $x = \pm\sqrt{3}$ and $y = (3)^2 - 18 \times 3 + 56 = +11$.

## Step 4

**Sketch the graph of y(x).**

A sketch of the function $y = x^4 - 18x^2 + 56$ must therefore pass through the following crucial points:

(A) $(-4, +24)$: a boundary point.
(B) $(-3.74, 0)$: an intercept with the x axis.
(C) $(-3, -25)$: a local minimum.
(D) $(-2, 0)$: an intercept with the x axis.
(E) $(-1.73, +11)$: a point of inflexion.
(F) $(0, +56)$: an intercept with the y axis and a local maximum.
(G) $(+1.73, +11)$: a point of inflexion.
(H) $(+2, 0)$: an intercept with the x axis.
(I) $(+3, -25)$: a local minimum.
(J) $(+3.74, 0)$: an intercept with the x axis.
(K) $(+4, +24)$: a boundary point.

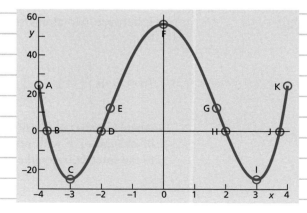

## End of Chapter Questions
(Answers to questions can be found at the end of the book)

### Basic

**1.** The table below contains information about the total number of cattle affected during an outbreak of foot and mouth disease.

| Time / weeks | 8 | 10 | 12 | 14 | 16 |
|---|---|---|---|---|---|
| Number affected | 1249 | 1501 | 1530 | 1598 | 1710 |

(a) Find the average rate of change in the number of cattle affected:
  (i) between week 8 and week 12;
  (ii) between week 10 and week 14; and
  (iii) between week 12 and week 16.
(b) Plot the data on a graph and estimate the instantaneous rate of change in week 12 by drawing a tangent line.

**2.** Use the first principles method to work out the derivative of $y$ with respect to $x$ for the following functions:
(a) $y = x^2 + x - 6$
(b) $y = x^3$

**3.** The electrical current $I$ passing through the core of an axon (nerve fiber) is defined as the flow of charge $q$ per unit time $t$ and can be summarized by the relationship:

$$I = \frac{dq}{dt}.$$

Use first principles differentiation to evaluate the current
(a) when $q = A$
(b) when $q = Bt$

where $A$ and $B$ are constants.

**4.** Differentiate the following expressions with respect to $x$ using the power function formula:
(a) $y = 4x^8$
(b) $y = -x^{12}$
(c) $y = x/\pi$
(d) $y = x^{-1/3}$
(e) $y = 4\sqrt{x}$

**5.** If $F(r) = \Delta P \pi r^4 / 8\eta L$, what is $dF/dr$?

**6.** Differentiate the following expressions with respect to $x$:
(a) $y = x^2 + 3x - 5$
(b) $y = x^{-1/3} - x - 6$
(c) $y = (x^3 + 1)/x^2$
(d) $y = (2/x) - 3\sqrt{x}$
(e) $y = 5x^4 - 1/x^3$

**7.**
(a) If the height of a child in meters is represented by $x$ and its age in years is $t$, what does $d^2x/dt^2$ represent and what units are needed to convert its numerical value to the corresponding physical quantity?
(b) If $x(t)$ denotes the position of a red blood cell in a vein, measured in mm at time $t$ ms, then what does $d^3x/dt^3$ represent and what units are needed to convert its numerical value to the corresponding physical quantity?

**8.** Consider the graph below. It shows the reaction coordinate diagram for a hypothetical enzyme-catalyzed reaction in which enzyme E converts a single substrate S into a single product P. Locate all minimum and maximum turning points and points of inflexion on the diagram.

Reaction coordinate diagram for an enzyme-catalyzed reaction: E, enzyme; S, substrate; ES, enzyme–substrate complex; ES$^‡$, transition state; EP, enzyme–product complex; P, product.

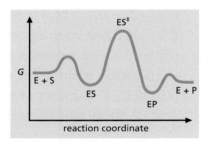

### Intermediate

**9.** The Lennard–Jones equation for the potential energy $V$ between two atoms separated by a distance $r$ is given by

$$V(r) = 4\varepsilon \left[ \left( \frac{\sigma}{r} \right)^{12} - \left( \frac{\sigma}{r} \right)^6 \right],$$

where $\varepsilon$ and $\sigma$ are positive constants. Determine the rate of change of $V$ with respect to $r$ (which is proportional to the force between the atoms).

**10.** According to Boyle's Law, if the temperature of a confined gas is held fixed, then the product of the pressure $P$ and the volume $V$ is a constant. Suppose that, for a certain gas sample, $PV = 8 \times 10^{-5}$ N m.
(a) Find the average rate of change of $P$ as $V$ is increased from 200 to 250 cm$^3$.
(b) Express $V$ as a function of $P$ and show that the instantaneous rate of change of $V$ with respect to $P$ is inversely proportional to the square of $P$.
(c) Evaluate the instantaneous rate of change of $P$ with respect to $V$ for $V = 225$ cm$^3$.

**11.** The population $N$ of rabbits on a small island is monitored by ecologists during an 11-year study:

| Time, $t$/yr | 0 | 1 | 2 | 3 | 4 | 5 |
|---|---|---|---|---|---|---|
| Population, $N$ | 100 | 157 | 233 | 333 | 452 | 578 |
| Time, $t$/yr | 6 | 7 | 8 | 9 | 10 | 11 |
| Population, $N$ | 692 | 789 | 861 | 910 | 947 | 967 |

(a) Plot a graph of $N$ against $t$ using a vertical axis range of $-800 \leqslant N \leqslant 1200$. Explain the significance of the shape of this curve.
(b) Calculate the average rate of change of population with respect to time ($\Delta N/\Delta t$) for each year of the study. Then plot $5 \times (\Delta N/\Delta t) / $ yr$^{-1}$ on the same axes as your original plot and account for the shape of this graph.
(c) Determine the rate of change of $\Delta N/\Delta t$ with respect to time, then plot $5 \times (\Delta^2 N/\Delta t^2) / $ yr$^{-2}$ on the same axes and, finally, explain the shape of this graph.

**12.** One hour after taking $x$ mg of a drug, the body temperature in degrees Celsius of a patient is given by the equation

$$T = T_0 - 0.00625x^2(18 - x),$$

where $T_0$ is the initial body temperature. Determine the value of $x$ that produces the greatest drop in body temperature, and the magnitude of the temperature change.

**13.** The concentration $c$ (in milligrams per liter) of phytoplankton in seawater varies with the depth $d$ (in meters) according to the equation:

$$c = -d(d^{1/2} - 3) + 4,$$

over the range $0 \leqslant d \leqslant 10$. What is the maximum concentration of phytoplankton?

**14.** The response $R$ of a body to a dose $D$ of a certain drug has been found to be

$$R = D^2\left(\frac{C}{2} - \frac{D}{3}\right),$$

where the maximum permitted dose, $C$, is a positive constant. If the sensitivity $S$ of the patient to the drug is defined as the rate of change of $R$ with respect to $D$, determine the value of $D$ at which the sensitivity is a maximum.

**15.** Hexokinase, an important enzyme in energy metabolism, catalyzes the conversion of glucose into glucose-6-phosphate (G6P). For the first hour after a meal that is rich in carbohydrates, the concentration, $c$, of G6P (measured in millimoles per liter) at time $t$ (measured in minutes) is given by:

$$c = 0.5t - 0.002t^2.$$

Determine the equation of the tangent to this time course curve at $t = 0$.

**16.**
(a) Find the equation of the tangent to the curve $y = x^3 - 1$ at the point where $x = -1$.
(b) A line that is perpendicular to the tangent to a curve and intersects with the curve at the same point of contact is called a 'normal'. Interestingly, when two straight lines are perpendicular, the product of their gradients is equal to $-1$. Find the equation of the normal to the curve $y = x^3 - 1$ at the point where $x = +1$.
(c) Find the coordinates of the point at which these two straight lines intersect.

**17.** Draw a diagram to show that there are two tangent lines to the parabola $y = x^2$ that pass through the point $(0, -4)$. Find the coordinates of the points where these tangent lines intersect the parabola.

**18.** In laminar flow of blood through a cylindrical artery, the resistance $R$ is inversely proportional to the fourth power of the radius $r$. Use a linear approximation to show that if $r$ is decreased by 2 %, $R$ will increase by approximately 8 %.

## Advanced

**19.** The ability to make rapid changes in direction plays an important role in helping fish to avoid predators. For two species, rainbow trout and green sunfish, the distance $x$ (measured in centimeters) covered at time $t$ (in seconds) after a stimulus is described by the equation:

$$x = kt^b,$$

with $k = 300$ cm s$^{-1.6}$ and $b = 1.60$ for trout, and $k = 210$ cm s$^{-1.71}$ and $b = 1.71$ for sunfish. Compare the distance traveled, the instantaneous velocity, and the instantaneous acceleration of the two fish at $t = 0.1$ s. Given that the lengths of the trout and sunfish are 14.4 and 8.0 cm, respectively, compare the velocities at $t = 0.1$ s in terms of fish lengths per second.

**20.** Predator–prey models are often used to study the interactions between species. Consider a population of lions at time $t$, given by $L(t)$, and zebra, given by $Z(t)$, in the Okavango Delta in Botswana. The interaction between the two species has been modeled by the equations $dZ/dt = aZ - bZL$ and $dL/dt = -cL + dZL$, where $a$, $b$, $c$, and $d$ are positive constants.

(a) What values of $dZ/dt$ and $dL/dt$ correspond to stable populations?
(b) How would the statement 'zebra go extinct' be represented mathematically?
(c) Suppose that $a = 0.05$, $b = 0.001$, $c = 0.05$, and $d = 0.00001$. Find all population pairs $(Z, L)$ that lead to stable populations. According to this model, is it possible for the two species to live in harmony, or will one or both species become extinct?

**21.** The body mass, $M$, of a polar bear is related to its length, $L$, by the equation

$$M = 7L^5,$$

where $M$ is measured in kilograms and $L$ in meters. Use a linear approximation to estimate the percentage error in the calculated body mass of a 2 m polar bear if the measurement of its length is only 99 % accurate.

# CHAPTER 6

# Integration

To get a clear picture of what integration is about, this chapter will treat it like a modern art installation that begins to make sense only after being gazed at from several points of view. The simplest approach is just to define integration as being *the inverse of differentiation*. This may give the impression that integration is rather like the 'undo' function common to many software packages.

**Figure 6.1**
*Calculus* is the Latin word for a pebble.

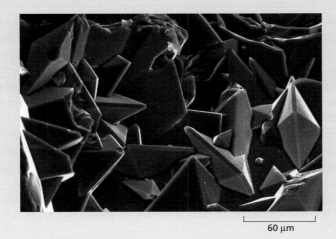

60 μm

**Figure 6.2**
In medicine, the word *calculus* refers to a hardened deposit, such as a kidney stone. Image courtesy of Tammy Bonstein, University of Manitoba.

The second and perhaps most useful way of thinking about integration is as *a process of adding up*. 'Calculus' is the Latin word for 'pebble' (Figure 6.1) and its root is the word 'calx', which means 'limestone'. The modern use of 'calculus' in mathematics seems to reflect the ancient Babylonian practice of arranging several pebbles in patterns, to teach arithmetic and to facilitate trade. In medicine, however, a 'calculus' indicates a hardened deposit within the body, such as a kidney stone (Figure 6.2). These are formed by the accumulation of many tiny crystalline flakes of calcium oxalate, which precipitate from urine as it passes through the kidney. Just as a whole kidney stone is the result of adding together all of its individual component flakes, useful physical quantities can be obtained by summing an appropriate collection of small contributions. For example, the total number of cattle that have been infected by the foot and mouth virus can be determined by summing together the number of cases confirmed on each day during an outbreak (Figure 6.3).

**Figure 6.3**
Number of confirmed cases of foot and mouth disease in cattle on each day during an outbreak.

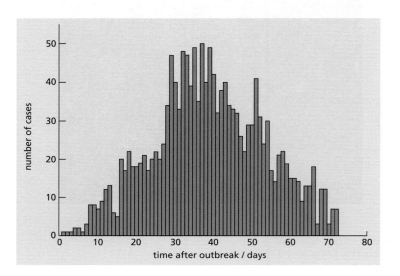

The graphical approach to differentiation introduced in Chapter 5 revealed that a derivative is the slope of the tangent to a curve at a particular point. By contrast, a third viewpoint on integration is to see it as a way of obtaining the *area under a curve*. An example from the field of pharmacokinetics shows how useful this process can be. A curve of the blood plasma concentration of a drug can be plotted as a function of time following an oral dose (Figure 6.4), and a similar curve can be obtained after an intravenous dose. The ratio of the areas under these two curves provides a way of measuring the fraction of the oral dose absorbed by the body. It is important that the total drug concentration is monitored carefully so that the correct dose is given to patients and toxic side-effects are minimized. Similarly, the area under a graph of the velocity of a vehicle against time can be analyzed to discover the distance that has been traveled at various times, as shown in Figure 6.5.

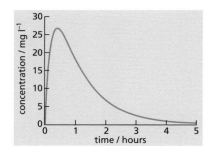

**Figure 6.4**
Concentration of gentamycin in blood plasma as a function of time after administration of an oral dose.

After examining the implications of all three perspectives on integration, this chapter also discusses how to understand the effects of cumulative change, as well as how to determine the mean value of a continuous function.

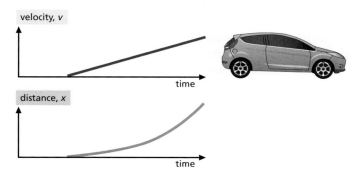

**Figure 6.5**
A graph of distance against time can be determined from the area under the velocity graph.

# 6.1 Undoing the effects of differentiation

Many mathematical operations can be matched with an inverse process, such as addition and subtraction, multiplication and division, or raising to a power and taking a root (for more examples, see Section 3.4). In the previous chapter, we found that by starting with a function of a certain variable and following the rules for differentiation we could obtain a new function, the derivative. The derivative represents the instantaneous rate of change, measuring how rapidly the initial function will alter when the value of a particular variable is perturbed.

However, in many situations it is easier to determine the rate at which a quantity changes than to measure how much of it has happened at a particular moment in time. For example, the population of a country is often difficult to monitor directly. It can be measured by a census, but citizens will only put up with intrusive questions once in each decade. Assuming that there is no immigration, the population between these data points can be determined by analyzing the birth rate and the death rate, which are easy to count if the law makes registering these events compulsory (Figure 6.6).

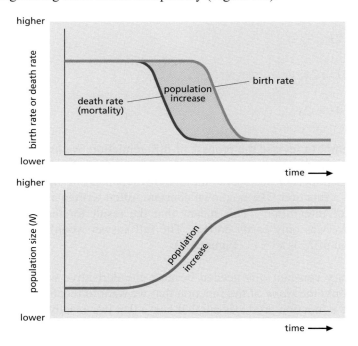

**Figure 6.6**
The relationship between rates of population change, due to births and deaths, and the population size.

Similarly, in a study of the ability of wetlands to trap nutrients it may be difficult to quantify the concentration of organic nitrogen species accumulating in a marsh. It is much less taxing to measure the rate at which water flows in from a polluted feeder stream and out through the river that drains the system. As in these examples, we often start out with information about the derivative of a function without prior knowledge of the equation from which the derivative is derived.

The name given to the process of reconstructing an original function from its derivative is **integration**. In other words, integration is the operation that must be performed to undo the effects of differentiation. We can start to unpack this concept by differentiating the function

$$y = x^2,$$

to achieve the familiar result that

$$\frac{dy}{dx} = 2x.$$

Integration is the reverse of this process, so alongside stating that 'the derivative of $x^2$ is $2x$', it looks like we should also be able to say 'the integral of $2x$ is $x^2$'.

Unfortunately, this picture is a little too simplistic. The true situation is more complex because an infinite number of functions can in fact be differentiated to get a result of $2x$. For example, we would get the same result if instead of $y = x^2$ we had chosen $y = x^2 + 3$ or even $y = x^2 - 14$. In each case, the derivatives of the second terms on the right-hand sides of the equations are zero, because they are all constants. Writing out these derivatives explicitly can emphasize the ambiguity of attempting to undo the effects of differentiation:

$$\frac{d}{dx}(x^2) = 2x;$$

$$\frac{d}{dx}(x^2 + 3) = 2x;$$

$$\frac{d}{dx}(x^2 - 14) = 2x.$$

In any attempt to reverse the process of differentiation, we have no idea what the value of the constant term in the original function may have been. Because of this, it is important to include an unknown constant in our answer for the integral of a function. The unknown constant, often written as $c$, is called the 'constant of integration'. This means that the result for the integral of $2x$ stated above was only partly correct. The full answer would be to state that 'the integral of $2x$ is $x^2 + c$' (Figure 6.7).

The arbitrary constant $c$ is needed because the derivative we start out with describes only the *slope* of the function that we want to reconstruct. Adding a constant to a function produces another curve that has exactly the same shape (that is, the same derivative or slope), but has just been displaced upward or

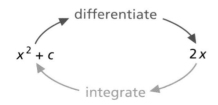

**Figure 6.7**
The inverse relationship between integration and differentiation.

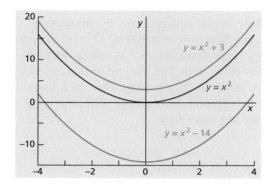

downward along the $y$ axis by a certain amount. Figure 6.8 illustrates this for $y = x^2$, $y = x^2 + 3$, and $y = x^2 - 14$. The slopes and shapes of these three functions are identical, but each has a different offset from the $y$ axis, so the value of each function will be different for any given value of the abscissa ('horizontal') variable. The constant of integration can only be evaluated if we are provided with additional information about a particular case.

The mathematical nomenclature for integration involves an unusual long s symbol, $\int$, otherwise known as an *integral sign*. Alongside the integral sign is another compulsory symbol, '$dx$', which must always be included because it indicates the name of the variable that is involved in the integration process – in this case $x$. Just as we saw with Leibniz's $dy/dx$ notation for the derivative in Section 5.3, the '$dx$' symbol here is the mathematical abbreviation for 'an infinitesimal increment in $x$'. In words, an equation of the sort shown in Figure 6.9 tells us that 'the integral of $2x$ with respect to $x$ is $x^2 + c$'. Note that the function being integrated, $2x$ in this case, is called the *integrand*. Technically, this whole expression is termed an *indefinite integral* to help distinguish it from the *definite integrals* that we will encounter shortly in Section 6.3.

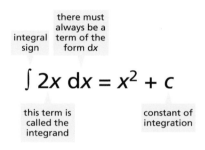

there must always be a integral term of the sign form d$x$

$$\int 2x \, dx = x^2 + c$$

this term is called the integrand

constant of integration

**Figure 6.9**
An indefinite integral consists of an integral sign, an integrand, and a differential.

Another way of describing the indefinite integral of a function $f(x)$ is as an 'antiderivative function', which is often written using the same label, but in upper case: $F(x)$. This allows us to write the defining feature of the antiderivative concisely as:

$$\frac{dF(x)}{dx} = f(x). \qquad (EQ6.1)$$

This means that an antiderivative is any function that when differentiated takes us back to the original function $f(x)$. The number of antiderivative functions is infinite, but they are all very closely related. If we know one of them, we can obtain any of the others simply by adding a constant to it. The new function will also be an antiderivative of $f(x)$ because the derivative of any constant term disappears. So, if $F(x)$ is one of the antiderivatives of the function $f(x)$, we can use it to write a general expression for an indefinite integral:

$$\int f(x) \, dx = F(x) + c. \qquad (EQ6.2)$$

We will apply these ideas straightaway by attempting to reconstruct the antiderivative of the function $r = 0.006\,t^{0.5}$, in which $r$ (measured in milligrams per hour) represents the rate at which the mass of a kidney stone increases with time $t$ (in hours). The integral of this expression with respect to time will make it possible to calculate the total mass of the crystalline aggregate. Using the general formula for differentiating a power function from Section 5.4, we can show that

$$\frac{d}{dt}(t^{1.5}) = 1.5t^{0.5}.$$

Differentiating $t^{1.5}$ makes a promising start, because it yields a term that contains $t^{0.5}$, which is what we need in our final answer. Unfortunately, $t^{0.5}$ is accompanied by a factor of 1.5, rather than the factor of 0.006 we should be aiming for. It therefore looks like we need to start with a $t^{1.5}$ term, multiply it by 0.006, and then divide by 1.5 before finally differentiating to get the proper result, because

$$\frac{d}{dt}\left(\frac{0.006}{1.5}t^{1.5}\right) = \frac{d}{dt}(0.004\ t^{1.5}) = 0.004 \times 1.5\ t^{0.5} = 0.006\ t^{0.5}.$$

We are now in a position to write out the correct form of the antiderivative, taking care to include a constant of integration:

$$\int 0.006t^{0.5}\,dt = 0.004t^{1.5} + c.$$

The arbitrary constant introduced by the process of integration can usually be determined if enough information is supplied. In this case, suppose we know that the mass $m$ of the kidney stone in milligrams is 50 at the start of the experiment when $t = 0$. If

$$m(t) = 0.004t^{1.5} + c,$$

then when $t = 0$, $m(0) = c = 50$. Knowing the value of the constant should therefore make it possible for us to predict the size of the stone at any time in the future, using the equation:

$$m(t) = 0.004t^{1.5} + 50.$$

## Box 6.1 Rules for integration

Because integration is the inverse of differentiation, we can use the basic rules of differentiation to write down the following simple rules for integration.

**Rule 1**

$$\int k\ f(x)\,dx = k\int f(x)\,dx.$$

**Rule 2**

$$\int [f(x) + g(x)]\,dx = \int f(x)\,dx + \int g(x)\,dx.$$

**Rule 3**

$$\int [f(x) - g(x)]\,dx = \int f(x)\,dx - \int g(x)\,dx.$$

## 6.2 Integrating simple expressions

By extending the same approach to determine the antiderivatives of a whole range of power functions of $x$, the following general formula can be deduced:

$$\int x^n \, dx = \frac{x^{n+1}}{n+1} + c,$$

provided that $n$ does not equal $-1$.

In other words, to integrate the function $x^n$, we must

(1) *increase* the value of the index ($n$) by 1, giving ($n + 1$);
(2) then *divide* the result by the new value of the index ($n + 1$);
(3) and, finally, *add* a constant of integration ($c$).

It is worth noting that this rule applies for all integer and non-integer values of the index $n$, except for $n = -1$. We will explore how to integrate functions involving $1/x$ in Chapter 7.

The best way to get the hang of the integration rule is by tackling some examples of different power functions. Note that it is sometimes useful to rewrite the original expression to produce a form that can be integrated more easily.

(1) *Positive integer values of n*:

If $y = x^{33}$, then

$$\int y \, dx = \int x^{33} \, dx = \frac{1}{33+1} x^{33+1} + c = \frac{1}{34} x^{34} + c.$$

(2) *Reciprocals* (*negative values of n*):

If $y = 1/x^2$ then we can write $y = x^{-2}$. Therefore

$$\int y \, dx = \int x^{-2} \, dx = \frac{1}{-2+1} x^{-2+1} + c = -x^{-1} + c = -\frac{1}{x} + c.$$

(3) *Roots* (*fractional values of n*):

If $y = 1/\sqrt[3]{x}$ then we can write $y = x^{-1/3}$. Therefore

$$\int y \, dx = \int x^{-1/3} \, dx = \frac{x^{1-1/3}}{1-1/3} + c = \frac{3}{2} x^{2/3} + c .$$

As we found for differentiation in Section 5.5, the general formula for integrating a power function needs to be supplemented by further rules that enable us to handle a wider range of functions (see Box 6.1). First, we will consider the integral of a function of $x$, $f(x)$, multiplied by a constant, $k$:

$$\int k\, f(x)\, \mathrm{d}x = k \int f(x)\, \mathrm{d}x.$$

A constant factor in an integral can be moved outside the integral sign, and the integrand can then be dealt with in the usual way by applying the general formula. We can demonstrate this by evaluating the integral $I = \int 12x^3\, \mathrm{d}x$:

$$I = \int 12x^3\, \mathrm{d}x = 12 \times \int x^3\, \mathrm{d}x = 12 \times \left\{ \frac{x^{3+1}}{3+1} + c \right\}$$

$$= 3x^4 + K,$$

where $K$ is a constant equal to $12c$. Because we do not know the value of the constant of integration $c$, it does not make much sense to include a term like '$12c$' in the final answer. Constants of integration are usually expressed in the simplest possible way and are often lumped together into a single term.

## Box 6.2 Incidence and mortality

Epidemiologists refer to both the *incidence* of disease and the *mortality* of disease.

*Incidence:*

The incidence of a disease is defined as the number of cases of the illness that are diagnosed in a specified population during a given period of time (typically a year).

*Mortality:*

The mortality of a disease is defined as the number of deaths that occur in a specified population during a given period of time (typically a year).

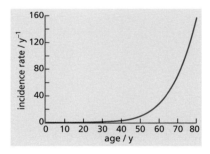

**Figure 6.10**
Graph of incidence of colon cancer per 100 000 women as a function of age.

An example from epidemiology would be to analyze the incidence rate (see Box 6.2) of a particular form of colon cancer (Figure 6.10). If $R(t)$, the incidence per 100 000 women per year as a function of age in years, $t$, is given by

$$R(t) = 7 \times 10^{-10} t^6,$$

then $I$, the integral of $R$ with respect to $t$, will be related to the total number of sufferers younger than a given age:

$$I = \int R\, \mathrm{d}t = \int 7 \times 10^{-10}\ t^6\, \mathrm{d}t$$

$$= 7 \times 10^{-10} \times \int t^6\, \mathrm{d}t = 7 \times 10^{-10} \times \frac{t^{6+1}}{6+1} + c$$

$$= 1 \times 10^{-10} t^7 + c.$$

Note that these are really numerical value equations, so we should have used curly brackets, such as $\{t\}$; however, this would have made the mathematics look *very* complicated!

This disease manifests itself late in life because a range of different somatic mutations are required for cancer to arise. Such mutations can occur whenever a cell divides, so older people who have undergone more cell divisions accumulate genetic damage and have a greater chance of developing cancer. The large positive value found for the power of $t$ here causes the number of cases to increase very slowly in the early years of life, but rapidly in later years. The chance of being diagnosed with colon cancer increases dramatically for women aged over 40 years.

A similar approach can be used to show that the antiderivative of a constant, such as $k$, is a linear function of the variable (for example $x$):

$$\int k \, dx = k \int 1 \, dx = k \int x^0 \, dx = k \left( \frac{x^{0+1}}{0+1} \right) + c = kx + c.$$

Here it is important to rewrite the integrand in a way that reveals it to be a function of $x$, by remembering that $x^0 = 1$. We will look again at the significance of this result in Section 6.4.

Next, to integrate a sum of two functions, such as $f(x)$ and $g(x)$, we need to integrate each term separately:

$$\int [f(x) + g(x)] \, dx = \int f(x) \, dx + \int g(x) \, dx.$$

This is the counterpart to the rule for differentiating the sum of two functions encountered in Section 5.5. To summarize the definition in words rather than symbols, we could say that the integral of a sum of functions is equal to the sum of the integrals of those functions.

For example, we might want to find the integral of the linear-quadratic dose–response relationship $n = Ar + Br^2$ that we came across in Section 5.5, where $n$ represents the average number of damaged chromosomes in a preparation of white blood cells that have been exposed to a dose of X-ray radiation $r$:

$$\int n \, dr = \int (Ar + Br^2) \, dr = \int Ar \, dr + \int Br^2 \, dr$$

$$= A \int r \, dr + B \int r^2 \, dr = \left( A \times \frac{r^{1+1}}{1+1} + c' \right) + \left( B \times \frac{r^{2+1}}{2+1} + c'' \right)$$

$$= \frac{Ar^2}{2} + \frac{Br^3}{3} + (c' + c'') = \frac{Ar^2}{2} + \frac{Br^3}{3} + c.$$

Although the original integral was separated into the sum of two individual integrals, only a *single* constant of integration, $c$, is needed in the final answer. All of the individual constants ($c'$ and $c''$) have been collected together into one overall constant ($c = c' + c''$).

A straightforward extension of this summation rule is to consider how to integrate the difference of two functions:

$$\int [f(x) - g(x)]\,dx = \int f(x)\,dx - \int g(x)\,dx.$$

This definition tells us that the integral of the difference between two functions is equal to the difference of the integrals of those functions.

**Figure 6.11**
A flock of migratory birds.

Suppose that birdwatchers have noticed that a certain species of bird gathers in its summer territory at a rate given approximately by $12t - 3t^2$, where $t$ is the number of months from the beginning of April (Figure 6.11). Given that the first bird arrives during April, we might want to know when all the mature birds have departed.

First we need to find an expression for the size of the mature bird population, so let us give this the label $p(t)$. The rate of change in the population is given by

$$\frac{dp(t)}{dt} = 12t - 3t^2.$$

This expression is a simple example of a *differential equation*, which we will consider in more detail in Chapter 12. We can determine $p(t)$, the population at any time, $t$, by integrating the derivative $dp/dt$ with respect to $t$. As before, we can tackle this integral by splitting the relationship into two separate terms:

$$p(t) = \int \frac{dp}{dt}\,dt = \int (12t - 3t^2)\,dt = \int 12t\,dt - \int 3t^2\,dt$$

$$= 12\int t\,dt - 3\int t^2\,dt = 12 \times \frac{t^2}{2} - 3 \times \frac{t^3}{3} + c = 6t^2 - t^3 + c.$$

After some practice, it should be possible to write down expressions for many integrals of various terms in a series of sums and differences directly, without having to spell out each step explicitly.

On this occasion, we have already been supplied with enough additional information to work out what the value of the constant of integration should be. At the beginning of April, when $t = 0$, no birds have arrived, so $p(0) = 0$. We can therefore write that:

$$0 = 6(0)^2 - (0)^3 + c.$$

As a consequence, $c = 0$, so the equation for the number of mature birds in the colony $t$ months after the beginning of April becomes

$$p(t) = 6t^2 - t^3.$$

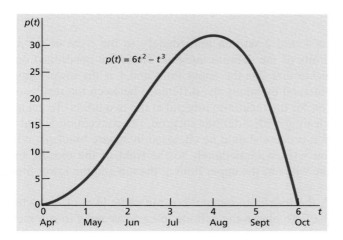

**Figure 6.12**
Graph of the population of migratory birds according to the model $p(t) = 6t^2 - t^3$.

A sketch of this function is shown in Figure 6.12. All the birds have left when

$$p(t) = 6t^2 - t^3 = t^2(6 - t) = 0,$$

in other words when $t = 0$ or when $t = 6$. Logic tells us that departures can only occur *after* the birds have arrived at $t = 0$. This means that all of the birds must have migrated by the beginning of October ($t = 6$). It is worth pointing out that the function $p(t)$ is not applicable for values of $t$ outside the range $0 \leqslant t \leqslant 6$, because a negative value for a population of birds would have no meaning.

Before progressing to the next section you should make sure you can do questions 1 and 2 from the End of Chapter Questions.

## 6.3 Definite and indefinite integrals

In Section 6.1 we encountered the idea of an 'indefinite integral':

Indefinite integral: $\displaystyle\int 2x \, dx = x^2 + c.$

The reason why this expression is given the label 'indefinite' is the inherent ambiguity of the integration process. There is no way of knowing what the actual value of the constant of integration $c$ will be unless additional information is provided, even if we know exactly how the antiderivative function depends on the variable.

This ambiguity can be resolved by specifying two values of the variable that represent limits between which the integral is to be evaluated. The result is called a 'definite integral', and some extra nomenclature is necessary to describe precisely what is going on. If we start at a lower limit of $x = 1$ and finish at an upper limit of $x = 2$, then we would write the definite integral as:

Definite integral:   $\displaystyle\int_1^2 2x\,\mathrm{d}x = \left[x^2 + c\right]_1^2.$

The numbers 1 and 2 written immediately to the right of the integral sign define the limits of the definite integral. These are positioned one above the other to indicate that 2 is the upper limit and 1 is the lower limit. A definite integral is obtained by taking the difference between the results of evaluating the expression for the indefinite integral at the two limits. To act as a reminder that we are dealing with a definite integral, the convention is that the result of the integration process should be enclosed in square brackets, with the limits of integration written immediately to the right of the closing bracket, once again with the value of the upper limit at the top and the lower limit below.

When $x = 2$, the indefinite integral has the value $[4 + c]$, but when $x = 1$ the value is $[1 + c]$. To obtain the definite integral we need to subtract the second expression (evaluated at the lower limit) from the first (evaluated at the upper limit). As we do this, the ambiguity inherent in the integration process disappears. Because the constant of integration appears in both of the expressions in square brackets, the actual value of $c$ no longer matters, because it disappears on taking the difference. Putting all of this together, we can now write down the process for obtaining the definite integral:

$$\int_1^2 2x\,\mathrm{d}x = \left[x^2 + c\right]_1^2 = [4 + c] - [1 + c] = 3.$$

When an indefinite integral is being solved, the constant of integration must always be included in the final answer. However, when evaluating a definite integral the constant will always cancel out in the way we have just seen. The constant of integration is therefore often omitted when the expression for a definite integral is written down between a pair of square brackets. It is therefore just a matter of personal preference whether or not you decide to include the constant of integration throughout a calculation or to omit it. It is never wrong to include a constant of integration, no matter what type of integral is being evaluated, so you could choose to make a point of always leaving it in, just to ensure that the constant is present when it has to be there (that is, when solving an indefinite integral).

If this attempt to define a definite integral seems confusing, do not worry! The following section gives a much more visual interpretation. However, in the meantime it is worth remembering that the indefinite integral of $f(x)$ can be expressed as:

$$\int f(x)\,dx = F(x) + c,$$

where $F(x)$ is any one of the infinite number of antiderivatives, chosen for convenience. If we now consider the definite integral of $f(x)$ between a lower limit of $x = a$ and an upper limit of $x = b$, we get the following result:

$$\int_a^b f(x)\,dx = \left[F(x) + c\right]_a^b = [F(b) + c] - [F(a) + c] = F(b) - F(a).$$

However, if the upper and lower limits of the integral are exchanged, then the answer is different:

$$\int_b^a f(x)\,dx = \left[F(x) + c\right]_b^a = [F(a) + c] - [F(b) + c] = F(a) - F(b).$$

This demonstrates that:

$$\int_a^b f(x)\,dx = -\int_b^a f(x)\,dx.$$

This is an important result: if you exchange the limits of integration in a definite integral, then the sign of the answer will change.

To close this section, it is worth pointing out that performing an indefinite integral produces an answer that is *a function*. This can have any value, not only due to the presence of the undefined constant of integration, but also because it can be evaluated at any value of the variable $x$. By contrast, if the upper and lower limits are both numbers, a definite integral essentially defines *a specific number* – the constant of integration cancels out when the difference is taken between the antiderivative evaluated at two particular values of the variable.

Before progressing to the next section you should attempt questions 3 and 4 from the End of Chapter Questions.

## 6.4 The area under a curve

Figure 6.13 shows once again the dependence of the population of migratory birds on the time after arrival at their colony. However, this time the graph is in disguise, acting as a more general example of the class of functions, $y(x)$, that remain positive for all values of their variable $x$ within the region of interest. Thus $y$ is used instead of $p$ and $x$ instead of $t$. The area of the region bounded by this curve, the $x$ axis, and the two vertical lines at $x = x_0$ and $x = x_1$ is designated $A$.

At the moment we have no way of knowing the value of area $A$, but we can ask ourselves a slightly easier question: what would be the change in this area, $\delta A$, if the value of $x_1$ is increased to $x_1 + \delta x$? This change is illustrated in Figure 6.14. Apart from the small, approximately triangular region that is

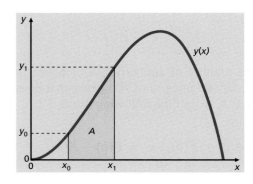

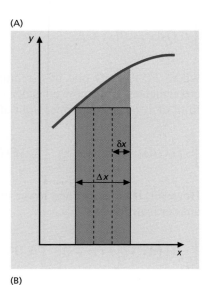

(A)

**Figure 6.13**
The area under the curve between $x_0$ and $x_1$ is shown in color (A).

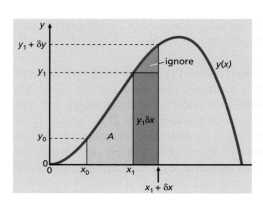

**Figure 6.14**
The change in area, $\delta A$, if the value of $x_1$ is increased to $x_1 + \delta x$, is approximately equal to $y_1 \delta x$.

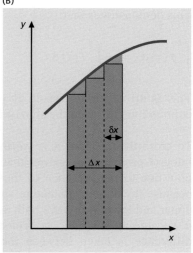

(B)

**Figure 6.15**
Replacing a single rectangular strip of width $\Delta x$ by three narrower strips of width $\delta x$ gives a better approximation to the correct area under the curve.

shaded in blue, all of the increase in area is accounted for by a tall, thin, green rectangle of height $y_1$ and width $\delta x$, which has an area of $y_1 \delta x$. We can therefore make an estimate of the increase in area:

$$\delta A \approx y_1 \, \delta x. \tag{EQ6.3}$$

If the increment $\delta x$ is made smaller, then the blue triangle will decrease rapidly in area – not only in absolute terms, but also as a fraction of the area of the green rectangle ($y_1 \delta x$). The approximation that $\delta A \approx y_1 \delta x$ is therefore rather accurate when $\delta x$ is very small, and in the limit when $\delta x$ approaches zero the approximation becomes exact. However, at first glance this might not appear to be much use to us, especially if we want to determine the value of $\delta A$ when $\delta x$ is large!

Fortunately, this is where integral calculus can come to the rescue. First we need to notice that if we divide both sides of EQ6.3 by $\delta x$ we get

$$\frac{\delta A}{\delta x} \approx y_1. \tag{EQ6.4}$$

In the limit when $\delta x$ approaches zero, this relationship becomes exact:

$$\frac{\mathrm{d}A}{\mathrm{d}x} = \lim_{\delta x \to 0} \left( \frac{\delta A}{\delta x} \right) = y_1. \tag{EQ6.5}$$

EQ6.5 tells us something important: that the rate of change of the area between the $x$ axis and the curve of any positive function, $y(x)$, is equal to the value of that function ($y_1$) determined at the particular abscissa value ($x_1$) at which the derivative is being evaluated. The implication of this result is that we should be able to use integration to calculate the area under any curve that can be described by an algebraic expression.

Before we do that, however, we need to discover how a large area can be treated as the sum of many smaller areas. One way of doing this is to divide the region we are interested in into a series of vertical strips, each of width $\delta x$. Figure 6.15 suggests that when the size of the increment $\delta x$ is made smaller, and the number of strips within the given region is increased, then the area excluded from the rectangle-based area calculation will diminish. We can therefore calculate an area under a curve more accurately if we treat it as being the sum of many thin, vertical strips.

As long as $\delta x$ is small enough for the relationship $\delta A \approx y_1 \delta x$ to be a good approximation for the area of each component strip, we can make an estimate for a large overall increase in area, $\Delta A$, brought about by increasing the value of the variable from $x_1$ to $x_1 + \Delta x$ in $n$ small increments of equal width, so that $\Delta x = n\delta x$. The overall increase in area is approximately equal to the sum of the areas of each of the component rectangular strips. Using the special notation for a summation over a large number of terms introduced in Chapter 2, the sum over all $n$ strips becomes:

$$\Delta A \approx \sum_{i=1}^{n} \delta A_i = \sum_{i=1}^{n} y_i \, \delta x. \tag{EQ6.6}$$

We can now attempt to improve the approximation by making $\delta x$, the width of each strip, smaller and smaller. However, to keep the total width of the area increment ($\Delta x$) the same size, we also have to make a concomitant increase in the number of strips $n$. In the limit when the width of each strip $\delta x$ tends toward zero, the relationship for $\Delta A$ becomes exact, because we are using an infinite number of strips with infinitesimally narrow widths. Figure 6.16 illustrates this process for the graph of $y(x) = 6x^2 - x^3$ between a lower limit of $x = 0$ and an upper limit of $x = 6$, using 3, 6, 12, and 24 vertical strips. The estimate of the area under the curve clearly becomes more accurate as the number of strips increases.

(A)

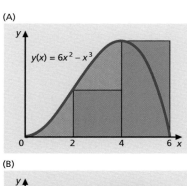

(B)

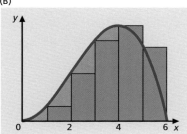

(C)

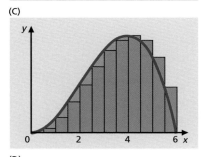

(D)

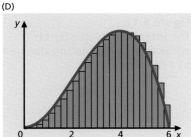

**Figure 6.16**
Taking more rectangular strips with decreasing width improves the accuracy of the estimate for the area under the curve. Note that the first strip in each panel has zero area.

Summarizing this procedure mathematically reveals the source of an important part of the jargon surrounding integration. In the limit as $n$, the number of strips, tends to infinity and $\delta x$, the width of each strip, tends to zero:

$$\Delta A = \lim_{\substack{n \to \infty \\ \delta x \to 0}} \sum_{i=1}^{n} y_i \, \delta x = \int_{x_1}^{x_1 + \Delta x} y \, dx. \qquad \text{(EQ6.7)}$$

In other words, the summation in the middle of EQ6.7 provides a definition of what the right-hand side means: the definite integral of a function can be regarded as the sum of the areas of an infinite number of infinitesimally narrow vertical strips, between two boundaries.

When the area in question lies entirely above the $x$ axis, as it does for the curve in Figure 6.13, its value is defined by the following definite integral:

$$\text{area} = \int_{x_0}^{x_1} y(x) \, dx \qquad \text{(EQ6.8)}$$

(provided that $x_1 > x_0$). For example, if $y = x^{-2}$, then the area under the curve between $x = 1$ and $x = 2$ is

$$\int_{1}^{2} x^{-2} \, dx = \left[ \frac{x^{-1}}{-1} \right]_{1}^{2} = \left[ -\frac{1}{x} \right]_{1}^{2} = \left[ -\frac{1}{2} \right] - \left[ -\frac{1}{1} \right] = \frac{1}{2}.$$

One of the examples in Section 6.2 demonstrated that the integral of a constant generates a linear function of the variable of integration. A graphical approach to integration can help to reveal some of the significance of this result. Figure 6.17 shows the area enclosed by the graph of $y = +10$, the $x$ axis, and vertical lines at $x = 2$ and $x = 8$. Using integration to determine this area, we find that:

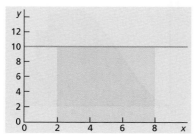

**Figure 6.17**
The integral of a constant is equivalent to evaluating the area of a rectangle.

$$\text{area} = \int_{2}^{8} 10 \, dx = \left[ 10x \right]_{2}^{8} = 80 - 20 = 60.$$

Integrating a constant is therefore equivalent to evaluating the area of a rectangle.

Often the area under a curve will have physical significance, such as the total amount of oxygen generated by photosynthesis in a pond over a given time period if the net rate of production of oxygen is already known. Thus we need to be able to interpret the physical significance of a definite integral. A classic example would be the area under a graph of velocity against time between times $a$ and $b$, which represents the distance traveled by an object between these two times. If the velocity $v$ m s$^{-1}$ of an object at time $t$ s is given by the formula $v(t) = 3t^2 + 5$, then the distance $s$ m traveled between times of 1 and 2 s is found from

$$s = \int_1^2 (3t^2 + 5) \, \mathrm{d}t = \left[\frac{3t^3}{3} + 5t\right]_1^2 = \left[(2)^3 + 5(2)\right] - \left[(1)^3 + 5(1)\right]$$

$$= [8 + 10] - [1 + 5] = 12.$$

That is, the distance traveled is 12 m.

## 6.5 Cumulative change

Figure 6.3 shows a day-by-day breakdown of diagnosed cases during an outbreak of foot and mouth disease. The total number of cattle affected during a particular period can be determined by summing together the number of cases reported on each day, which is equivalent to the sum of the areas of the rectangles that convey this information in the histogram. However, suppose we have developed a mathematical model that uses a continuous function to describe the infection rate as a function of time, rather than returning a set of discrete population measurements at specific time intervals (Figure 6.18). In this case, the total number of cases recorded during a particular period is represented by the area under the curve, which could be obtained from a definite integral.

To account for the effects of cumulative change in more detail, let us suppose that $R(t)$ is the instantaneous rate of change with respect to time of a population $N(t)$, which can be represented in the form:

$$R = \frac{\mathrm{d}N}{\mathrm{d}t}. \tag{EQ6.9}$$

The change in the population between an initial time $t = T_0$ and a later time $t = T_1$ is represented by

$$\Delta N = N(T_1) - N(T_0). \tag{EQ6.10}$$

The change in population that occurs during this interval is described by the area under the graph of $R(t)$ plotted as a function of time:

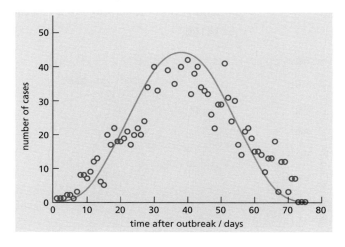

**Figure 6.18**
Number of new confirmed cases of foot and mouth disease in cattle on each day during an outbreak plotted as points on a graph with a possible model (the smooth curve).

$$\Delta N = \int_{T_0}^{T_1} R(t)\, dt. \tag{EQ6.11}$$

This equation indicates that the cumulative change in the size of a population is given by the integral with respect to time of the instantaneous rate of population change. Another useful way of stating this relationship is:

$$N(T_1) = N(T_0) + \int_{T_0}^{T_1} R(t)\, dt. \tag{EQ6.12}$$

This link between the area under a curve, the definite integral that describes it, and a cumulative rate of change is a feature of integration that is important to grasp.

For example, if the population of a wasp nest starts at 500 wasps and increases at a rate of $W(t)$ wasps per week, then $500 + \int_0^{26} W(t)\, dt$ represents the number of wasps present after 26 weeks.

As another example, Figure 6.19 shows the mortality rate for lung cancer in older men aged between 64 and 75 years plotted against the time since the start of a national 'no smoking' campaign. The anti-smoking message apparently took more than 20 years to affect the health of men in this age range. Over a 35-year period, the statistics indicate that the mortality rate ($y$, in deaths per 100 000 of population per year) varies with time ($t$, in years) according to the equation:

$$y = 200 + 30t - 0.6t^2.$$

During the first 30 years of the campaign, the number of deaths due to lung cancer per 100 000 of population can be obtained from the following integral:

$$\begin{aligned}
\text{Total number of deaths} \atop \text{(per 100 000)} &= \int_0^{30} (200 + 30t - 0.6t^2)\, dt \\
&= \left[ 200t + \frac{30}{2} t^2 - \frac{0.6}{3} t^3 \right]_0^{30} \\
&= [200 \times 30 + 15 \times 900 - 0.2 \times 27000] - [0] \\
&= 14100.
\end{aligned}$$

**Figure 6.19**
Mortality rate in terms of deaths per 100 000 population per year due to lung cancer in men aged 64–75 years.

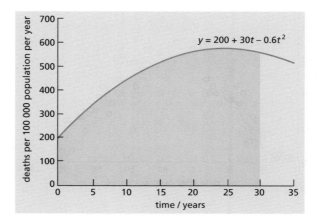

# 6.6 Composite, zero, and negative areas

The realization that a definite integral is related to the area bounded by a curve provides a useful insight into many important properties. For example, in the same way that an area can be split up into several fragments, a definite integral can be divided into smaller portions. If $a < c < b$, then we can rewrite the definite integral of a function of $x$ as the sum of two component integrals:

$$\int_a^b f(x)\,dx = \int_a^c f(x)\,dx + \int_c^b f(x)\,dx. \qquad \text{(EQ6.13)}$$

This relationship can be useful for integrating certain difficult functions by dividing the problem up into portions that are more manageable. It is also needed when a model is made up of different functions, each of which is valid for a different interval.

For example, consider a curve that contains a discontinuity (a sudden change of gradient), such as the time dependence of the concentration of luteinizing hormone (LH), which governs the excretion of progesterone during the menstrual cycle of a female mammal (Figure 6.20). We could determine the area under a discontinuous curve like this by interrupting our integral at the discontinuity, splitting it up into two component integrals that deal separately with the concentration changes beforehand and afterward.

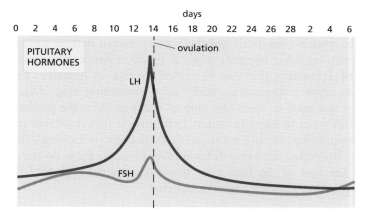

**Figure 6.20**
The concentration of luteinizing hormone (LH) and follicle-stimulating hormone (FSH) during the menstrual cycle of a female mammal.

A simpler demonstration of this technique would be to use integral calculus to find the area of a triangle OBA that has vertices at coordinates (0, 0), (1, 1), and (3, 0), as sketched in Figure 6.21. The graph that forms the boundary of the area inside the triangle has the equation:

$$y = \begin{cases} x & \text{for } 0 \leqslant x \leqslant 1 \quad \text{(OB)} \\ \dfrac{3}{2} - \dfrac{x}{2} & \text{for } 1 \leqslant x \leqslant 3 \quad \text{(BA)}. \end{cases}$$

This function has a sharp corner or discontinuity at $x = 1$. The area formed by the two triangular regions is given by

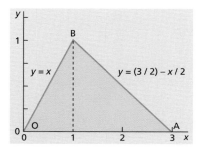

**Figure 6.21**
The area under the triangle can be determined by integrating $y = x$ (between 0 and 1) and $y = (3/2) - x/2$ (between 1 and 3).

$$\text{Area} = \int_0^3 y\,dx = \int_0^1 x\,dx + \int_1^3 \left(\frac{3}{2} - \frac{x}{2}\right) dx$$

$$= \left[\frac{x^2}{2}\right]_0^1 + \left[\frac{3x}{2} - \frac{x^2}{4}\right]_1^3$$

$$= \left[\frac{(1)^2}{2} - \frac{(0)^2}{2}\right] + \left[\frac{3\times(3)}{2} - \frac{(3)^2}{4} - \left\{\frac{3\times(1)}{2} - \frac{(1)^2}{4}\right\}\right]$$

$$= \left[\frac{1}{2}\right] + \left[\frac{9}{2} - \frac{9}{4}\right] - \left[\frac{3}{2} - \frac{1}{4}\right] = \frac{3}{2}.$$

The result of this calculation can be checked by using the familiar $1/2 \times$ base $\times$ height formula for the area of a triangle, which confirms that the final answer should indeed be $(1/2) \times 3 \times 1 = 3/2$.

A graphical understanding of integration is also useful for illustrating the effect of setting the upper and lower limits of a definite integral equal to each other. This is equivalent to determining the area enclosed by a curve, the $x$ axis, and two vertical lines that are superimposed on top of each other. Figure 6.22 attempts to illustrate this situation for the function $y = 10$, with the upper and lower boundaries set to $x = 5$. The enclosed area is clearly zero:

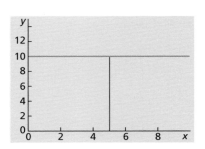

**Figure 6.22**
When the lower and upper limits of a definite integral are the same, the integral corresponds to a vertical line with zero area.

$$\text{Area} = \int_5^5 10\,dx = \left[10x\right]_5^5 = 50 - 50 = 0.$$

So far in this chapter, we have used integration to determine areas that are bounded by positive functions and the $x$ axis between defined limits. We now need to consider what happens when a function becomes negative. As an example of a function that can be either positive or negative, Figure 6.23 shows data from a kinematics study of a swimming frog. Digital image technology was used to study the flow of water in the wake produced by the frog's feet in order to identify factors that provide forward thrust in different styles of swimming. During a stroke of 'asynchronous kicking', the body of the frog is propelled forward, defining the direction for a positive velocity. However, the foot of the frog must move backward for part of the stroke, during which it possesses a negative velocity.

How can we describe the area bounded by a curve that possesses a portion that dips below the horizontal axis? We will approach this by considering a simple example, such as the area between the function $y = -\sqrt{x}$ and the $x$ axis within the range $1 \leqslant x \leqslant 4$. As shown in Figure 6.24, this region is completely below the $x$ axis. On attempting to evaluate this area using the normal definite integral method, we see that:

$$\text{Area} = \int_1^4 (-x^{1/2})\,dx = \left[-\frac{2}{3}x^{3/2}\right]_1^4 = -\frac{2}{3} \times \left[(4)^{3/2} - (1)^{3/2}\right]$$

$$= -\frac{2}{3}[8 - 1] = -\frac{14}{3}.$$

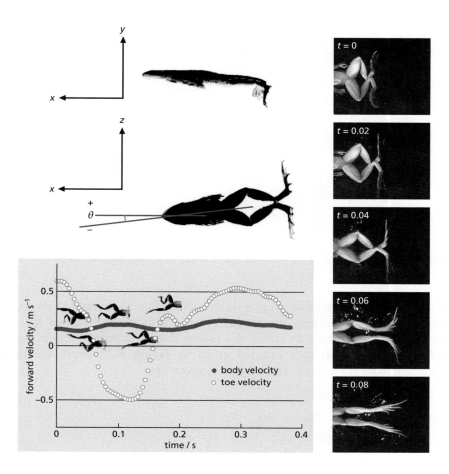

**Figure 6.23**
Kinematics study of a frog swimming. Adapted with permission from Johansson, L.C. and Lauder, G.V. (2004) *Journal of Experimental Biology*, 207: 3945.

It is perhaps no surprise to discover that this procedure yields a negative value for the area, because the function $y(x)$ is itself negative throughout the range $1 \leqslant x \leqslant 4$. The concept of a negative area may be rather perturbing, but the minus sign here is by convention simply an indication that the area we have calculated lies below the $x$ axis. This property is called the 'polarity' of the area under a curve: a positive polarity indicates that the region under consideration is above the horizontal axis, while a negative polarity indicates that the required area is below it.

Taking this line of thought one step further, we also need to consider the area enclosed by a function that possesses portions that are above the horizontal axis as well as sections that are below it. For example, we might want to determine an area enclosed by the curve $V(t) = 2t^3 - 28t^2 + 98t - 72$ in Figure 6.25, which models the action potential voltage (measured in milli-volts) in a nerve cell axon plotted as a function of time (in milliseconds) after a stimulus. A close look at the graph shows that it starts at the resting potential voltage of $-72$ mV and then rises, cutting the horizontal axis at $t = 1$, goes through a maximum and then decreases, cutting the $t$ axis again at $t = 4$. If we choose to determine the area bounded by vertical lines at $t = 1$ and $t = 6$, there are two sections to consider: the portion corresponding to the

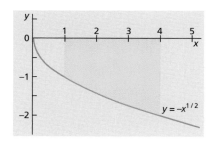

**Figure 6.24**
The integral of $y = -\sqrt{x}$ for $1 \leqslant x \leqslant 4$ corresponds to an area entirely below the $x$ axis; it has a negative value.

**Figure 6.25**
The integral of
$V(t) = 2t^3 - 28t^2 + 98t - 72$ for
$1 \leqslant t \leqslant 6$ corresponds to the sum
of an area above the $t$ axis
(positive contribution) and an
area below the $t$ axis (negative
contribution).

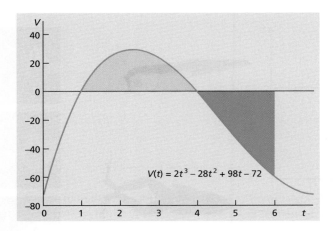

range $1 \leqslant t \leqslant 4$, which lies above the $t$ axis, and the region for $4 \leqslant t \leqslant 6$, lying below it.

If no account is taken of the sign of the function when we attempt to integrate over the range of interest, the quantity that is determined is called the *net area*, or the *signed area*. In effect, this means that we subtract the area found below the horizontal axis (shaded green in Figure 6.25) from the area enclosed above it (yellow). The net area is given by:

$$\text{Net area} = \int_1^6 (2x^3 - 28x^2 + 98x - 72)\, dx$$

$$= \left[ \frac{2x^4}{4} - \frac{28x^3}{3} + \frac{98x^2}{2} - \frac{72x^1}{1} \right]_1^6 = \left[ \frac{x^4}{2} - \frac{28x^3}{3} + 49x^2 - 72x \right]_1^6$$

$$= \left[ \frac{(6)^4}{2} - \frac{28 \times (6)^3}{3} + 49 \times (6)^2 - 72 \times 6 \right] - \left[ \frac{1}{2} - \frac{28}{3} + 49 - 72 \right]$$

$$= -36 + 31.833 = -4.167.$$

The net area is negative, indicating that the green portion enclosed below the $t$ axis is larger than the lobe that protrudes above it. Note that the net area corresponds to a quantity of $-4.167$ ms mV.

On the other hand, if we are interested in determining the *total area* or *unsigned area* enclosed by a curve, we would have to consider the absolute value (the unsigned part) of the function. To do this, we must examine the function over the required range and evaluate a separate integral for each portion found above or below the horizontal axis. To perform this calculation for the action potential problem in Figure 6.25 we need to split the area up into two parts: the yellow section $A_1$ and the green component $A_2$. First, for the region that lies above the axis, we have $1 \leqslant t \leqslant 4$:

$$A_1 = \int_1^4 (2x^3 - 28x^2 + 98x - 72)\, dx = \left[ \frac{x^4}{2} - \frac{28x^3}{3} + 49x^2 - 72x \right]_1^4$$

$$= 26.666 + 31.833 = +58.5.$$

For the second component, spanning $4 \leqslant t \leqslant 6$:

$$A_2 = \int_4^6 (2x^3 - 28x^2 + 98x - 72)\,\mathrm{d}x = -36 - 26.666 = -62.667.$$

This calculation corresponds to the region in which the curve dips below the horizontal axis and therefore gives an area with negative polarity. As we would expect, the net area is given by the sum of these two components:

Net area $= A_1 + A_2 = +58.5 - 62.667 = -4.167.$

However, to determine the total area enclosed by the curve over the region of interest, we would need to sum the absolute values of the areas of the two portions:

$$\begin{aligned}
\text{Total area} &= \int_1^6 |V(t)|\,\mathrm{d}t = |A_1| + |A_2| \\
&= |+58.5| + |-62.667| = +121.167.
\end{aligned}$$

The quantity corresponding to the unsigned area is therefore 121.167 ms mV.

Whether it is appropriate to determine the net area or the total area will depend on the context of each new case. It might not be readily apparent why one would want to determine the area under the voltage curve in Figure 6.25 (although a good reason to do this is given in the next section). However, the velocity profile of the toe of the swimming frog shown in Figure 6.23 provides an example that is easier to justify, because the integral of a plot of velocity against time yields the distance traveled. The net area under the body velocity curve reports on how far the whole frog moves during a given interval, whereas the unsigned area under the toe velocity curve indicates the length of the path traced out by the frog's foot during a stroke.

## 6.7 The mean value of a function

**Figure 6.26**
An ocean ecosystem.

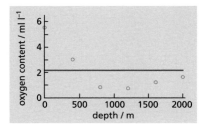

**Figure 6.27**
Plot of oxygen concentration against depth showing a high concentration near the surface but a depleted zone around 1000 m. The horizontal line represents the average oxygen concentration.

**Table 6.1  Oxygen concentration (in milliliters of gas per liter of water) for depths from 0 to 2000 m**

| Depth / m | 0 | 400 | 800 | 1200 | 1600 | 2000 |
|---|---|---|---|---|---|---|
| Oxygen / ml l$^{-1}$ | 5.5 | 3.0 | 0.8 | 0.7 | 1.2 | 1.6 |

Dissolved oxygen is essential to the survival of all ocean ecosystems (Figure 6.26). The oxygen concentration in seawater varies with temperature, salinity, and depth, but can be locally enriched or depleted by many factors, such as underwater currents, photosynthesis, and the breakdown of organic material. Table 6.1 contains measurements of the oxygen concentration (in milliliters of gas per liter of water) at different depths. As Figure 6.27 shows, the concentration is high near the surface, owing to exchange with oxygen in the atmosphere, and there is a depleted zone at a depth of approximately 1000 m. However, what is the average oxygen concentration over the depth range studied?

Statistical analysis is dealt with in detail in Chapters 9–11, but for now we will focus on $\bar{a}$, the arithmetic mean (or average value) of a series of $n$ numbers $a_1, a_2, \ldots, a_n$, which is given by the expression

$$\bar{a} = \frac{a_1 + a_2 + \ldots + a_n}{n} = \frac{1}{n} \sum_{k=1}^{n} a_k. \tag{EQ6.14}$$

The mean oxygen concentration, $\bar{c}$, is therefore:

$$\bar{c} = \frac{5.5 + 3.0 + 0.8 + 0.7 + 1.2 + 1.6}{6} = 2.1 \text{ ml l}^{-1} \text{(to 2 significant figures)}.$$

This value is indicated by the level of the blue bar in Figure 6.27.

**Figure 6.28**
The mean value of the bird population over the range $0 \leqslant t \leqslant 6$ is equivalent to the height of a rectangle that has the same area as that under the population curve.

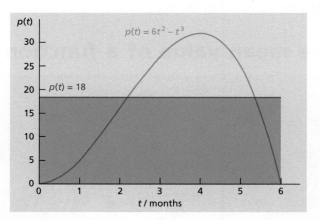

It is sometimes useful to find a way of calculating the mean value of a continuous function $f(x)$, rather than of a set of discrete numbers. A graphical way of thinking about the mean value of a function is presented in Figure 6.28, which shows the number of adult birds in a nature reserve during the spring and summer months (see Figure 6.12). The mean value of the population over the range $0 \leqslant t \leqslant 6$ is equivalent to the height of a rectangle that has the same area as that bounded by the function over the same range of abscissa coordinates. In Figure 6.28 the equivalent rectangular area is shaded in green.

In other words, the mean value of a function, $\bar{f}$, multiplied by the difference between the abscissa coordinates at the start and end of the range, $(b - a)$, is equal to the area under the curve $f(x)$ over the same range:

$$\bar{f} \times (b - a) = \int_a^b f(x)\,dx. \tag{EQ6.15}$$

The recipe for calculating the mean value of a function is therefore:

$$\bar{f} = \frac{1}{b - a} \int_a^b f(x)\,dx. \tag{EQ6.16}$$

For the range $0 \leqslant t \leqslant 6$, the population of the colony is given by $p(t) = 6t^2 - t^3$. The average number of adult birds present during this period is therefore:

$$\bar{p} = \frac{1}{6 - 0} \int_0^6 (6t^2 - t^3)\,dt = \frac{1}{6}\left[ \frac{6t^3}{3} - \frac{t^4}{4} \right]_0^6$$

$$= \frac{1}{6}\left[ 2 \times 216 - \frac{1296}{4} \right] - \frac{1}{6}[0] = 18.$$

On average, 18 birds are present during the period under consideration, compared with the maximum population of 32, which occurs at the beginning of August (when $t = 4$). The light purple-shaded area above the horizontal mean population line at $p = 18$ and enclosed by the population curve is equal to the area enclosed below it (shaded yellow).

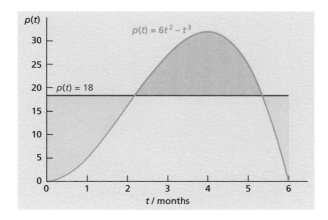

**Figure 6.29**
The area enclosed by the population curve above the mean value is equal to the area enclosed below it.

If we are interested in the mean value of a function that dips below the horizontal axis, it is worth noting that this approach relies on a calculation of the net area: portions below the axis may cancel out those above it. For example, in a study of the time dependence of the action potential in an axon (Figure 6.30), the mean voltage between 0 and 6 ms is given by:

**Figure 6.30**
The mean of the action potential of an axon has negative and positive contributions with an overall negative value (−6 mV).

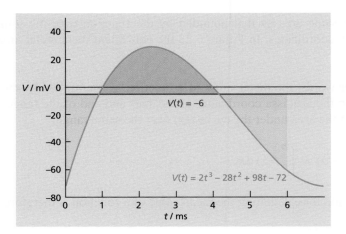

$$\overline{V} = \frac{1}{6-0} \int_0^6 (2t^3 - 28t^2 + 98t - 72)\, dt$$

$$\overline{V} = \frac{1}{6-0} \left[ \frac{2t^4}{4} - \frac{28t^3}{3} + \frac{98t^2}{2} - \frac{72t^1}{1} \right]_0^6 = \left[ \frac{6^3}{2} - \frac{28 \times 6^2}{3} + 49 \times 6 - 72 \right] - [0]$$

$$= -6.$$

It is worth emphasizing here that we don't need to split this integral into three parts (for the positive and negative regions) because we are focusing on the *net* area. Over this range the area enclosed between the portions of the curve with negative voltage and the *t* axis is larger than the positive lobe above it, giving a mean voltage of −6 mV. As before, the light purple area above the line representing the mean voltage ($V = -6$) is equal to the sum of the yellow areas below it.

## Presenting Your Work

### QUESTION A

The velocity $v$ of blood in a cylindrical blood vessel of radius $R$ and length $l$ is given by $v(r) = P(R^2 - r^2)/4\eta l$, where $\eta$ and $P$ are constants and $r$ is the distance from the axis of the cylinder. Find the average velocity over the interval $0 \leqslant r \leqslant R$ and compare this with the maximum velocity.

---

If $v(r) = \dfrac{P}{4\eta l}(R^2 - r^2)$, then in the interval $0 \leqslant r \leqslant R$, then

$$\bar{v} = \frac{1}{R-0}\int_0^R \frac{P}{4\eta l}(R^2 - r^2)\,dr = \frac{P}{4\eta lR}\left[R^2 r - \frac{r^3}{3}\right]_0^R = \frac{P}{4\eta lR}\left[R^3 - \frac{R^3}{3} - 0 + 0\right] = \frac{P}{4\eta lR} \times \frac{2R^3}{3}$$

$$= \frac{PR^2}{6\eta l}.$$

The maximum velocity occurs when $\dfrac{dv}{dr} = 0$, i.e. when $\dfrac{dv}{dr} = \dfrac{P}{4\eta l}(0 - 2r) = -\dfrac{Pr}{2\eta l} = 0$,

which is when $r = 0$. This point must be a maximum because $\dfrac{d^2v}{dr^2} = -\dfrac{P}{2\eta l} = $ negative.

When $r = 0$, $v_{max} = \dfrac{PR^2}{4\eta l}$, so $\dfrac{\bar{v}}{v_{max}} = \dfrac{4}{6} = \dfrac{2}{3}$.

**QUESTION B**

A species of bird gathers in its summer territory and then disperses at a rate given approximately by

$$\frac{dN}{dt} = 24t^2 - 4t^3$$

where $N$ is the number of mature birds in the colony at time $t$, which is measured in months from the beginning of February. Assume that there are no mature birds present at the beginning of February, that no mature birds die, and that birds born in the summer territory are not included in the count.

(a) Find an expression for the subsequent size of the mature bird population.

(b) Determine when all of the mature birds have departed.

(c) Calculate the maximum number of birds in the colony and indicate when this occurs.

(d) Determine the mean population of the colony over the first four months.

---

$$\frac{dN}{dt} = 24t^2 - 4t^3$$

(a) Integrating with respect to $t$ gives:

$$N = \int \frac{dN}{dt}\, dt = \int (24t^2 - 4t^3)\, dt = \frac{24}{3}t^3 - \frac{4}{4}t^4 + c = 8t^3 - t^4 + c.$$

When $t = 0$, $N = 0$, therefore

$$0 = 8 \times (0)^3 - (0)^4 + c,$$

i.e. $c = 0$, so

$$N(t) = 8t^3 - t^4.$$

(b) All the birds will have departed when $N = 0$,
that is, when $0 = 8t^3 - t^4 = t^3(8 - t)$
that is, when $t = 0$ or when $t = 8$ months.
Therefore all the birds will have departed by the beginning of October.

(c) The turning points occur when $\frac{dN}{dt} = 0$

that is, when $0 = 24t^2 - 4t^3 = 4t^2(6 - t)$
that is, when $t = 0$ or when $t = 6$ months.
When $t = 0$, $N = 0$, so this cannot be the maximum population.

$$\frac{d^2N}{dt^2} = 48t - 12t^2,$$

so when $t = 6$

$$\frac{d^2N}{dt^2} = 48 \times 6 - 12 \times (6)^2 = 288 - 432 = -144.$$

Because the curvature is negative, there is a maximum turning point at $t = 6$.

When $t = 6$, $N = 6^3(8 - 6) = 432$.

Therefore the maximum number of birds is 432, and this occurs at the beginning of August.

(d) The mean population over the first four months is given by

$$\bar{N} = \frac{1}{4-0} \int_0^4 (8t^3 - t^4)\, dt = \frac{1}{4} \times \left[ \frac{8t^4}{4} - \frac{t^5}{5} \right]_0^4$$

$$= \frac{1}{4} \times \left[ 2 \times (4)^4 - \frac{(4)^5}{5} \right] - \frac{1}{4} \times [0] = 128 - \frac{256}{5} = 76.8.$$

Therefore the mean population is 77 birds (to the nearest whole bird).

## End of Chapter Questions
(Answers to questions can be found at the end of the book)

### Basic
**1.** Integrate the following expressions with respect to $x$:
(a) $y = 7$
(b) $y = 4x^6$
(c) $y = x^{-2}$
(d) $y = 2x + 1$
(e) $y = 21x^{2/5}$
(f) $y = 2x - 3x^{1/2} - x^{-4/3}$

**2.** Integrate each of the following expressions by first rewriting in a suitable form:
(a) $y = (x + 3)^2$

(b) $y = \dfrac{(x^3 - 1)}{x^2}$

(c) $y = \sqrt[3]{x^2} - \sqrt{x^3}$

**3.** Evaluate the following definite integrals and sketch the areas that you have determined:

(a) $\displaystyle\int_0^4 x^3 \, dx$

(b) $\displaystyle\int_{-2}^{-8} \frac{1}{x^3} \, dx$

(c) $\displaystyle\int_3^{12} \sqrt[3]{x} \, dx$

(d) $\displaystyle\int_0^4 (\sqrt{x} - 2)(\sqrt{x} - 4) \, dx$

**4.** Evaluate the definite integral $\int_a^b f(x) \, dx$ when:
(a) $f(x) = 9 - 2x, a = -1, b = 3$
(b) $f(x) = 5x - x^2/2, a = -10, b = 10$
(c) $f(x) = (1 - x)/\pi, a = 0, b = \pi$

**5.** Find the area bounded by the $x$ axis, the curve $y = x^{-2}$, and vertical lines at $x = 1$ and
(a) $x = 2$

(b) $x = 3$

(c) $x = 5$

(d) $x = +\infty$

**6.** Determine the net area bounded by the function $y = t^2 - 2t + 3$ and the $x$ axis:
(a) over the range $0 \leqslant t \leqslant 3$
(b) over the range $1 \leqslant t \leqslant 6$

(c) over the range $-5 \leqslant t \leqslant -2$
(d) What is the mean value of this function over the range $0 \leqslant t \leqslant 3$?

**7.**
(a) A hive of honeybees starts with a population of 100 which increases at a rate of $R(t)$ bees per week.

What does $100 + \displaystyle\int_0^{15} R(t) \, dt$ represent?

(b) If $x$ is a length measured in m and $a(x)$ is the mass per unit length measured in $\text{kg m}^{-1}$, then what should be the units for

$$\int_2^8 a(x) \, dx?$$

(c) If $G(t)$ is the rate of growth of a child in kilograms per year, and $t$ is the time in years,

what does $\displaystyle\int_5^{10} G(t) \, dt$ represent?

**8.** Find the average value of the square root of $x$ between $x = 1$ and $x = 4$.

### Intermediate
**9.** Ice is forming on a pond at a rate given by $dy/dt = k\sqrt{t}$, where $k$ is a positive constant and $y$ is the thickness of the ice in centimeters at time $t$ measured in hours since the ice started forming. Find $y$ as a function of $t$.

**10.** The rate of change $R$ of the population $p$ of a colony of geckos is given as a function of time $t$ (in months) by $R = 1/(3\sqrt{t})$. Determine an expression for $p$ if the population is 30 when $t = 0$.

**11.** At each point $(x, y)$ on a curve, the slope is equal to the square of the distance between the point and the $y$ axis. If it is known that the point $(-1, 2)$ lies on the curve, find its equation.

**12.** If $A(t)$, the area in square centimeters of a healing wound, changes at a rate given approximately by

$$\frac{dA(t)}{dt} = -4t^{-3},$$

where $t$ is the time in days, $1 \leqslant t \leqslant 10$, and $A(1) = 2$, then what will be the area of the wound after 10 days?

**Advanced**

**13.** An insect population, initially numbering 100, increases to a population of size $p(t)$ after a time $t$ (measured in days). If the rate of growth at time $t$ is given by $dp/dt = 2t + 3t^2$, determine the size of the population after (a) 1 day and (b) 10 days.

**14.** The instantaneous velocity, $v(t)$, of a body moving along a straight line is given by the equation $v(t) = \int a(t) \, dt$, where $a(t)$ is the instantaneous acceleration, whereas its position, $s(t)$, is given by $s(t) = \int v(t) \, dt$. Freefall motion can be modeled by setting $s$ equal to zero when an object is on the ground, choosing the positive direction to be up, and also setting $a(t)$ equal to $-g$, where $g$, the acceleration due to gravity, is $9.8 \text{ m s}^{-2}$. If $v(0) = v_0$ and $s(0) = 0$, derive equations for $v(t)$ and $s(t)$. When a kangaroo rat jumps vertically off the ground with an initial velocity of $3 \text{ m s}^{-1}$, what is the maximum height it will reach?

**15.** Find the area of the region enclosed by the curves $y = x^2$ and $y = 2x - x^2$.

**16.** Find the area of the region enclosed by the graphs of $x = y^2$ and $y = x - 2$.

(Hint: think about this problem sideways!)

# CHAPTER 7

# Calculus: Expanding the Toolkit

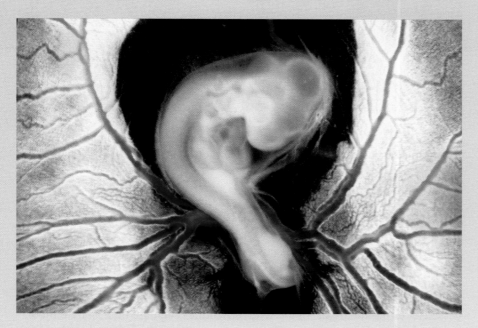

**Figure 7.1**
A 3-day-old chick embryo. Image courtesy of Nobue Itasaki, Division of Developmental Neurobiology, MRC National Institute for Medical Research.

Chapters 5 and 6 introduced the basic equipment you need to construct new theories by analyzing the behavior of curves. To push the analogy with a builder's toolkit rather too far, we could view differentiation as supplying a metaphorical 'spirit level' that is sensitive to the gradient of a function, whereas integration provides a trowel for spreading mortar over a range of different areas. The aim of this chapter is to describe a set of 'power tools' that will make it possible for you to perform tasks that couldn't be accomplished 'by hand' using the methods encountered up to this point. For differentiation, the chain rule, the product rule, and the quotient rule are vital for slicing your way through an array of more complex functions. Similarly, the rule for changing the variable of an integral provides just the sort of heavy lifting gear that is necessary for shifting weighty integration problems.

These new techniques are put to work to investigate a collection of biological topics, from the metabolism of reptiles to the size of rabbits, by way of weak acids, population dynamics, and the forces experienced by the foot of a

**Figure 7.2**
In the wild, crocodiles regulate their body temperature by basking in water when they need to cool off, and then shuttling back to the land to warm up in the sunshine. Image courtesy of Mister-E under Creative Commons Attribution 2.0 Generic.

sprinting athlete. Periodic functions appear regularly as applications, culminating in a discussion of simple harmonic motion at the end of the chapter.

# 7.1 Sinusoidal functions

The trigonometric functions discussed in Chapter 4 are useful for modeling biological data that vary in a repetitive way as a function of time. For example, the body temperature of a cold-blooded reptile, such as a crocodile, would be expected to increase through the day owing to solar heating, but to decrease during the night. In the wild, crocodiles regulate their body temperature by basking in water when they need to cool off, and then shuttling back to dry land to warm up in the sunshine. Despite this activity, the core body temperature of a small crocodile is far from stable and will fluctuate in a sinusoidal manner during the course of a day (Figure 7.3). By contrast, the body temperature of a warm-blooded mammal (for example a human) is much more tightly regulated, owing to differences in metabolism and adaptations such as sweating or shivering.

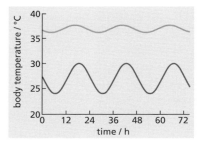

**Figure 7.3**
Diurnal variation in the body temperature of a small crocodile (blue) and a human (red). The human body temperature is much more tightly regulated, owing to differences in metabolism and adaptations such as sweating or shivering.

The curves plotted in Figure 7.3 can serve as a reminder of the basic properties of a sinusoidal function, which are encapsulated by the general equation:

$$y(t) = A \sin\left(\frac{2\pi}{T} t + \phi\right) + y_0, \qquad \text{(EQ7.1)}$$

where $A$ is the amplitude of the oscillation; $T$ is the time period; $\phi$ is the phase angle; and $y_0$ is the mean value of the sine curve or its vertical offset from the horizontal axis. (Note that EQ7.1 is an extension of EQ4.35.) The amplitude of the crocodilian body temperature oscillation is much larger than that of the human, but the mean temperature of the human is greater than that of the crocodile. The time periods and phase angles of the two curves are similar, both producing minimum temperatures in the early hours of the morning and maxima in the afternoon.

Moving our focus back to calculus, we will now exploit the periodic nature of these sinusoidal curves. To characterize the whole function, it is only necessary to think about the slope in a region that spans one complete time period, which for the crocodile's body temperature could be from 0 to 24 hours. Portions outside this range simply repeat the behavior that takes place inside it. For the simpler function $y = \sin(x)$ shown in Figure 7.4, the appropriate range is $0 \leqslant x \leqslant 2\pi$. In addition, we can see that the slope of a tangent to this curve would clearly be positive from 0 to $\pi/2$, negative from $\pi/2$ to $3\pi/2$, positive again between $3\pi/2$ and $2\pi$, and zero at $x = \pi/2$ and $x = 3\pi/2$. The plot is at its steepest, and so the slope possesses its maximum and minimum values, at $x = 0$, $\pi$, $2\pi$, and so on. From this portrait it is clear that the slope of $y = \sin(x)$ must also be periodic. The important question here is, which other sinusoidal function has similar characteristics?

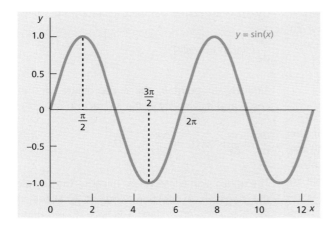

**Figure 7.4**
Plot of the function sin(x) for values of x between 0 and 12.

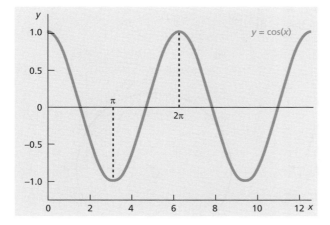

**Figure 7.5**
Plot of the function cos(x) for values of x between 0 and 12.

As Figure 7.5 demonstrates, the cosine function has all of the features mentioned above that describe the derivative of $\sin(x)$: it possesses maxima at $x = 0$ and $2\pi$, with a minimum at $\pi$; it cuts the $x$ axis at $\pi/2$ and again at $x = 3\pi/2$; it is positive from 0 to $\pi/2$, negative from $\pi/2$ to $3\pi/2$, and positive again between $3\pi/2$ and $2\pi$. These properties suggest that, if $x$ is in radians,

$$\frac{\mathrm{d}}{\mathrm{d}x}\sin(x) = \cos(x).$$

(EQ7.2)

A simple geometrical approach for proving this relationship is outlined in Box 7.1.

## Box 7.1 The derivatives of the sine and cosine functions

Imagine a centrifuge that contains a sample at the end of a rotor of length 1 dm, which is rotating at an angular velocity of 1 rad μs⁻¹ in an anticlockwise direction in a horizontal plane defined by the $x$ and $y$ axes shown in Figure 1A. If the sample is at point $A$ at time zero, then at time $t$ (measured in microseconds) later it has traveled along an arc of length $t$ (in decimeters) to point $B$, having swept through an angle of $t$ (in radians) with respect to the center of the centrifuge at point $O$. The $y$ coordinate of the sample now corresponds to the length $CB$, which is $\sin(t)$ dm; its $x$ coordinate is represented by $OC$, which is $\cos(t)$ dm. The velocity of the sample along the direction of the $y$ axis is given by:

$$\frac{\mathrm{d}y}{\mathrm{d}t} = \frac{\mathrm{d}}{\mathrm{d}t}(\sin(t)).$$

Next, we need to consider what would happen if the sample breaks off from the rotor at time $t$ μs. Ignoring the effects of gravity, air resistance, and damage to other laboratory equipment, we would expect the sample to continue moving in the same direction at the same velocity. After a delay $\Delta t$ of 1 μs, it will have traveled 1 dm from point $B$ to point $D$, as illustrated in Figure 1B. In the $y$ direction, the sample will have traveled from $B$ to $E$, a distance we can call $\Delta y$. The velocity of the sample in this direction is $\Delta y/\Delta t = BE/1$. However, $OBC$ forms a right angle triangle, so the angle $OBC$ must be $((\pi/2) - t)$. Also, the three angles $OBC$, $OBD$, and $DBE$ must add up to $\pi$ radians and the tangent $BD$ is perpendicular to the radius line $OB$, so we know that the size of angle $DBE$ must be $t$ (in radians). Because $BDE$ is also a right angle triangle, this means that the length of side $BE$ has to be $\cos(t)$ (in decimeters). Because the velocity of the sample has not changed since time $t$ μs, meaning that $\mathrm{d}y/\mathrm{d}t = \Delta y/\Delta t = BE$, we can state that:

$$\frac{\mathrm{d}}{\mathrm{d}t}(\sin(t)) = \cos(t).$$

Similarly, at point $B$, the $x$ coordinate of the sample is $\cos(t)$ and its velocity in the $x$ direction is:

$$\frac{\mathrm{d}x}{\mathrm{d}t} = \frac{\mathrm{d}}{\mathrm{d}t}(\cos(t)) = \frac{\Delta x}{\Delta t} = -ED.$$

The minus sign is used here because the sample moves in the negative $x$ direction when it is above the $x$ axis. Because $ED = BC = \sin(t)$, we have shown that:

$$\frac{\mathrm{d}}{\mathrm{d}t}(\cos(t)) = -\sin(t).$$

(A)

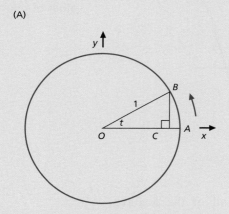

(B)

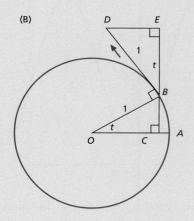

**Figure 1**
Diagrams showing the motion of a sample in a centrifuge:
(A) rotating anticlockwise from point $A$ at time 0 to point $B$ at time $t$; and
(B) traveling in a straight line from point $B$ to point $D$ after breaking away from the rotor.

However, what about the derivative of the function $y = \cos(x)$? Inspection of Figure 7.5 reveals that the slope of this curve is zero at $x = 0$, then becomes negative, reaching its steepest at $\pi/2$; the gradient then becomes more positive (i.e. less negative), passing through zero at $x = \pi$, increasing to a maximum at $3\pi/2$, and then returning once more to zero at $2\pi$. This behavior has the same form as the function obtained when $\sin(x)$ is multiplied by minus one (Figure 7.6), implying that:

$$\frac{d}{dx}\cos(x) = -\sin(x). \tag{EQ7.3}$$

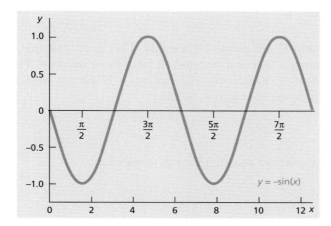

**Figure 7.6**
Plot of the function $-\sin(x)$ for values of $x$ between 0 and 12.

We have seen how the sine and cosine functions can be transformed into each other by way of differentiation. However, if we wanted to determine the mean body temperature of a crocodile over a given time period, we would need to match these relationships with a complementary set of rules that describe the inverse process, namely integration:

$$\int \sin(x)\,dx = -\cos(x) + c, \tag{EQ7.4}$$

and

$$\int \cos(x)\,dx = \sin(x) + c. \tag{EQ7.5}$$

The basic sine and cosine functions have an intriguing complementary relationship, but to perform calculus with more useful forms of trigonometric functions, such as EQ7.1, our toolkit of methods needs to be expanded to include a few more sophisticated techniques.

Before progressing to the next section, ensure that you can do question 1 from the End of Chapter Questions.

## 7.2 Differentiating a pair of nested functions

The first new tool makes it possible to differentiate a pair of 'nested functions' – but to demonstrate what this term actually means, a concrete application is needed. Acetic acid is the active component of table vinegar and the major biological precursor of cholesterol. It is also ejected by whip scorpions as a chemical defense against attackers. As a weak acid, it only partly dissociates into ions when it dissolves in water:

$$CH_3CO_2H \text{ (aq)} \rightleftharpoons CH_3CO_2^- \text{ (aq)} + H^+ \text{ (aq)}.$$

If the total concentration of acetic acid species present, $[CH_3CO_2H] + [CH_3CO_2^-]$, is represented by $x$, then the concentration of hydrogen ions in solution, $y$, is given by:

$$y = \sqrt{A + Bx} - C,$$

where $A$, $B$, and $C$ are positive constants. This is an example of a pair of nested functions, a fact that becomes a little more apparent by inventing a new function of $x$, which we will call $u(x)$. If we set $u(x)$ equal to the expression inside the square root, so that

$$u = A + Bx,$$

then we can rewrite the equation for $y$ in terms of $u$:

$$y = u^{1/2} - C.$$

By applying the rules encountered in Sections 5.4 and 5.5, we can work out the derivative of $u$ with respect to $x$, and the derivative of $y$ with respect to the new variable $u$:

$$\frac{du}{dx} = B \qquad \text{and} \qquad \frac{dy}{du} = \frac{1}{2}u^{-1/2}.$$

However, to get an expression for the derivative of $y$ with respect to the original variable $x$, we need help, which is where the *chain rule* comes in (Box 7.2). If $y$ is a function of $u$ (that is, $y = f(u)$), and $u$ is itself a function of the variable $x$ (that is, $u = g(x)$, so that $y = f(u) = f(g(x))$), then the chain rule states that:

$$\frac{dy}{dx} = \frac{dy}{du} \times \frac{du}{dx}. \tag{EQ7.6}$$

Returning to the acetic acid problem, we can make use of the chain rule to work out the rate of change of the concentration of hydrogen ions at any given concentration of acid:

$$\frac{dy}{dx} = \frac{dy}{du} \times \frac{du}{dx} = \frac{1}{2}u^{-1/2} \times B.$$

Substituting $u = A + Bx$ back into this equation will give an expression written in terms of $y$ and $x$ only:

$$\frac{dy}{dx} = \frac{1}{2}(A + Bx)^{-1/2} \times B = \frac{B}{2\sqrt{A + Bx}}.$$

## Box 7.2 The chain rule

If $y$ is a function of $u$ and $u$ is a function of $x$, then a small increment $\delta x$ in the variable $x$ will cause corresponding increments $\delta u$ and $\delta y$ in $u$ and $y$, respectively. Using the same kind of approximation introduced in Section 5.11, $\delta y$ can be expressed in terms of $\delta u$ by using the derivative of $y$ with respect to $u$. Similarly, $\delta u$ can also be expressed in terms of $\delta x$:

$$\delta y \approx \frac{dy}{du}\delta u \quad \text{and} \quad \delta u \approx \frac{du}{dx}\delta x.$$

So, substituting for $\delta u$, we get

$$\delta y \approx \frac{dy}{du} \times \frac{du}{dx}\,\delta x.$$

Dividing both sides of this equation by $\delta x$ therefore gives:

$$\frac{\delta y}{\delta x} \approx \frac{dy}{du} \times \frac{du}{dx}.$$

In the limit where $\delta x$ tends toward zero, the relation becomes exact:

$$\frac{dy}{dx} = \lim_{\delta x \to 0}\left(\frac{\delta y}{\delta x}\right) = \frac{dy}{du} \times \frac{du}{dx}.$$

Note that different rules are needed for second or higher derivatives.

---

This function always remains positive, but decreases as $x$ becomes larger. It enables us to predict that the solution will become more acidic if further acetate is added, although increasing the overall concentration of acetic acid species will cause ever-smaller effects.

By using the chain rule, we can construct derivatives for more complex functions, such as a trig function of the form $y = \sin(ax + b)$, where $a$ and $b$ are constants. Making a substitution for the term in brackets, $u = ax + b$, should enable us to deduce that:

$$y = \sin(u) \qquad \text{and} \qquad \frac{dy}{du} = \cos(u).$$

Because $u = ax + b$, we also know that $du/dx = a$. Applying the chain rule therefore gives:

$$\frac{dy}{dx} = \frac{d}{dx}(\sin(ax + b)) = \frac{du}{dx} \times \frac{d}{du}(\sin(u)) = a \times \cos(u).$$

Substituting $u = ax + b$ back into the final answer shows us that

$$\frac{dy}{dx} = a\cos(ax + b). \tag{EQ7.7}$$

We can use this general result to write down the answers to related problems. For example, the slope of the general sinusoidal function:

$$y(t) = A\sin\left(\frac{2\pi}{T}t + \phi\right) + y_0 \tag{EQ7.8}$$

is given by:

$$\frac{dy}{dt} = \frac{2\pi A}{T}\cos\left(\frac{2\pi}{T}t + \phi\right), \tag{EQ7.9}$$

because the derivative with respect to $t$ of $\left(\dfrac{2\pi}{T}t + \phi\right)$ is simply $\dfrac{2\pi}{T}$.

It may take a little practice to spot the most convenient substitution for applying the chain rule. With more complex expressions it will not always be the best strategy to 'substitute for the term in brackets', as we did here. For example, to differentiate $y = \cos^3(x)$, the best approach would be to make the substitution $u = \cos(x)$, so that $y = u^3$ and $du/dx = -\sin(x)$. Next, applying the chain rule we find that:

$$\frac{dy}{dx} = \frac{d}{dx}\left(\cos^3(x)\right) = \frac{du}{dx} \times \frac{d}{du}(u^3) = -\sin(x) \times 3u^2$$

$$= -3\sin(x)\cos^2(x).$$

Of course, there is no reason why a nested function should not also contain a function that is itself nested, which could contain yet another nested function, and so on. To obtain the derivative of a ridiculously complicated function, such as

$$y = \ln\left(\frac{1}{\sin(x^2 - 3)}\right),$$

we need an extended form of the chain rule. If $f_1$ is a function of $f_2$, $f_2$ is a function of $f_3$, $f_3$ is a function of $f_4$, $\ldots$ and $f_{n-1}$ is a function of $f_n$ which is itself a function of $x$, then

$$\frac{df_1}{dx} = \frac{df_1}{df_2} \times \frac{df_2}{df_3} \times \frac{df_3}{df_4} \times \cdots \times \frac{df_n}{dx}. \tag{EQ7.10}$$

Setting $\quad t = x^2 - 3 \quad$ gives $\quad \dfrac{dt}{dx} = 2x.$

Setting $\quad u = \sin(t) \quad$ gives $\quad \dfrac{du}{dt} = \cos(t).$

Setting $\quad v = \dfrac{1}{u} \quad$ gives $\quad \dfrac{dv}{du} = -u^{-2}.$

Setting $\quad y = \ln(v) \quad$ gives $\quad \dfrac{dy}{dv} = \dfrac{1}{v}$ (see Table 5.3).

Putting all of this together, we find that

$$\frac{dy}{dx} = \frac{dy}{dv} \times \frac{dv}{du} \times \frac{du}{dt} \times \frac{dt}{dx}$$

$$= \frac{1}{v} \times -\frac{1}{u^2} \times \cos(t) \times 2x = -\frac{1}{u} \times \cos(t) \times 2x = -\frac{\cos(t)}{\sin(t)} \times 2x.$$

After tidying up, so that the invented functions $v$, $u$, and $t$ are written in terms of the original variable $x$, this gives

$$\frac{dy}{dx} = \frac{-2x}{\tan(x^2 - 3)}.$$

# 7.3 Differentiating a product of two functions

Products of simple functions are frequently encountered in mathematical models of biological systems. In glycolysis, at the start of the pathway that generates ATP from glucose, we find that glucose is converted into glucose-6-phosphate (G6P), which is then transformed into fructose-6-phosphate (Figure 7.7):

$$\text{glucose} \xrightarrow{\text{hexokinase}} \text{glucose-6-phosphate} \xleftrightarrow{\text{G6P isomerase}} \text{fructose-6-phosphate}.$$

According to one model, at time $t$ the concentration $c$ of the intermediate G6P can be described by the physical value equation:

$$c(t) = At^2 e^{-Bt}, \tag{EQ7.11}$$

where $A$ and $B$ are positive constants. To analyze the rate at which the G6P concentration varies with time, we need a rule for working out the derivative of a product of two functions.

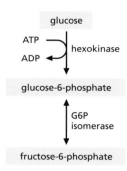

**Figure 7.7**
The first two steps of the glycolytic pathway.

## Box 7.3 Deriving the product rule

To derive the product rule formula, we need to start by allowing an increment $\delta x$ in $x$ that produces a corresponding increment $\delta y$ in the function $y$, which can be written as the product of $u$ and $v$, which are also functions of $x$:

$$y + \delta y = (u + \delta u)(v + \delta v) = uv + v\delta u + u\delta v + \delta u\delta v.$$

Because $y = uv$, we can subtract $y$ from the left-hand side of this last equation and $uv$ from the right-hand side, to get an expression for $\delta y$:

$$\delta y = v\delta u + u\delta v + \delta u\delta v.$$

As we make $\delta x$ get smaller, *all* of the terms in this last equation decrease in size, but the last term, $\delta u\delta v$,

approaches zero *more rapidly* than any of the others. We can see this in pictorial form in Figure 1. We can therefore make an approximation by setting the $\delta u\delta v$ term equal to zero. After dividing the remaining terms by $\delta x$, we have:

$$\frac{\delta y}{\delta x} \approx \frac{\delta u}{\delta x}v + u\frac{\delta v}{\delta x}.$$

At the limit when $\delta x$ tends to zero, the approximation will become exact, so we can replace all of the ratios by derivatives:

$$\frac{dy}{dx} = \lim_{\delta x \to 0}\left(\frac{\delta y}{\delta x}\right) = \frac{du}{dx}v + u\frac{dv}{dx}.$$

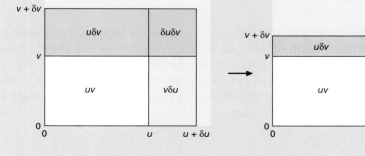

**Figure 1**
Showing how the product $(u + \delta u)(v + \delta v)$ changes with the size of $\delta u$ and $\delta v$. As $\delta x$ gets smaller, *all* of the terms in $v\delta u + u\delta v + \delta u\delta v$ decrease in size, but $\delta u\delta v$ approaches zero *more rapidly* than any of the others.

If both $u(x)$ and $v(x)$ are functions of the same variable $x$, then

$$\frac{\mathrm{d}}{\mathrm{d}x}(uv) = \frac{\mathrm{d}u}{\mathrm{d}x}v + u\frac{\mathrm{d}v}{\mathrm{d}x}.$$    (EQ7.12)

In other words, to differentiate the product of two functions we need to *multiply the derivative of the first function by the second function left alone, then add to this the first function left alone multiplied by the derivative of the second.* Further details of where the product rule comes from are outlined in Box 7.3.

To apply the product rule to the time dependence of the G6P concentration, we start by setting:

$$u = At^2 \qquad \text{and} \qquad v = e^{-Bt}.$$

Using the power function rule introduced in Section 5.4, we can show that

$$\frac{\mathrm{d}u}{\mathrm{d}t} = A \times 2 \times t^{2-1} = 2At.$$

The properties of the exponential function will be discussed in more detail in the following chapter, but for now just applying the appropriate entry from the table of standard derivatives in Table 5.3 reveals that:

$$\frac{\mathrm{d}v}{\mathrm{d}t} = 1 \times (-B) \times e^{-Bt} = -Be^{-Bt}.$$

Assembling these components together produces an expression for the rate of change of $c$, the concentration of G6P, with respect to time, $t$:

$$\frac{\mathrm{d}c}{\mathrm{d}t} = \frac{\mathrm{d}u}{\mathrm{d}t}v + u\frac{\mathrm{d}v}{\mathrm{d}t} = (2At) \times (e^{-Bt}) + (At^2) \times (-Be^{-Bt})$$

$$= 2Ate^{-Bt} - ABt^2e^{-Bt} = At\,(2 - Bt)\,e^{-Bt}.$$

This derivative could be used in several ways, for example to identify when the maximum concentration of G6P occurs, using the methods described in Section 5.7.

# 7.4 Differentiating a ratio of two functions

Ratios of simple functions are also quite common in biology. For example, changes in the metabolism of a boa constrictor as it digests and absorbs food can be monitored by measuring the rate at which it consumes oxygen after a large meal. A mathematical model (see Figure 7.8) predicts that if $R$ is the rate of oxygen consumption in milliliters per gram of reptile per hour and $t$ is the time in hours since the meal, then:

$$R = \frac{at}{1 + bt^2},$$

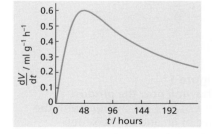

**Figure 7.8**
Graph showing the rate of oxygen consumption of the boa constrictor as a function of time after its meal.

where $a$ and $b$ are constants. To determine when the snake consumes oxygen most rapidly, we need to deploy yet another tool for differentiation: the quotient rule.

Setting $u(t) = at$ and $v(t) = 1 + bt^2$, then $R(t) = u(t)/v(t)$ and we have:

$$\frac{du}{dt} = a \text{ and } \frac{dv}{dt} = 2bt.$$

Applying the quotient rule (see Box 7.4):

$$\frac{d}{dt}\left(\frac{y}{v}\right) = \frac{\dfrac{du}{dt}v - u\dfrac{dv}{dt}}{v^2} \qquad\qquad \text{(EQ7.13)}$$

we find that:

$$\frac{dR}{dt} = \frac{a \times (1 + bt^2) - at \times 2bt}{(1 + bt^2)^2} = \frac{a + abt^2 - 2abt^2}{(1 + bt^2)^2}$$

$$= \frac{a(1 - bt^2)}{(1 + bt^2)^2}.$$

The snake consumes oxygen at its maximum rate when $dR/dt = 0$, which can only occur when $1 - bt^2 = 0$, that is, when $t = \sqrt{1/b}$.

Returning to the theme of trigonometric applications, recall that the tangent function is a ratio of the sine and cosine functions. The tangent function is periodic (see Figure 7.9A), but unlike the sine and the cosine it is not continuous. Instead, the slope regularly shoots off toward $\pm\infty$, with vertical asymptotes at $x = \pi/2$, $3\pi/2$, $5\pi/2$, and so on. These asymptotes occur because of the relationship:

$$\tan(x) = \frac{\sin(x)}{\cos(x)}. \qquad\qquad \text{(EQ7.14)}$$

## Box 7.4 Deriving the quotient rule

A ratio can be rewritten as a product in which the second term is raised to a negative power, so the equation $y = u/v$ is the same as $y = uv^{-1}$. We can use the product rule to differentiate this as follows:

$$\frac{dy}{dx} = \frac{d}{dx}\left(\frac{u}{v}\right) = \frac{d}{dx}(uv^{-1}) = \frac{du}{dx} \times (v^{-1}) + u \times \frac{d}{dx}(v^{-1}).$$

If we set $v^{-1}$ in the second term on the far right-hand side equal to a new function $w$, it should become clear that we can differentiate the expression in brackets using the chain rule, because $w$ is a function of $v$, which itself is a function of $x$. The differentiation goes as follows:

$$\frac{d}{dx}(v^{-1}) = \frac{dw}{dx} = \frac{dw}{dv} \times \frac{dv}{dx} = \frac{d}{dx}(v^{-1}) \times \frac{dv}{dx} = -v^{-2} \times \frac{dv}{dx}.$$

Substituting this into the first expression, we get the following expression for the derivative of $y$ with respect to $x$:

$$\frac{dy}{dx} = \frac{du}{dx} \times (v^{-1}) - u \times v^{-2} \times \frac{dv}{dx} = \frac{\dfrac{du}{dx} \times v - u \times \dfrac{dv}{dx}}{v^2}.$$

**Figure 7.9**
Plots of the functions (A) tan(x) and (B) 1/cos²(x) for values of x between 0 and 4π.

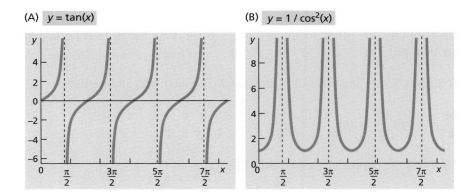

(A) $y = \tan(x)$

(B) $y = 1 / \cos^2(x)$

When the graph of the cosine function periodically cuts the $x$ axis, the denominator becomes zero, so the ratio $\sin(x)/\cos(x)$ becomes infinite. The slope of $\tan(x)$ is positive throughout the range $0 \leqslant x \leqslant 2\pi$, but tends to infinity at $x = \pi/2, 3\pi/2, 5\pi/2$, and so on. These features produce yet another interesting periodic function (see Figure 7.9B) – but how can we quantify its shape?

To make progress we need to deploy the quotient rule. Setting $u = \sin(x)$ and $v = \cos(x)$ means that:

$$\frac{du}{dx} = \frac{d}{dx}(\sin(x)) = \cos(x) \qquad \text{and} \qquad \frac{dv}{dx} = \frac{d}{dx}(\cos(x)) = -\sin(x).$$

The quotient rule now indicates that:

$$\frac{d}{dx}(\tan(x)) = \frac{d}{dx}\left(\frac{u}{v}\right) = \frac{\frac{du}{dx}v - u\frac{dv}{dx}}{v^2}$$

$$= \frac{\cos(x) \times \cos(x) - \sin(x) \times (-\sin(x))}{\cos^2(x)} = \frac{\cos^2(x) + \sin^2(x)}{\cos^2(x)}.$$

We can tidy up this result by recalling the version of Pythagoras' theorem summarized in EQ4.15:

$$\sin^2(x) + \cos^2(x) = 1. \qquad \text{(EQ7.15)}$$

Therefore

$$\frac{d}{dx}(\tan(x)) = \frac{1}{\cos^2(x)}.$$

In trigonometry, the reciprocal of the cosine of an angle is referred to as a secant, leading to the relationship:

$$\sec(x) = \frac{1}{\cos(x)}.$$

Hence we can write

$$\frac{\mathrm{d}}{\mathrm{d}x}(\tan(x)) = \sec^2(x). \tag{EQ7.16}$$

Figure 7.9B indicates that the value of $\sec^2(x)$ is always greater than zero but shoots off to infinity at $x = \pi/2,\ 3\pi/2,\ 5\pi/2,\ \ldots$.

## 7.5 Analyzing asymptotes

The graphs of $\tan(x)$ and $\sec^2(x)$ in Figure 7.9 both contain obvious vertical asymptotes, so in this section we take a detour to examine asymptotic behavior more closely. A simpler example of a function that contains a vertical asymptote is a so-called 'doomsday equation', such as that obtained after fieldwork which determined $n$, the average population per square meter of flatworms, as a function of time $t$ (in months) after they had colonized a small island:

$$n(t) = \frac{60}{120 - t}. \tag{EQ7.17}$$

Flatworms devour common earthworms, which are ecologically important because they aerate and mix the soil, making it easier for plants to absorb nutrients. A close analysis of EQ7.17 should help us to understand the time range over which the population model is valid, as well as demonstrating the level of damage that the invading species could potentially wreak.

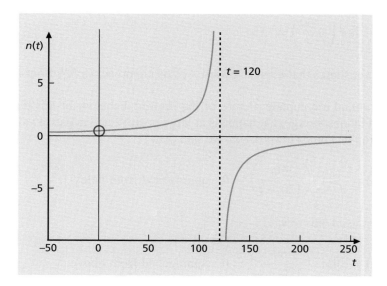

**Figure 7.10**
Plot of the doomsday function: $y = 60/(120 - t)$ for values of $t$ between $-50$ and $250$.

The plot intersects the $n$ axis at $t = 0$, where $n(0) = 60/(120 - 0) = 0.5$. There is no finite value of $t$ that could produce a population of zero, so the curve does not have an intercept with the $t$ axis. In addition, analysis of the first and second derivatives of $n$ would demonstrate that the curve lacks many typical features, such as maximum or minimum turning points, or points of inflexion. The main feature is that as $t$ approaches 120, the denominator approaches zero and the value of $n$ zooms off to infinity, creating a vertical asymptote about the line $t = 120$ (see Figure 7.10).

We can investigate the behavior of the doomsday function in the vicinity of the asymptote by making two useful substitutions. First, to examine the low $t$ side of the asymptote we will replace $t$ with $(120 - \varepsilon)$, where $\varepsilon$ is a small positive number, giving:

$$n = \frac{60}{120 - (120 - \varepsilon)} = \frac{60}{\varepsilon}.$$

If we set $\varepsilon$ to a small value, such as 0.001, then $n$ will be large (that is, $n = 60/0.001 = 60\,000$). If even smaller values are selected for $\varepsilon$, $n$ remains positive but becomes increasingly huge. Using the nomenclature introduced in Section 5.2, the limit as $\varepsilon$ tends toward zero can be expressed as:

$$\lim_{\varepsilon \to 0}(n) = \lim_{\varepsilon \to 0}\left(\frac{60}{\varepsilon}\right) = +\infty.$$

In other words, as $t$ approaches 120 from below, $n$ tends toward $+\infty$.

Similarly, to examine the high $t$ side of the asymptote, we should replace $t$ with $(120 + \varepsilon)$, giving:

$$n = \frac{60}{120 - (120 + \varepsilon)} = -\frac{60}{\varepsilon}.$$

In the limit as $\varepsilon$ tends toward zero:

$$\lim_{\varepsilon \to 0}(n) = \lim_{\varepsilon \to 0}\left(-\frac{60}{\varepsilon}\right) = -\infty.$$

This demonstrates that $n$ tends toward $-\infty$ as $t$ approaches 120 from above.

We can obtain a complete picture of the limiting behavior of this doomsday equation by observing what happens at large positive and negative values of $t$. In the limit as $t$ tends toward $+\infty$, we find that:

$$\lim_{t \to +\infty}(n) = \lim_{t \to +\infty}\left(\frac{60}{120 - t}\right) = 0 \quad \text{approached from below.}$$

Similarly, as $t$ tends toward $-\infty$:

$$\lim_{t \to -\infty}(n) = \lim_{t \to -\infty}\left(\frac{60}{120 - t}\right) = 0 \quad \text{approached from above.}$$

These results show that the doomsday equation has a horizontal asymptote around the line $n = 0$.

All of the information about the doomsday function is assembled on an axis system in Figure 7.10. In terms of evaluating this function as a mathematical model for the flatworm population, it is worth emphasizing the obvious point that a negative population has no meaning. It would therefore be inappropriate to apply the model when values of $t$ are greater than 120. When data were first collected, the flatworm population was quite low and was increasing only slowly. According to the model, the colonization process appears not to be too serious during the first few years. However, if growth continues unchecked and no other factors come into play, the model predicts that the population of flatworms will explode during the tenth year. As there are no proven chemical methods of control, now would therefore be a good time for the government to start funding research into how to avoid this potential ecological disaster.

# 7.6 Changing the variable of an integral

The last few sections have expanded the repertoire of tools that we have available for differentiation. In particular, the chain rule has equipped us to differentiate nested functions, such as $y(t) = A \sin[(2\pi t / T) + \phi] + y_0$, by making judicious substitutions. It would be useful to match this ability with an equivalent approach for integration.

Fortunately, it is also possible to introduce substitutions that can simplify an expression, and so make it easier to integrate. There are many different ways for doing this, but one common strategy is to set a new function, $u$, equal to the most complicated part of the function that we want to integrate. Often, but not always, this corresponds to an expression in brackets.

Suppose that the population of toads in a wetland ecosystem is 100 at the beginning of a study and that the rate of growth of the population, $p(t)$ toads, after $t$ years is given by the equation

$$\frac{\mathrm{d}p(t)}{\mathrm{d}t} = (9 + 2t)^{1/2}.$$  (EQ7.18)

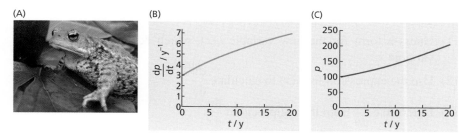

**Figure 7.11**
(A) A toad. Image courtesy of timitalia under Creative Commons Attribution 2.0 Generic. (B) Plot of rate of change of toad population against time. (C) Plot of toad population against time.

This relationship could be used to estimate the number of toads 10 years later (Figure 7.11). To accomplish this, as we found in Section 6.5, we must solve the integral:

$$p(t) = \int (9 + 2t)^{1/2} \, \mathrm{d}t.$$  (EQ7.19)

If the bracketed expression in the integrand was raised to a simple integer power, for example $(9 + 2t)^2$, we might consider expanding the brackets to produce three terms (that is, $81 + 36t + 4t^2$) and then integrating each term separately. Unfortunately, in EQ7.19 the term in brackets is raised to a non-integer power, namely 1/2, indicating that a square root is required, so the approach of expanding the brackets is not appropriate. However, from Section 6.2, we know how to use the power function formula to integrate simpler expressions, such as $x^{1/2}$:

$$\int x^{1/2}\,\mathrm{d}x = \frac{x^{(1/2)+1}}{(1/2)+1} + c = \frac{2}{3}x^{3/2} + c. \tag{EQ7.20}$$

It is therefore tempting to make a substitution with a new variable $u$ by setting $u = 9 + 2t$, which would transform EQ7.19 to:

$$p(t) = \int u^{1/2}\,\mathrm{d}t. \tag{EQ7.21}$$

This has certainly simplified the integrand, but sadly it has also introduced a new problem: we now need to discover how to integrate a function of one variable ($u$) with respect to a different variable ($t$).

The appropriate procedure is summarized in the rule for 'changing the variable of an integral', which is the counterpart of the chain rule for differentiation introduced in Section 7.2. If $f(x)$ can be written in terms of a new variable $u$, such that $u$ is also a function of $x$ (that is, $u = u(x)$), then $x$ can be written as a function of $u$ (that is, $x = x(u)$). The change of variable rule states that:

$$\int f(x)\,\mathrm{d}x = \int f(x(u)) \times \frac{\mathrm{d}x}{\mathrm{d}u}\,\mathrm{d}u = \int g(u)\,\mathrm{d}u \tag{EQ7.22}$$

where $g(u)$ represents the function $f(x)$ expressed in terms of the new variable $u$. The recipe for deploying the rule is then:

(1) Guess a form for the new variable $u(x)$, which depends on the original variable $x$.

(2) Use the expression for $u(x)$ to calculate $\dfrac{\mathrm{d}u}{\mathrm{d}x}$.

(3) Determine $\dfrac{\mathrm{d}x}{\mathrm{d}u}$ from the reciprocal of $\dfrac{\mathrm{d}u}{\mathrm{d}x}$.

(4) In the integral, rewrite $f(x)$ in terms of $u$.

(5) In the integral, replace '$\mathrm{d}x$' by '$\dfrac{\mathrm{d}x}{\mathrm{d}u}\,\mathrm{d}u$'.

(6) Solve the integral with respect to the new variable $u$.

The hardest part of applying the change of variable rule is that it relies on choosing a suitable substitution; that is, the best possible form for $u(x)$. The appropriate substitution can be difficult to spot, so the best way to learn is by trying a few examples (such as questions 4, 5, and 6 from the End of Chapter Questions).

First, returning to the population of toads in the wetland study, we found that:

$$p(t) = \int (9 + 2t)^{1/2}\, \mathrm{d}t.$$

Choosing to make a substitution for the expression in brackets gives $u = 9 + 2t$. Differentiating $u$ with respect to $t$ then yields:

$$\frac{\mathrm{d}u}{\mathrm{d}t} = 2, \qquad \text{so therefore} \qquad \frac{\mathrm{d}t}{\mathrm{d}u} = \left(\frac{\mathrm{d}u}{\mathrm{d}t}\right)^{-1} = \frac{1}{2}.$$

Placing this information into the change of variable formula, we find that:

$$p = \int u^{1/2} \times \frac{\mathrm{d}t}{\mathrm{d}u}\, \mathrm{d}u = \frac{1}{2}\int u^{1/2}\, \mathrm{d}u = \frac{1}{2} \times \frac{1}{3/2} \times u^{3/2} + c$$

$$= \frac{1}{2} \times \frac{2}{3} \times u^{3/2} + c = \frac{1}{3} u^{3/2} + c.$$

It is usually best to substitute back for $u$ to produce an answer in terms of the original variable $t$:

$$p(t) = \frac{1}{3}(9 + 2t)^{3/2} + c. \tag{EQ7.23}$$

We were provided with extra information about the initial population of the colony, meaning that $p(0) = 100$, so in this case the value of the constant of integration $c$ can be determined:

$$100 = \frac{1}{3}(9)^{3/2} + c = \frac{1}{3}(3)^3 + c = 9 + c,$$

that is,

$$c = 100 - 9 = 91,$$

so that

$$p(t) = \frac{1}{3}(9 + 2t)^{3/2} + 91.$$

This solution makes it possible for us to predict that after 10 years the population of toads will be:

$$p(10) = \frac{1}{3}(9 + 2 \times 10)^{3/2} + 91 = 143.06.$$

It would obviously be ridiculous to predict a fraction of a toad, so our final answer for the population after 10 years should be rounded to the nearest integer, giving 143.

More general functions, such as $y = (ax + b)^n$, where $a$, $b$, and $n$ are constants, can be integrated using the same strategy. If we set $u = ax + b$, then it should now be simple to show that

$$\frac{du}{dx} = \frac{d}{dx}(ax + b) = a, \qquad \text{and so} \qquad \frac{dx}{du} = \left(\frac{du}{dx}\right)^{-1} = \frac{1}{a}.$$

Changing the variable of integration from $x$ to $u$ therefore gives:

$$\int (ax + b)^n \, dx = \int u^n \times \frac{dx}{du} \, du = \int \frac{u^n}{a} \, du$$

$$= \frac{1}{a(n + 1)} u^{n+1} + c = \frac{1}{a(n + 1)} (ax + b)^{n+1} + c. \tag{EQ7.24}$$

If we set $a = 2$, $b = 9$, and $n = 1/2$, this expression predicts the result we obtained in EQ7.23. With a bit of practice, the same rule can be used to integrate functions of the form $(a + bx)^n$ by inspection, enabling the answer to be written down directly without too many lines of tedious working in between. The trick is to notice that the value of the power $n$ has been increased by 1, while the result is divided by the new index $(n + 1)$, and then once again by the derivative of the bracketed expression.

The change of variable method can be applied to more than just variations on the theme of power functions: it also makes it possible to integrate modified trigonometric functions, such as $y = \cos(ax + b)$, where $a$ and $b$ are constants. If we substitute for the expression in brackets once again, setting $u = ax + b$, so that

$$y = \cos(u), \qquad \text{then} \qquad \frac{du}{dx} = a, \qquad \text{and} \qquad \frac{dx}{du} = \left(\frac{du}{dx}\right)^{-1} = \frac{1}{a}.$$

The integral of $y$ with respect to $x$ is therefore given by:

$$\int y \, dx = \int \cos(ax + b) \, dx = \int \cos(u) \times \frac{dx}{du} \, du = \int \frac{\cos(u)}{a} \, du$$

$$= \frac{\sin(u)}{a} + c = \frac{1}{a} \sin(ax + b) + c. \tag{EQ7.25}$$

Deciding on the appropriate substitution is not always so obvious, although choosing a convenient function gets easier with practice. An example of a more challenging substitution would be to solve the integral:

$$I = \int \sin(x) \cos(x) \, dx.$$

Here the integrand is made up of the product of two trig functions: $\sin(x)$ and $\cos(x)$, which we already know have an interesting connection, in that $\cos(x)$ is the derivative of $\sin(x)$. If we start out by making the substitution $u = \sin(x)$, we can show that:

$$\frac{du}{dx} = \cos(x) \quad \text{and} \quad \frac{dx}{du} = \left(\frac{du}{dx}\right)^{-1} = \frac{1}{\cos(x)}.$$

The next task is to change the variable of the integral:

$$I = \int \sin(x)\cos(x)\,dx = \int u \times \cos(x) \times \frac{dx}{du}\,du$$

$$= \int u \times \cos(x) \times \frac{1}{\cos(x)}\,du = \int u\,du = \frac{1}{2}u^2 + c.$$

Finally, substituting back for $u$ gives:

$$I = \frac{1}{2}\sin^2(x) + c.$$

In this case, a term in the integrand that contained the variable $x$ (that is, $\cos(x)$) remained behind after we substituted $\sin(x)$ with $u$. Fortunately, this did not spoil the solution, because the $\cos(x)$ term canceled with the $dx/du$ term.

It is worth noting that no substitution we choose could be described as 'wrong'; it is just that some are more helpful than others. Signs that you may have made an unhelpful substitution include: terms that contain the original variable surviving in the integrand; or being left with an expression that is even more difficult to integrate than the one you started with. When the whole process shudders to a halt, the best course of action is to pause, take a deep breath, go back to the start, and consider whether a different choice of substitution would make progress easier.

The change of variable method works for definite integrals as well as the indefinite integrals used as examples in this section until now. However, if a substitution is to be made in a definite integral, special care must be taken to ensure that the upper and lower limits of a definite integral are applied correctly. There are two possible approaches:

(1) Leave the values of the upper and lower limits unchanged, but make sure that the result is written only in terms of the original variable before these limits are applied.
(2) Write the result only in terms of the new variable, but redefine the upper and lower limits to reflect the fact that they now refer to the new variable.

To emphasize this, the change of variable rule can be rewritten in a slightly different form for a definite integral:

$$\int_{x=a}^{x=b} f(x)\,dx = \int_{u=u(a)}^{u=u(b)} f(x(u)) \times \frac{dx}{du}\,du = \int_{u=u(a)}^{u=u(b)} g(u)\,du. \qquad \text{(EQ7.26)}$$

For example, returning to the wetland study set out in EQ7.18, the mean rate of increase in the toad population over ten years is described by the definite integral:

$$\bar{R} = \frac{1}{10-0} \int_{t=0}^{t=10} (9+2t)^{1/2} \, dt.$$

Using the first approach, we decide to leave the values of the upper and lower limits written in terms of the original variable $t$, but then have to make sure that only terms that contain $t$ remain before finally evaluating the square brackets:

$$\bar{R} = \frac{1}{10} \times \left[ \frac{1}{3}(9+2t)^{3/2} \right]_{t=0}^{t=10} = \frac{1}{30} \times \left[ (29)^{3/2} - (9)^{3/2} \right] = 4.31.$$

Alternatively, adopting the second approach, after performing the integral we leave the expression written in terms of the substitute variable $u = 9 + 2t$, but must then convert the upper and lower limits into boundaries for $u$ before evaluating the square brackets:

$$\bar{R} = \frac{1}{10} \times \left[ \frac{1}{3} u^{3/2} \right]_{t=0}^{t=10} = \frac{1}{30} \times \left[ u^{3/2} \right]_{u=9}^{u=29}$$

$$= \frac{1}{30} \times \left[ (29)^{3/2} - (9)^{3/2} \right] = 4.31.$$

Either way, we have discovered that the population increases by an average of 4.31 toads per year.

## 7.7 Using a table of standard integrals

The method of differentiation from first principles was introduced in Section 5.3 to provide a way of obtaining the derivative of any given function, but no equivalent protocol exists for working out any given indefinite integral. Fortunately, there is no need to despair, because we can rely instead on tables of standard integrals that have been compiled by experts. An example of such a table is shown in Table 7.1.

If the problem we need to tackle is a perfect match with an entry in the table of standard integrals, then the appropriate result should simply be plugged in. For example, to solve:

$$I = \int_{p=0}^{p=3} 3^p \, dp,$$

we simply need to recognize that the integrand has exactly the same form as the entry for '$a^x$' in Table 7.1. Following that, we can work our way through the evaluation of the upper and lower limits:

$$I = \int_{p=0}^{p=3} 3^p \, dp = \left[ \frac{3^p}{\ln(3)} \right]_{p=0}^{p=3} = \frac{1}{\ln(3)} \left[ (3)^3 - (3)^0 \right] = 23.67.$$

## Table 7.1  A table of standard integrals

| function, $f(x)$ | integral, $\int f(x)\,dx$ |
|---|---|
| $x^n$ | $\dfrac{x^{n+1}}{n+1} + c$   (for $n \neq -1$) |
| $\dfrac{1}{x}$ | $\ln(x) + c$ |
| $e^x$ | $e^x + c$ |
| $a^x$ | $\dfrac{a^x}{\ln(a)} + c$ |
| $\ln(x)$ | $x\ln(x) - x + c$ |
| $\sin(x)$ | $-\cos(x) + c$ |
| $\cos(x)$ | $\sin(x) + c$ |
| $\tan(x)$ | $\ln\left(\dfrac{1}{\cos(x)}\right) + c$ |
| $\sin^2(x)$ | $\dfrac{1}{2}x - \dfrac{1}{2}\sin(x)\cos(x) + c$ |
| $\cos^2(x)$ | $\dfrac{1}{2}x + \dfrac{1}{2}\sin(x)\cos(x) + c$ |
| $\dfrac{1}{\sqrt{1-x^2}}$ | $\sin^{-1}(x) + c$ |
| $\dfrac{-1}{\sqrt{1-x^2}}$ | $\cos^{-1}(x) + c$ |
| $\dfrac{1}{x^2+1}$ | $\tan^{-1}(x) + c$ |

Alternatively, we might be presented with an integral that *almost* corresponds to a standard form, but must be handled using the normal rules for manipulating integrals described in Section 6.2. Problems of this sort include sums and differences of standard functions, or standard forms multiplied by a constant. For example:

$$I = \int \{12\ln(x) + 33\sin(x)\}\,dx = 12\int \ln(x)\,dx + 33\int \sin(x)\,dx$$

$$= 12x\ln(x) - 12x - 33\cos(x) + c.$$

On frequent occasions, an integrand must be transformed into a more tractable form, for example by using the change of variable rule. As an application, if the rate of growth of a rabbit is modeled by:

$$\frac{dL}{dt} = \frac{A}{1 + (t-m)^2},$$

where $L$ is the nose-to-tail length, $A$ and $m$ are both constants, and $t$ is the time since birth in months, then the size of a rabbit of a given age can be determined from the integral:

$$L = \int \frac{A}{1 + (t-m)^2}\,dt.$$

(A)

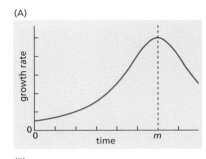

(B)

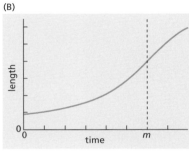

**Figure 7.12**
Figure showing (A) the growth rate and (B) the nose-to-tail length of a rabbit as functions of time.

If we make the substitution $u = t - m$, then $du/dt = 1$ and $dt/du = (du/dt)^{-1} = 1$. Feeding this information into the change of variable formula gives:

$$L = \int \frac{A}{1 + u^2} \times \frac{dt}{du}\, du = \int \frac{A}{1 + u^2}\, du.$$

Next, inserting the appropriate standard integral from Table 7.1 indicates that the answer contains an inverse tangent function:

$$L = A \tan^{-1}(u) + c = A \tan^{-1}(t - m) + c.$$

As Figure 7.12 shows, the rabbit experiences a growth spurt that produces a point of inflexion at month $m$ in a plot of its length as a function of time, corresponding to the maximum in the growth rate curve.

A table of standard integrals can also serve quite well as a table of standard derivatives. All that is required here is to follow the appropriate row in Table 7.1 from right to left instead of from left to right. For example, in a study of the forces exerted on the foot of a sprinting athlete, the leg can be modeled simply as two segments jointed at the knee (Figure 7.13). The angle $\alpha$ between the vertical and the line that joins the foot to the hip ($FH$) is given by:

$$\alpha = \tan^{-1}\left(\frac{x}{h}\right).$$

To obtain the derivative of $\alpha$ with respect to $x$, we first need to make the substitution:

$$u = \frac{x}{h}, \qquad \text{so that} \qquad \frac{du}{dx} = \frac{1}{h}.$$

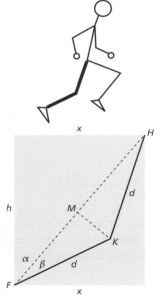

**Figure 7.13**
Diagram of a sprinting athlete, showing a two-segment model of the leg (*F*, foot; *K*, knee; *H*, hip). Image courtesy of Mark Sadowski under Creative Commons Attribution – Share Alike 2.0 Generic.

Applying the chain rule now gives:

$$\frac{d\alpha}{dx} = \frac{d\alpha}{du} \times \frac{du}{dx} = \frac{d}{du}\left\{\tan^{-1}(u)\right\} \times \frac{1}{h} = \frac{1}{u^2 + 1} \times \frac{1}{h}.$$

And substituting back in for $x$ gives:

$$\frac{d\alpha}{dx} = \frac{1}{(x/h)^2 + 1} \times \frac{1}{h} = \frac{h}{x^2 + h^2}.$$

Analysis of the stress exerted on the ankle during motion can lead to improvements in sprinting technique, as well as minimizing the risk of injuring the athlete.

The table of integrals given in Table 7.1 is quite short, but more comprehensive lists are available, such as those in *CRC Standard Mathematical Tables and Formulae*™. In addition, several computer programming environments, including *Mathematica*™ and *Maple*™, can use symbolic logic to evaluate definite and indefinite integrals.

# 7.8 Simple harmonic motion

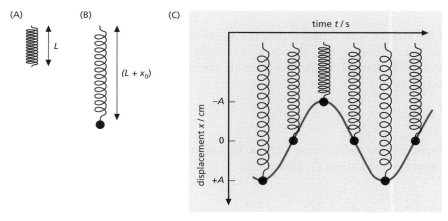

**Figure 7.14**
Figure illustrating simple harmonic motion by the motion of a loaded spring oscillating about its point of equilibrium. (A) Length of the unloaded spring; (B) length of the loaded spring at equilibrium; (C) after extending and releasing, the displacement of the loaded spring varies with time in a sinusoidal manner.

Simple harmonic oscillations, in which a variable changes sinusoidally (see Section 4.6), can occur whenever a system possesses inertia alongside a restoring force that is proportional to the displacement from equilibrium conditions. The simplest examples of this type of motion are a pendulum (see Section 4.6) or a weight bobbing up and down on the end of a spring (see Figure 7.14), but similar principles apply to the wing-beats of a hovering hummingbird and to the atoms vibrating at either end of a covalent bond in a biological metabolite or a protein molecule. The net force on the weight, determined by gravity pulling down (a positive force in this case) plus the spring pulling upward (a negative force), is zero when the weight is at its rest position. If we assume that the spring has negligible mass, the downward gravitational force is given by

$$F_{\text{grav}} = mg, \tag{EQ7.27}$$

where the $m$ is the mass of the bob in kilograms and $g$ is the acceleration due to gravity. As the spring is stretched further, the strength of the negative restoring force increases according to Hooke's law:

$$F_{\text{spring}} = -k(x_0 + x), \tag{EQ7.28}$$

where $k$ is the stiffness constant (in units of per centimeter) and $x$ (in centimeters) is the displacement from $x_0$, the distance that the loaded spring at equilibrium extends from its unloaded length $L$ (see Figure 7.14). The net force $F$ on the bob is therefore:

$$F = F_{\text{grav}} + F_{\text{spring}} = mg - k(x_0 + x).$$

At equilibrium, the net force is zero, so

$$mg - k(x_0 + 0) = 0; \quad \text{that is,} \quad x_0 = \frac{mg}{k}. \tag{EQ7.29}$$

The equilibrium displacement $x_0$ is therefore determined by both the mass of the bob and the properties of the spring (with $k$ representing its stiffness). The net force on the bob will be negative for positive values of the displacement, but $F$ will be positive when $x$ is negative.

If the weight is now pulled downward to $x = A$, held still for a few seconds, and then released, its velocity will initially be zero. However, the bob will accelerate upward in the direction of the net force according to Newton's second law:

$$F = ma, \tag{EQ7.30}$$

where $a$ is the acceleration (that is, the second derivative of the displacement $x$ with respect to time $t$). Therefore

$$ma = m\frac{\mathrm{d}^2 x}{\mathrm{d}t^2} = mg - k(x_0 + x) = (mg - kx_0) - kx.$$

But from EQ7.29 we know that

$$x_0 = \frac{mg}{k},$$

which gives

$$mg - kx_0 - kx = -kx,$$

so that

$$\frac{\mathrm{d}^2 x}{\mathrm{d}t^2} = -\frac{k}{m}x. \tag{EQ7.31}$$

EQ7.31 is an example of a differential equation, a topic that will be discussed further in Chapter 12. In this case, the differential equation defines simple harmonic motion (SHM).

As the weight moves upward, its velocity increases, but the restoring force diminishes, becoming zero when the spring is at its equilibrium length (that is, $x = 0$). At this point the bob possesses a negative velocity, so it will continue to move, entering the zone in which $x$ is negative. The spring is now being compressed, so it produces a restoring force in the downward (positive) direction, which gradually decelerates the bob until it comes to rest at $x = -A$. However, the restoring force continues to act, accelerating the weight in the downward direction, so that it now gains a positive velocity. The velocity continues to increase until the bob passes through $x = 0$, at which point the restoring force changes sign again, slowing it down until it comes to rest at $x = +A$. The whole cycle then repeats itself.

It can be shown that solutions to the SHM master equation (EQ7.31) all have the same general form:

$$x(t) = C_1 \cos(\omega t) + C_2 \sin(\omega t), \tag{EQ7.32}$$

where $C_1$, $C_2$, and $\omega$ are constants. To check that this is a genuine solution, we must differentiate twice with respect to $t$:

$$\frac{dx}{dt} = -C_1 \omega \sin(\omega t) + C_2 \omega \cos(\omega t). \tag{EQ7.33}$$

$$\frac{d^2x}{dt^2} = -C_1 \omega^2 \cos(\omega t) - C_2 \omega^2 \sin(\omega t) \tag{EQ7.34}$$

$$= -\omega^2 \{C_1 \cos(\omega t) + C_2 \sin(\omega t)\} = -\omega^2 x.$$

EQ7.31 and EQ7.34 are equivalent if:

$$\omega = \sqrt{\frac{k}{m}}.$$

The constant $\omega$ represents the 'angular frequency' of the oscillating system and is measured in units of radians per second (rad s$^{-1}$). The solution to the SHM master equation presented in EQ7.32 makes use of the unique calculus properties of the sine and cosine functions: the second derivative of $\sin(x)$ is $-\sin(x)$, whereas the second derivative of $\cos(x)$ is $-\cos(x)$.

For the case shown in Figure 7.14, we have enough information to determine the values of the constants $C_1$ and $C_2$. If $x = +A$ when $t = 0$, then from EQ7.32:

$$A = C_1 \cos\left(\sqrt{\frac{k}{m}} \times 0\right) + C_2 \sin\left(\sqrt{\frac{k}{m}} \times 0\right),$$

that is,

$C_1 = A.$

We also know that the bob is at rest at $t = 0$, so it has zero velocity. From EQ7.33, the equation for velocity, we find that:

$$0 = -A\sqrt{\frac{k}{m}}\sin\left(\sqrt{\frac{k}{m}} \times 0\right) + C_2\sqrt{\frac{k}{m}}\cos\left(\sqrt{\frac{k}{m}} \times 0\right),$$

that is,

$C_2 = 0.$

The motion of the bob is therefore completely described by the equation:

$$x(t) = A\cos\left(\sqrt{\frac{k}{m}}\, t\right).$$

Comparison with the master equation for sinusoidal motion (EQ7.1) and remembering that $\sin(\theta + (\pi/2)) = \cos(\theta)$ reveals that the amplitude of the oscillation is $A$, whereas the time period $T$ is given by:

$$\frac{2\pi}{T} = \sqrt{\frac{k}{m}} = \omega,$$

that is,

$$T = 2\pi\sqrt{\frac{m}{k}} = \frac{2\pi}{\omega}.$$

The frequency of the oscillation is given by the reciprocal of the time period:

$$f = \frac{1}{T} = \frac{1}{2\pi}\sqrt{\frac{k}{m}} = \frac{\omega}{2\pi}.$$

This relationship indicates that the weight will bob up and down slowly if it is heavy (that is, if $m$ is large), but move rapidly if the spring is very stiff (that is, if $k$ is large). In terms of molecular motions, this means that strong covalent bonds vibrate rapidly, but weaker hydrogen bonds oscillate more slowly.

# Presenting Your Work

### QUESTION

In a temperate climate, the length $L$ (measured in centimeters) of a freshly hatched river fish (a brown trout) depends on time $t$ according to the equation:

$$L(t) = t + \cos(t).$$

Plot this function for the range $0 \leqslant t \leqslant +13$ and identify the points of inflexion. What do you think the units of time should be? What time of year do you think the fish was hatched, assuming the fish grows more rapidly in warm water?

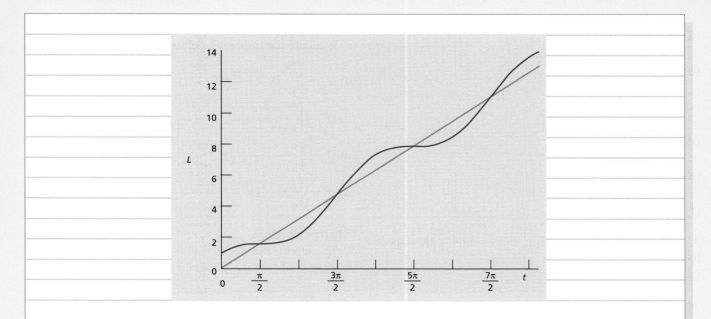

$L(t) = t + \cos(t)$, so $\dfrac{dL}{dt} = 1 - \sin(t)$ and $\dfrac{d^2L}{dt^2} = -\cos(t)$.

Stationary points occur when $\dfrac{dL}{dt} = 0$, that is, when $\sin(t) = +1$,

that is, when $t = \dfrac{\pi}{2}, \dfrac{5\pi}{2}, \cdots$.

Points of inflexion occur when $\dfrac{d^2L}{dt^2} = 0$,

that is, when $t = \dfrac{\pi}{2}, \dfrac{3\pi}{2}, \dfrac{5\pi}{2}, \cdots$.

It looks like $2\pi$ time units should represent 1 year, with the fish growing only when the water temperature exceeds a certain threshold. If this is the case, then the fish hatches in late autumn and growth stops shortly afterward in mid-winter. With spring and summer comes a growth spurt, which slows down again as autumn approaches.

## End of Chapter Questions    (Answers to questions can be found at the end of the book)

### Basic

**1.** Differentiate the following expressions with respect to $x$:
(a) $y = x + \sin(x)$
(b) $y = 0.5\cos(x) + x^2 + 3$
(c) $y = 3\sin(x) + 4\cos(x)$

**2.** Differentiate the following expressions with respect to $x$:

(a) $y = (3x - 1)^{12} - \dfrac{5}{x}$

(b) $y = 4 + \sin\left(\dfrac{x}{2}\right)$

(c) $y = x\sin(x)$

(d) $y = \dfrac{x}{x + 1}$

(e) $y = \dfrac{\sin^2(x)}{x^2}$

(f) $y = (2 + 3x^2)^4$

**3.** Differentiate the following expressions with respect to $x$:
(a) $y = (x - 1)(2x + 1)^{10}$

(b) $y = \dfrac{x^3 + p^3}{x + p}$

(c) $y = \dfrac{x^2 + 1}{x^2 - 1}$

(d) $y = y_0 + \cos\left(\dfrac{2\pi x}{T} + \phi\right)$

(e) $y = \dfrac{x}{\cos(x^3)}$

(f) $y = \dfrac{p^2 + 2pqx + q^2x^2}{p + qx}$

**4.** Integrate the following expressions with respect to $x$:
(a) $y = \cos(x) - \sin(x)$

(b) $y = 4 + \cos\left(\dfrac{x}{2}\right)$

(c) $y = y_0 + \sin\left(\dfrac{2\pi x}{T} + \phi\right)$

(d) $y = x\sin(x^2)$

**5.** Integrate the following expressions with respect to $x$:

(a) $y = \dfrac{\cos(x)}{\sin^4(x)}$

(b) $y = \dfrac{\cos(x)}{\sqrt{2 + \sin(x)}}$

(c) $y = \dfrac{x^2 - p^2}{x - p}$

(d) $y = (3x - 1)^{11} - 5/x^2$

**6.** Evaluate the following definite integrals:

(a) $I = \displaystyle\int_2^{10} 3\sqrt{x - 2} \, dx$

(b) $I = \displaystyle\int_0^8 \dfrac{4}{\sqrt{x + 3}} \, dx$

(c) $I = \displaystyle\int_0^1 \dfrac{x}{(x + 1)^3} \, dx$

(d) $I = \displaystyle\int_0^3 x \sqrt[3]{3x^2 + 2} \, dx$

### Intermediate

**7.** The length $l$ of a chord that connects any two points on a circle of radius $r$ is given by:

$$l = r\sqrt{2(1 - \cos(\alpha))},$$

where $\alpha$ is the angle obtained by connecting the two ends of the chord to the center of the circle. Determine the instantaneous rate of change of $l$ with respect to $\alpha$.

**8.** Use the linear approximation method to derive approximations for the following functions, by setting $x$ to zero and making $\delta x$ very small:
(a) If $y = \sqrt[3]{1 + x}$, show that $\sqrt[3]{1 + \delta x} \approx 1 + \delta x/3$;
(b) If $y = \cos(x)$, show that $\cos(\delta x) \approx 1$;
(c) If $y = 1/(1 + x)$, show that $1/(1 + \delta x) \approx 1 - \delta x$.
Determine the percentage error introduced by these approximations for $\delta x = 0.01, 0.1$, and $0.5$.

**9.** Sketch the function of the damped oscillation $y = \cos(t) \, e^{-t}$ within the range $0 \leqslant t \leqslant 2\pi$, indicating the coordinates of any boundary points, intercepts with the $t$ and $y$ axes, vertical or horizontal asymptotes, local maxima, local minima, or points of inflexion. (NB: $t$ is measured in radians.)

**10.** Environmental health officers monitoring an outbreak of food poisoning after a wedding banquet were able to model the time course of the recovery of guests using the equation:

$$r = \frac{100t}{1+t},$$

where $t$ represents the number of days since infection and $r$ is the percentage of guests who no longer display symptoms. Determine an expression for the rate of recovery.

**11.** Sketch the function

$$y = (x-1) + \frac{1}{(x-2)},$$

in the range $0 \leqslant x \leqslant +5$ and $-5 \leqslant y \leqslant +5$, indicating the coordinates of any boundary points, intercepts with the $x$ or $y$ axes, vertical or horizontal asymptotes, maximum or minimum turning points, or points of inflexion. What are the largest and smallest values of $y$ over this range?

**12.** In a kinematics study of the swimming motion of a trout, the velocity ($v$ in centimeters per second) of the anal fin tip depends on time ($t$ in milliseconds) according to the equation:

$$v(t) = 90 \cos\left(\frac{2\pi}{0.3}t\right).$$

Determine the total (unsigned) distance traveled by the fin tip during a single oscillation.

**13.** Find the mean values of the function $y = \sin(2t)$ over the ranges $0 \leqslant t \leqslant \pi/2$ and $0 \leqslant t \leqslant 2\pi$.

**14.** A population of frogs was estimated to be 100 000 at the beginning of the year 2000. Suppose that the rate of growth of the population $p(t)$ (in thousands) after $t$ years is given by the equation

$$\frac{dp(t)}{dt} = (4 + 0.15t)^{3/2}.$$

Estimate the population at the beginning of the year 2015.

**15.** In an experiment to measure elasticity, a tendon is suspended vertically with a mass $m$ hanging at the bottom. The mass is observed to perform simple harmonic motion according to the equation:

$$y = y_m \times \sin(\omega t + \pi),$$

where $y_m = 5$ mm and $\omega = 15$ s$^{-1}$.
(a) What are the amplitude, period, and frequency of the oscillation?
(b) Draw a graph showing the variation of $y$ with $t$ for two cycles of the oscillation.
(c) At what times is the displacement increasing most rapidly?
(d) The acceleration $a$ of the mass is given by:

$$a = -\frac{k}{m} \times y,$$

where $k$ is a measure of elasticity, and $m$ is 0.30 kg. What is the value of $k$, in appropriate units?

**Advanced**
**16.** The sine and cosine functions can be written in the form of the following infinite series:

$$\sin(x) = x - \frac{x^3}{3 \times 2 \times 1} + \frac{x^5}{5 \times 4 \times 3 \times 2 \times 1}$$
$$- \frac{x^7}{7 \times 6 \times 5 \times 4 \times 3 \times 2 \times 1} + \cdots;$$
$$\cos(x) = 1 - \frac{x^2}{2 \times 1} + \frac{x^4}{4 \times 3 \times 2 \times 1}$$
$$- \frac{x^6}{6 \times 5 \times 4 \times 3 \times 2 \times 1} + \cdots;$$

Differentiate these series term by term to verify the standard expressions for

$$\frac{d}{dx}(\sin(x)) \text{ and } \frac{d}{dx}(\cos(x)).$$

**17.** A contaminated lake is treated with a bactericide. The rate of change of harmful bacteria $t$ days after treatment is given by

$$\frac{dN}{dt} = -\frac{2000t}{1+t^2},$$

where $N(t)$ is the number of bacteria in 1 ml of water.

(a) State with a reason whether the count of bacteria increases or decreases during the period $0 \leqslant t \leqslant 10$.

(b) Find the minimum value of $dN/dt$ during this period.

**18.** The mean square value of a function $f(t)$ over the interval $a \leqslant t \leqslant b$, is defined as:

$$\text{msv}(f(t)) = \frac{1}{b-a} \int_a^b \{f(t)\}^2 dt.$$

Determine the mean square values of $y = \sin(2t)$ over the ranges (a) $0 \leqslant t \leqslant \pi/2$ and (b) $0 \leqslant t \leqslant 2\pi$.

**19.** Analyze the behavior of the Michaelis–Menten equation

$$v_0 = \frac{V_{\max}[S]}{K_M + [S]},$$

in the limit as the concentration of substrate $[S]$ tends to infinity. (Hint: divide both the numerator and the denominator of the fraction by $[S]$ before analyzing the limit.)

**20.** The flow of water pumped upward through the xylem of a tree, $F$, is given by the relationship:

$$F = M_0(p + qt)^{3/4},$$

where $t$ is the age of the tree in days, $p$ and $q$ are positive constants, and $M_0 p^{3/4}$ is the mass of the tree at the time of planting, when $t = 0$. If $p = 10$, $q = 0.01$ day$^{-1}$, and $M_0 = 0.92$ l day$^{-1}$, determine the total volume of water pumped up the tree in the tenth year (that is, between $t = 3285$ days and $t = 3650$ days).

# CHAPTER 8

# The Calculus of Growth and Decay Processes

The aim of this chapter is to show the reader how to steal the family jewels of two functions that have important applications in all branches of biology: the natural logarithm and the exponential function. This master plan for larceny unfolds by cracking open a window and sneaking into a discussion about how to integrate reciprocal functions. Solving this mystery will unlock the door for a deeper understanding of the logarithms, unmasking their secret identity, and revealing how they can be manipulated. After ransacking the belongings of logarithms and pocketing valuable rules for performing differentiation and integration, the whole house gets turned upside-down as we move our focus to the inverse process of taking an exponential. Here, we uncover a dangerous history of predators, uncontrolled growth, and unstoppable decay.

After the heist, you will be able to use your ill-gotten gains to analyze a range of important processes, taking in acid–base titrations, how to survive when lost at sea, fluorescence microscopy, working out the age of a tiger shark, and how drugs are cleared from the body.

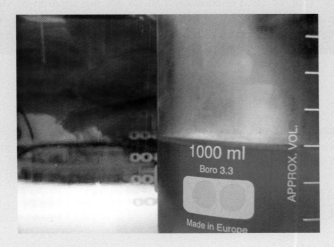

**Figure 8.1**
An acid–base titration using the indicator methyl orange. Image courtesy of Cudmore under Creative Commons Attribution – Share Alike 2.0 Generic.

**Figure 8.2**
Fluorescence microscopy of bovine pulmonary artery endothelial cells, with microtubules stained green.

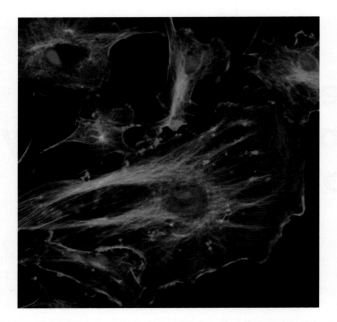

## 8.1 Integrating a reciprocal function

When the rule for integrating power functions of $x$ was introduced in Section 6.2, we noted that the rule does not apply for $n = -1$:

i.e. $\displaystyle\int x^n \, dx = \frac{x^{n+1}}{n+1} + c$    [for $n \neq -1$].    (EQ8.1)

It is easy to see why this restriction is applied: setting $n = -1$ would cause the denominator to become zero, and dividing by zero is never a good idea; in mathematical jargon this means that the result is 'undefined'. The power functions $x^n$ that we have encountered so far are not capable of yielding $1/x$ when they are differentiated. However, expressions that contain reciprocals are important and widespread in biology, such as the relationship between the energy $E$ and the wavelength $\lambda$ of light:

$$E = \frac{hc}{\lambda},$$

(where $h$ and $c$ are constants), or between the flow rate $\Phi$ and the viscosity $\eta$ of blood coursing through an artery of radius $R$ and length $L$ for a given pressure difference $\Delta P$:

$$\Phi = \frac{\pi R^4 \Delta P}{8\eta L}.$$

It would therefore be very useful to find a way of integrating reciprocal functions.

We can sneak up toward an answer by remembering that, as explained in Section 6.4, an integral represents the area under a curve. Figure 8.3 shows a plot of the positive portion of the function $y = 1/t$, which remains finite for all

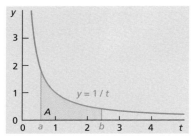

**Figure 8.3**
The integral $\int_a^b (1/t)\,dt$ represented as the area under the curve $y = 1/t$ between $t = a$ and $t = b$.

values of $t \neq 0$. The definite integral of a positive function gives the area bounded by the curve, the axis of the variable of integration, and vertical lines at the limits of integration. Because the function $y = 1/t$ is positive if $t$ is positive, the shaded area $A$ must correspond to the following definite integral:

$$A = \int_a^b \frac{1}{t}\, dt. \qquad \text{(EQ8.2)}$$

If we fix the lower limit $a$ of this integral equal to 1 and allow the upper limit to float, by setting $b$ equal to a variable $x$, then the area becomes a function that depends on the value of the upper limit $x$ (see Figure 8.4):

$$A(x) = \int_1^x \frac{1}{t}\, dt \quad \text{[for } x > 0]. \qquad \text{(EQ8.3)}$$

The area $A(x)$ is used as the mathematical definition of the *logarithmic function* for positive values of $x$. As mentioned in Section 3.6, the logarithmic function is denoted by $\ln(x)$, which is short for 'the natural logarithm of $x$'. We can therefore rewrite EQ8.3 in the form:

$$\ln(x) = \int_1^x \frac{1}{t}\, dt \quad \text{[for } x > 0]. \qquad \text{(EQ8.4)}$$

We should now examine this new definition to see if it agrees with what we already know about logarithms from Section 3.6. First, because the function $y = 1/t$ is positive for all values of $t$ that are greater than 1, the yellow area that represents $\ln(x)$ in Figure 8.4 must also be positive for $x > 1$. This fits in with what we already know about ordinary logarithms, because $\log_{10}(x) > 1$ for $x > 1$; for example, $\log_{10}(100) = +2$.

However, what about values of $x$ that are in the range $0 < x < 1$? The graphical representation of this situation is shown in Figure 8.5. In Section 6.3, we found that swapping the limits of integration for a definite integral causes the result to change sign, so

$$\ln(x) = \int_1^x \frac{1}{t}\, dt = -\int_x^1 \frac{1}{t}\, dt.$$

For $0 < x < 1$, the logarithmic function is therefore equal to the area enclosed by the curve $y = 1/t$ and the $t$ axis between a lower limit of $t = x$ and an upper limit of $t = 1$, all multiplied by $-1$. Because $y = 1/t$ remains positive for $0 < x < 1$, $\ln(x)$ must therefore be negative. This is again consistent with what we already know, because $\log_{10}(x) < 0$ for $0 < x < 1$. For example, $\log_{10}(0.01) = -2$.

To complete the picture, the integral-based definition of the logarithmic function implies that when $x = 1$, the region of interest reduces to a single vertical line, which has zero area (see Figure 8.6).

$$\ln(1) = \int_1^1 \frac{1}{t}\, dt = 0.$$

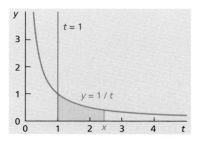

**Figure 8.4**
The function $A(x)$ defined by $A(x) = \int_1^x (1/t)\, dt$ is represented by the area under the curve $y = 1/t$ between $t = 1$ and $t = x$. This area represents the natural logarithmic function $\ln(x)$.

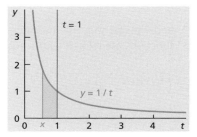

**Figure 8.5**
Illustration to show that $\ln(x)$ is negative for $0 < x < 1$. The upper limit of the integral $\ln(x) = \int_1^x (1/t)\, dt$ is less than the lower limit, hence the area under the curve is in the reverse direction.

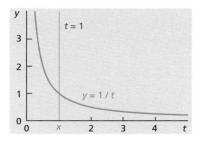

**Figure 8.6**
Illustration to show that $\ln(1) = 0$. The upper limit of the integral $\ln(x) = \int_1^x (1/t)\, dt$ is equal to the lower limit, hence the area under the curve is zero.

For comparison, $\log_{10}(1) = 0$; in fact, $\log_a(1) = 0$ for any $a \neq 0$, because $a^0 = 1$.

Overall, we have shown that the definition of the logarithmic function given by EQ8.4 predicts that $\ln(x)$ will behave in a way that is broadly similar to what we would expect for common base 10 logarithms. This goes some way to justifying the name of the function, but the argument in Box 8.1 goes further, deriving the following expression for the derivative of the logarithmic function:

$$\frac{d}{dx}(\ln(x)) = \frac{1}{x}. \tag{EQ8.5}$$

## Box 8.1 Differentiating the logarithmic function

Section 6.1 introduced the idea of $F(t)$, the antiderivative of a function $f(t)$. For any antiderivative function:

$$\frac{dF(t)}{dt} = f(t),$$

and we saw that the indefinite integral and antiderivative functions are related by the equation:

$$\int f(t)\, dt = F(t) + c.$$

When a definite integral is expressed using the same nomenclature we get:

$$\int_a^b f(t)\, dt = [F(b) + c] - [F(a) + c] = F(b) - F(a).$$

Applying this notation to the limits introduced in the definition of the logarithmic function in EQ8.4 gives:

$$\int_1^x f(t)\, dt = [F(x) + c] - [F(1) + c] = F(x) - F(1).$$

This kind of expression is referred to as an 'improper integral' because one of the limits of the definite integral is not defined precisely. By contrast, for a 'proper integral', both limits are defined and finite. If we now differentiate this last equation with respect to $x$ we obtain:

$$\frac{d}{dx} \int_1^x f(t)\, dt = \frac{d}{dx}(F(x) - F(1))$$

$$= \frac{d}{dx}(F(x)) - \frac{d}{dx}(F(1)) = \frac{d}{dx}(F(x)) = f(x).$$

The term containing $F(1)$ becomes zero because 1 is just a number, but the term containing $F(x)$ remains because it is a function of the variable $x$. If $f(t)$ is a reciprocal function $(1/t)$, this argument has demonstrated that:

$$\frac{d}{dx} \int_1^x \frac{1}{t}\, dt = \frac{1}{x}.$$

Mixing this result with the definition given in EQ8.4 produces an expression for the derivative of the logarithmic function:

$$\frac{d}{dx}(\ln(x)) = \frac{d}{dx}\left(\int_1^x \frac{1}{t}\, dt\right) = \frac{1}{x}.$$

Viewing integration as the inverse of differentiation, we can convert this relationship into the appropriate entry from the table of standard integrals (see Table 7.1):

$$\int \frac{1}{x}\, dx = \ln(x) + c. \tag{EQ8.6}$$

This is the result we were aiming for: an answer to the question of how to integrate the power function $x^n$ for $n = -1$. However, if $\ln(x)$ really does define some kind of logarithm, some questions spring to mind, prompted by Section 3.6. What is the base of this logarithm? And what is so 'natural' about it?

The base of a logarithm is revealed by the number that yields a result of 1. For example, we know that $\log_{10}(10) = 1$ and that $\log_2(2) = 1$. The value of $e$ is therefore defined by the number that makes a true statement out of the equation $\ln(e) = 1$. In terms of the definition given in EQ8.4, the value of $e$ is equal to the upper limit required for the area bounded by the graph of $y = 1/t$, the $t$ axis, and vertical lines at $t = 1$ and $t = e$ to be equal to 1 (see Figure 8.7):

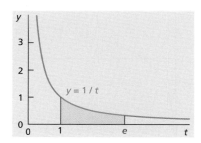

$$\ln(e) = \int_1^e \frac{1}{t}\, dt = 1. \tag{EQ8.7}$$

Determining this area yields a value for $e$ that is an irrational number: 2.71828 (to six significant figures). The value of $e$ is therefore selected 'naturally' by the shape of the reciprocal function $y = 1/t$.

The behavior of the logarithmic function is illustrated in Figure 8.8. The value of $\ln(x)$ is negative for $0 < x < 1$ and quickly becomes very large and negative as $x$ decreases, with a vertical asymptote at $x = 0$. It is worth reiterating here that we have not defined the logarithmic function for values of $x$ that are less than or equal to zero. This is why calculators return an error

**Figure 8.7**
Illustration to show that $\ln(e) = 1$. The average height of the region under the curve $y = 1/t$ between $t = 1$ and $t = e$ is slightly greater than 0.5, whereas the base length of the region is slightly less than 2, making the region itself exactly equal to 1.

## Box 8.2 Proving that ln(x) behaves like a true logarithm

Section 3.6 suggested that a function, $f(x)$, will possess all the properties of a logarithm as long as

$$f(x^a) = a\ f(x).$$

To prove that EQ8.4 defines a function that is truly a logarithm, we must therefore demonstrate that the expression above follows as a logical consequence.

To start, we will define two functions: $p(x) = \ln(x^n)$ and $q(x) = n \times \ln(x)$, where $n$ is a constant. To differentiate the first function, we need to make the substitution $u = x^n$, so that

$$\frac{du}{dx} = nx^{n-1} \quad \text{and} \quad \frac{dp}{du} = \frac{d}{du}\{\ln(u)\} = \frac{1}{u}.$$

We can now use the chain rule from Section 7.2 to obtain the derivative of $p$ with respect to $x$:

$$\frac{dp}{dx} = \frac{dp}{du} \times \frac{du}{dx} = \frac{1}{u} \times nx^{n-1} = \frac{1}{x^n} \times nx^{n-1} = \frac{n}{x}.$$

Next, we need to find the derivative of $q$ with respect to $x$:

$$\frac{dq}{dx} = \frac{d}{dx}\{n \times \ln(x)\} = \frac{n}{x}.$$

The derivatives of $p(x)$ and $q(x)$ are therefore equal. Following the argument in Section 6.1, this means that the two functions differ only by a constant, so we can write:

$$p(x) = q(x) + C,$$

that is,

$$\ln(x^n) = n\ln(x) + C.$$

To determine that value of the constant $C$, we need to set $x = 1$:

$$\ln(1^n) = n\ln(1) + C.$$

However, in Section 8.1 we demonstrate that $\ln(1) = 0$, so:

$$0 = 0 + C, \quad \text{that is,} \quad C = 0.$$

Therefore a logical consequence of the relation $d\{\ln(x)\}/dx = 1/x$ is that $\ln(x^n) = n\ln(x)$. All the other properties of normal logarithms that must apply to $\ln(x)$ can be shown in a similar manner.

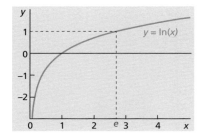

**Figure 8.8**
Plot of ln(x) for 0 < x < 5.

message if we ask them to determine the logarithm of a negative number. The ln(x) curve cuts the horizontal axis just once, at x = 1, and possesses a gentle positive but ever decreasing gradient for x > 1.

It should not be a surprise to discover that the shape of the logarithmic function y = ln(x) is consistent with its first derivative dy/dx being equal to 1/x, as this is the implication of the statement made in EQ8.5. For x > 0, the function 1/x is always positive, and so ln(x) always increases with increasing x. However, its gradient is shallow for large values of x because 1/x is then a very small positive number. Because 1/x reaches zero only at x = +∞, the graph of ln(x) has no stationary points. As x tends toward zero, 1/x becomes large and positive, so the gradient of the logarithmic function becomes very steep.

The effect of these properties is to expand the scale of ln(x) for the range 0 < x < 1, but to contract the scale for 1 < x < +∞. As a result, logarithms are very useful for displaying data that are spread over a large range of values. For example, Figure 8.9 compares the genome sizes of different classes of organisms, from viruses (down to $10^3$ base pairs ('bp') of DNA) to plants (up to $10^{11}$ bp).

**Figure 8.9**
Comparison of genome sizes for various classes of organism. Because of the great range of values, even within some classes, it is better to use a logarithmic scale.

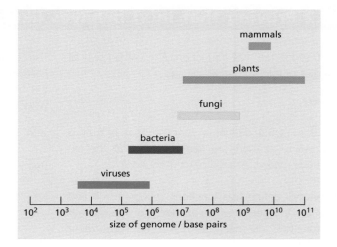

Another interesting application comes in the area of allometric modeling, which is used to study how characteristics of organisms vary with body size. The relationships between many parameters can be summarized by the basic equation:

$$Y = Y_0 X^b, \tag{EQ8.8}$$

where $Y$ represents a dependent variable (such as heart beat rate), $Y_0$ is a normalization constant, $X$ is the independent variable (for example body mass), and the exponent $b$ is a scaling constant. As Figure 8.10A shows, it is difficult to discern the relationship between average resting heart rates and average body mass for mammals, ranging from horses down to mice. However, taking the natural logarithm of both sides of EQ8.8 and applying the appropriate log laws gives:

$$\ln(Y) = \ln(Y_0 X^b) = \ln(Y_0) + b \ln(X). \qquad \text{(EQ8.9)}$$

If the allometric equation is obeyed, this indicates that a plot of $\ln(Y)$ against $\ln(X)$ should produce a straight line, with the gradient yielding the value of the scaling constant, $b$. The appropriate log–log plot for mammalian heart rates is displayed in Figure 8.10B. It turns out that all of the data points lie rather close to a straight line defined by $\ln(Y) = 5.44 - 0.29 \ln(X)$. In this case, the scaling constant is less than 1, indicating that the heart rates of larger animals will be similar, and less than 0, suggesting that the heart rate is inversely proportional to body mass. Further information about fitting procedures and the use of models in biology is provided in Chapter 12.

## 8.2 Calculus with logarithms

Clotting is a vital part of the process for making cheese. It involves adding enzymes to milk that disrupt micelles made from the protein casein. The molecules of casein are cleaved into water-loving and water-hating fragments, which self-assemble into a three-dimensional network, forming a viscous gel. Some experiments on a newly discovered protease measured the 'clotting time' – the time required for milk to reach a defined viscosity after the addition of enzyme. The clotting time $t_c$ was found to be inversely proportional to the amount of enzyme added, $x$, such that:

$$t_c = \frac{K}{x}.$$

To estimate the mean clotting time (see Section 6.7) expected after addition of an amount of enzyme in the range $a \leqslant x \leqslant b$, we would have to solve the integral:

$$\bar{t_c} = \frac{1}{b-a} \int_a^b \frac{K}{x} \, \mathrm{d}x.$$

We can use the standard integral for a reciprocal function given by EQ8.6 to show that:

$$\bar{t_c} = \frac{K}{b-a} \Big[ \ln(x) \Big]_a^b = \frac{K}{b-a} [\ln(b) - \ln(a)] = \frac{K}{b-a} \ln\left(\frac{b}{a}\right).$$

If $K = 300$ mg min l$^{-1}$, then the enzyme concentrations $a = 5$ mg l$^{-1}$ and $b = 10$ mg l$^{-1}$ correspond to clotting times of 60 min and 30 min, respectively. The value predicted by the equation for the mean clotting time $\bar{t_c}$ is 41.6 min, which is shorter than the result of 45 min that can be obtained from a simple average of the upper and lower boundary times.

In addition to direct applications of the standard formulae given by EQ8.5 and EQ8.6, we can deploy all of the extra tools for differentiation and integration introduced in Chapter 7. For example, the chain rule (Section 7.2) can be used to extend EQ8.5 to more complicated expressions. To differentiate the general

The properties of logarithms are summarized in a series of 'log laws':

(i) $\log_n(a) + \log_n(b) = \log_n(ab)$
(ii) $\log_n(a) - \log_n(b) = \log_n(a/b)$
(iii) $\log_n(a^b) = b\log_n(a)$
(iv) $\log_a(b) = \log_c(b)/\log_c(a)$

(A)

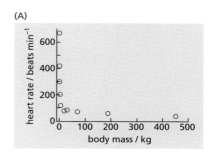

(B)

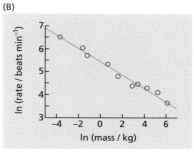

**Figure 8.10**
A normal plot (A) and a log–log plot (B) for allometric modeling of the relationship between resting heart beat rate and body mass for adult mammals, ranging in size from mice to horses.

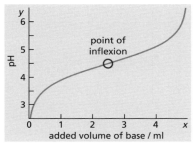

**Figure 8.11**
The pH of a solution of acetic acid after the addition of sodium hydroxide solution. In the vicinity of the point of inflexion, the mixture of acetic acid and acetate ions acts as a pH buffer system that is relatively insensitive to the concentration of base.

The rule for changing the base of a logarithm:

$$\log_a(b) = \frac{\log_c(b)}{\log_c(a)}$$

function $y = \ln(ax + b)$ with respect to $x$, we should set $u = ax + b$, giving $y = \ln(u)$ and $du/dx = a$. Therefore

$$\frac{dy}{dx} = \frac{dy}{du} \times \frac{du}{dx} = \frac{1}{u} \times a = \frac{a}{ax + b}. \qquad \text{(EQ8.10)}$$

The simple rule given by EQ8.10 could be used to obtain derivatives for a wide range of more specific examples. Setting $a = -3$ and $b = 4$ gives the result:

$$\frac{dy}{dx} = \frac{d}{dx}(\ln(4 - 3x)) = \frac{-3}{4 - 3x} = \frac{3}{3x - 4}.$$

The product rule (Section 7.3) can also be used in combination with logarithms. To differentiate the expression $y = x \ln(x) - x$ with respect to $x$, we need the product rule to solve the derivative of the first term on the right-hand side of the equation:

$$\frac{dy}{dx} = \frac{d}{dx}(x\ln(x) - x) = 1 \times \ln(x) + x \times \frac{1}{x} - 1 = \ln(x) + 1 - 1 = \ln(x).$$

Writing this result in reverse puts the focus on integration rather than differentiation and yields the standard result from Table 7.1 for integrating a logarithmic function:

$$\int \ln(x)\,dx = x\ln(x) - x + c.$$

The 'log laws' can be deployed to simplify an expression before we attempt to differentiate it. For example, the Henderson–Hasselbalch equation (mentioned in Section 3.8 in the subsection on pH) can be adapted to show that the pH of 100 ml of a solution containing 50 mM acetic acid being titrated with 1 M sodium hydroxide is given by:

$$y = 4.75 + \log_{10}\left(\frac{x}{5 - x}\right),$$

where $y$ is the pH value, and $x$ represents the volume of base added in milliliters. Can we use this relationship to predict how much base must be added for the titration curve (shown in Figure 8.11) to reach a point of inflexion, where the gradient is at its shallowest?

First, we should convert from base 10 common logarithms to natural logarithms:

$$y = 4.75 + \frac{\ln\left(\dfrac{x}{5 - x}\right)}{\ln(10)} = 4.75 + 0.43 \times \ln\left(\frac{x}{5 - x}\right).$$

Next, we can split the logarithmic term into two components:

$$y = 4.75 + 0.43 \times \ln(x) - 0.43 \times \ln(5 - x).$$

This equation can now be differentiated with respect to $x$ to obtain the first derivative by applying the result of EQ8.10:

$$\frac{dy}{dx} = \frac{0.43}{x} - \frac{0.43 \times -1}{5 - x} = \frac{0.43}{x} + \frac{0.43}{5 - x}.$$

For the second derivative, we need to differentiate again with respect to $x$:

$$\frac{d^2 y}{dx^2} = -\frac{0.43}{x^2} + (-1) \times (-1) \times \frac{0.43}{(5 - x)^2}.$$

At a point of inflexion, the second derivative is equal to zero:

$$-\frac{0.43}{x^2} + \frac{0.43}{(5 - x)^2} = 0,$$

that is,

$$\frac{0.43}{(5 - x)^2} = \frac{0.43}{x^2}.$$

Dividing both sides of this equation by 0.43 and multiplying by $x^2 (5 - x)^2$ gives:

$$x^2 = (5 - x)^2 = 25 - 10x + x^2.$$

Deleting $x^2$ from both sides of the equation indicates that:

$$0 = 25 - 10x \qquad \text{so that} \qquad x = 2.5.$$

The calculation reveals that the point of inflexion occurs at the midpoint of the titration, when 2.5 ml of base has been added, as illustrated in Figure 8.11.

We can also make use of the change of variable rule from Section 7.6 to solve an integral that has the general form:

$$I = \int \frac{1}{ax + b} \, dx, \qquad\qquad \text{(EQ8.11)}$$

where $a$ and $b$ are constants. If we set $u = ax + b$, so that

$$\frac{du}{dx} = a \qquad \text{and} \qquad \frac{dx}{du} = \left(\frac{du}{dx}\right)^{-1} = \frac{1}{a},$$

then

$$I = \int \frac{1}{ax + b} \, dx = \int \frac{1}{u} \times \frac{dx}{du} \, du = \int \frac{1}{u} \times \frac{1}{a} \, du$$

$$= \frac{1}{a} \times \int \frac{1}{u} \, du = \frac{1}{a} \times \ln(u) + c.$$

Substituting back for $u$ gives us a handy new rule:

$$I = \int \frac{1}{ax+b} \, \mathrm{d}x = \frac{1}{a} \ln(ax+b) + c. \qquad \text{(EQ8.12)}$$

If only they had the opportunity to apply it, this relationship would be very useful to humans who find themselves immersed in cold water without proper insulation, such as passengers stranded in the open sea after a shipwreck. The risk of hypothermia due to prolonged exposure to the cold increases as heat is lost to the environment, with death certain to occur when the core body temperature drops below 30 °C. A crude method for predicting how the survival time ($t_s$, measured in hours) depends on the surface temperature of the sea ($T$, in degrees Celsius) uses the equation:

$$t_s = \frac{0.2}{0.1 - 0.004\,T}.$$

**Figure 8.12**
(A) Cold-water survival strategy after a shipwreck is to stay close to minimize heat loss.
(B) Survival times as a function of water temperature. At temperatures between 10 °C and 15 °C there is a large variation in survival times, with an average of 4 hours. Image courtesy of Deep Silence under Creative Commons Attribution – Share Alike license version 3.0.

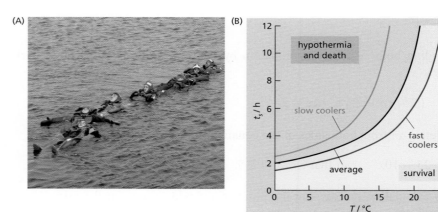

This relationship indicates that when the water is warm, the chance of a successful rescue is good, but the outlook is bleak when the sea is cold. As Figure 8.12B shows, significant corrections are required for individuals who cool rapidly (for example, owing to age or lack of insulation from a layer of fat) or slowly (perhaps because they huddle together to conserve heat). Despite these uncertainties, the mean predicted survival time (see Section 6.7) for an average swimmer in water temperatures within the range 10 °C $\leq T \leq$ 15 °C is given by:

$$\overline{t_s} = \frac{1}{15-10} \int_{10}^{15} \frac{0.2}{0.1-0.004\,T} \, \mathrm{d}T = \frac{0.2}{5} \times \left[ -\frac{1}{0.004} \ln(0.1-0.004\,T) \right]_{10}^{15}$$

$$= -\frac{0.04}{0.004} \times [\ln(0.1-0.004\times 15) - \ln(0.1-0.004\times 10)]$$

$$= -10 \times \ln\left(\frac{0.04}{0.06}\right) = +4.05.$$

The mean predicted survival time is therefore approximately 4 hours. Note that units were left out throughout the calculation for the sake of simplicity.

The rule given in EQ8.12 is really a special case of a more general result, which can be applied to expressions that have the form of the reciprocal of a function multiplied by its derivative:

$$\int \frac{(df(x)/dx)}{f(x)} \, dx = \ln(f(x)) + c. \qquad \text{(EQ8.13)}$$

To illustrate where this rule comes from, we can investigate the energetic cost of the digestion and absorption of food by a boa constrictor. One way to monitor the metabolic response of such an animal is to measure the rate at which it consumes oxygen after a large meal, as shown in Figure 8.13. If $V$ is the volume of oxygen consumed in milliliters per gram of reptile and $t$ is the time in hours since the meal, then the rate of change can be modeled as:

$$\frac{dV}{dt} = \frac{0.025t}{1 + 4.3 \times 10^{-4} \times t^2}.$$

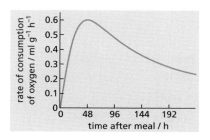

**Figure 8.13**
Rate of oxygen consumption of the boa constrictor as a function of time after the meal.

To calculate the total volume of extra oxygen consumed as a result of the meal within the first 48 hours, we must evaluate the integral:

$$V = \int_0^{48} \frac{0.025t}{1 + 4.3 \times 10^{-4} \times t^2} \, dt.$$

If we set $u = 1 + 4.3 \times 10^{-4} \times t^2$, so that

$$\frac{du}{dt} = 8.6 \times 10^{-4} \times t, \qquad \text{then} \qquad \frac{dt}{du} = \left(\frac{du}{dt}\right)^{-1} = \frac{1162.8}{t}.$$

Applying the change of variable rule gives us an integrand that contains a term closely related to the derivative of $u$ with respect to $t$ ($0.025t$ instead of $8.6 \times 10^{-4} \times t$) multiplied by the reciprocal of $u$, multiplied again by the derivative of $t$ with respect to $u$. The two derivative terms in this expression ($du/dt$ and $dt/du$) should therefore cancel out, leaving us to solve simply the integral of a standard function multiplied by a constant. For this example, we will use approach (1) from Section 7.6, rewriting the answer in terms of the original variable $t$ before applying the limits:

$$V = \int_{t=0}^{t=48} \frac{0.025t}{u} \times \frac{dt}{du} \, du = \int_{t=0}^{t=48} \frac{0.025t}{u} \times \frac{1162.8}{t} \, du = 29.07 \times \int_{t=0}^{t=48} \frac{1}{u} \, du$$

$$= 29.07 \times \left[\ln(u)\right]_{t=0}^{t=48} = 29.07 \times \left[\ln(1 + 4.3 \times 10^{-4} \times t^2)\right]_{t=0}^{t=48}$$

$$= 29.07 \times [\ln(1.991) - \ln(1)] = 20.0.$$

The snake therefore consumes an additional 20 ml of oxygen per gram of body weight in the two days after its meal. Notice again that numerical value equations have been used to avoid repeatedly writing out the units and cluttering up the appearance of these equations.

# 8.3 Calculus with exponential functions

The polymerase chain reaction (PCR) has revolutionized biology by providing a way to amplify selected DNA sequences. The method involves adding nucleotide triphosphates and a DNA polymerase enzyme to a biological sample along with a pair of 'primers', short DNA fragments that will bind to the beginning and end of the target sequence. The double-stranded DNA in the biological sample becomes single stranded at high temperature (melting), joins with a new partner strand when the temperature is lowered (annealing), and new DNA is made at an intermediate temperature (extension). This melting–annealing–extension temperature cycle is repeated many times in order to prepare large amounts of the desired DNA fragment. At any point, the number of molecules of target DNA produced, $N(c)$, is given by the formula:

$$N(c) = N_0 (E + 1)^c,$$

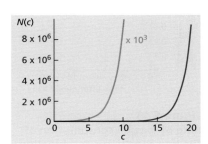

**Figure 8.14**
Increase in the number of target DNA molecules, $N(c)$, present in successive cycles ($c$) of a PCR experiment. For the red trace, the scale has been expanded by a factor of $10^3$.

where $N_0$ represents the number initially present, $c$ represents the number of cycles, and $E$ is a number between 0 and 1, representing the efficiency of DNA synthesis during each cycle. Figure 8.14 demonstrates that as the number of cycles increases, the amount of product grows exponentially: initially the number of copies is small, but it increases dramatically in successive cycles.

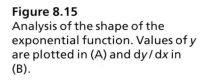

The exponential function is the inverse of the logarithmic function:
if $\ln(x) = y$, then $x = e^y$.

This section revisits the exponential function, $e^x$, which is the mirror image or inverse of the logarithmic function, $\ln(x)$. A plot of the function $y = e^x$ is given in Figure 8.15A. The curve is always positive, with a horizontal asymptote at $y = 0$ for large negative values of $x$. The graph possesses a value of 1 when $x = 0$ and increases very rapidly for $x > 0$. Interestingly, the slope of $y = e^x$ is also always positive, tends toward zero as $x$ tends toward $-\infty$, possesses a value of 1 when $x = 0$, and grows ever steeper as $x$ tends to positive infinity (Figure 8.15B). In fact, the exponential function has the unique property of being equal to its own derivative (see Box 8.3), such that:

$$\frac{\mathrm{d}}{\mathrm{d}x}(e^x) = e^x. \tag{EQ8.14}$$

This unusual feature provides another reason why $e$ is the mathematician's favorite base.

**Figure 8.15**
Analysis of the shape of the exponential function. Values of $y$ are plotted in (A) and $\mathrm{d}y/\mathrm{d}x$ in (B).

(A)

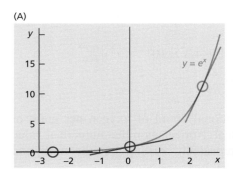

(B)

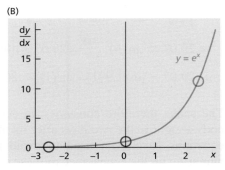

## Box 8.3 The exponential function is equal to its own derivative

We start by considering the function obtained by raising any base $a$ to the power of $x$, giving $y = a^x$, where $a$ is a constant. Taking the natural logarithm of both sides of this equation gives:

$$\ln(y) = \ln(a^x) = x \times \ln(a).$$

We will now differentiate both sides of this equation with respect to $x$. The right-hand side gives:

$$\frac{d}{dx}\{x\ln(a)\} = \ln(a).$$

The derivative of the left-hand side can be solved by using the chain rule:

$$\frac{d}{dx}\{\ln(y)\} = \frac{dy}{dx} \times \frac{d}{dy}\{\ln(y)\} = \frac{dy}{dx} \times \frac{1}{y}.$$

We have now demonstrated that

$$\frac{dy}{dx} \times \frac{1}{y} = \ln(a).$$

Multiplying both sides of this equation by $y$ gives:

$$\frac{dy}{dx} = y\ln(a).$$

To examine the specific case of the exponential function, we need to set $a = e$, recalling that $\ln(e) = 1$:

$$\frac{dy}{dx} = y\ln(e) = y,$$

which means that

$$\frac{d}{dx}(e^x) = e^x.$$

We are now in a position to derive expressions for the derivatives of more complex functions. The intensity of light is attenuated by water, by any dissolved compounds, and by particulate matter. These properties can be used to estimate the abundance of plankton algae in a lake by measuring the depth to which sunlight can penetrate (see Figure 8.16). The intensity of light $I(a)$ measured by a probe at a given depth $z$, is given by the equation:

$$I(a) = I_0 \exp\{-(K_w + K_a a)z\},$$

where $a$ is the concentration of algae, $I_0$ is the intensity of sunlight at the surface of the lake, $K_w$ is an attenuation coefficient due to absorption of light by water, and $K_a$ is an attenuation coefficient due to absorption of light by the chlorophyll molecules in the algae. To consider how the intensity of light changes with the concentration of algae, we need to differentiate $I(a)$ with respect to $a$. Making the substitution $u = -(K_w + K_a a)z$ yields $I = I_0 e^u$ and $du/da = -K_a z$, so applying the chain rule gives:

$$\frac{dI}{da} = \frac{dI}{du} \times \frac{du}{da} = I_0 e^u \times -K_a z$$

$$= -I_0 K_a z \times \exp\{-(K_w + K_a a)z\}.$$

The negative sign for the derivative indicates that if the concentration of algae increases, less light will penetrate down to depth $z$; however, as $a$ continues to increase, the intensity changes at different depths become less significant.

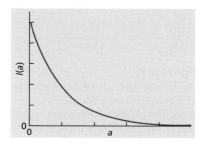

**Figure 8.16**
How the intensity of light $I(a)$ at a depth $z$ in a lake depends on the concentration of algae $a$.

A vital component of technology required for PCR is the DNA polymerase, an enzyme that joins individual nucleic acids together to make long polynucleotide polymers. *Taq* polymerase, an enzyme obtained from the bacterium *Thermus aquaticus*, is often used for this job. This extremophile micro-organism grows best at high temperatures and was originally discovered in samples taken from hot springs in Yellowstone National Park. *Taq* polymerase is ideal for performing PCR reactions because it has a high degree of thermal stability: its optimum activity is at around 80 °C and it still functions after repeated exposure to temperatures up to 95 °C.

The effect of temperature on the rate constant $k$ of a chemical reaction is predicted by the Arrhenius equation:

$$k = Ae^{-E_a/RT}, \tag{EQ8.15}$$

where $E_a$ is the activation energy (which could be measured in joules), $T$ is the absolute temperature (which, as always, must be in Kelvin), $R$ is the universal gas constant ($8.314 \, \text{J K}^{-1} \, \text{mol}^{-1}$), and $A$ is a temperature-independent factor (typically with units of 'per second'). For the DNA synthesis reaction of *Taq* polymerase, the pre-exponential factor has a value of $1.1 \times 10^{17}$ nucleotides per second ($\text{nt s}^{-1}$), and the activation energy is $99.5 \, \text{kJ mol}^{-1}$.

We can use this information to deduce how the rate constant of a reaction will change with temperature. This requires some differentiation, which is made easier by performing the substitution

$$u = -\frac{E_a}{RT},$$

so that

$$\frac{du}{dT} = -\frac{E_a}{R} \times -T^{-2} = \frac{E_a}{RT^2}.$$

It is then straightforward to show that:

$$\frac{dk}{du} = \frac{d}{du}(Ae^u) = Ae^u,$$

so application of the chain rule now gives:

$$\frac{dk}{dT} = \frac{dk}{du} \times \frac{du}{dT} = Ae^u \times \frac{E_a}{RT^2} = \frac{AE_a}{RT^2}e^{-E_a/RT}.$$

The temperature dependencies of $k$ and $dk/dT$ are shown in Figure 8.17. The theoretical curve for the rate constant increases continuously as the temperature is raised, but possesses a point of inflexion, which corresponds to a maximum turning point in the plot of the first derivative.

(A)

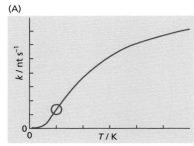

(B)

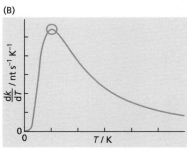

**Figure 8.17**
(A) The temperature dependence of a reaction rate constant according to the Arrhenius equation. The circle indicates a point of inflexion. (B) The temperature dependence of the derivative of $k$ with respect to $T$. The circle indicates a maximum turning point.

The Arrhenius equation is often converted into a linear form to interpret the results of reaction kinetics experiments performed at different temperatures. Because $k$ and $A$ have the same units, we can take the natural logarithm of both sides of EQ8.15, which gives:

$$\ln(k) = \ln(Ae^{-E_a/RT}) = \ln(A) + \ln(e^{-E_a/RT}) = \ln(A) - \frac{E_a}{R} \times \frac{1}{T}.$$

$$(EQ8.16)$$

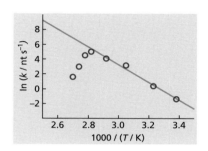

An 'Arrhenius plot' of $\ln(k)$ against $1/T$ should therefore produce a straight line with a gradient of $-E_a/R$. The data shown in Figure 8.18 verify that *Taq* polymerase obeys the Arrhenius equation for a wide range of temperatures. However, the nonlinear features in the high-temperature region of the Arrhenius plot indicate that reaction mechanism has changed. In this case, deviation from the straight line is due to the enzyme becoming inactive, perhaps because of unfolding or aggregation at high temperatures.

Because integration is the inverse process of differentiation, EQ8.14 indicates that the standard integral for the exponential function should be:

$$\int e^x \, dx = e^x + c.$$

$$(EQ8.17)$$

**Figure 8.18**
Fit of experimental data (circles) for *Taq* polymerase to the Arrhenius equation (solid curve) in which $\ln(k)$ is plotted against $1/T$. The Arrhenius equation fits the experimental data well for lower temperatures but breaks down at higher temperatures; that is, when $1000/T$ is small.

To integrate the more complex function $y = e^{ax+b}$ with respect to $x$, we can once again use the substitution, $u = ax + b$. Because $du/dx = a$, we know that

$$\frac{dx}{du} = \left(\frac{du}{dx}\right)^{-1} = \frac{1}{a}.$$

Applying the change of variable rule enables us to solve the integral:

$$\int e^{ax+b} dx = \int e^u \times \frac{dx}{du} \, du = \int e^u \times \frac{1}{a} \, du$$

$$(EQ8.18)$$

$$= \frac{1}{a}e^u + c = \frac{1}{a}e^{ax+b} + c.$$

This result can be applied in many practical situations. For example, ornithologists discovered that the change in stomach temperature ($\Delta T$, measured in degrees Celsius) of an African penguin after ingesting $50 \text{ cm}^3$ of seawater (at 14.5 °C) obeyed the equation:

$$\Delta T = 14 \times (1 - e^{-0.13t}),$$

where time $t$ was measured in minutes. The average temperature change in the range 10–20 minutes after swallowing the water can therefore be determined as follows:

$$\overline{\Delta T} = \frac{1}{20 - 10} \int_{10}^{20} 14 \times (1 - e^{-0.13t})\, dt = \frac{14}{10} \int_{10}^{20} (1 - e^{-0.13t})\, dt$$

$$= 1.4 \times \left[ t - \frac{e^{-0.13t}}{-0.13} \right]_{10}^{20} = 1.4 \times \left\{ \left[ 20 + \frac{e^{-0.13 \times 20}}{0.13} \right] - \left[ 10 + \frac{e^{-0.13 \times 10}}{0.13} \right] \right\}$$

$$= 11.9.$$

The average temperature change is therefore 11.9 °C.

## 8.4 Decay processes

Radioactive isotopes decay at a rate that is proportional to the quantity of substance present. As the process continues, fewer nuclei of this particular type exist in the sample, so the rate of decay diminishes. In more mathematical jargon, we can set $N$ equal to the number of radioactive nuclei at time $t$ and then construct an equation in which the rate of decay (that is, the first derivative of $N$ with respect to $t$) is directly proportional to the amount of substance present:

$$\frac{dN(t)}{dt} = -\lambda N(t), \tag{EQ8.19}$$

where $\lambda$ is a proportionality constant.

The minus sign on the right-hand side of EQ8.19 indicates that the slope of a plot of $N$ against $t$ must be negative. This is exactly what we need, because if the number of radioactive nuclei decreases with time, the gradient of this curve must be negative. EQ8.19 contains a derivative, so it is yet another example of a 'differential equation'. Chapter 12 presents a detailed strategy for solving this simple kind of differential equation, but for now we can demonstrate that the expression

$$N(t) = N_0 e^{-\lambda t}, \tag{EQ8.20}$$

is a valid solution of EQ8.19, where $N_0$ is the value of $N$ at time $t = 0$.

**Figure 8.19**
The shape of the exponential function depends on the sign of the constant multiplying the independent variable $t$. Negative exponent terms give a decreasing function of $t$ (decay), whereas positive values produce an increasing function of $t$ (growth).

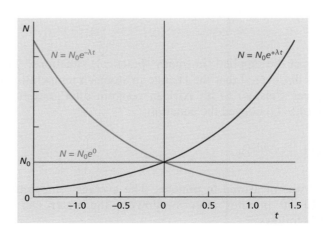

Figure 8.19 indicates that EQ8.20 has the correct time dependence: if the term in the exponent is negative, then the number of radioactive nuclei decreases from $N_0$ as $t$ increases. When the function is traced backward along the negative time direction, we find that much more of the radioactive substance was present in the past. The value of the exponent has a major effect on the shape of the exponential function. When $\lambda$ is zero, the value of $N$ does not change with time. However, when the exponent is positive, the number of nuclei will increase with time. In the latter case, the equation no longer describes decay, but instead a situation of explosive growth.

A straightforward way to prove that an expression is a genuine solution of a particular differential equation is to calculate its derivative and show that it can be written in the same form. Differentiating EQ8.20 with respect to $t$ reveals that:

$$\frac{dN(t)}{dt} = \frac{d}{dt}\left(N_0 e^{-\lambda t}\right) = N_0 \times -\lambda \times e^{-\lambda t} = -\lambda \times N_0 e^{-\lambda t}$$

$$= -\lambda N(t),$$

as required by EQ8.19.

Knowing that the rate of decay is proportional to the current amount of substance gives us the ability to predict the amount that will be present at any time in the future. This is crucial information for handling compounds that are labeled with rapidly decaying radioactive isotopes, such as fluorine-18 ($^{18}$F)-tagged fluorodeoxyglucose ($^{18}$F-FDG), which is used in positron emission tomography (PET) studies to image the rapid uptake of glucose by tumors or by areas of the brain that consume a lot of fuel (see Figure 8.20).

EQ8.20 also enables scientists to calculate how much of an isotope would have been present at some time in the past, providing the basis for using radioactive dating techniques in archaeology and geology. For example, the carbon isotopes in living matter are in dynamic equilibrium with those in the environment, but are no longer able to exchange after death. By measuring the current amount of the carbon-14 ($^{14}$C) isotope in a once-living sample and comparing it with the background concentration, the time of death can be estimated.

Each radioactive substance is specified by a characteristic half-life, $t_{1/2}$, which describes the kinetics of its decay process. The half-life is defined as the time taken for the amount of the radioactive element present at any time $T$ to be reduced by one half. According to EQ8.20, when $t = T + t_{1/2}$,

$$\frac{1}{2} N_0 e^{-\lambda T} = N_0 e^{-\lambda(T + t_{1/2})} = N_0 e^{-\lambda T} e^{-\lambda t_{1/2}}.$$

After canceling out a factor of $N_0 e^{-\lambda T}$ and taking logarithms of both sides, we get

$$\ln\left(\frac{1}{2}\right) = \ln\left(e^{-\lambda t_{1/2}}\right). \tag{EQ8.21}$$

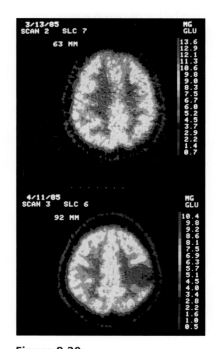

**Figure 8.20**
PET scans of a normal brain and the brain of a patient with an astrocytoma (tumor of the central nervous system).

The left-hand side of EQ8.21 can be simplified using the appropriate log law:

$$\ln\left(\frac{1}{2}\right) = -\ln(2).$$

For the right-hand side, we need to recall that the logarithmic function is the inverse of the exponential function, meaning that:

$$\ln\left(e^{-\lambda t_{1/2}}\right) = -\lambda t_{1/2}.$$

Therefore

$$-\lambda t_{1/2} = -\ln(2),$$

and

$$t_{1/2} = \frac{\ln(2)}{\lambda}. \qquad\qquad (EQ8.22)$$

Note that for a simple exponential decay, the value of the half-life, $t_{1/2}$, does not depend on the amount of radioactive substance present or on the time $T$ at which the experiment began. This is illustrated in Figure 8.21, which shows three consecutive half-lives for the function $y = e^{-2t}$.

**Figure 8.21**
For a simple exponential decay, the time required for a decrease by a factor of two is always the same, and is known as the half-life.

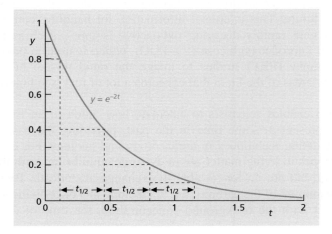

The $^{18}$F label in $^{18}$F-FDG has a half-life of only 110 min, so this tracer compound has to be synthesized in a nearby cyclotron shortly before being used in a PET scan. Suppose that a sample of $^{18}$F-FDG with an activity of 400 MBq (that is, $4 \times 10^8$ nuclei decaying per second) is prepared at a cyclotron 5 hours drive away from the center where a PET scan is to be performed. After being injected into a patient through a saline drip, there is a delay of 1 hour while the tracer compound is taken up by the body's tissues. If the half-life of $^{18}$F-FDG is 110 min, the maximum possible activity at the start of the PET experiment can be calculated as follows. Because $t_{1/2} = 110$ min, then from EQ8.22,

$$\lambda = \frac{\ln(2)}{t_{1/2}} = \frac{0.693}{110} \text{ min}^{-1} = 0.00630 \text{ min}^{-1}.$$

If $N_0 = 400$ MBq and $t = 6 \times 60$ min $= 360$ min, then

$$N = N_0 e^{-\lambda t} = 400 e^{-0.00630 \times 360} = 41.4 \text{ MBq.}$$

The activity at the start of the PET experiment will have fallen to just over 41 MBq.

There is nothing magic about the level of radioactivity decreasing by a factor of two after we have waited for a half-life to elapse. The $t_{1/2}$ parameter is just a convenient tool for thinking about the reciprocal of the rate constant for decay by giving it a meaningful time scale. An alternative approach is used to characterize electronically excited molecules. After being irradiated by light at a certain wavelength, many compounds fluoresce, emitting photons with a longer wavelength (see Figure 8.22). The intensity $I$ of this fluorescence depends on the time $t$ after excitation by a short pulse of light in the following way:

$$I(t) = I_0 e^{-kt}, \tag{EQ8.23}$$

where $I_0$ is the intensity when $t = 0$ and $k$ is a constant. The decay of the electronically excited state is characterized by the mean fluorescence lifetime, $\tau$, defined as the time required for the fluorescence intensity to reduce by a factor of $e$. From the definition of $\tau$, we can write:

$$I(\tau) = I_0 e^{-k\tau} = \frac{1}{e} I_0.$$

Therefore $e \times e^{-k\tau} = 1 = e^{1-k\tau}$. Taking logs gives $1 - k\tau = 0$, so that

$$\tau = \frac{1}{k}. \tag{EQ8.24}$$

This is illustrated in Figure 8.23. For a group of electronically excited molecules, the mean lifetime corresponds to the mean of the lifetimes of each individual molecule.

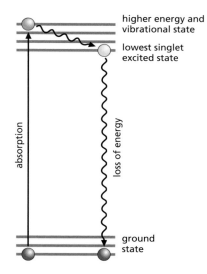

**Figure 8.22**
The mechanism of fluorescence. An electron absorbs light energy and moves to a higher energy (excited) state. The electron subsequently loses energy until it finally falls back to the ground state, emitting light at a longer wavelength.

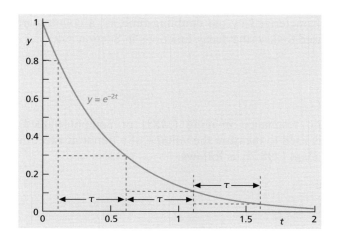

**Figure 8.23**
For a simple exponential decay, the time required for a decrease by a factor of $e$ is always the same, and is known as the lifetime.

# 8.5 Growth processes

The process that is complementary to exponential decay is unrestricted growth, in which a parameter such as volume $V$ is governed by the differential equation:

$$\frac{dV}{dt} = +\gamma V,$$  (EQ8.25)

which has the solution:

$$V = V_0 e^{+\gamma t},$$  (EQ8.26)

where $\gamma$, the rate constant for growth, is positive and $V_0$ is the volume when $t = 0$.

Unrestricted exponential growth is often dangerous, because the numbers involved can rapidly become frighteningly large. For example, on average each nucleus of uranium-235 ($^{235}U$) that undergoes nuclear fission produces 2.5 neutrons, and two or more lightweight elements. Each of these neutrons can interact with another $^{235}U$ nucleus, which will then experience its own fission reaction. If a lump of uranium is greater than the critical mass and this process is left unchecked, the result will soon be a thermonuclear explosion.

Unrestricted growth occurs in biology when competition for resources is not a significant factor. Such situations occur when an organism that has no natural predators is introduced to an ecosystem, or when a fresh batch of culture medium is inoculated with bacteria. The PCR reaction described in Section 8.3 also exhibits exponential growth: if the process is efficient, then during each cycle the number of DNA fragments can double.

To become a malignant cancer, a human cell must gain several abilities: to divide without limit, to ignore intracellular and extracellular signals that might curtail its growth, to develop its own nutrient supply, and to invade other tissues. Tumors start with a single cell and continue to divide at a constant rate, their volume growing with time. As mentioned in Section 3.8 in the first subsection on plotting the logarithm of a number of cells, cell division is characterized by the doubling-time, $t_2$, a counterpart to the half-life that we used to describe decay functions in Section 8.4:

$$t_2 = \frac{\ln(2)}{\gamma}.$$  (EQ8.27)

If a magnetic resonance imaging (MRI) or computer-aided tomography (CAT) scan is used to measure the volume of a tumor at two times, $t_a$ and $t_b$, then we can adapt EQ8.26 as follows:

$$\frac{V_b}{V_a} = \frac{V_0 e^{+\gamma t_b}}{V_0 e^{+\gamma t_a}} = e^{+\gamma(t_b - t_a)}.$$  (EQ8.28)

Taking logs of both sides and rearranging, we find that:

$$\gamma = \frac{1}{t_b - t_a} \ln \left( \frac{V_b}{V_a} \right).$$

Applying this to EQ8.27 and setting $\Delta t = t_b - t_a$ gives an expression for the doubling-time in terms of measurable quantities:

$$t_2 = \frac{\ln(2) \times \Delta t}{\ln(V_b/V_a)}. \tag{EQ8.29}$$

In lung cancer, an inoperable tumor usually becomes fatal when it reaches a volume of approximately 500 cm$^3$. If the tumor has volume $V_d$ (in cubic centimeters) when it is detected, EQ8.29 can be rearranged to make $\Delta t$ the subject, allowing us to predict that the patient will survive for a time $t_s$,

$$t_s = \frac{t_2 \times \ln(500\,\text{cm}^3/V_d)}{\ln(2)}.$$

If no further treatment is possible, Figure 8.24 shows the dependence of the predicted survival time on the doubling-time for a series of spherical tumors with different diameters at the time of detection.

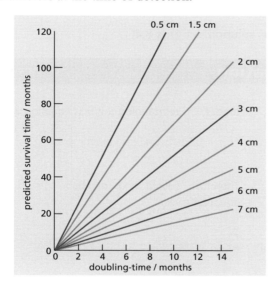

**Figure 8.24**
Predicted survival time after the detection of lung tumors of different sizes as a function of the doubling-time. Reprinted from Geddes, D.M. (1979) *British Journal of Diseases of the Chest* 73:1. The natural history of lung cancer: a review based on rates of tumour growth p1–17. Used with permission from Elsevier.

By contrast, environmental factors restrict the growth of most cells, organisms, and populations. The simplest treatment of restricted growth is provided by the von Bertalanffy equation, which can be applied to organisms whose body shapes remain the same throughout their lives when there is a constant supply of food. For example, the rate of growth of a sand tiger shark (Figure 8.25) of length $L(t)$ (in meters) and age $t$ (in years) is found to decrease as it approaches its maximum length ($L_\infty$). If the growth rate ($dL/dt$) is proportional to the difference between the current and the maximum possible length of the shark, we can construct the equation:

$$\frac{dL}{dt} = k(L_\infty - L), \tag{EQ8.30}$$

**Figure 8.25**
A sand tiger shark. Image courtesy of Daniel HP under Creative Commons Attribution – Share Alike 2.0 Generic.

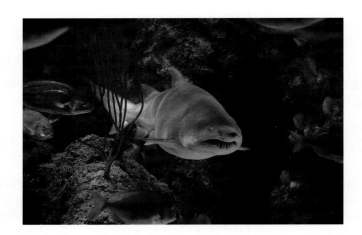

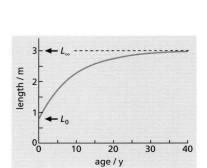

**Figure 8.26**
The von Bertalanffy growth curve for a sand tiger shark.

where $k$ is a positive constant. This relationship predicts that the shark will experience rapid growth when it is a juvenile (that is, when $L << L_\infty$), but the older it gets, the slower it grows (Figure 8.26). If $L_0$ is the length of the shark at birth (that is, when $t = 0$), the solution to the von Bertalanffy differential equation is given by:

$$L = L_\infty - (L_\infty - L_0)e^{-kt}. \tag{EQ8.31}$$

For a proof of this solution, see Box 8.4.

## Box 8.4 The von Bertalanffy equation

If

$$L = L_\infty - (L_\infty - L_0)e^{-kt},$$

then differentiating both sides with respect to $t$ gives:

$$\frac{dL}{dt} = -(L_\infty - L_0) \times -k \times e^{-kt} = k(L_\infty - L_0)e^{-kt}.$$

This equation gives the growth rate as a function of

time, $t$. On rearranging the first equation we find that:

$$(L_\infty - L_0)e^{-kt} = (L_\infty - L).$$

Therefore, substituting for $(L_\infty - L_0)e^{-kt}$ in the second equation gives:

$$\frac{dL}{dt} = k(L_\infty - L).$$

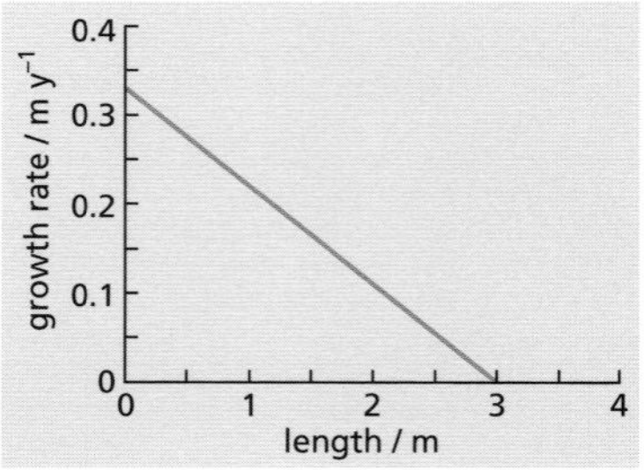

Predictions of the future size of the shark can be made if values are obtained for three parameters: $L_0$, $L_\infty$, and $k$. If a shark is held in captivity from birth it can be measured regularly, so its age and initial length will be known quantities. The constant $k$ and the maximum size can therefore be determined from a linear plot of the growth rate ($dL/dt$) as a function of length $L$ (Figure 8.27); the horizontal intercept yields $L_\infty$, whereas the gradient gives a value for the von Bertalanffy constant $k$.

To reflect the idea that a new organism begins to grow when a fertilized egg divides, rather than at birth, the restricted growth equation EQ8.31 is sometimes restated in an equivalent form:

$$L = L_\infty(1 - e^{-k(t-t_0)}). \tag{EQ8.32}$$

**Figure 8.27**
The growth rate of a sand tiger shark as a function of length.

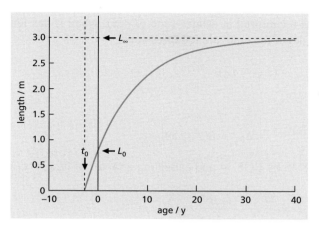

**Figure 8.28**
The von Bertalanffy growth curve showing the meaning of $t_0$.

Here $t_0$ represents the hypothetical time at which a sand tiger shark is predicted to have zero size (Figure 8.28). To obtain the relationship between $t_0$, $L_0$, and $L_\infty$, we can equate the right-hand sides of EQ8.31 and EQ8.32:

$$L_\infty - L_\infty e^{-k(t-t_0)} = L_\infty - L_\infty e^{-kt} + L_0 e^{-kt}.$$

The $L_\infty$ terms cancel out and we can divide all of the remaining terms by $e^{-kt}$ to leave:

$$L_\infty e^{kt_0} = L_\infty - L_0.$$

Changing the subject of this formula to $t_0$ gives:

$$t_0 = \frac{1}{k} \ln \left( \frac{L_\infty - L_0}{L_\infty} \right). \tag{EQ8.33}$$

For a female sand tiger shark, $L_0$ is 0.8 m, $L_\infty$ is 3.0 m, and $k$ is 0.11 m y$^{-1}$, which correspond to a $t_0$ value of $-2.82$ years. This is somewhat longer than the actual gestation period of a shark pup, which is at most 2 years. The von Bertalanffy model breaks down here because growth before birth is more rapid than predicted, probably because the supply of nutrients is more efficient and the shape of the fetus is somewhat different from that of the adult shark.

Applying the von Bertalanffy method to an animal in the wild is not always so easy. For example, there is currently no straightforward, non-invasive method for determining the age of a wild green turtle. For example, counting the growth rings on the scales of its shell is not a reliable indicator of age, because a turtle may form several growth rings in a good year. A more successful approach is to capture, measure, release, and then recapture the same animal at a later date. If a turtle is caught for the first time and tagged at age $T$, when it has length $L(T)$, we can state that:

$$L(T) = L_\infty - (L_\infty - L_0)e^{-kT}.$$

Rearranging this a little produces the expression:

$$(L_\infty - L_0)e^{-kT} = L_\infty - L(T). \tag{EQ8.34}$$

If the turtle is recaptured at a later time $(T + x)$, when it has length $L(T + x)$, we can write down the equivalent expressions:

$$L(T + x) = L_\infty - (L_\infty - L_0)e^{-k(T+x)},$$

and

$$(L_\infty - L_0)e^{-k(T+x)} = L_\infty - L(T + x). \tag{EQ8.35}$$

Dividing EQ8.35 by EQ8.34 and canceling a factor of $(L_\infty - L_0)e^{-kT}$ gives:

$$\frac{L_\infty - L(T + x)}{L_\infty - L(T)} = \frac{(L_\infty - L_0)e^{-k(T+x)}}{(L_\infty - L_0)e^{-kT}} = e^{-kx}. \tag{EQ8.36}$$

Finally, rearranging EQ8.36 to make $L(T + x)$ the subject gives

$$L(T + x) = L_\infty - (L_\infty - L(T))e^{-kx}. \tag{EQ8.37}$$

This version of the restricted growth equation has a very similar form to EQ8.31, but shows that the length of a turtle and the time since it was last captured can be used to determine the rate constant $k$ and the ultimate length $L_\infty$ without knowing either its age or its length at birth.

For example, when it is first captured, tagged, and measured, a green turtle has length 77.2 cm, but on recapture 19 years later it has grown to 93.0 cm (Figure 8.29). EQ8.37 can be rearranged to show that:

$$L_\infty = \frac{L(T + x) - L(T)\ e^{-kx}}{(1 - e^{-kx})}.$$

If the von Bertalanffy constant $k$ for this species is known to be 0.075 cm y$^{-1}$, the ultimate length of the turtle is predicted to be 98.0 cm.

EQ8.31 describes a process in which a quantity (length) changes from a specified initial value $(L_0)$ to a final equilibrium value $(L_\infty)$, with the difference between the current and the equilibrium values decaying exponentially to zero. Similar differential equations can be used to model a range of interesting phenomena, including the rate at which an object approaches the temperature of its surroundings, the decay of leaves falling on to the floor of a forest, predicting the equilibrium concentration of a pollutant released into a lake, or analyzing how the attainment level of a new skill improves with practice. For further details, see Chapter 12.

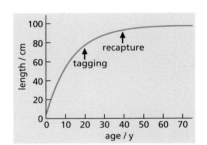

**Figure 8.29**
Turtle tagging and recapture experiments.

# Presenting Your Work

## QUESTION A

In genetics, the uncertainty or randomness associated with the expression of a gene can be quantified in terms of its 'informational entropy'. A particular gene has only two states, switched on or switched off, which can be detected in samples from different tissues using a DNA microarray. If $x$ is the probability that the gene is switched on, then the binary entropy function $H(x)$ is given by

$$H(x) = -x \ln(x) - (1 - x) \ln(1 - x).$$

Determine the value of $x$ that gives the greatest value of $H(x)$.

---

We need to use the product rule to differentiate the two terms on the right-hand side of the entropy equation:

$$\frac{d}{dx}(x \ln(x)) = 1 \times \ln(x) + x \times \frac{1}{x} = \ln(x) + 1$$

$$\frac{d}{dx}((1 - x)\ln(1 - x)) = -1 \times \ln(1 - x) - (1 - x) \times \frac{1}{(1 - x)} = -\ln(1 - x) - 1$$

$$\frac{dH}{dx} = -\ln(x) - 1 + \ln(1 - x) + 1 = \ln\left(\frac{1 - x}{x}\right).$$

Turning points occur when $\frac{dH}{dx} = 0$, i.e. when $\ln\left(\frac{1 - x}{x}\right) = 0$,

which occurs when $\frac{1 - x}{x} = 1$.

Multiplying both sides of this equation by $x$ gives: $1 - x = x$,

that is, $x = \frac{1}{2}$.

To work out whether this is a maximum or a minimum, we should check the sign of the second derivative:

$$\frac{d^2 H}{dx^2} = \frac{d}{dx}\left\{\ln\left(\frac{1 - x}{x}\right)\right\} = \frac{d}{dx}\{\ln(1 - x) - \ln(x)\} = -\frac{1}{1 - x} - \frac{1}{x}.$$

When $x = 0.5$, $\frac{d^2 H}{dx^2} = -\frac{1}{1 - 0.5} - \frac{1}{0.5} = -4$, so this is a local maximum.

The entropy $H$ therefore has its greatest value at $x = 0.5$. This corresponds to the situation where the probability that the gene is switched on ($x = 0.5$) is equal to the probability that it is switched off ($1 - x = 0.5$).

**QUESTION B**

A human femur and some fragments of charcoal are discovered next to each other at an archaeological site and are found to contain 55 % and 51 % of the original level of $^{14}C$, respectively. If the half-life of $^{14}C$ is 5730 years, determine whether the human could have started the fire that produced the charcoal.

Assuming that $^{14}C$ decay is exponential

$$N(t) = N_0 e^{-\lambda t}, \text{ so } t = -\frac{1}{\lambda} \ln \left( \frac{N}{N_0} \right).$$

But $t_{1/2} = \dfrac{\ln(2)}{\lambda}$, so $\lambda = \dfrac{\ln(2)}{t_{1/2}}$, giving $t = -\dfrac{t_{1/2}}{\ln(2)} \ln \left( \dfrac{N}{N_0} \right).$

The age of the femur is

$$t_{femur} = -\frac{5730}{\ln(2)} \ln(0.55) = 4942 \text{ years.}$$

The age of the artifact is

$$t_{artifact} = -\frac{5730}{\ln(2)} \ln(0.51) = 5566 \text{ years.}$$

Because these ages are apparently separated by 624 years, it is unlikely that the human started this fire — unless he or she was burning an ancient heirloom!

## End of Chapter Questions

*(Answers to questions can be found at the end of the book)*

### Basic

**1.** Differentiate the following expressions with respect to $x$:

(a) $y = \ln(4 - 3x)$

(b) $y = \exp(3 - 2x)$

(c) $y = \exp(x^2 + 1)$

(d) $y = x \times \ln(x)$

(e) $y = \left(x + \dfrac{1}{x}\right) \times e^x$

(f) $y = \dfrac{e^{-3x}}{3x}$

**2.** Differentiate the following expressions with respect to $x$. Where possible, rewrite the expression first to make the differentiation easier.

(a) $y = \left(\dfrac{e^{2x}}{3}\right)^3$

(b) $y = (e^x + 1)^7$

(c) $y = \ln\left\{\dfrac{(1 - 3x)}{(x + 4)(1 - x)^2}\right\}$

(d) $y = \ln(1 + x^2)$

(e) $y = x\, e^x \sin(x)$

(f) $y = \cos(\ln(x))$

**3.** Integrate the following expressions with respect to $x$:

(a) $y = e^{+3x} + e^{-3x}$

(b) $y = \dfrac{1}{1 - 2x}$

(c) $y = xe^{-x^2}$

**4.** Integrate the following expressions with respect to $x$. Where possible, rewrite the expression first to make the integration easier.

(a) $y = \dfrac{x}{x - 1}$

(b) $y = \ln(e^{3x})$

(c) $y = \dfrac{\exp(\ln(x^2))}{x}$

**5.** Evaluate the following definite integrals:

(a) $I = \displaystyle\int_1^e \dfrac{2}{x}\, dx$

(b) $I = \displaystyle\int_{-1}^{\ln(10)} e^{-x}\, dx$

**6.** Evaluate each of the following definite integrals by making an appropriate change of variable.

(a) $I = \displaystyle\int_0^1 \dfrac{x^3}{x + 1}\, dx$

(b) $I = \displaystyle\int_{-1}^{1} \dfrac{e^x - e^{-x}}{e^x + e^{-x}}\, dx$

### Intermediate

**7.** At time $t$ (in minutes) the concentration $c$ (in millimoles per liter) of a reaction intermediate is expressed by the equation

$$c(t) = 10t^3 e^{-t}.$$

Sketch the form of this function for positive values of $t$.

**8.** In an experiment called 'the reptilian drag race', the distance traveled on a horizontal surface when an agamid lizard accelerates from a standing start has been modeled by the equation:

$$x = v_{max}\left(t + \dfrac{e^{-kt}}{k} - \dfrac{1}{k}\right).$$

In this equation $v_{max}$ is the maximum velocity, $t$ is the time after the start, and $k$ is a rate constant. Obtain expressions for the velocity $v$ and the acceleration $a$ of the lizard as a function of time. If $v_{max}$ is 3 m s$^{-1}$ and $k$ is 10 s$^{-1}$, make a rough sketch of $x / \text{m}$, $v / \text{m s}^{-1}$, and $(a / 10)$ m s$^{-2}$ on the same set of axes for $0 \leqslant t \leqslant 1$ s.

**9.** The Clausius–Clapeyron equation links the vapor pressure $p$ and temperature $T$ of a gas with the gas constant $R$ and the latent heat of vaporization, $\Delta H_{vap}$ (which can also be treated as a constant). Show that the expressions

$$\dfrac{d\{\ln(p)\}}{dT} = \dfrac{\Delta H_{vap}}{RT^2} \quad \text{and} \quad \dfrac{dp}{dT} = \dfrac{p\Delta H_{vap}}{RT^2}$$

are equivalent.

**10.** Use the linear approximation method to derive approximations for the following functions, by setting $x$ to zero and making $\delta x$ very small:

(a) If $y = \ln(1 + x)$, show that $\ln(1 + \delta x) \approx \delta x$

(b) If $y = \exp(x)$, show that $\exp(-\delta x) \approx 1 - \delta x$. Determine the percentage error of these approximations for $\delta x = 0.01, 0.1$, and $0.5$.

**11.** Sketch the function $y = (\ln(x))/x$ within the range $0 \leqslant x \leqslant 10$ and $-1 \leqslant y \leqslant +0.5$ indicating the coordinates of any boundary points, intercepts with the $x$ and $y$ axes, vertical or horizontal asymptotes, local maxima, local minima, or points of inflexion.

## Advanced

**12.** The philosopher Aristotle died in 322 BC, but the earliest known copies of his work date from 850 AD. If fragments of a newly discovered manuscript are analyzed, what proportion of the original amount of $^{14}C$ would you expect to find (a) if it was written in 850 AD; or (b) if it was written by Aristotle himself. (Note that the half-life of $^{14}C$ is 5730 years.)

**13.** The half-life of cocaine in human blood plasma is 1.5 hours. If initially there is 0.2 mg/kg of cocaine in the bloodstream, determine when the amount drops below 0.01 mg/kg, and what percentage will remain after 1 day.

**14.** Between 1950 and 1990, the human world population increased approximately exponentially at a rate of 1.86 % per year.
(a) Estimate the doubling-time of the world population.
(b) Assuming that this approximation is still valid and that the mean lifespan of a human is 25 years, calculate the current world population as a percentage of $N_{total}$, the total number of people who have lived during the last million years.

(Hint: integrate the equation for the world population with respect to time to find the total number of person-years for everyone who has lived during the last million years. Divide this result by the mean lifespan to obtain an expression for $N_{total}$.)

**15.** The Gompertz growth curve, given by the equation

$$x(t) = a \exp(-b \exp(-ct)),$$

where $a$, $b$, and $c$ are positive constants, can be used to model the size of a tumor as a function of time. Show that the curve possesses a point of inflexion at $t = \ln(b)/c$.

# CHAPTER 9

# Descriptive Statistics and Data Display

Biology experiments rarely yield a single number; rather, they produce a set of measurements. Many scientific experiments and observational studies yield huge amounts of data: techniques such as brain scanning may produce many millions of numbers for a single scan. So, what do we do with all these data? Just staring at a long list of numbers is not very informative. It is best if we can find a sensible way to draw a meaningful picture of the data, like a three-dimensional visualization of our brain scan.

(A)

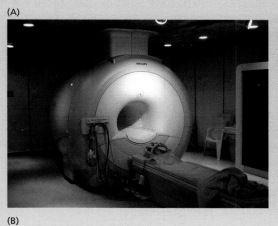

(C)

**Figure 9.1**
A modern brain scanner produces many millions of numbers in a single scanning session (A). The numbers themselves (B) are no use unless we have a way of visualizing and summarizing them, such as this three-dimensional image of the differences in brain activity between normal individuals and schizophrenic patients (C).

(B)

| | | | | | | | | | |
|---|---|---|---|---|---|---|---|---|---|
| 92.27232 | 27.59873 | 68.27159 | 16.45766 | 61.99743 | 97.93327 | 23.74764 | 24.70277 | 27.00433 | 57.62179 |
| 72.96887 | 48.36348 | 98.89933 | 11.45828 | 81.54283 | 53.30085 | 88.74505 | 52.20137 | 90.81593 | 29.56227 |
| 58.38643 | 34.29861 | 67.80288 | 13.25921 | 23.36585 | 97.16449 | 73.59811 | 77.2467 | 60.94119 | 71.36301 |
| 62.63576 | 66.90434 | 88.65279 | 80.19005 | 71.7458 | 36.61784 | 44.95184 | 56.85336 | 77.94018 | 89.21044 |
| 63.26059 | 3.361856 | 37.91634 | 97.46252 | 24.66708 | 76.48902 | 19.80798 | 9.472042 | 46.05279 | 56.70383 |
| 5.986422 | 13.78547 | 8.19881 | 41.64089 | 99.59356 | 29.49696 | 10.73063 | 7.200171 | 34.01174 | 22.78648 |
| 23.20195 | 87.51052 | 61.98893 | 14.4239 | 53.75503 | 49.25431 | 47.71024 | 24.73094 | 3.697212 | 62.6111 |
| 8.085785 | 31.62249 | 85.57363 | 39.76502 | 97.68176 | 53.69371 | 47.07124 | 29.46008 | 48.55319 | 4.134846 |
| 45.91201 | 22.99768 | 81.37854 | 39.80351 | 34.69779 | 88.10599 | 10.04007 | 47.2223 | 91.34105 | 73.72224 |
| 68.35866 | 71.57353 | 44.76168 | 25.60956 | 95.19553 | 36.47399 | 32.19412 | 80.49207 | 33.78497 | 37.38513 |
| 76.59016 | 13.11525 | 26.0191 | 54.6943 | 64.92567 | 56.09012 | 82.22509 | 56.51576 | 62.38169 | 70.37322 |
| 56.40326 | 57.56379 | 55.10071 | 65.25372 | 52.14133 | 1.176595 | 76.47659 | 84.08017 | 72.50476 | 17.43844 |
| 3.39358 | 89.96927 | 9.381817 | 59.95629 | 51.87232 | 59.17203 | 49.60708 | 15.95584 | 23.44681 | 55.91028 |
| 56.09719 | 5.260781 | 87.26575 | 58.03758 | 8.240363 | 49.73513 | 69.88527 | 80.13369 | 14.06698 | 75.04928 |
| 51.64076 | 33.9103 | 54.20902 | 7.908238 | 87.93817 | 61.91476 | 86.01317 | 9.292247 | 80.29093 | 3.511675 |
| 85.05463 | 59.84094 | 19.01969 | 13.49588 | 64.1395 | 35.16145 | 41.71987 | 68.57865 | 32.02744 | 68.74529 |
| 42.6446 | 61.90268 | 67.37326 | 93.93701 | 76.47176 | 50.24529 | 80.15839 | 17.65467 | 7.640986 | 83.76912 |
| 33.71358 | 79.57247 | 2.376451 | 60.25503 | 85.784 | 5.808636 | 67.74646 | 66.83327 | 90.64096 | 40.29587 |
| 22.24921 | 31.51778 | 57.20564 | 37.68627 | 37.97682 | 91.43793 | 33.01351 | 83.20688 | 95.85603 | 29.48397 |
| 30.61237 | 47.99463 | 76.35992 | 51.91269 | 55.64647 | 25.10737 | 61.07285 | 83.48908 | 67.42621 | 13.9276 |
| 65.18154 | 99.60292 | 21.4301 | 86.14157 | 17.51436 | 29.62082 | 27.20651 | 16.26169 | 27.21718 | 21.05209 |
| 69.03907 | 8.251522 | 59.32826 | 83.99454 | 31.50223 | 4.487203 | 89.95593 | 78.35896 | 84.49923 | 88.09859 |
| 53.82024 | 63.49776 | 32.66586 | 32.55293 | 23.44242 | 49.17725 | 90.89192 | 70.64258 | 1.884067 | 24.55445 |
| 89.66551 | 29.12665 | 74.00128 | 87.71271 | 47.61983 | 95.52565 | 78.53784 | 41.23409 | 81.17832 | 73.37301 |
| 30.69607 | 87.26133 | 14.60693 | 75.7596 | 56.16886 | 55.54137 | 92.89115 | 5.077133 | 83.65748 | 21.25392 |
| 24.35928 | 53.65804 | 96.72502 | 57.56316 | 48.92942 | 56.94031 | 31.89628 | 74.51281 | 11.4204 | 80.8832 |
| 24.537 | 98.98097 | 13.2443 | 93.45118 | 5.131202 | 14.66745 | 70.13449 | 11.91343 | 22.92332 | 51.08262 |
| 71.62712 | 71.07652 | 63.66945 | 18.7923 | 80.31483 | 93.10207 | 76.57894 | 2.895843 | 61.20741 | 39.06401 |
| 5.569961 | 23.72177 | 92.58949 | 42.50456 | 58.4579 | 36.14749 | 25.66358 | 63.15894 | 5.058649 | 2.124288 |
| 50.43553 | 48.45663 | 13.68348 | 6.546351 | 12.84572 | 78.96522 | 38.57635 | 79.43858 | 87.40821 | 57.18554 |

So far in this book, we have been looking at mathematics and numbers as precise measurements or abstract symbols for scientific concepts, but many of these concepts may vary across cells or individuals (for example height). Even if we are measuring something that we do not think is variable, our measurement process may have some errors, so that if we try to measure a concept repeatedly, the measurements will be slightly different every time we run an experiment.

In this chapter, we will look at techniques for dealing with *sets of data*, rather than single numbers. To make sense of the results of any real study, we need techniques to organize, summarize, and produce a visual representation of our data. These techniques are known as exploratory data analysis or descriptive statistics. We will also see how we can describe a simple relationship between two sets of observations.

# 9.1 Measurement scales

When we have finished an experiment, we will have a set of data: these may be detailed audio-video recordings, entire organisms, or simply a collection of numbers in a notebook. However proud we may be to have collected these data, they are not much good on their own. Our results are only useful if we can share them with others – and we certainly cannot simply give everybody else a copy of all the data, and expect them to read or understand them all.

What we need is a systematic way to summarize, or describe, the results of an experiment. This type of analysis is called *descriptive statistics*, and is often distinguished from the techniques that are used when we wish to draw conclusions from our data (these are called *inferential statistics*, which we will meet in Chapter 11). The purpose of descriptive statistics is to summarize a set of data in a few numbers, so that we can convey a brief, but accurate, impression of the results.

Our first consideration when exploring the results of an experiment has to be the type of data we have collected. We can express a single measurement in the form of a number (or a set of numbers) and some units. However, not all numbers are the same: we have already seen the importance of the units of measured physical qualities, and obviously specifying the units of measurement we are using is important if we want to measure a physical quality such as the mass, or height, of an organism.

However, when we come to measuring some of the biological values we are interested in, there are issues concerning the different types of numerical measurement that go beyond the problem of picking the correct SI units. Imagine a colleague returns from a whale-watching experiment. Looking at her notebook, which she has left open, we see a list of dates, each accompanied by a set of values between 1 and 7. The number by each date could represent the number of whales that were spotted on each sighting, the estimated length in meters of a spotted whale, or the species of whale spotted (1 = humpback, 2 = bottlenose dolphin, 3 = blue whale, and so on). Alternatively, our poor colleague may never have seen a whale, and was just recording the number of times she was seasick!

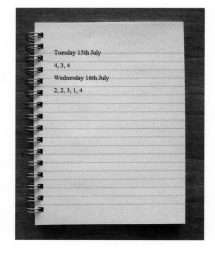

**Figure 9.2**
A data notebook of numbers cannot be interpreted unless we know the measurement scale that relates these numbers to the physical quantities.

It should be obvious that before we try to summarize the data, we need some kind of data 'key': that is, something that tells us what each number stands for. This might simply be some units for the numbers (for example whales per trip, or meters), or it might be a table of which species corresponds to which number.

The key that is used to relate a pure number to a 'measurement' (or vice versa) is known as a measurement scale. Talking about the species key above as a 'scale' may seem peculiar, but it is really just the same as if we were talking about a temperature scale. We might hear the weather report give the statement: 'today the air temperature is 35'. Whether we wrap up warm, or turn on the air conditioner, will depend on whether we think that the number 35 should be converted into a temperature using the Celsius or Fahrenheit measurement scales.

It is traditional to distinguish between four different types of scale for scientific measurements: *nominal*, *ordinal*, *interval*, and *ratio* scales.

## Nominal scales

A purely **nominal scale** relates each value to a different label, in no particular order, so that the *size* (numerical value) of the label does not carry any meaning. These scales are used with measurements of category, which do not need to have any specific order. We can assign a different number (or word) for each category in any way we choose. Examples might be sex ($1 =$ male, $2 =$ female), or species, as we saw above.

## Ordinal scales

An **ordinal scale** is one in which a larger value always relates to a larger measurement. To be expressed on such a scale, our measurements of category must have a meaningful order, such that they can be ranked (that is, put into ascending, or descending, order). We can assign any number we like to each category, so long as the order of the numbers reflects the ranking of the categories. As larger measurements are given larger values on the scale, the *order* of two values on the scale tells us which is larger. The exact numerical difference between points on the scale does not necessarily tell us anything, as a gap of two units may not measure the 'same amount' of difference all along the scale. Some examples are the Beaufort scale of wind power, military rank ($1 =$ general, $2 =$ captain, $3 =$ private), or undergraduate degree ($1 =$ first class, $2 =$ upper second, $2.5 =$ lower second, $3 =$ third class, $4 =$ without honors).

## Interval (and ratio) scales

An **interval scale** is one in which an increase in value of one unit corresponds to a particular increase in a measurement. Measurements of 'amount' are generally taken on interval scales. Because each unit of increase corresponds to the same amount of change in measurement, the difference in value between any two measurements (the *interval* along the scale) is a meaningful measure of change.

Generally, a measurement of the 'amount' of some property such as length is also taken on a **ratio scale**, an interval scale with a meaningful 'zero point' (corresponding to an absence of the property being measured). For ratio-scale measurements, it is not only the differences in value that are meaningful: a value that is twice as large reflects a measurement of 'twice as much' of the amount. In other words, the *ratio* of any two values on the scale is meaningful.

## Continuous and discrete data

We first met numbers as the counting numbers, and expanded these to give us the integers, rational, and real numbers (see the Introduction). Recall that the integers have 'gaps' in them (there is no integer between 1.2 and 1.4) whereas the real numbers are continuous, in the sense that there is always one of them in any interval we can think of. We can make the same distinction between measurements on interval scales. Some measurements are made on **discrete scales**, for which only certain values are meaningful (that is, there are values that are not possible measurements). An example of this would be the number of whales seen (must be a non-negative integer). Other scales are **continuous scales**, for which any value in a range is a possible measurement. An example would be the length of a whale spotted.

Most continuous measurements that we take are also ratio-scale measurements, but not quite all: time and temperature (measured in degrees Celsius) are measurements on an interval scale, not a ratio scale (because a measurement of zero degrees Celsius, or zero seconds since midnight, does not mean an *absence* of heat or time).

It is worth noting that the different types of measurement scale are not really exclusive categories, so much as a *hierarchy*: for example ratio and interval scales have ordinal properties (the size order of two values is meaningful), and ratio scales have interval properties (the difference between two values is meaningful).

Most interval and ratio scales are *always positive or zero* (often because they are measures of amount), and are sometimes called *positive definite* scales. The concepts of positive definite or non-ratio interval scales are rarely very important to us, as any interval scale will provide a corresponding scale of *differences between measurements* which is a ratio scale (differences in Celsius temperature, or in time, can be twice as large) and can take negative values.

Why do we bother to distinguish between the types of measurement scale? We wish to calculate statistics to describe our data, in order to understand, communicate, or draw conclusions from them. What makes a sensible summary or description of our numbers will depend on what the numbers are measuring, not just on what the values are. This is a very important point, and easy to forget.

To see why this is true, let us consider the numbers in our whale-spotting notebook. Imagine two notebooks for which we want to summarize the data, which happen to contain the same numbers. The first notebook contains numbers that correspond to the estimated length of each whale spotted (in meters); the second contains numbers corresponding to the species of each whale. The *numbers* in the different notebooks are exactly the same, but the way in which we should summarize them would be completely different because the two notebooks were recording different types of data. For example, it would make sense to add the numbers for the first notebook (estimated length of whales), but not for the other (a humpback + a bottlenose does not equal a blue whale!). This type of consideration relates to the type of scale on which our numbers are measured. There are several different techniques for analyzing and summarizing data, and not all are suitable for each type of scale.

Obviously, we should never be confronted with a situation where we do not know the meaning of numbers in the data we have collected. However, this example should remind you how important it is to think about what kind of data we have collected, before we try to summarize or analyze.

For the remainder of this chapter we will consider a data set based on an imaginary study of the impact of temperature change on amphibian species. This shows data for 50 captures of adult pickerel frogs, *Rana palustris*, in an area of North American wetland in July and September of the same year. The weather (full sun, dry with cloud cover, rain, sleet/hail/snow), temperature at time of capture, body length (snout to anal vent), and mass are recorded for each frog captured.

**Figure 9.3**
An adult pickerel frog, *Rana palustris.* Image courtesy of Sam Hopewell under Creative Commons Attribution Share Alike 3.0 Unported license.

As these frogs have no visible sexual characteristics outside the breeding season (March–May), they are difficult to sex reliably. However, the researcher in this study knows that the sex is usually clear from the mass (adult males are never above 20 g, and adult females are never below 15 g). Where the sex of the adult captured can be determined on this basis, it is noted.

Before going to the next section you should tackle question 1 from the End of Chapter Questions.

## 9.2 Summarizing a data set

Glancing at the frog data in Table 9.1 is not very informative: we get no real sense of what was observed in this study. The most useful approach to

| Table 9.1 Example data set | | | | | |
|---|---|---|---|---|---|
| **Capture number** | **Sex** | **SVL / cm** | **Mass / g** | **Temperature / °F** | **Weather** |
| 1 | 1 | 6.6 | 29.3 | 81.2 | 1 |
| 2 | 3 | 5.0 | 12.4 | 77.2 | 1 |
| 3 | 2 | 5.7 | 15.2 | 77.6 | 3 |
| 4 | 3 | 5.6 | 12.1 | 79.3 | 1 |
| 5 | 1 | 5.8 | 26.3 | 80.6 | 2 |
| 6 | 3 | 5.3 | 11.6 | 77.9 | 1 |
| 7 | 3 | 4.4 | 8.3 | 77.4 | 1 |
| 8 | 1 | 6.2 | 22.4 | 76.3 | 1 |
| 9 | 3 | 5.0 | 12.3 | 76.3 | 2 |
| 10 | 1 | 6.6 | 35.7 | 76.6 | 1 |
| 11 | 3 | 5.1 | 12.4 | 78.4 | 1 |
| 12 | 2 | 5.4 | 16.2 | 75.8 | 2 |
| 13 | 3 | 5.2 | 12.1 | 77.5 | 1 |
| 14 | 1 | 6.8 | 42.5 | 72.0 | 2 |
| 15 | 3 | 5.3 | 14.1 | 72.1 | 2 |
| 16 | 3 | 4.7 | 10.2 | 78.6 | 2 |
| 17 | 1 | 6.8 | 38.0 | 74.2 | 1 |
| 18 | 3 | 5.8 | 13.9 | 71.7 | 2 |
| 19 | 3 | 4.7 | 12.1 | 78.4 | 1 |
| 20 | 2 | 6.4 | 19.8 | 76.3 | 1 |
| 21 | 1 | 6.7 | 33.3 | 75.0 | 2 |
| 22 | 1 | 6.3 | 28.0 | 72.9 | 2 |
| 23 | 3 | 5.1 | 14.2 | 70.6 | 1 |
| 24 | 1 | 6.4 | 25.7 | 71.0 | 3 |
| 25 | 2 | 5.2 | 15.7 | 77.3 | 1 |
| 26 | 2 | 5.5 | 17.5 | 76.8 | 1 |
| 27 | 3 | 5.2 | 13.1 | 71.4 | 2 |
| 28 | 3 | 4.6 | 9.4 | 73.8 | 2 |
| 29 | 3 | 4.9 | 12.6 | 69.1 | 2 |
| 30 | 3 | 4.8 | 11.0 | 75.1 | 1 |
| 31 | 2 | 5.3 | 16.7 | 76.5 | 1 |
| 32 | 3 | 5.3 | 12.8 | 67.5 | 3 |
| 33 | 2 | 5.7 | 17.3 | 72.2 | 2 |
| 34 | 3 | 5.0 | 13.5 | 69.7 | 3 |
| 35 | 3 | 4.3 | 7.7 | 71.2 | 3 |
| 36 | 3 | 4.6 | 10.3 | 72.8 | 1 |
| 37 | 1 | 6.6 | 29.7 | 69.6 | 3 |
| 38 | 2 | 6.0 | 17.4 | 73.9 | 3 |
| 39 | 1 | 5.8 | 21.1 | 72.5 | 2 |
| 40 | 1 | 6.5 | 30.8 | 70.0 | 3 |
| 41 | 3 | 5.7 | 13.3 | 76.3 | 1 |
| 42 | 2 | 5.5 | 19.7 | 73.9 | 2 |
| 43 | 2 | 6.0 | 18.8 | 70.1 | 3 |
| 44 | 3 | 4.9 | 11.2 | 76.3 | 1 |
| 45 | 1 | 6.5 | 27.0 | 68.1 | 4 |
| 46 | 2 | 5.9 | 16.1 | 70.0 | 3 |
| 47 | 1 | 5.5 | 23.2 | 69.5 | 3 |
| 48 | 3 | 4.9 | 10.2 | 72.1 | 3 |
| 49 | 1 | 6.0 | 29.0 | 74.2 | 2 |
| 50 | 1 | 6.1 | 29.9 | 76.2 | 3 |

A data set describing the capture of 50 pickerel frogs. These data are constructed for the purposes of demonstration, based on published data on the same species.
Sex: 1, female; 2, unknown; 3, male.
Weather: 1, sunshine; 2, dry, cloudy; 3, rain; 4, hail/sleet/snow.
SVL, snout to vent length (that is, length of body excluding tail).

describing a data set is to create a graphical summary: this will allow you to grasp its important features quickly. Of course, it is crucial that any graphical display conveys an accurate impression!

At this stage, we will focus on how we summarize a single set of measurements, where data points are equivalent measurements in no particular order. Sometimes, of course, we need to summarize how one measured variable changes with respect to another (such as time, if we take a measurement at different time points to get a 'time-series' of data). We will return to this type of situation later in the chapter.

## Pie charts and column graphs

If data are measured on a nominal scale, they can only be frequency counts. It is often sensible to express data as proportions or ratios, and the best visual presentation in this case is usually the pie chart. These are circular graphs, where the circle is divided up into sectors that show the proportion of scores in each category. The angle (as a proportion of $2\pi$, or 360°) for a sector is set to the proportion of scores. In general, pie charts should only be used when you want to show all of a data set, in terms of the proportion that fall into each category. It is common, but not required, to order the categories in decreasing size, starting with the largest category (clockwise or anticlockwise) from the top of the circle.

Ordinal measurements are generally better presented in the form of a column graph, where the number or proportion of values in each range is shown by the height of a column. The main difference between a column graph and a pie chart is that the categories in a column graph are presented in order along the $x$ axis. Because of this, a column graph allows rapid comparison of the proportion of scores that have adjacent values more easily than does a pie chart.

The sex data in Table 9.1 are measured on a nominal scale, and the weather data are measured on an ordinal scale (from sunny to heavy precipitation). We can illustrate pie charts and column graphs using these data.

The weather data are expressed on an ordinal scale (in terms of how 'bad' the weather is: sun, dry, rain, hail/sleet/snow). Thus a column graph is suitable, as we can order the columns. To generate a column graph we simply sum the number of observations in each category, and draw a bar whose height corresponds to that number, as shown in Table 9.2 and Figure 9.4.

The sex data are expressed on a nominal scale (male, female, unknown) with no particular order. A pie chart is the most suitable graphical method.

To generate a pie chart, we sum the number within each category, and express each as a proportion of the total observations. A circle is drawn and divided into sectors, one for each category. The internal angle for each sector is calculated as $p \times 360°$, where $p$ is the proportion of scores in the category (we express the angle here in degrees, as this is often more useful when drawing).

| Table 9.2  Weather distribution (number of captures for each weather category) | |
|---|---|
| Sun | 20 |
| Dry/cloudy | 16 |
| Rain | 13 |
| Hail/sleet/snow | 1 |

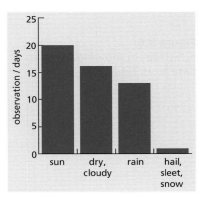

**Figure 9.4**
A column graph showing the weather data from Table 9.2.

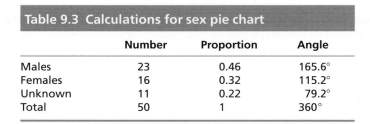

| Table 9.3 Calculations for sex pie chart | | | |
| --- | --- | --- | --- |
| | **Number** | **Proportion** | **Angle** |
| Males | 23 | 0.46 | 165.6° |
| Females | 16 | 0.32 | 115.2° |
| Unknown | 11 | 0.22 | 79.2° |
| Total | 50 | 1 | 360° |

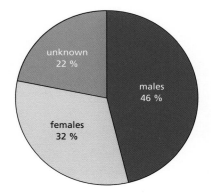

**Figure 9.5**
A pie chart showing the sex data from Table 9.3.

Following the simple convention of starting with the largest 'slice', and working clockwise, we get Figure 9.5.

## Histograms

If our data comprise interval measurements, we can summarize them in the form of a histogram. This is a special name for a column graph when the $x$ axis is an interval scale. Unlike a column graph for ordinal data, it makes sense to talk about the 'width' of each interval in a histogram. Usually, to create a histogram, we simply divide the range of the data into several equal intervals, and count how many values fall into each one. Then we draw a bar for each interval, with height equal to the number (or proportion) of values that fall within it. The heights of all the bars will sum to the total number of observations (or 100 %); and if you want to know the number (or proportion) of the data that fall into a certain range, you can simply sum the heights of the bars for that range.

The mass, snout to vent length (SVL), and temperature data from Table 9.1 are all measured on interval scales (SVL and mass are also ratio scales), so we can illustrate the process of generating a histogram using the mass data. We have 50 observations in the interval 7.7−42.5. For simplicity, we can choose the edges of our scale to be 5 g and 45 g, giving a range of 40 g. Choosing eight categories gives a simple width of 5 g for each category, as shown in Table 9.4.

| Table 9.4 Frequency distribution of mass for the frogs | |
| --- | --- |
| **Mass / g** | **Frequency** |
| 5–10 | 3 |
| 10–15 | 20 |
| 15–20 | 11 |
| 20–25 | 3 |
| 25–30 | 8 |
| 30–35 | 2 |
| 35–40 | 2 |
| 40–45 | 1 |

We then plot bars of heights equal to the frequency in each interval to get Figure 9.6.

**Figure 9.6**
A histogram showing the mass data from Table 9.4.

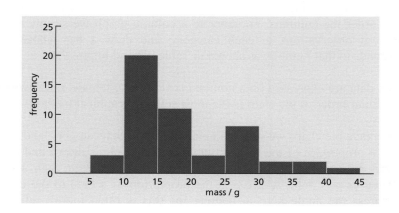

However, how many intervals should we use? If we choose too few or too many, the histogram will be relatively uninformative, as we can see from the two (rather extreme) examples in Figure 9.7.

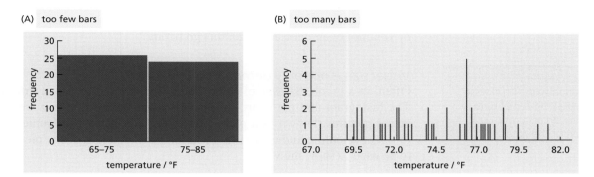

**Figure 9.7**
Two histograms showing the temperature data from Table 9.1, drawn with an inappropriate number of intervals.

A useful rule when choosing the number of intervals is that for $N$ data points, we choose about $\sqrt{N}$ intervals of the same width. In the example above, we had 50 data points, suggesting that we should use approximately $\sqrt{50} = 7.07$ categories. As we saw, dividing the range into eight equal intervals gave a suitable number of columns to show the pattern in the data. You do not need to stick rigidly to this principle, but it usually works well for data sets of 200 items or fewer.

It is the areas of the bars in a histogram that give the visual sense of 'size', and thus the area should reflect the proportion of values in each interval, rather than the height of the bar. If all your intervals are of the same width, then this distinction is unimportant: the relative heights of your bars will correspond to their relative areas. However, if for some reason the intervals you are plotting differ in width, you must make sure that it is the area of each bar, not the height, that reflects the proportion of values in that interval. To do this you have to plot values on the $y$ axis that correspond to the 'values per unit in the interval' known as the **frequency density** (if plotting frequencies) or **proportion density** (if plotting proportions).

In our example above we could have divided each of these frequency values by 50 to express them as proportions, then divided each proportion by 5 g (the width of each interval), to give proportion density values, rather than plain frequencies. Why did we not plot proportion density in this case as well? Is it not 'better' to do the same thing in every case?

If all the intervals are equal, then plotting proportion density (rather than frequency) is only a change of scale on the $y$ axis: it makes no difference to the shape of the graph. The only thing we achieve by plotting the density when we have equal intervals is to make the graph slightly more difficult to understand for the reader. As the point of the graph is to communicate with the reader, this is not really a 'better' approach!

## Lies, damn lies, and graphs

Graphical displays of information are very helpful in terms of quickly conveying information. However, if not drawn carefully (and honestly) such figures can just as quickly give a misleading impression. There is one key point to look out for when drawing or inspecting figures. The *relative values* we are plotting should be reflected in the *relative apparent size* (height, area, or volume) of an item. How can this go wrong?

### Visual size is not only determined by height

Often, in a graph, the *size* of some visual item is used to depict information. The histogram and pie chart are obvious examples of this, but we may use actual pictures (known as a pictogram) especially when communicating data to a non-scientific audience. It is common for the height of the pictorial item to be used to show the value.

> The most important consideration when producing a graphical summary of data is that it gives an accurate impression of the data.

However, we must be careful that the visual impression of size for an item corresponds to the size it is representing. Scaling both the width and height of a picture in a pictogram breaks this rule, because if we double the height and width a two-dimensional object will look four times as big (as the area is four times larger). The problem gets even worse if our picture is of a three-dimensional object: a doubling in size now conveys an eight-fold increase in volume!

This will also be a problem if we plot a histogram using unequal interval widths on the *x* axis, and simply plot the number or proportion in each category. When we do this, the relative area for each category will not show the proportion in each category: wider intervals will seem to have a higher number. We get round this by plotting the proportion density (that is, proportion per unit width of category) or frequency density.

### Changing scales will change relative height

The scale of any graph is important, and we should not choose a scale that highlights any differences between items unduly. Some data may look very different if the zero point of a scale is excluded. This also applies to pie charts: we should always make sure our categories include all possible observations: by convention, readers will expect this.

Imagine that another study was conducted in which 1000 frogs were observed at a particular site of special scientific interest. The sex data were as follows: males, 320; females, 460; not recorded, 220. All of the graphs in Figure 9.8 are attempts to summarize these data.

Looking at the figure, it is not obvious that all the graphs are showing 'the same data': some give very different impressions than others. Graph A is most informative, as the area of each 'bar' represents the proportion of scores. The others all misrepresent the data to some degree, by excluding one category (B), giving a false zero point (D), or, in the case of graph C, by using an image of a three-dimensional object such that apparent differences in 'size' seem larger than differences in 'height'.

(A)

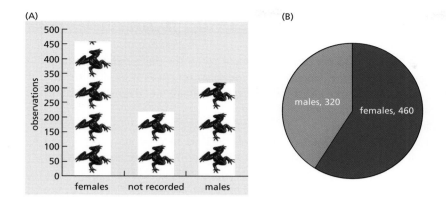

(B)

males, 320    females, 460

(C)

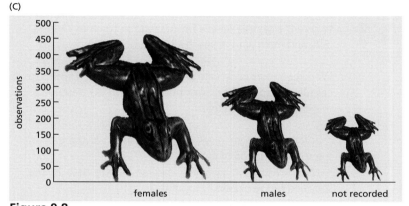

(D)

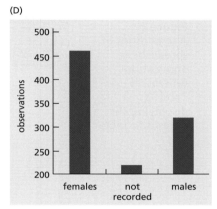

**Figure 9.8**
A variety of graphics displaying the sex data for the data described in the text. Some of these figures are very misleading.

# 9.3 Numerical summary of a data set

In addition to graphing your data, you can summarize the key features using *descriptive sample statistics*. These are numbers that reflect the distribution of values in your data. The properties you will usually be most interested in are the central tendency and dispersion of your data. The central tendency describes how large the measurements typically are (*where* on the measurement scale the data tend to lie): in other words, how big are the values? The dispersion describes how spread out the data are along the scale: in other words, are the values tightly or loosely clustered?

### Interval data: sample mean, variance, and standard deviation

Most continuous numeric measurements in biology are measured on interval or ratio scales, with units of measurement that have the same meaning all along the scale. This means that the sizes of differences, or intervals, between measurements are meaningful. In this situation, we can describe our sample in terms of the 'average' value observed. The data have to be interval for this type of averaging to be sensible, because calculating an average relies on adding values together.

# Box 9.1 Descriptive analysis using a spreadsheet program

Numerical descriptions of data tables such as Table 9.1 are often created using computer programs called *spreadsheet* applications such as Microsoft Excel®, or OpenOffice Calc®. These programs store data in tables called *worksheets*. The cells in these tables are usually referred to by a coordinate such as 'A3', which refers to the 'cell' in the third row of column A (the first column).

Entering data into a worksheet is usually a matter of typing (or copying and pasting) the data into the cells.

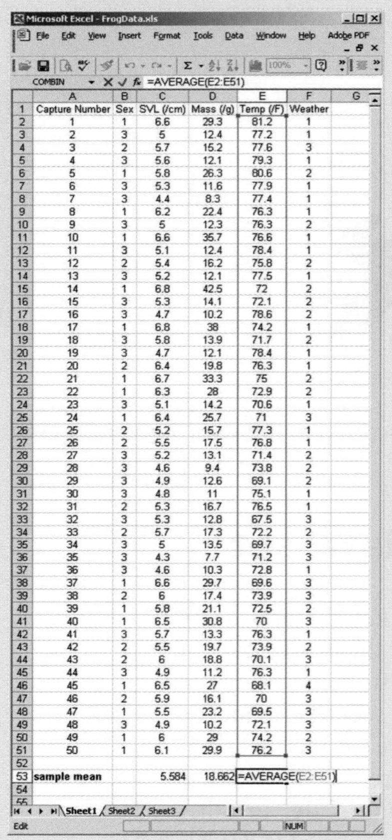

**Figure 1**
A screenshot of a spreadsheet application (Microsoft Excel®) displaying the data from Table 9.1.

| | A | B | C | D | E | F |
|---|---|---|---|---|---|---|
| 1 | Capture Number | Sex | SVL (/cm) | Mass (/g) | Temp (/F) | Weather |
| 2 | 1 | 1 | 6.6 | 29.3 | 81.2 | 1 |
| 3 | 2 | 3 | 5 | 12.4 | 77.2 | 1 |
| 4 | 3 | 2 | 5.7 | 15.2 | 77.6 | 3 |
| 5 | 4 | 3 | 5.6 | 12.1 | 79.3 | 1 |
| 6 | 5 | 1 | 5.8 | 26.3 | 80.6 | 2 |
| 7 | 6 | 3 | 5.3 | 11.6 | 77.9 | 1 |
| 8 | 7 | 3 | 4.4 | 8.3 | 77.4 | 1 |
| 9 | 8 | 1 | 6.2 | 22.4 | 76.3 | 1 |
| 10 | 9 | 3 | 5 | 12.3 | 76.3 | 2 |
| 11 | 10 | 1 | 6.6 | 35.7 | 76.6 | 1 |
| 12 | 11 | 3 | 5.1 | 12.4 | 78.4 | 1 |
| 13 | 12 | 2 | 5.4 | 16.2 | 75.8 | 2 |
| 14 | 13 | 3 | 5.2 | 12.1 | 77.5 | 1 |
| 15 | 14 | 1 | 6.8 | 42.5 | 72 | 2 |
| 16 | 15 | 3 | 5.3 | 14.1 | 72.1 | 2 |
| 17 | 16 | 3 | 4.7 | 10.2 | 78.6 | 2 |
| 18 | 17 | 1 | 6.8 | 38 | 74.2 | 1 |
| 19 | 18 | 3 | 5.8 | 13.9 | 71.7 | 2 |
| 20 | 19 | 3 | 4.7 | 12.1 | 78.4 | 1 |
| 21 | 20 | 2 | 6.4 | 19.8 | 76.3 | 1 |
| 22 | 21 | 1 | 6.7 | 33.3 | 75 | 2 |
| 23 | 22 | 1 | 6.3 | 28 | 72.9 | 2 |
| 24 | 23 | 3 | 5.1 | 14.2 | 70.6 | 1 |
| 25 | 24 | 1 | 6.4 | 25.7 | 71 | 3 |
| 26 | 25 | 2 | 5.2 | 15.7 | 77.3 | 1 |
| 27 | 26 | 2 | 5.5 | 17.5 | 76.8 | 1 |
| 28 | 27 | 3 | 5.2 | 13.1 | 71.4 | 2 |
| 29 | 28 | 3 | 4.6 | 9.4 | 73.8 | 2 |
| 30 | 29 | 3 | 4.9 | 12.6 | 69.1 | 2 |
| 31 | 30 | 3 | 4.8 | 11 | 75.1 | 1 |
| 32 | 31 | 2 | 5.3 | 16.7 | 76.5 | 1 |
| 33 | 32 | 3 | 5.3 | 12.8 | 67.5 | 3 |
| 34 | 33 | 2 | 5.7 | 17.3 | 72.2 | 2 |
| 35 | 34 | 3 | 5 | 13.5 | 69.7 | 3 |
| 36 | 35 | 3 | 4.3 | 7.7 | 71.2 | 3 |
| 37 | 36 | 3 | 4.6 | 10.3 | 72.8 | 1 |
| 38 | 37 | 1 | 6.6 | 29.7 | 69.6 | 3 |
| 39 | 38 | 2 | 6 | 17.4 | 73.9 | 3 |
| 40 | 39 | 1 | 5.8 | 21.1 | 72.5 | 2 |
| 41 | 40 | 1 | 6.5 | 30.8 | 70 | 3 |
| 42 | 41 | 3 | 5.7 | 13.3 | 76.3 | 1 |
| 43 | 42 | 2 | 5.5 | 19.7 | 73.9 | 2 |
| 44 | 43 | 2 | 6 | 18.8 | 70.1 | 3 |
| 45 | 44 | 3 | 4.9 | 11.2 | 76.3 | 1 |
| 46 | 45 | 1 | 6.5 | 27 | 68.1 | 4 |
| 47 | 46 | 2 | 5.9 | 16.1 | 70 | 3 |
| 48 | 47 | 1 | 5.5 | 23.2 | 69.5 | 3 |
| 49 | 48 | 3 | 4.9 | 10.2 | 72.1 | 3 |
| 50 | 49 | 1 | 6 | 29 | 74.2 | 2 |
| 51 | 50 | 1 | 6.1 | 29.9 | 76.2 | 3 |
| 52 | | | | | | |
| 53 | sample mean | | 5.584 | 18.662 | =AVERAGE(E2:E51) | |

Figure 1 shows the data from Table 9.1 in a Microsoft Excel® worksheet. Different types of data (sex, SVL, mass, temperature, and weather) are stored in separate columns, to form a table where each row describes one capture.

As well as simply displaying and storing our data table, a spreadsheet program allows us to calculate the descriptive statistics such as the sample mean, or average, for each set of measurements by means of a *worksheet formula*. A worksheet formula is a formula that can be inserted into a cell of a worksheet to calculate something based upon the data stored within the worksheet. The figure shows how the worksheet formula '=AVERAGE(*range*)' is being used

to calculate the sample mean for the temperature data. The '*range*' in the formula expression is used to tell the spreadsheet which cells contain the data you wish to describe.

In Figure 1, the cell 'E53' is selected for editing, and '=AVERAGE(E2:E51)' has been typed into that cell. Once 'Enter' is typed on the keyboard, the cell E53 will show the value of the sample mean, or average, of the data in the cells from E2 to E51 (shown with the blue boundary). Similar formulae have been typed into cells C53 and D53 to give the sample means for the data in columns C and D (SVL and mass data, respectively).

Worksheet formulae can be used in Microsoft Excel® to calculate many of the descriptive statistics we describe in this chapter. For example:

'=AVERAGE(*range*)' gives the sample mean of the data in the specified *range* of cells.
'=MEDIAN(*range*)' gives the median of the data in the specified *range* of cells.

'=QUARTILE(*range*, *Q*)' gives one value of the five-point summary of the data in the *range*. The value of *Q* is used to determine which summary value is calculated ($Q = 0$ for the minimum value, 1 for the LQB, 2 for the median, 3 for the UQB, or 4 for the maximum value).

'=STDEV(*range*)' gives the sample standard deviation of the data in the specified *range* of cells.
'=VAR(*range*)' gives the sample variance of the data in the specified *range* of cells.
'=COVAR(*range_x*, *range_y*)' gives the literal covariance (not the sample covariance) for the data in the two ranges.
'=SKEW(*range*)' gives the skewness of the data in the specified *range* of cells.
'=KURT(*range*)' gives the kurtosis of the data in the specified *range* of cells.
'=CORREL(*range_x*, *range_y*)' gives the correlation coefficient, *r*, for the data in the two *ranges*; the data are treated as being matched pairs.
'=RSQ(*range_x*, *range_y*)' gives the coefficient of determination ($r^2$) for the data in the two ranges.

Similar functions are available in all spreadsheet programs, and many have tools (such as the Microsoft Excel® 'data analysis toolpack' add-in) which can enable us to calculate a whole range of descriptive statistics for each column in a data table automatically.

The **sample mean** (sometimes called the average) is probably the most familiar of all concepts in statistics. The sample mean of a data set, written $\bar{x}$, is calculated by summing all the values in the sample, and dividing by the number of values. Formally, this is written:

$$\text{sample mean } \bar{x} = \frac{\sum x}{n}. \tag{EQ9.1}$$

Here, $x$ denotes a value or measurement, $n$ denotes the number of data values in your sample, and the capital sigma, $\sum$, denotes summation. Thus this formula can be read (in English) as 'add up all the measurements, and divide by the number of measurements'. The sample mean is a good measure of central tendency for most interval data, but it is quite sensitive to extreme values (values much larger or smaller than the majority); a few extreme values, or 'outliers', can drag the mean below or above the majority of measurements in your sample.

The most useful measure of dispersion in statistics is the **sample variance**, which we write as $s_x^2$. The sample variance is obtained by summing up the squared differences between each value and the sample mean, and dividing by $n - 1$ (that is, by 1 less than the number of values in your data set). In other words,

$$\text{sample variance } s_x^2 = \frac{\sum(x - \bar{x})^2}{n - 1}. \tag{EQ9.2}$$

The sample variance is, therefore, a sort of 'typical squared deviation'. It may seem odd that you divide by 1 less than the number of values in your data set when calculating the sample variance, but there is a very good reason for this, which we will return to at the end of the chapter.

It is possible to express the sum of squared deviations in a different way, which can make calculation a little easier:

$$\sum(x - \bar{x})^2 = \sum x^2 - \frac{\left(\sum x\right)^2}{n}. \tag{EQ9.3}$$

We often express the sample variance in terms of its positive square root, known as the **sample standard deviation**, $s_x$. This has the advantage of having units that correspond to the units of measurement on our measurement scale.

$$\text{standard deviation } s_x = \sqrt{s_x^2} = \sqrt{\frac{\sum(x - \bar{x})^2}{n - 1}}. \tag{EQ9.4}$$

Let us see how we calculate the sample mean and sample variance using the sample of frog mass data shown in Table 9.1. The calculation of the mean and variance is based upon the sum of all the values in the data set, and the sum of all the squares of the values. In this case:

$$\sum x = 933.1 \text{ g}; \quad \sum x^2 = 20\,984.95 \text{ g}^2.$$

The sample mean can be calculated by dividing the sum by $n$:

$\bar{x} = 933.1$ g$/50 = 18.66$ g.

The sample variance is given by the sum squared deviation divided by 49. Now, the sum squared deviation can be calculated by EQ9.3 as

$$\sum(x - \bar{x})^2 = \sum x^2 - \frac{(\sum x)^2}{n} = 20985 \text{ g}^2 - (933.1 \text{ g})^2/50 = 3571.4 \text{ g}^2.$$

We divide this value by $(n - 1)$ to get the sample variance, which is 72.89 g$^2$.

Of course, to perform this type of analysis, we would almost certainly use a hand calculator or computer (the computer or calculator will perform all the calculations outlined above for us, and simply give the correct result).

Using a computer, we could enter the data into a range of cells in a spreadsheet such as Microsoft Excel®. In Excel®, the formula '=AVERAGE(*range*)' will calculate the sample mean of a range of cells, and the formula '=VAR(*range*)' will calculate the sample variance.

Using a typical statistical calculator, we would enter each of the data points, and read off the sample mean and sample variance. If using a calculator, remember two things. First, most calculators give the sample standard deviation, rather than the sample variance. If so, we have to square this value. Second, the sample variance formula divides the sum of squared differences by $n - 1$. An alternative formula, in which the total sum of squared differences is divided by $n$, is also available on most calculators. (This alternative formula is only appropriate when the data are a *complete population* of numbers, rather than a representative sample; this is an important point and we will return to it later.)

There is one other statistic that is worth mentioning at this stage, namely the **standard error** (or, more fully, the *standard error of the mean*). This statistic describes how 'accurate' we should believe our experiment result to be. However, understanding this statistic requires a little bit more than we have yet discussed, so we will leave it until later in the chapter.

## Ordinal or interval data: median, range, and interquartile range

The statistics discussed above (for describing interval data) are all based on adding or subtracting the values in our sample. However, we noted at the start of this chapter that if our data are not on an interval scale, it does not make sense to add or subtract the numbers. Also, if our data are interval but very strangely distributed with extreme values, the sample mean and variance might be a poor summary of most of the data points (because they are influenced by extreme values). In such a case, we might wish to have an alternative way of summarizing data.

The **median** (sometimes written $\tilde{x}$) provides an alternative measure of central tendency to the sample mean that can be calculated for ordinal data or interval data. The median is the middle measurement in a set of data; that is, there are just as many values larger than the median as there are smaller. You can easily calculate the median by ranking your data, and taking the middle value if you have an odd number of measurements, or taking the midpoint between the two middle values if you have an even number of measurements.

Returning to our frog mass data, we have 50 numbers. To calculate the median and the quartiles, we need to arrange the numbers into ascending order. This is easily done by hand (and even more simply by using a spreadsheet program) and gives the data shown in Table 9.5.

The median is the number that comes at position $(n + 1)/2$ in the ordered list; that is, the mean of the 25th and 26th number. In this case, the 25th and 26th highest numbers are 15.7 g and 16.1 g, so the median is 15.9 g.

Another way of expressing this is to say that the median is the $m$th largest value where $m = (n + 1)/2$ is known as the 'median location'. This is a roundabout way of saying that the median is 'the middle number!'. Although a formal expression is nice, it can sometimes lead to a mistake when remembering this formula for the median location, $m$, and thinking that it *is* the median. The median value is **not** equal to $(n + 1)/2$!

Similarly, we can measure dispersion in terms of its **range**, which is just as it sounds (the smallest to largest values in the data set). The range of our frog mass data is from 7.7 g to 42.5 g, i.e. 34.8 g.

Additional information about dispersion is provided by the **interquartile range (IQR)** or **H-spread**. This is the range of the upper and lower **quartile boundaries** (also called **hinges**) of our data set. What are quartile boundaries? These are the values that divide our data, when placed in ascending order, into four equally sized sets, or **quartiles**. Now, as the median splits the data into halves, we need two other values to split each half into two quarters: these are known as the *upper and lower* quartile boundaries (UQB and LQB). For most samples, there are no values that *precisely* split the data into four quarters, and a variety of techniques exists to generate the quartile boundaries. The commonest, and simplest, is as follows.

Split your data at the median into two sets. If you started with an odd number of items, include the median in both of these sets. The lower quartile boundary is the median of the bottom half of the data set, and the upper quartile boundary is the median of the top half of the data set (because the lower quartile boundary is halfway into the first half, it is one quarter of the way from the bottom; because the upper quartile boundary is halfway into the top half, it is three quarters of the way from the bottom). By definition, one half of the data points lie within the IQR (i.e. between upper and lower quartile boundaries).

In our frog mass data, there are 50 data points. The lower quartile boundary is the median of the lower 25 numbers; that is, the 13th number from the bottom, and the upper quartile boundary is the 13th number up from the middle (38th in the overall ascending list). The 13th number in the list is 12.3, and the 38th is 25.7, thus the IQR is 12.3 g to 25.7 g, i.e. 13.4 g.

| Rank | Mass / g |
|------|----------|
| 1 | 7.7 |
| 2 | 8.3 |
| 3 | 9.4 |
| 4 | 10.2 |
| 5 | 10.2 |
| 6 | 10.3 |
| 7 | 11.0 |
| 8 | 11.2 |
| 9 | 11.6 |
| 10 | 12.1 |
| 11 | 12.1 |
| 12 | 12.1 |
| 13 | 12.3 |
| 14 | 12.4 |
| 15 | 12.4 |
| 16 | 12.6 |
| 17 | 12.8 |
| 18 | 13.1 |
| 19 | 13.3 |
| 20 | 13.5 |
| 21 | 13.9 |
| 22 | 14.1 |
| 23 | 14.2 |
| 24 | 15.2 |
| 25 | 15.7 |
| 26 | 16.1 |
| 27 | 16.2 |
| 28 | 16.7 |
| 29 | 17.3 |
| 30 | 17.4 |
| 31 | 17.5 |
| 32 | 18.8 |
| 33 | 19.7 |
| 34 | 19.8 |
| 35 | 21.1 |
| 36 | 22.4 |
| 37 | 23.2 |
| 38 | 25.7 |
| 39 | 26.3 |
| 40 | 27.0 |
| 41 | 28.0 |
| 42 | 29.0 |
| 43 | 29.3 |
| 44 | 29.7 |
| 45 | 29.9 |
| 46 | 30.8 |
| 47 | 33.3 |
| 48 | 35.7 |
| 49 | 38.0 |
| 50 | 42.5 |

Table 9.5  Mass data in ascending order

We could have used a spreadsheet function to calculate the median and quartile boundaries. For example, Microsoft Excel® provides the functions '=MEDIAN(*range*)' and '=QUARTILE(*range*,*Q*)', in which $Q = 1$ for the lower quartile boundary and $Q = 3$ for the upper quartile boundary. However, if we use this QUARTILE function, we get slightly different values for the quartile boundaries than those given above, as Excel® uses a slightly different approach to the one given in the text for calculating quartiles of small samples. Many statistical calculators use yet another technique slightly different to the one above for samples containing an odd number of data points (calculating the quartile boundaries from each half *not* including the median). In general, the thing you need to remember is that if you use different computer packages to calculate quartile boundaries, you will get slightly different answers. These differences are usually very small, and not worth worrying about: there is no generally accepted 'correct' way of calculating quartile boundaries.

### The five-point summary (quartile boundaries) of a data set

The combination of the range (maximum and minimum), the upper and lower quartile boundaries, and the median gives quite detailed summary information about the distribution of a data set, in that it shows the range spanned by each quartile of the data set. This set of values is known as the **five-point summary** of the data set.

Combining the values for our frog mass data, above, we can report a five-point summary as follows:

Minimum = 7.7 g
Lower quartile boundary = 12.3 g
Median = 15.9 g
Upper quartile boundary = 25.7 g
Maximum = 42.5 g

## All data: the mode

Finally, there is a summary statistic that can be used with any type of data: the mode. This is the commonest value in a set of nominal data, or the value corresponding to the highest point on the histogram or column graph – the most common interval – for ordinal or interval data. If a data set is bimodal with two roughly equal 'peaks' on its histogram, then we would report both of these modes. The mode is not a particularly useful measure unless there is a very large number of measurements, and we will not discuss it further.

Before continuing, you should tackle questions 4 and 5 from the End of Chapter Questions.

## 9.4 Exploratory summaries of a data set

Simple graphical or numerical summarizing is only a small part of the techniques that we can use to investigate the properties of a large data set. These techniques are generally known as 'exploratory data analysis', as they are used to explore, rather than simply to summarize key properties of, a data set for communication, as with a histogram or pie chart. We will not look at exploratory analysis in much detail, but there are a couple of techniques that are useful to discuss, because they are easy to do and they are quite commonly used throughout biology.

The first of these techniques is a simple way to sketch a histogram-like plot of our data when we are exploring our results. It is known as a **stem-and-leaf plot**, and is useful because it is quick to draw, and keeps all the information from our data sample.

To see how it is done, look at the two samples below, and the corresponding stem-and-leaf plot shown in Figure 9.9:

Sample 1 {61, 71, 57, 63, 83, 54, 79, 85, 69, 68, 66, 75, 62, 99, 53},

Sample 2 {62, 92, 64, 51, 81, 49, 77, 53, 83, 78, 67, 78, 73, 62, 72}.

How is this plot constructed? We start by looking at the numbers and deciding which decimal digit (units, tens, tenths, and so on) should represent the intervals for our histogram. This is called the stem digit: in this case it is the digit corresponding to 'tens', but we could choose any digit, on either side of the decimal point. If our data were in the range 125−195, for example, our stems would be 12, 13, 14, ... 19 (tens); if our data were between 0.710 and 0.839, our stems would be 71, 72, ... 82, 83 (hundredths).

We then write the stem numbers in a column, in order. For each number in the data set, the *next digit* that follows the stem (the 'leaf') is then written down next to its stem. When this is done for all numbers, we end up with a kind of rotated histogram.

The stem-and-leaf plot is very similar to a histogram and has very much the same uses and advantages. Unlike the histogram, it is generally very quick to do by hand for a small data set, and it is easy to check if it has been done correctly. We can also compare two samples very quickly, by plotting them 'back to back' as in the figure above.

The second technique that is commonly used is the **box-and-whiskers plot**, sometimes simply called a **boxplot**. This is a graphical extension of the five-point summary we described earlier, with some extra conventions that are useful when individual data points have relatively extreme values.

Figure 9.10 shows both a box-and-whiskers plot and a stem-and-leaf plot of the frog mass data from Table 9.1. To construct the box-and-whiskers plot, a box is drawn from the upper to lower quartile boundaries. This box therefore shows the IQR (half the dataset is 'in the box'). A line is drawn showing the median, which splits the box into two regions corresponding to the two central quartiles. To complete the box-and-whiskers plot, we can draw lines from the edges of the box to the maximum and minimum values, thus showing the top and bottom quartiles. It is possible to add a 'cross' representing the position of the sample mean.

| sample 1 | | sample 2 |
|---:|:---:|:---|
| | 40 | 9 |
| 3 4 7 | 50 | 1 3 |
| 2 6 8 9 3 1 | 60 | 2 4 7 2 |
| 5 9 1 | 70 | 7 8 8 3 2 |
| 5 3 | 80 | 1 3 |
| 9 | 90 | 2 |

**Figure 9.9**
A stem-and-leaf plot of two samples of numbers. Each value is plotted as a 'leaf' in one row. The value described is the sum of the 'leaf digit' and the 'stem' against which it is plotted.

**Figure 9.10**
A box-and-whiskers plot (A) and a stem-and-leaf plot (B) of the mass data from Table 9.1. Here the stem values are written twice, once for leaves 0 to 4, and again for 5 to 9; leaves are rounded to the nearest gram.

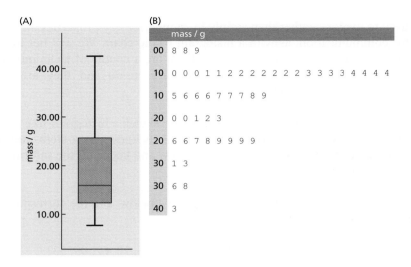

However, there is another important aspect to the box-and-whiskers plot concerning extreme values. Before drawing the whiskers, we check how long they would be. If we would be drawing whiskers more than one and a half times the length of the box, we decide that some values are too extreme, and we exclude the extreme values from the whiskers. We do this by imagining a cut-off that is 1.5 times the box length (that is, 1.5 times the IQR) outside each quartile boundary. Any points that lie outside these cut-offs (known as *fences*) are temporarily excluded, and the whiskers are drawn from the edge of the box to the highest and lowest values that are inside the fences.

We therefore have a box showing the IQR, and whiskers showing the range of data that lie *inside* our imaginary cut-off points or *fences*. The points we excluded from the whiskers are called **outliers**, and we plot these data points individually with a dot or circle. It is common to use a different symbol (an asterisk) to denote **extreme outliers** that are more than twice as far from the quartile boundaries as our cut-offs; that is, points that lie more than three times the IQR beyond the upper or lower quartile boundary. A final convention that we recommend, although it is ignored by most computer packages, is to put a small bar at the end of the whisker only if it shows true maximum or minimum value (and thus has no outliers beyond it).

These conventions can be slightly confusing, so we will go through a complete example to show how it is done.

Taking the sixteen measurements of mass data for female frogs, in ascending order, we get Table 9.6.

We start by calculating the sample mean, median, and quartile boundaries. The sample mean is 29.49 g; the median is the average of the 8th and 9th value = 29.15 g; the lower quartile boundary is the average of the 4th and 5th value = 26.0 g; and the upper quartile boundary is 32.05 g.

The 'box' on our graph goes from 26.0 g to 32.05 g, with a line for the median at 29.15 g. In addition, we will add a cross for the sample mean at 29.49 g (this is a useful component of a plot, but frequently omitted.). The plot at this stage is shown in Figure 9.11A.

| Table 9.6 Ranked masses for females | |
|---|---|
| **Rank** | **Mass / g** |
| 1 | 21.1 |
| 2 | 22.4 |
| 3 | 23.2 |
| 4 | 25.7 |
| 5 | 26.3 |
| 6 | 27.0 |
| 7 | 28.0 |
| 8 | 29.0 |
| 9 | 29.3 |
| 10 | 29.7 |
| 11 | 29.9 |
| 12 | 30.8 |
| 13 | 33.3 |
| 14 | 35.7 |
| 15 | 38.0 |
| 16 | 42.5 |

Next, we work out where the fences (imaginary cut-off values) should go. The IQR is 32.05 g − 26.0 g = 6.05 g. The fences lie 1.5 times this distance from the quartile boundaries. Thus the upper fence is 32.05 g + 1.5 × 6.05 g = 41.13 g (two decimal places), and the lower fence is 26.0 g − 1.5 × 6.05 g = 16.93 g (two decimal places). We do not draw these lines on our figure, but instead compare them with the values in our data set. No data points are lower than the lower fence, so we draw the whisker from the box to the minimum value (21.1 g), and mark the edge of the whisker with a line, as shown in Figure 9.11B.

The largest data point, 42.5 g, *does* exceed the upper fence. Therefore we temporarily exclude this point and draw the upper whisker to the largest remaining data point, 38.0 g, and do not mark the edge of the whisker. Finally, we draw a mark showing the data points outside the fences individually (in this case, we mark a point for 42.5 g). We will use a circle to mark this point as it is not very far outside the fence (if it was 'two fences', that is three times the IQR, beyond the box, we would use an asterisk). The finished plot is shown in Figure 9.11C.

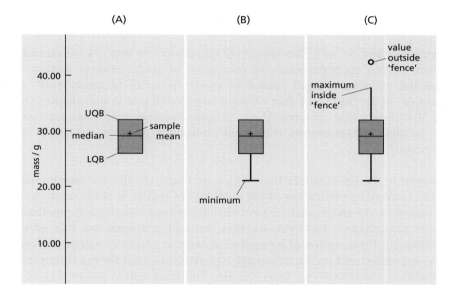

**Figure 9.11**
An illustration of the stages involved in drawing a boxplot. See text for details.

There are several useful things about a box-and-whiskers plot. First, it shows the general pattern of data, especially the symmetry of the data, clearly, often more clearly than a histogram. Second, it is helpful in identifying the presence of outliers and extreme values. It might be sensible to exclude extreme values before calculating our descriptive statistics, such as the sample mean or variance, as extreme values may have a very large distorting influence on these statistics. Third, we can visually compare two or more groups that are measured on the same scale, by drawing adjacent boxplots, as shown in Figure 9.12.

## Population parameters

We have been careful throughout this chapter to use the terms sample variance, sample mean, and sample covariance. Why not just say 'mean',

**Figure 9.12**
The boxplots comparing the mass data for the three sex groups from Table 9.1.

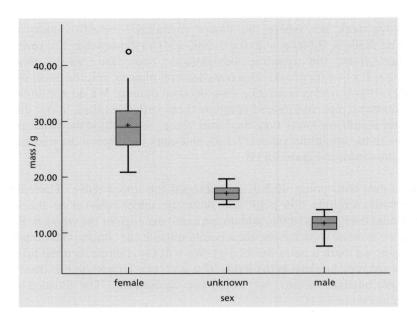

'variance', and so on? The answer is related to the concept of **sampling**. When we run an experiment we take a set of measurements that can be regarded as a very small subset of many 'possible measurements'. For example, in the frog study that we have been looking at in this chapter, there are 50 different frog captures. These 50 measurements are a small fraction of the possible measurements (the many thousands of frogs we *might have* measured).

We refer to the set of values from which we might sample our measurements as the *underlying* population of our sample. We will look at the mathematics that allows us to understand the relationship between samples and populations in the next chapter. However, for now, we will just point out that given a mathematical description of a population, we can generally calculate a mean (average value) and variance (average squared deviation) for the values in the population, just as we can for a sample. These are called parameters of the population, and they correspond to descriptions of, for example, 'frogs in general' rather than a description of 'the frogs we found in a particular experiment'.

The reason we chose to use the sample mean and sample variance as descriptions of our data is that these values tend to be good **estimators** of the corresponding parameters, namely the mean and variance of the underlying population. What do we mean by a good estimator? The obvious answer would be 'one that gives the right value'. We know that if we repeated the experiment with 50 more frog captures, we would not get exactly the same sample mean and sample variance. Different samples from the same population will contain different values, and thus have different sample means and sample variances. So the sample mean and variance cannot give us *exactly the right answer every time*.

The criteria we use to decide what is a good estimator are:

1. Does it give the right answer, on average? (If we take a large enough number of samples, and took an estimate from each sample, would the average of the estimates be the parameter we are estimating?) Statisticians call this being *unbiased*. An estimator that is, on average, larger or smaller than the value being estimated is called a *biased* estimator.
2. Does it give an answer close to the right answer? (How small is the variability of estimates taken from different samples?) This is called the efficiency of the estimator.

One of the reasons that the sample mean and sample variance are so widely used is that they are always unbiased estimators of the population mean and variance (so, if we took a large enough set of samples, the average of all these sample means and variances would be the true population mean and variance). This is the reason that we divide by $n - 1$ in the formulae for the sample variance: when calculated in this way, the sample variance is an unbiased estimator of the true population parameter.

When we sample from a distribution that is approximately 'normal' (see Figure 9.14), it is also true that the sample mean and sample variance are the most efficient estimators of the population mean and variance. However, in certain circumstances (generally, if the population is symmetrical but very *heavy-tailed*), the sample median can be a more efficient estimator of the population mean than the sample mean.

We said above that different samples will not have exactly the same sample mean and sample variance (because they will not contain the same values). This variability between samples is called **sampling error**. There are two different sources of sampling error in any real experiment: first, there is the variability in whatever we are measuring (for example, we could catch different frogs, or measure at different times, as the mass of a frog varies owing to growth, excretion, and ingestion). This type of variation is called *sampling variation*. Second, there are errors that we could make in measuring a particular individual (*measurement error*).

The sample variance we have calculated in this chapter represents our estimate of sampling error between different measures, which therefore includes measurement error. For convenience, we usually do not try to distinguish between the sources of variability in our measurements. Instead, we simply describe the 'set of possible *measurements* of frog mass' as our population (rather than the set of possible 'true frog mass values'; that is, values measured with no measurement error).

It may be that we are very interested in the sampling variation (that is, variation in the thing we are measuring), but not in the measurement error. In such a case it would be possible to estimate the size of measurement error directly (for example by repeatedly measuring values that are known to be constant).

## The standard error and accuracy of estimation
Looking at the frog mass data in Table 9.1, we have seen how to summarize it in terms of its average (sample mean) and its dispersion (sample standard

deviation). However, there is one further aspect to these data that we can report. Imagine that someone else ran a similar study involving 5000 frog captures. Because this new study is so much larger, it is therefore more informative. We would probably feel that we could trust the sample mean from the larger study as being more accurate as a description of all possible frog measurements than the sample mean from the original study involving only 50 captures.

Is there a way we can describe the result of the larger experiment that reflects the fact that we think the results are more trustworthy? Yes: there is a statistic called the *standard error*, which reflects how accurate an estimate of the population parameter a statistic like the sample mean will generally be.

A set of repeated experiments will all give slightly different sample means. The standard error of the mean is an estimate of the variability (standard deviation) of the sample means of all possible samples of the same size we could take from the population. The derivation of the standard error is very simple, and we will go through it systematically at the end of the chapter. For now, we will just give the formula:

$$s_{\bar{x}} = \frac{s_x}{\sqrt{n}}. \qquad \text{(EQ9.5)}$$

Let us return to our two different-sized studies of pickerel frogs to see exactly what this tells us. The two studies ($n = 50$, and $n = 5000$) each involve sampling values from the same population, and calculating the sample mean. It should be obvious that the two samples will not both have a sample mean exactly equal to the population mean. Owing to sampling error, sometimes a sample mean will be greater than the population mean, or sometimes it will be less than the population mean. However, how far from the population mean would we expect the sample mean to be? This is what the standard error of the mean tells us.

Let us assume that both the large and small studies gave us the same sample standard deviation for snout to vent length data (4 mm). The standard error of the mean for the sample of $n = 50$ is 0.566 mm (4 mm divided by $\sqrt{50}$). The standard error of the mean for the sample of $n = 5000$ pickerel frogs is 0.0566 mm. Thus the study that is 100 times larger gives us an estimate of the population mean that is *ten times* as accurate.

As we can see, the formula for the standard error shows that bigger samples (larger values of $n$) give smaller standard errors. In other words, the more values you have in a sample, the closer the sample mean is likely to be to the mean of the population.

A schematic illustration of this is given in Figure 9.13. The figure shows a set of possible population values (top), and a series of possible samples from the population. The mean of each possible sample is marked with a red line. The figure shows that the possible sample means from small samples tend to be more 'spread out' than the sample means from large samples (the samples chosen in the figure were deliberately chosen to exaggerate this effect).

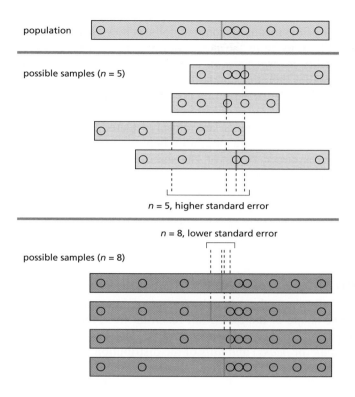

**Figure 9.13**
An illustration of the concept of the standard error of the sample mean. Different sample means have different values (red lines). For larger samples, the possible values of the sample mean get closer together. This is reflected in the size of the standard error of the sample mean.

Because the standard error of the mean is the standard deviation of all possible sample means drawn from the population in question, it will be larger for smaller samples.

Before continuing, you should tackle question 7 from the End of Chapter Questions.

## Difference between standard error and standard deviation

The distinction between the standard deviation and the standard error often causes some confusion for those learning statistics. It is worth going through the difference quite carefully. We will look at it in terms of the familiar SVL measurements from our frogs.

There is variability in our frog length measurements. Remember that this will be caused by the actual variability of frogs (sampling variability), and by the variability in our measuring device (measurement error). As we cannot mathematically distinguish between these sources of variability in a data set, we describe the total variability in measured frog size as sampling error. The amount of sampling error we have in measuring frogs is essentially fixed, and the same for all experiments.

The standard deviation is a measure of sampling error in individual measurements. It reflects how far apart the individual, randomly selected, measured values are (in this case, the measured length of frogs). The standard deviation is therefore not influenced by the size of experiment we choose to run.

The standard error of the mean, by contrast, tells us how close an estimate the sample mean is likely to be to the population parameter being estimated. It reflects the variability in *experimental conclusions* rather than in *frog measurement*, and is smaller than the standard deviation. This reflects the fact that we have overcome some of the sampling error by measuring repeatedly.

Obviously the two measurements are closely related. The value of the standard error will be larger if the variability between real measurements (the standard deviation) is larger. However, the standard error gets smaller as we increase the number in the sample.

In summary, the *standard deviation* describes the population, the variability in possible individual measurements. The *standard error* describes the *estimate of the sample mean*, in terms of how variable such estimates are likely to be.

## Distribution shape and transformations

Scientists often talk about the 'shape' of a set of interval data. What they really mean is the shape of the histogram resulting from the data. The shape can be described qualitatively, and by numbers. The key characteristics of interest are the *peak* (or *peaks*) and *tails* of the distribution.

A symmetrical distribution is one in which, if you were to draw a vertical line through the mean, the distribution on one side is a mirror image of the picture on the other side. A histogram will rarely be precisely symmetrical, but the term is not used in an all-or-nothing way. For example, we would describe the histogram in Figure 9.15 as roughly symmetrical. A distribution is *more* symmetrical if each side is *more closely approximated* by a mirror image of the other.

A non-symmetrical distribution is generally *skewed*, in that it has a **skewness** that is not zero. Positive skewness is the tendency for there to be fewer, and more spread-out, data points above the mean than below. A negative skewness indicates the reverse (fewer, and more spread-out points below the mean). A data set with positive skewness is said to be *positively skewed*, or *skewed to the right*, as it tends to have a longer 'tail' on the right-hand side, containing the more spread-out higher values. Similarly, a data set with negative skewness is said to be *negatively skewed*, or *skewed to the left*, and has a longer 'tail' on the left-hand side.

As well as the tails of the distribution, we can also describe its peaks: that is, intervals in which there are a high proportion of scores. The **modality** of a data set or population describes how many distinct peaks there are on a histogram of the distribution. A unimodal distribution is one with a single prominent peak; a bimodal distribution has two such peaks. If a distribution has more than two peaks, we generally stop counting, and just describe it as *multimodal*.

Finally, we might be interested in whether a histogram is generally 'peaky' or 'flat'. A *tail-heavy* histogram may have a narrow peak, with longer, flatter tails than a normal distribution curve (that is, a higher proportion of values concentrated in the center and the tails). A *light-tailed* histogram is one in which the peak is broader and flatter than in the normal distribution, with fewer values in the tails. Statisticians have a name for this property (**kurtosis**), and it is usually expressed relative to a particular bell-shaped curve known as the normal distribution (see Figure 9.14; we will discuss this distribution in more detail in Chapter 10).

You do not need to worry about exactly how skewness and kurtosis are calculated, as you will never do it by hand. They can be calculated easily in a spreadsheet program like Microsoft Excel® using the worksheet formulae '=SKEW(*range*)' and '=KURT(*range*)'. However, being aware of their definitions may help us to remember what these statistics tell us.

Any data point in our sample can be expressed as a deviation $d$ from the sample mean (we saw above that the sample variance is a form of average squared deviation). Items in the 'positive tail' correspond to large positive deviations, and those in the negative tail correspond to large negative deviations. The skewness of a sample is proportional to the average cubed deviation, and the kurtosis is proportional to the average fourth power of the deviations.

Thus:

- If the skewness is positive, it must be because there are more extreme values on the positive side, so positive skewness indicates a long tail on the positive side.
- Negative skewness indicates a long tail on the negative side.
- If the kurtosis is positive, the distribution is *heavy-tailed*, and has more very extreme values in the tails than does a normal curve.
- If the kurtosis is negative, the distribution is *light-tailed*, and has fewer very extreme values in the tails than does a normal curve.

So what can we do with the knowledge about the distribution of a data set? Well, if our distribution shape is peculiar in some way, it is often useful to express the data on a different scale. Re-expressing our data on a different scale is known as a **data transformation**. The simplest kind of data transformation is a change of units (such as meters to millimeters, or degrees Celsius to degrees Fahrenheit). This type of change of units can be described in terms of multiplying each data point by a constant (which could be 1, for no change), and/or adding a constant value (which could be 0, for no change).

This type of transformation – multiplying all the data by a positive constant value, and/or adding any constant value – is known as a **linear transformation** and has no effect on the histogram 'shape'. Such a transformation will change the position and/or scale of the $x$ axis, and thus change the mean and/or variance. However, the shape (i.e. both the skewness and kurtosis) of the distribution will remain the same. The only thing to bear in mind is that if we multiply by a negative value, then this will produce a histogram that is a 'mirror image' of the histogram we started with: the kurtosis of the transformed sample will be the same, but the skewness will change sign; that is, the transformed skewness will be the negative of the untransformed skewness.

**Figure 9.14**
The tendency of a distribution to differ from the normal distribution (top left) is measured by the modality, skewness, and kurtosis of the data set.

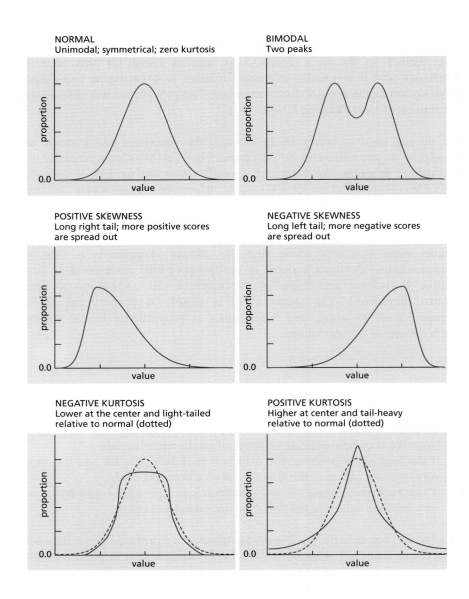

We have noted that the histogram may differ a good deal from the normal distribution. Some of the analysis techniques we will meet in Chapter 11 are useful only for data that are approximately normal in distribution, so this deviation from the normal shape may be a problem. However, we often find that our data can be made more normal in distribution (that is, have a curve closer to the normal curve, with skewness and kurtosis closer to zero), by means of a **nonlinear transformation** (such as taking the reciprocal or logarithm of each data point).

Nonlinear transformations influence different regions of the interval scale by different amounts, and can therefore change the shape of the distribution. Thus a nonlinear transformation can help make a data sample a closer fit to the normal curve. The commonest transformations are the reciprocal transformation, logarithmic transformation, and rank transformation.

Let us look at the frog mass data one more time, and see what happens to the histogram if we apply a logarithmic transformation. The first thing we need to do is calculate the logarithm of each of our numbers. We could use any base for our logarithm, so we will use the natural (base $e$) logarithm. Each value is replaced by the logarithm of that value.

| Table 9.7 Taking logarithmic values | | | | | | |
|---|---|---|---|---|---|---|
| Mass / g | 29.25 | 12.38 | 15.16 | 12.10 | ... | 29.90 |
| ln(mass / g) | 3.38 | 2.52 | 2.72 | 2.49 | ... | 3.40 |

This gives a series of new data points, as illustrated in Table 9.7.

We have 50 observations in the range 2.03–3.75. We need around $\sqrt{50} = 7.07$ categories. We can choose the edges of our scale to be 2.00 and 3.75, giving a range of 1.75. Choosing seven categories gives us a relatively easy-to-use value of 0.25 for the width of each interval. The corresponding frequency distribution is given in Table 9.8.

The shape of the histogram (Figure 9.15) is now considerably more symmetrical than that plotted earlier for the raw data (Figure 9.6).

| Table 9.8 Distribution of the logarithms of the mass data | |
|---|---|
| ln(mass / g) | Frequency |
| 2.00–2.25 | 3 |
| 2.25–2.50 | 9 |
| 2.50–2.75 | 12 |
| 2.75–3.00 | 10 |
| 3.00–3.25 | 4 |
| 3.25–3.50 | 8 |
| 3.50–3.75 | 4 |

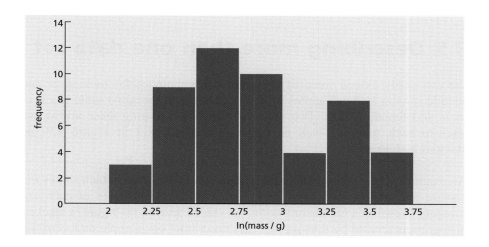

**Figure 9.15**
A histogram showing the distribution of the logarithm of the mass data in Table 9.1.

Another approach we could apply to obtain this histogram is to transform the *boundaries* of the cells in our frequency table. This approach also works, provided we remember that we will end up with *unequal interval widths*, so we would have to take care to plot *frequency density* in our histogram.

It is important to be careful using these transformations, because when they change the shape of the distribution, they also change the 'meaning' of the summary statistics we met before (such as the sample mean and sample variance). For example, imagine we transform two data sets by taking the base 10 logarithm, and we find that the transformed scores in group A are roughly 2 units larger, on average, than the scores in group B. We might conclude that, because $\log_{10}(100) = 2$, this means that group A scores are roughly 100 units larger than group B. However, this is *not* correct.

Why not? Because increasing the logarithm of the data is essentially the same as multiplying the untransformed data. Thus an increase of 2.0 in the base 10 logarithm corresponds to an untransformed score that is 100 times as large, not 100 units larger!

The reciprocal transformation sometimes produces data that are very difficult to interpret (the mean reciprocal of the mass of a frog does not have an obvious interpretation, whereas the reciprocal of the time taken for a frog to cross a pond simply reflects the speed of the frog's movement). The most important thing to remember for reciprocal data is that they have the opposite *sense* – the smallest values before transformation become the largest values after transformation.

Finally, there is one more transformation that is commonly used when we have more than one data set: the rank transformation. The lowest value in a sample is recoded as 1, the second lowest as 2, and so on. (If two or more values are the same, they are given the average of the ranks they would get if they were very slightly different.) For example 10, 10, 12, 12, 12, 13 becomes 1.5, 1.5, 4, 4, 4, 6.

This simply recodes the interval data (the values) on an ordinal scale; the transformed data points represent only the *order* (in increasing size) of the untransformed data.

# 9.5 Describing more than one data set

We can easily extend some of the graphical techniques for looking at one sample of data to two samples of data measured on the same scale (such as male and female SVLs). Two common ways of exploring these are to use **back-to-back graphs** (such as the stem-and-leaf plot in Figure 9.9) or **stacked histograms**.

An example of a stacked histogram is shown in Figure 9.16, which shows the capture temperature data from Table 9.1, with the contribution of each sex category shown in a contrasting color. As can be seen in the figure, a stacked histogram is created by drawing the bars describing one sample on top of the bars describing another sample (using the same vertical scale). The total height of each bar will thus represent the total number in each range, and the colors will reflect the proportion of each range that is in each category.

Back-to-back graphs are useful if you wish to see differences between two samples measured on the same scale; stacked histograms are useful if we are interested in the combined effect of the two or more samples.

## Covariance and correlation

Sometimes we need to describe data in a way that *combines* information from two different types of measurement. Looking back at Table 9.1, we can see that each capture of a frog involved more than one property being measured. If we have two samples of data measured on different scales (such as SVL and mass) then it makes no sense to compare their values using back-to-back graphs or stacked histograms. It would not be appropriate to ask if the average

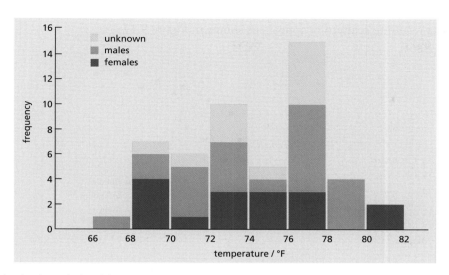

**Figure 9.16**
A stacked histogram showing the distribution of the temperature data in Table 9.1, with the contribution of each sex shown separately.

body length is 'bigger' than the average mass, for example. However, we might be interested in the *relationship* between the measurements on the two scales: do *more massive* frogs have *longer* bodies (SVL)?

How can we determine whether two measurements go together in this way? Each row of Table 9.1 refers to one sample (that is, one frog capture), and contains both a mass data point and an SVL data point. Let us call the variables we are measuring $X$ (mass) and $Y$ (SVL). Each row thus contains a value of mass, $x$, measured in grams and a corresponding SVL value, $y$, in centimeters.

The best thing to do with this type of data is to plot them on a scatterplot. We draw axes corresponding to the ranges of each variable, and then plot a point corresponding to each $x$, $y$ pair. The $x$ and $y$ coordinates of the point we plot represent a pair of values describing one combination of measurements. In this case, the combination is the mass value and the SVL value for one of the 'capture' events described by a row of Table 9.1.

Drawing a pair of axes for mass and SVL (both of these are ratio scales), we then mark each capture as a single point, where the coordinates correspond to the mass and SVL observed on that particular capture.

| Table 9.9  An extract from Table 9.1 | | | |
| --- | --- | --- | --- |
| **Capture number** | **Sex** | **Mass / g** | **SVL / cm** |
| 1 | 1 | 29.3 | 6.6 |
| 2 | 3 | 12.4 | 5.0 |
| 3 | 2 | 15.2 | 5.7 |

Table 9.9 shows an extract from Table 9.1. Thus capture one is indicated by the point with coordinates $x = 29.3$, $y = 6.6$; capture 2 is indicated by the point $x = 12.4$, $y = 5.0$, and so on. Finally, we can include the sex data by plotting each sex group by using a different symbol (square for female, triangle for male, circle for unknown).

The resulting scatterplot appears as shown in Figure 9.17.

**Figure 9.17**
A scatterplot showing the relationship between mass and SVL for the data in Table 9.1. Each sex group is shown with a different symbol.

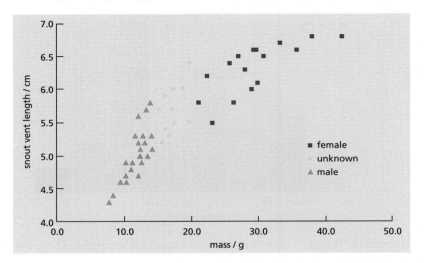

Here we are combining different groups of data, and we can indicate which group an observation is in by changing the marker plotted for that point on the scatterplot. This is always worth doing, as the overall scatterplot may be misleading if there are different relationships within particular subgroups. This becomes very clear when they are plotted separately!

Looking at the scatterplot for SVL and mass data, we note that the data show a tendency to 'slope upward': above-average SVL measurements tend to go with above-average mass values. This tendency is called a **positive correlation**, and it is measured by a statistic called the **sample covariance**. The sample covariance measures the degree to which above-average scores for one measurement go along with above-average scores for another measurement.

$$s_{xy}^2 = \sum \frac{(x - \bar{x})(y - \bar{y})}{n - 1} = \frac{\sum xy - (\sum x \sum y)/n}{n - 1} \qquad \text{(EQ9.6)}$$

**Variance and covariance**
The sample variance reflects the typical squared deviation in a single set of data, and is a measure of dispersion. Covariance is the average product of paired deviations in two variables, and is a measure of the degree to which they tend to vary together.

From this formula we can see that the covariance is the average product of the *x* and *y* deviations: it is the degree to which *above-average values of x* go with *above-average values of y.*

As an example, we can calculate the sample covariance for the relationship between the mass and SVL data for the whole set of frogs in Table 9.1.

**Table 9.10 Calculations for the covariance**

| Capture number | x Mass / g | y SVL / cm | x × y |
|---|---|---|---|
| 1 | 29.3 | 6.6 | 193.38 |
| 2 | 12.4 | 5.0 | 62.00 |
| 3 | 15.2 | 5.7 | 86.64 |
| 4 | 12.1 | 5.6 | 67.76 |
| . . . | . . . | . . . | . . . |
| 50 | 29.9 | 6.1 | 182.39 |
| Total | 933.1 | 279.2 | 5466.69 |

The calculation of the covariance is based upon the sum of all the $x$ and $y$ values, and upon the sum of their products, as illustrated in Table 9.10.

Thus $\sum x = 933.1$ g;   $\sum y = 279.2$ cm;   and   $\sum xy = 5466.7$ g cm.

Using the calculation formula for the sample covariance, we have:

$$s_{xy}^2 = \frac{\sum xy - (\sum x \sum y)/n}{n - 1}$$

$$= \frac{5466.7 \text{ g cm} - \dfrac{933.1 \text{ g} \times 279.2 \text{ cm}}{50}}{49} = 5.23 \text{ g cm}.$$

We can see here that the value of the sample covariance is not always easy to understand, and is in the units of $x$ and $y$ scores multiplied. This means that the sample covariance is not the best statistic to work with. Instead, we use the **correlation coefficient, $r$.**

$$r = \frac{s_{xy}^2}{s_x s_y} = \frac{1}{(n - 1)} \sum \frac{(x - \bar{x})}{s_x} \times \frac{(y - \bar{y})}{s_y}. \hspace{2cm} \text{(EQ9.7)}$$

This represents the covariance in sensible units because the $x$ and $y$ deviations have been divided by their respective standard deviations. The terms being multiplied in this formula are called *standardized* values (distances from the sample mean, divided by the sample standard deviation), which are unitless ratios.

The correlation coefficient therefore gives us a measure of the relationship that is independent of the units of measurement.

Applying this to our frog data, we divide the covariance by the two standard deviations. These are 8.53 g for $x$ and 0.687 cm for $y$; thus $r$ is given by:

$$r = \frac{s_{xy}^2}{s_x s_y} = \frac{5.23 \text{ g cm}}{8.53 \text{ g} \times 0.687 \text{ cm}} = 0.892.$$

The value of $r$ we obtained for the relationship between the SVL and mass data confirms the tendency for the scores to show a positive correlation. It can be shown that $r$ ranges from $+1$ (perfect positive correlation: all points lie on a line with positive slope) to $-1$ (perfect negative correlation: all points lie on a line with a negative slope).

The key interpretations of $r$ are in terms of what a change in one variable tells us about the other:

- The typical value of $x$ tends to increase by $r$ standard deviations ($s_x$) for every standard deviation ($s_y$) that $y$ increases, and vice versa.
- Small (close to zero) values of $r$ do not imply a lack of relationship. There may be a 'U-shaped' relationship, so we can only interpret a zero correlation by looking at a scatterplot.

- Correlation is symmetrical, saying 'large $x$ is associated with increasing $y$' is exactly the same as saying 'large $y$ is associated with increasing $x$', and thus it does not tell us about whether $x$ causes $y$ or vice versa. Indeed, two numbers may be highly correlated (for example your own age and the price of crude oil increased between 1998 and 2008) without any simple causal relationship!

> A correlation between two variables does not imply that one causes the other.

Using a computer, we could enter the data into a range of cells in a spreadsheet such as Microsoft Excel®. The formula '=CORREL(*range*)' will calculate the value of $r$. Excel® also has a formula '=COVAR(*range*)' for the covariance. However, this calculates what is sometimes called the *literal covariance* ($\sigma_{xy}^2$), rather than the sample covariance ($s_{xy}^2$). This would be the more appropriate formula if a data set reflected the entire population, rather than a sample, and it is given by:

$$\sigma_{xy}^2 = \frac{\sum xy}{n} - (\mu_x \times \mu_y) = \frac{\sum xy - (\sum x \sum y)/n}{n}. \qquad \text{(EQ9.8)}$$

If the sample covariance is required, it can be obtained from the literal covariance using a simple correction formula:

$$s_{xy}^2 = \frac{\sum xy - (\sum x \sum y)/n}{n-1} = \frac{n}{n-1}\sigma_{xy}^2. \qquad \text{(EQ9.9)}$$

Although $r$ is a very convenient description of a correlation, it also has some disadvantages because it is a very *sample-specific* statistic: the exact value of $r$ depends a lot upon the nature of the sample data from which it is calculated, as well as the underlying relationship.

The first problem is that (even more than the sample mean and variance) $r$ can be very influenced by one or two extreme values. The second, and less well-known problem, is that if our sample contains only a restricted range of values for $x$ or $y$, then we will tend to see a smaller correlation than we would for samples that contain a larger range. If the linear relationship holds over a wider range, then the value of the correlation coefficient will increase as we increase the range in our sample. For example, if we were to calculate the correlation between SVL and mass for the three sex groups separately (giving us three restricted ranges of mass, as shown in Figure 9.17), we would find $r$ values of 0.782, 0.508, and 0.750 for the male, unknown, and female groups, respectively. All of these are substantially lower than the overall combined correlation of $r = 0.892$, which was calculated over the whole range.

What can we do about this? The answer is, sadly, not much, and therefore we cannot generalize the value of a correlation in a sample to a larger or smaller range of a variable.

Probably the most important interpretation of $r$ is in terms of predicting one variable from another, by means of linear regression. When we have a correlation, we can use one measure to judge the likely value of another. For example, imagine that (for some reason) we knew the mass of a particular pickerel frog, and had to guess the body length of that frog.

Obviously, knowing that there is a relationship like that in the scatterplot, we should be able to predict the SVL better if we know the mass of the frog. The value of $r^2$ tells us *how much better* the prediction will be (the proportion of sampling error in $y$ which is removed when we know $x$). We will look at this idea in a lot more detail when we consider curve fitting in Chapter 12.

## Covariance and the variance sum law

The sample covariance is useful in other ways than simply describing the relationship of two sets of data. It allows us to answer a lot of interesting questions about the sample mean and sample variance of properties we derive by adding up, or calculating the difference between, two measurements. This means that the covariance will be very useful later on (especially in Chapter 11).

Imagine that we have measured leg length, $x$, and torso length, $y$, in a sample of adult humans. For each individual, we could calculate a new measure $h$ (shoulder height) by summing these data. Thus each value of $h$ represents the sum of two sampled values, $h = x + y$. Alternatively, we might be interested in some score $d$ that represents the *difference* between the two sampled values for each individual, giving $d = x - y$.

It turns out the sample means and sample variances of $h$ and $d$ can be obtained if we know the sample means, sample variances, and the covariance of $x$ and $y$.

The sample mean and variance of $h$ values are:

$$\bar{h} = \bar{x} + \bar{y},$$
$$s_h^2 = s_x^2 + s_y^2 + 2s_{xy}^2.$$

The sample mean and variance of differences ($d$ values) are:

$$\bar{d} = \bar{x} - \bar{y},$$
$$s_d^2 = s_x^2 + s_y^2 - 2s_{xy}^2.$$

Do these results make sense? Well, if the covariance is positive (high scores tend to go together) then the variance of the difference $(x - y)$ is less than the sum of the two variances. This is sensible, as there is going to be less difference on average between the scores if the similar ones go together! If the covariance is negative, then the variance of the sum $(x + y)$ will be less than the sum of the two variances, which also makes sense, as these scores tend to 'cancel out' when added.

## Box 9.2 Proof of the relationship between variances and covariance

The proof of the relationship between variances and covariance is so simple, it is worth going through in full. We have measured leg length, $x$, and torso length, $y$, in a sample of adult humans. For each individual, we can calculate the shoulder height $h$ by summing their leg length ($x$) and torso length ($y$). Thus each sampled value of $h$ represents the sum of two sampled values, $h = x + y$.

The sample mean of the shoulder height ($h$) values can be shown to be simply related to the sample means of $x$ and $y$ as follows:

$$\bar{h} = \frac{\sum h}{n} = \frac{\sum(x+y)}{n} = \frac{\sum x}{n} + \frac{\sum y}{n} = \bar{x} + \bar{y}.$$

The sample variance of the $h$ values is a bit more complicated:

$$s_h^2 = \sum \frac{(h - \bar{h})^2}{n-1} = \sum \frac{((x+y) - (\bar{x} + \bar{y}))^2}{n-1}.$$

We can multiply this expression out, although it gets a bit messy ...

$$s_h^2 = \sum \frac{\begin{array}{c} x^2 + xy - x\bar{x} - x\bar{y} + yx + y^2 - y\bar{x} - y\bar{y} \\ -\bar{x}x - \bar{x}y + \bar{x}^2 + \bar{x}\bar{y} - \bar{y}x - \bar{y}y + \bar{y}\bar{x} + \bar{y}^2 \end{array}}{n-1}$$

... but we can regroup ...

$$s_h^2 = \sum \frac{\begin{array}{c} (x^2 - 2x\bar{x} + \bar{x}^2) + (y^2 - 2y\bar{y} - \bar{y}^2) \\ + 2(xy - x\bar{y} - \bar{x}y + \bar{x}\bar{y}) \end{array}}{n-1}$$

... and these terms can all be rewritten:

$$s_h^2 = \sum \frac{(x - \bar{x})^2 + (y - \bar{y})^2 + 2(x - \bar{x})(y - \bar{y})}{n-1}.$$

Separating these terms gives ...

$$s_h^2 = \sum \frac{(x - \bar{x})^2}{n-1} + \sum \frac{(y - \bar{y})^2}{n-1} + 2 \sum \frac{(x - \bar{x})(y - \bar{y})}{n-1}$$

... which we can write in a much simpler form.

$$s_h^2 = s_x^2 + s_y^2 + 2s_{xy}^2.$$

Thus the sample variance of $h$ values (that is, the sample variance of the sum of the $x$ and $y$ values) is the sum of the sample variances of $x$ and $y$ values, plus twice their sample covariance.

What if we were interested in the difference $d = x - y$ between these values, rather than their sum? If we did the whole thing again (replacing $y$ with $(-y)$, above) we would find that the sample mean of the differences is given by:

$$\bar{d} = \frac{\sum d}{n} = \frac{\sum(x-y)}{n} = \frac{\sum x}{n} - \frac{\sum y}{n} = \bar{x} - \bar{y},$$

and the sample variance of the differences is given by: $s_d^2 = s_x^2 + s_y^2 - 2s_{xy}^2$. Thus the sample variance of the differences between $x$ and $y$ is the sum of their sample variances *minus* twice their sample covariance.

Finally, if we repeat this process substituting '$a \times x$' for $x$, and '$b \times y$' for $y$ (where $a$ and $b$ are constants), we can find the general solution for the variance sum law, as given in the main text.

In fact, we can find a very general solution for what happens when adding or subtracting values by substituting $ax$ for $x$ and $by$ for $y$ throughout the equations, in a similar manner to that shown in Box 9.2. If we do this, we find a general solution for adding sampled values together, known as the **variance sum law**:

Sample variance $(ax + by) = s_{(ax+by)}^2 = a^2 s_x^2 + b^2 s_y^2 + 2ab \times s_{xy}^2.$

In Section 9.4, we introduced the standard error of the mean, and said that there was a simple derivation of the formula given in EQ9.5. This derivation is based upon the variance sum law, and is very simple, so we can go through it here.

Remember that the standard error of the mean describes the variability between the different possible sample means we could observe from a population. When we calculate a sample mean, we are sampling $n$ items from the population, adding them, and dividing by $n$. The variance sum law tells us that the variance of the sum of $n$ independent observations from a population will be $n$ times the population's variance. So, if we repeatedly took samples of size $n$ from the population, and added them all up, the variance of these values would be around $n$ times the variance of the population. Our best estimate of the population variance is the sample variance, $s_x^2$, so our estimate of the variance of possible sums from $n$ observations is $ns_x^2$; that is, the standard deviation of possible sums is $s_x\sqrt{n}$.

Why does this help? Well, the sample means from all these samples are the sum of each sample, scaled down by a factor of $n$. So we can scale down this standard deviation of different sums, to get the standard deviation of the different sample means, which is the *standard error of the mean*:

$$s_{\bar{x}} = \frac{s_x\sqrt{n}}{n} = \frac{s_x}{\sqrt{n}}.$$

## Presenting Your Work

### QUESTION A

The table below shows the mass of each of 30 meerkats, taken from a field study of meerkat groups in the Kalahari. Data shown are the mass of each organism (in grams). Produce a graphical summary of these data, report the five-point summary, and give the sample mean along with its standard error.

| | | | | | | | | | | | | | | |
|---|---|---|---|---|---|---|---|---|---|---|---|---|---|---|
| 459 | 592 | 595 | 511 | 633 | 510 | 569 | 582 | 605 | 585 | 522 | 526 | 629 | 575 | 481 |
| 513 | 520 | 575 | 473 | 555 | 466 | 561 | 535 | 510 | 497 | 680 | 583 | 516 | 521 | 695 |

Graphical summary:

Data are measured on a ratio scale, so a histogram. 30 numbers, so 5 or 6 intervals.

Range is 459–695, take 451–700 for convenience (5 intervals of 50):

| interval | frequency |
|---|---|
| 451–500 | 5 |
| 501–550 | 10 |
| 551–600 | 10 |
| 601–650 | 3 |
| 651–700 | 2 |

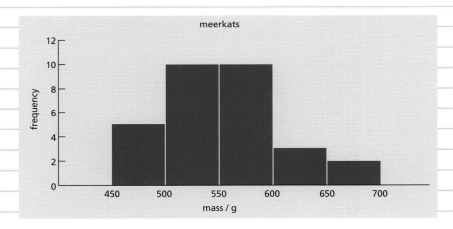

Five-point summary: first, we arrange the data into ascending order:

| | | | | | | | | | | | | | | |
|---|---|---|---|---|---|---|---|---|---|---|---|---|---|---|
| 459 | 466 | 473 | 481 | 497 | 510 | 510 | 511 | 513 | 516 | 520 | 521 | 522 | 526 | 535 |
| 555 | 561 | 569 | 575 | 575 | 582 | 583 | 585 | 592 | 595 | 605 | 629 | 633 | 680 | 695 |

$n = 30$, so the median is the $(30 + 1)/2 = 15.5$th number (the average of the 15th and 16th).

The lower quartile boundary is the median of the bottom 15 numbers, i.e. the 8th smallest.

The upper quartile boundary is the 8th largest.

Min = 459 g, LQB = 511 g, Median = 545 g, UQB = 585 g, Max = 695 g.

Sample mean = 552.5 g, sample standard deviation = 59.4 g.

Standard error of the mean = $59.4 \text{ g}/\sqrt{30} = 10.8$ g.

The sample mean = 552.5 g, and its standard error is 10.8 g.

## QUESTION B

Two different scientific techniques are used to estimate the nerve-firing rate in equivalent samples of nervous tissue. One technique gives a sample of 10 000 measurements in ten minutes, with a standard deviation between measurements of 15 spikes per second. The other technique gives fewer measurements, and less measurement error, producing 2500 measurements in the same interval with a standard deviation of only 8 spikes per second between measurements.

Assuming that both techniques are calibrated to avoid systematically too high or too low estimates, which should be recommended as giving the more accurate estimate of the true firing rate in the tissue sample?

The precision of the estimate is measured by the standard error of the mean.

SEM = standard deviation/$\sqrt{n}$.

First technique: SEM = 15/100 = 0.15 spikes per second.

Second technique: SEM = 8/50 = 0.16 spikes per second.

Thus the first technique seems to give the more accurate estimate of population mean.

## End of Chapter Questions (Answers to questions can be found at the end of the book)

### Basic

**1. Measurement scales**

(a) For each column of data presented in Table 9.1, decide which type of measurement scale corresponds with each of the properties.

(b) For each of the following types of biological data, decide whether the quantity being measured is discrete or continuous: number of legs; body mass; blood oxygen concentration; litter size; lever press responses.

(c) Are the following measurements best described as being measured on nominal, ordinal, interval, or ratio scales? Blood group; payroll number; employment start date; number of days sick leave this year.

**2. Pie charts and column graphs**

(a) A sample of color data shows that 60 % of observed items are green, 15 % are blue, and the remaining 25 % are red. Draw a pie chart to summarize these data.

(b) 100 observations of wind strength are taken on different days, in different locations. The measurements are on the Beaufort scale of wind strength (an ordinal scale from 0 to 12, with larger values indicating increasing wind speed). The number of observations of each wind strength were as follows:

| force 0 | force 1 | force 2 | force 3 | force 4 |
|---------|---------|---------|---------|---------|
| 10      | 12      | 30      | 26      | 9       |

| force 5 | force 6 | force 7 or more |
|---------|---------|-----------------|
| 6       | 6       | 1               |

Summarize these data in a column graph.

(c) In a genotype survey of a particular gene site in 800 adult humans, 400 individuals were found to have genotype A, 127 were found to have genotype B, 17 were found to have genotype C, and 256 were found to have genotype D. Summarize these data with a suitable figure.

**3. Histograms**

(a) In an experiment, 100 mice are weighed. The number observed within each weight range is shown below.

| 100–119 g | 120–139 g | 140–159 g | 160–179 g |
|-----------|-----------|-----------|-----------|
| 10        | 12        | 30        | 26        |

| 180–199 g | 200–219 g | 220–239 g | 240–259 g |
|-----------|-----------|-----------|-----------|
| 9         | 6         | 6         | 1         |

Summarize these data in a histogram.

(b) Which of the following descriptions is best for the histogram in part (a)?
(i) unimodal and symmetrical;
(ii) bimodal and negatively skewed;
(iii) unimodal and positively skewed;
(iv) bimodal and symmetrical;
(v) unimodal and negatively skewed?

(c) Plot a histogram of the following data, using a sensible number of categories:

| 68 | 65 | 74 | 64 | 79 | 69 |
|----|----|----|----|----|----|
| 70 | 64 | 62 | 56 | 68 | 68 |
| 70 | 66 | 63 | 64 | 60 | 62 |
| 65 | 65 | 53 | 56 | 62 | 56 |
| 64 | 61 | 57 | 68 | 76 | 60 |
| 66 | 50 | 58 | 60 | 54 | 62 |

**4. Interval data: sample mean, variance, and standard deviation**

(a) Calculate the sample mean of the following values:

| 13 | 18 | 16 | 16 | 10 |
|----|----|----|----|----|
| 13 | 13 | 15 | 13 | 18 |

(b) Calculate the sample variance and sample standard deviation of the numbers in part (a).

(c) Calculate the sample mean, sample variance, and sample standard deviation of the following numbers:

| 12 | 16 | 24 | 30 | 82 |
|----|----|----|----|----|
| 28 | 83 | 12 | 44 | 9  |

### 5. Ordinal or interval data: median, range, and IQR

(a) Calculate the median of the data sets in questions 4(a) and 4(c).

(b) Calculate the range and IQR for the data sets in questions 4(a) and 4(c).

(c) Give a five-point summary (minimum, lower quartile boundary, median, upper quartile boundary, and maximum) for the following numbers:

20   38   40   22   24   54
92   105   24   34   114   92

### 6. Exploratory summaries of a data set

(a) What are the maximum, minimum, and median values in the data set shown in the following stem-and-leaf plot?

```
 10 : 146
 20 : 2
 30 : 37
 40 : 23
 50 : 2
 60 : 224
 70 : 35
 80 : 09
 90 : 5
100 :
```

(b) Produce a boxplot comparing the data sets in questions 4(c) and 5(c).

(c) Produce a stem-and-leaf plot comparing the data sets in questions 4(c) and 5(c).

### 7. The standard error and accuracy of estimation

(a) A random sample of 10 items is drawn from a population. The sample mean of the 10 items is 7, and the standard deviation is 20. What is the standard error of this sample mean?

(b) A random sample of 30 items is drawn from a population. The sample mean of the items is 7, and the sample variance is 400. What is the standard error of this sample mean?

(c) Provide the sample mean and standard error of the sample mean for the data sets in questions 4(c) and 5(c).

### 8. Covariance and correlation

(a) What is the correlation between these two sets of paired measurements?

| A | 10 | 13 | 9 | 5 | 10 | 13 | 8 | 14 | 8 |
|---|----|----|---|---|----|----|---|----|---|
| B | 21 | 19 | 32 | 60 | 25 | 18 | 64 | 23 | 40 |

(b) What are the sample covariance and correlation between length and height in the following sample of measurements?

| length / mm | 12 | 24 | 18 | 21 | 7 | 19 | 17 |
|-------------|----|----|----|----|---|----|----|
| height / mm | 251 | 299 | 178 | 230 | 105 | 183 | 224 |

| length / mm | 23 | 9 |
|-------------|----|---|
| height / mm | 260 | 112 |

(c) In a sample of 20 pairs of $x$, $y$ measurements, the sample covariance between the $x$ and $y$ values is 15. If the sample standard deviations of $x$ and $y$ are 20 and 3, respectively, what is the correlation between the $x$ and $y$ values?

**9.** Before a mayoral election, 200 people were asked their voting intention. The following responses were obtained: 70 said candidate A, 30 said candidate B, and 82 said candidate C. The remainder failed to give an answer. Provide an informative graphic to summarize these data.

**10.** Draw a histogram of the following data set, and describe the shape of the distribution.

$$\begin{bmatrix} 30 & 36 & 13 & 50 & 20 \\ 9 & 19 & 46 & 20 & 17 \\ 11 & 42 & 27 & 46 & 25 \\ 18 & 19 & 26 & 9 & 21 \\ 28 & 33 & 14 & 34 & 15 \end{bmatrix}$$

**11.** Give a five-point summary of the data given in question 10.

**12.** Calculate the sample mean, sample variance, standard deviation, and standard error of the mean for the data in question 10.

**13.** If samples were taken for each of the measurements in question 1(c), for which would a sample mean have a sensible value?

## Intermediate

**14.** The histogram below shows the different lengths of a species of grass found in a field study. Determine what can be concluded about the value of the skewness, sample median, and IQR of the dataset.

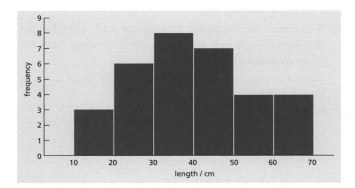

**15.** The figure below appears in a newspaper article 'Bananas are the new apples', which describes changes in fruit consumption in a town during the first 5 years of the 21st century. Knowing that consumption of the two fruits was equal at the start of the century, do you feel that the graphic chosen gives a reasonable summary of the data?

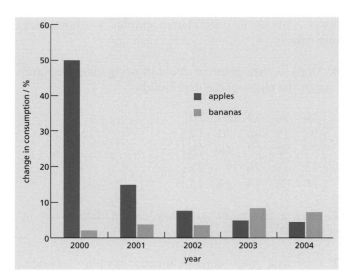

**16.** Redraw the column graph in Figure 9.4 as a stacked graph, showing the contribution of each sex group to the captures in each weather condition. (Use the data from Table 9.1.)

**17.** Calculate the magnitude of the correlation between mass and capture temperature, for the data presented in Table 9.1.

**18.** In an observational study of a rodent colony, the number of behavioral signs of illness shows a strong positive correlation with the amount of a certain plant that the animal has eaten. Can we conclude that the plant is making these animals sick?

## Advanced

**19.** In Meanland, the average height of an $n = 100$ sample of married men is 175 cm, and that of their wives is 160 cm. The sample standard deviation of heights for both genders is 15 cm. People in Meanland seem to choose their partners independently of their height, as the covariance between the heights of married couples in the sample is zero. What is the sample mean and variance of (a) the difference in height between, and (b) the average height of, the married couples in the sample?

**20.** In the neighboring country of Modenia, a similar sample is obtained. The sample mean height of 100 married men is 175 cm, that of their wives is 160 cm, and the standard deviation of height for both genders is 15 cm. However, in this country, people do not choose their partners independently of their height, and there is a sample covariance of $+30$ cm$^2$ between the heights of married couples. What is the sample mean and variance of (a) the difference in height between, and (b) the average height of, the married couples in this sample?

**21.** What are the correlation coefficients between married couples' heights in the samples in questions 19 and 20?

# CHAPTER 10

# Probability

The language of biology rarely contains terms such as 'always' or 'certainly': we may observe that a baboon *will usually* receive more grooming if it has groomed others, that camouflage makes an organism *less likely* to fall victim to predation, or that smoking puts an individual at *higher risk* of developing lung cancer. All of these statements relate to **probabilities**. To understand any of these statements, we need to understand a little of the theory of probability.

This chapter will cover the basic concepts of probability: what a probability value means, and how to calculate and combine probabilities. We will also explore conditional probabilities, which we commonly use, but often fail to understand: conditional probabilities are vital to the understanding of all the statistical tests most commonly used by biologists.

Finally, we will look at the concept of probability distribution functions, including two distributions that are frequently used to describe events in the biological world: the normal and binomial distributions.

**Figure 10.1**
Baboons grooming one another. Grooming may change the probability of being groomed in return.

**Figure 10.2**
The melanic (black) and original 'peppered' variety of moths on lichen- (left) and soot- (right) covered tree trunks. There are two moths on each tree, with very different probabilities of being spotted. Photographs from the estate of G. B. Ford courtesy of J. S. Haywood.

# 10.1 Probability

There is an old saying that the only certainties in life are death and taxes. Whether or not we agree with the sentiment, we should notice that this implies that everything else is uncertain: there is a chance that it will, or will not, happen. Probability is the branch of mathematics that tries to describe these uncertainties. We use a simple form of probability mathematics whenever we say something is 'more likely' than anything else, 'inevitable', or 'impossible'.

This is probably easiest to understand by example. A biologist studying plants is interested in the different properties of flowers she is breeding. She is considering a particular seedling, and wants to describe how that seedling might develop. She thinks that the seedling will probably grow into a plant that is less than 20 cm tall, and that it is quite likely to have red flowers. She thinks it is almost impossible that it will have blue flowers. We will use this example to introduce some terminology.

What does it mean for our biologist to believe that red flowers are probable? The statement 'the seedling produces red flowers' is a description of the world that may, or may not, *turn out to be true*. This type of statement (a description of what might happen) is known as an **event**. When an event turns out to be true (is an accurate description of what happens) the event is said to **occur**. Every probability can be thought of as the probability of a certain event occurring; that is, turning out to be an accurate description of what actually happens.

The state of affairs that is finally observed is known as the **outcome**. To begin with, the biologist believes that a plant will probably be less than 20 cm tall. The 'probably' means she does not know whether the statement 'less than

20 cm tall' will be true about the outcome. Although she does not know exactly what the outcome will be, she does know that there are many different *possibilities*: the seedling may grow 170 mm tall, with blue flowers; it may be 251.2 mm tall with red flowers; it may be 212 mm tall with white flowers, and so on.

Suppose our biologist now waits to observe the growing plant and measures the height of the plant that actually grows from the seedling. Of course, she might record the height as a rounded value, for example 21 cm, or even describe it relative to a standard, for example writing 'tall' to indicate that it is taller than 200 mm. Note that her recording of observed height is a description of the outcome: all observations, being descriptions, are *events that have occurred*. Events can thus be considered the basic element in probability.

When we come to make our observation, exactly one of the possibilities has turned out to be the outcome we observe (mathematicians call this *sampling* the outcome). It should be obvious that once we know what the outcome is, we also know which events have occurred. Imagine that the seedling grew to 205 mm tall, and produced red flowers. The *outcome* is the plant itself, and the following *events* have occurred: 'Red flowers', 'A tall plant', 'A tall plant with red flowers', 'Flowers that are *not* blue', and even 'A short plant *or* red flowers'.

An event that must always occur (is true of every possibility) is said to be *inevitable*, whereas one that can never occur (is true of no possible outcome) is said to be *impossible*. All other events might be said to have a chance of occurring that is somewhere in between these two extremes. We call the chance of the event occurring its **probability**, written as **P(event)**.

We are able to assign values to probabilities. The value we assign may be thought of as '*the proportion of all possible, equally likely, outcomes for which the event is true*'. From this definition, it follows that the probability of an *inevitable* event is 1 (certainty: it occurs for all possible outcomes), and the probability of *impossible* events is zero (they occur for none of the outcomes). Given that all other events must occur for a proportion of outcomes somewhere between these extremes, then each probability value must be in this range: a probability can never be greater than one, nor less than zero.

Of course, if P(A) is the proportion of possible (equally likely) outcomes for which A occurs (that is, for which A is true), then for the remaining proportion of outcomes it must be that A does not occur. Thus if we define a new event $\sim$**A**, which is the event of 'A not happening', then **P($\sim$A)** is the probability that A does not occur, and must be equal to $1 - P(A)$. This is known as the *converse* probability of A.

$$P(\sim A) = 1 - P(A).\hspace{4cm}\text{(EQ10.1)}$$

For example, if we know that it rains on 25 % of days (so the probability of rain on a randomly selected day is 0.25), then we also know that it *does not rain* on 75 % of days; thus the probability of no rain on a randomly selected

The probability that A does not occur, $P(\sim A) = 1 - P(A)$, is sometimes easier to calculate than P(A).

## Box 10.1 Alternative views of probability

The interpretation of probability as a 'proportion of possible outcomes' is known as the *classical* view of probability. It dates back to around 1800 and the work of the great French mathematician Pierre Simon Laplace. However, it is no longer the most commonly used interpretation of probability. Imagine a common situation, such as estimating the probability that it will rain tomorrow. The 'classical' view is difficult to apply here, because we cannot easily construct a set of 'equally likely possible tomorrows' and then agree what proportion of them are rainy days.

Two other views of probability are now most commonly used. *Frequentist* probability is very similar to the classical view. This interpretation of probability is 'the proportion of times an event would be observed in the future, if observations were repeated indefinitely'. How does this help with the probability of rain? Imagine that in our town, it rains on 25 % of all days throughout the year, in the long run. This would allow us to agree that 'the probability of rain tomorrow is 25 %', based on the frequentist interpretation of probability.

An alternative view of probability is the *Bayesian* view. This interprets probability as the degree of certainty we have that a statement is true. This can be defined to be the same thing as the frequentist probability for future events (if I believe it rains on 25 % of days, I should be '25 % certain' it will rain tomorrow).

One advantage of the Bayesian view of probability is that it can be applied to past events. If I have no way of knowing whether it rained yesterday, I might say that the probability that it *did* rain yesterday is the same as the probability of rain for any other day, namely 25 %. This

kind of statement seems reasonable, but if you think about it carefully, you will see that it *requires* probability to be regarded as 'the degree of certainty' we have about something. It cannot be interpreted using either a classical or frequentist view of probability: 'rain yesterday' either happened or did not, so we cannot talk about 'the proportion of observations in the long run', nor about a set of 'equally likely possible outcomes'.

The main *disadvantage* of Bayesian probability is that it is 'subjective': unless we have exactly the same information available, we will not have the same certainty, and thus not the same probability. If someone was walking outside yesterday, and got rained on, they would not agree with us that the probability of rain yesterday was only 25 %!

The exact interpretation of probability is not important for understanding the rules of probability in this chapter (all the interpretations result in the same rules for combining and calculating probability), so we will largely ignore it! However, it is a little more important when interpreting the kind of statistical tests we will meet in Chapter 11: these techniques are all based on *frequentist* views of probability (what would happen in the long run).

Frequentist probabilities, although useful, have to be interpreted carefully. It is common to interpret the probability from a statistical test in terms of 'how likely a theory is to be true'. Saying that a theory is true is like saying 'it rained yesterday': a theory is either true, or it is not. From our data, we may well become more or less confident about the theory (Bayesian probability), but it does not make sense to claim that a theory 'will be true a certain proportion of times in the future'.

day is 0.75. It is worth remembering that the probability of something *not* happening is often much easier to work out than the probability that it does happen.

## An example of probability in biology

One of the areas of biology in which the principles of probability are most evident is in reproduction and simple genetics. It has been known for a long time that offspring tend to resemble their parents and we now know many of the mechanisms that lead to such similarity. We will use this example throughout the chapter, so if you are unsure of the terminology of simple genetics you should review Box 10.2 before continuing.

## Box 10.2 The terminology of genetics

Throughout this chapter, we use the classic theory of genetic inheritance first postulated by Gregor Mendel as an illustration of probability in a biological context. The terminology of this example is reviewed here, in case any of it is unfamiliar. A more thorough treatment is given in *Biology Today*, Chapter 2 (Minkoff and Baker, 2004).

According to Mendel's account of genetic transmission, each parent has two *genes* (i.e. is *diploid*). Each gene can exist in different alternative forms, or *alleles*. The pair of alleles that a parent has is fixed, and called the parent's *genotype*. A genotype that contains two identical alleles is called *homozygous*, whereas one that contains two different alleles is called *heterozygous*.

For sexual reproduction, each parent produces a *gamete*, or germ cell, which only contains one gene (i.e. is *haploid*). The single gene in the gamete will be a copy of one of the two genes of the parent, and thus will be one of the alleles in the parent's genotype. When the two gametes are combined, the offspring's genotype is therefore a combination of one allele from the maternal genotype, and one allele from the paternal genotype (see Figure 10.3).

The set of physical characteristics that arise from a particular genotype is called the *phenotype*. One allele of a gene may be dominated by another, such that the phenotype of a heterozygous individual is the same as that of an individual who is homozygous with two copies of the *dominant* allele. The other allele in this case is called *recessive*, and the recessive phenotype is only observed in individuals who are homozygous with two copies of the recessive phenotype.

The alleles in each gamete are determined randomly, and each instance of reproduction involves the unpredictable selection of a single male gamete and a single female gamete. We can talk about each of these selection processes in the language of probability.

In terms of selecting the allele within a gamete, an *observation* or *outcome* will be the allele contained in the gamete, and an *event* that might occur is a description of the observed gamete. For example, an event that might occur during reproduction could be 'the male gamete contains the dominant allele'.

In terms of the selection of a pair of gametes, our observations or outcomes will concern the *genotype* of the offspring as a whole. For example, an event that might occur could be 'the offspring is heterozygous'.

A simplified theory of diploid genetic inheritance states that each offspring has two alleles of each gene: one is a copy of one allele from the mother, and the other a copy of one allele from the father. Consider a simple case where there are only two alleles of a specific gene, and that one of the alleles is recessive (meaning that only those who are homozygous – having two copies of the relevant allele – express the trait for that allele rather than that for the dominant allele).

Using the tools of probability we can answer questions based on this simple theory.

(a) What proportion of the offspring of heterozygous parents will be homozygous?
(b) If, say, 4 % of the population express a recessive trait, how many of the population are likely to be heterozygous (that is, have one of each allele)?

We will look at different ways to find the answers to these two simple questions as we introduce some basic concepts.

**Figure 10.3**
Simple inheritance of a dominant trait. The first generation is created by cross-fertilization. All members of this generation are heterozygous. The second generation may have three different genotypes, with different probabilities. The violet-red allele, V, is dominant and the white allele, v, is recessive.

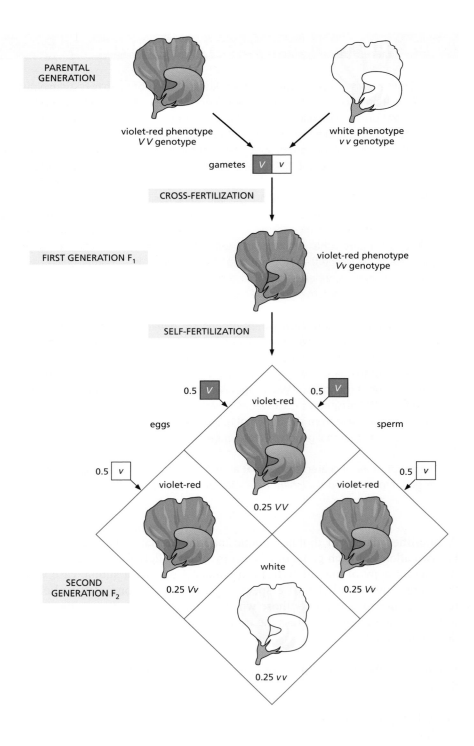

## Types of event

Some events may have an impact on one another; that is, one event makes the other more, or less, likely. Other events may have no such impact: these are known as **independent events**. In our reproduction example, if a woman is equally likely to conceive a male offspring as a second child, whatever the gender of the first child, then we would say that the genders of first and second offspring are independent.

Some events are alternatives: if one has happened then the other has not. These are known as **mutually exclusive** events. In the case of our reproduction example, the alleles of a gene in a gamete are mutually exclusive events, because if the gamete contains the recessive allele of a gene, it does not contain the dominant allele.

Finally, we label any set of events that includes all possible outcomes as an **exhaustive** set. The pair of alleles of the gene make up an exhaustive set, similarly the set of events 'male' and 'female' make up an exhaustive set of possible events for the gender of offspring.

Now consider the events A and ~A, as we defined above. Try to answer the following questions before reading on. Are events A and ~A independent? Are the events exhaustive? Are they mutually exclusive?

When you think you have the answers, read on to see how well you have done.

Either A or ~A must occur, by our definition of ~A. So A and ~A are exhaustive events.

If A occurs, this makes it impossible that ~A has occurred. So A and ~A are mutually exclusive events. The probability of ~A occurring if A occurs is 0, whereas the probability of ~A occurring if A does not is 1. These are not the same, so A and ~A are not independent events.

## Probability diagrams

Sometimes the quickest way to calculate a probability is to think of all the possible outcomes, and simply count the proportion for which the event in question occurs. In Figure 10.3 there are four possible combinations of gametes from heterozygous parents, and for only one of these does the event of homozygous recessive offspring occur. So the probability of getting a recessive phenotype from heterozygous parents is 25 %. However, we must be careful, because 'listing all possible outcomes' will only work if all the outcomes we list are equally likely. For this reason, in anything but the very simplest cases, we use a different method, known as a *probability tree diagram*.

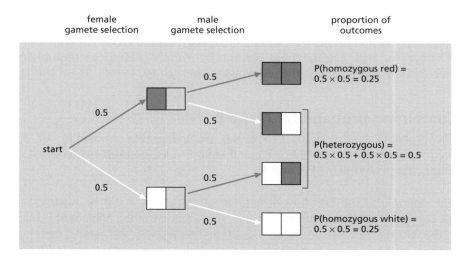

**Figure 10.4**
A simple probability tree for the genetics example shown in Figure 10.3. The first branch shows the possible alleles in the female gamete and the second branches show the possible alleles in the male gamete (red lines show dominant, red alleles; white lines show recessive, white alleles). The probability of a particular sequence, such as homozygous recessive, is given by multiplying the probabilities along each branch.

A probability tree is a good way of considering the probability of different events when the overall result depends upon a combination of different outcomes (such as one dice roll followed by another, or the combination of gene selection in male and female gametes). A simple probability tree for the cross-breeding example is shown in Figure 10.4. Working across from left to right (or, if the tree is drawn vertically, top to bottom) we consider all the exclusive alternatives of the first event and draw a branch for each alternative along with its probability. There are two branches, one for each allele within the maternal gamete, and these have equal probabilities of 50 %. Then, for each individual branch, we consider all the exclusive alternatives of the next event. When we have finished, the probability of each possible sequence is given by multiplying all of the probabilities along the path. In this case, only one sequence (the bottom one) results in a homozygous recessive offspring. So the probability of being homozygous recessive is given by multiplying the probabilities along this path: that is, 25 %. This confirms the result we saw above.

What do we do if an event is true for the outcome of more than one possible sequence? Remember that a probability is the proportion of possible outcomes for which the event occurs. We know that the probability, namely the proportion of possible outcomes, for each sequence of branches in the tree is the result of multiplying along that sequence. Therefore we simply need to add up the probability values for all sequences where the event is true.

For example, the probability of heterozygous offspring is given by the sum of the probabilities of the two central sequences in Figure 10.4. Each of these sequences has probability of 25 %, and adding these probabilities gives 25 % + 25 % = 50 %. So this tells us that the probability of observing a heterozygous offspring (from two heterozygous parents) is one half.

From this we can see some basic principles of combining probabilities:

- If we are looking at the probability of **both of two events** happening in sequence, we **multiply some probabilities**.
- If we are looking at the probability of **either of a set of alternatives** happening, we **add some probabilities**.

Knowing these principles can help us to use simple arithmetic to calculate the probability of a combination of events, but only under certain circumstances. Whether we will get the right answer by simply adding or multiplying probabilities depends upon the relationship between the events, as we will see now.

> $P(A \wedge B)$ is the **joint** probability that both events A and B occur (are true of the outcome).

## Combining probability: A and B

The probability of both of two events occurring (the probability of co-occurrence) is written as $P(A \wedge B)$, and is known as the **joint probability**, or simply the **probability of A and B**.

Our example of the breeding of heterozygous parents could be regarded as the combination of two different events: the selection of the allele within each of the two gametes. We create an event F that represents 'the female, maternal gamete contains a copy of the recessive allele', and another event M that

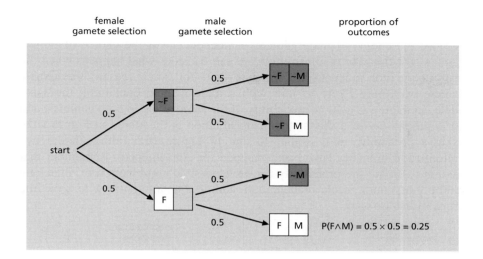

female gamete selection    male gamete selection    proportion of outcomes

start

0.5

~F

0.5

0.5    ~F  ~M

0.5    ~F  M

0.5

F

0.5    F  ~M

0.5    F  M    P(F∧M) = 0.5 × 0.5 = 0.25

**Figure 10.5**
The same probability tree as in Figure 10.4, labeled with the events F (female gamete contains recessive, white allele) and M (male gamete contains the recessive, white allele). The joint probability of F and M is found by multiplying the probabilities of F and M.

states 'the male, paternal gamete is recessive'. Therefore the female gamete containing a dominant allele is the event ~F, whereas ~M is the event that the male gamete contains a dominant allele.

All the possible outcomes for the genotype of an offspring are a combination of these four events: M, ~M, F, and ~F. We can put these labels into our probability tree diagram, as shown in Figure 10.5. The bottom path in this tree shows the joint occurrence of F and M, namely recessive alleles from both parents. This is the only path in which the offspring shows the recessive phenotype, and has a probability equal to the probability of the maternal gene containing the recessive allele P(F) *multiplied by* the probability of the paternal gene also containing the recessive allele P(M).

We have already seen that the probability of an offspring being recessive phenotype is 0.25. This event is, of course, the co-occurrence of events F and M (a recessive offspring has *both* a recessive maternal allele *and* a recessive paternal gamete), so we can rewrite this as $P(F \wedge M) = 0.25$. Looking at Figure 10.5, we can see that this value comes from multiplying two different probabilities: the probability of a maternal recessive allele P(F), and the probability of a male recessive gamete P(M), namely $P(F \wedge M) = P(F) \times P(M)$. This principle of multiplying probabilities always applies for independent events, and is sometimes called the *multiplicative rule*:

---

The joint probability of two *independent* events both being true is the product of their individual probabilities. Thus, if A, B are independent events, the joint probability of them *both* occurring is given by:

$P(A \wedge B) = P(A) \times P(B).$                    (EQ10.2)

---

This multiplicative rule **applies only if the events are independent**. If the occurrence of one event influences the probability of another, then their joint probability is **not** equal to the product of their individual probabilities.

In Figure 10.5, the probabilities for the 'male allele branches' are all 0.5, because the choice of male gamete is independent of the choice of female gamete. For the sake of illustration, we can consider what happens when the choices are not independent. Imagine that, for some reason, the female gamete was more likely to join with a male gamete containing the same allele, so there was a 60 % probability of the same allele in both gametes, and 40 % probability of a mis-match. This situation is illustrated in Figure 10.6, using a probability tree similar to that in Figure 10.5 (although the two bottom-right branches have been swapped for convenience). Remember that we can calculate the probability for each possible combination of alleles by multiplying along each branch, and adding the alternatives to give the probabilities we want.

**Figure 10.6**
The figure shows a probability tree similar to that in Figure 10.5 (note that the two lower-right branches have been swapped). This tree shows a situation where the probabilities are not independent: a recessive allele in a female gamete (F) makes the male gamete more likely to contain the recessive allele.

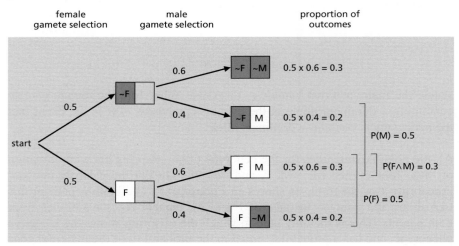

In this situation, the probability of *both* M and F occurring is 0.3, although the simple probabilities P(M) and P(F) are once again 0.5. This confirms that multiplying the probabilities *does not work* for events that are not independent. We will look at this kind of situation in more detail later in the chapter when we discuss conditional probability in Section 10.2.

## Combining probability: A or B

The probability of either one of two events, or both of them, occurring is written as P(A ∨ B), and is known as the **probability total** of the events, or simply the **probability of A or B**.

P(A ∨ B) is the **combined** probability that either of events A or B (or both) occur.

In the section on constructing a probability tree, above, we added the probabilities from different branches to obtain the probability of either branch being followed. This follows from our definition of probability as the proportion of outcomes for which an event is true. Adding the probability always gives the right answer for the total probability of mutually exclusive events. This is sometimes called the *additive rule*:

> The total probability of either one, or both, of two mutually exclusive events occurring is the sum of their simple probabilities. If A, B are exclusive events:
>
> $$P(A \lor B) = P(A) + P(B).$$
> (EQ10.3)

For example, if we are breeding two heterozygous plants, we know that there is a 25 % chance of the offspring being homozygous recessive, and a 25 % chance that it will be homozygous dominant. What is the total probability that it will be either of these two phenotypes? The sum of the two probabilities is 50 %, and they are mutually exclusive, as both could not happen at once. So the probability that the offspring is one of the two homozygous genotypes is 50 %.

What about non-exclusive events? Remember that 'having a recessive allele from the female parent' is event F, and 'getting a recessive allele from the male parent' is event M. From our theory of genetics, $P(M) = 0.5$, and $P(F) = 0.5$. These are *non-exclusive* events, because it is possible for them both to occur: we know that the joint probability that they will both occur is 25 %.

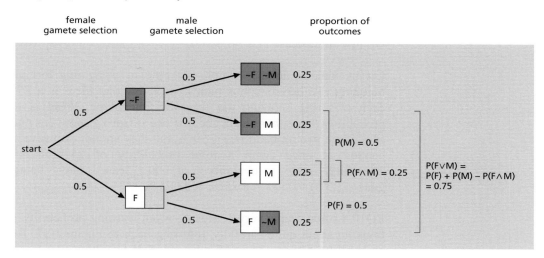

**Figure 10.7**
Total probability. The figure shows the probability tree from Figure 10.5 redrawn with the two lower-right branches swapped. The lower three paths make up the outcomes for which at least one of events M and F is true, that is the outcomes containing at least one recessive, white allele. The middle two outcomes are counted as part of P(M), as M (paternal allele recessive) is true for these outcomes. The bottom two outcomes are counted as part of P(F), as F (maternal allele recessive) is true for these outcomes. We wish to know the proportion of outcomes for which M *or* F is true. If we simply add P(M) and P(F) we count one of these outcomes twice. To obtain the correct proportion, we must subtract the joint probability.

Let us look one more time at a probability tree for offspring of heterozygous parents, Figure 10.7. We will return to the simple situation where the selection of the gametes is independent, although the same approach would work for the non-independent case. If we want to know the total probability of M or F (the probability of an offspring having *at least one recessive gene*), a look at the probability tree shows that the answer is 0.75, or 75 %: three of the four paths give us one or more recessive alleles, and each path has a probability of 0.25.

There are four possible distinct outcomes (branches on the tree). M is true for two of the outcomes, and F is true for two outcomes. However, when we combine these two 'M' outcomes and the two 'F' outcomes, we do not get four different branches because there is one possible outcome for which both events M and F are true. To get the correct proportion of possible outcomes, we have to avoid counting this possibility twice. The outcome that could get counted twice is one for which M and F are both true, namely any outcome that is part of $P(M \wedge F)$.

So, in general, we need to subtract this from the total of $P(M)$ and $P(F)$ to get the correct answer. Thus $P(M \vee F) = P(M) + P(F) - P(M \wedge F) = 75\,\%$. This is the general additive rule for combined probability:

> The total probability of at least one of a pair of non-exclusive events A, B occurring is the sum of their individual probabilities minus their joint probability:
>
> $$P(A \vee B) = P(A) + P(B) \ - \ P(A \wedge B). \qquad \text{(EQ10.4)}$$

Note that this formula also gives us EQ10.3 for mutually exclusive events because in this case $P(A \wedge B) = 0$.

Let us look at another simple example of using this formula. If we are breeding two heterozygous plants, we know that there is a 25 % chance of each offspring showing the recessive phenotype. What is the combined probability that either one of the first two offspring show this phenotype? The sum of the two probabilities is 50 %, but they are not exclusive, as both could happen at once (with a joint probability of $25\,\% \times 25\,\% = 6.25\,\%$). Therefore, the probability of at least one being heterozygous $= 25\,\% + 25\,\% - 6.25\,\% = 43.75\,\%$.

EQ10.4 often seems very confusing when learning probability, so it is worth thinking about quite carefully, referring back to Figure 10.7. To get the total probability of events M and F, we need to do two things:

- Start with the proportion of outcomes that correspond to M (this is $P(M)$).
- Add any outcomes *we have not counted yet* that correspond to F (this is $P(F) - P(M \wedge F)$).

So, in total, we have $P(M) + P(F) - P(M \wedge F)$.

We can use these rules for probability arithmetic to answer simple probability questions without the need for diagrams. For example, suppose that a given gene site is not sex-linked, and has two alleles, one of which is recessive. If 4 % of a randomly interbreeding population shows the recessive phenotype, what is the probability that an individual in the population with the dominant phenotype is heterozygous?

We defined M and F as the events of male and female gametes being recessive, so we can keep these labels. We will now use the event R to refer to any gamete being recessive (so $\sim\!R$ occurs for any dominant gamete). Because the gene is not sex-linked and the population is randomly interbreeding, $P(M) = P(F) = P(R)$.

The probability of a recessive phenotype is the probability that both gametes are recessive. Therefore:

$P(M) \times P(F) = 4 \%$, which implies that $(P(R))^2 = 0.04$, i.e. that $P(R) = 0.2$.

The probability that a gamete contains the dominant allele is: $P(\sim R) = 1 - P(R) = 0.8$.

Now we can work out the probability of each genotype:

P(homozygous dominant) $= P(\sim R) \times P(\sim R) = 0.64$,
P(homozygous recessive) $= P(R) \times P(R) = 0.04$.

These are mutually exclusive, so using additive rule:

P(homozygous either) $= 0.64 + 0.04 = 0.68 = P(\sim \text{heterozygous})$.
Therefore P(heterozygous) $= 1 - 0.68 = 0.32$.

The probability of an individual in the population being heterozygous is 32 %.

In this example, we can see a couple of points worth noting about probability calculations.

- The fact that the probability for a single offspring showing the recessive phenotype is the product of the maternal and paternal recessive probability is familiar. However, using the arithmetic rules (the multiplicative rule), we can see that the answer is obtained very much more simply than by drawing a probability tree.
- The answer to the question about the probability of any person in the population being heterozygous is most easily answered by calculating the converse probability. Look out for this as a way of answering questions that seem difficult.

When learning probability, it is very easy to get confused over the additive and multiplicative rules. A simple mnemonic to help with this is:

**A**dd **A**lternatives
**M**ultiply co**M**pounds.

In other words, when we want the probability of either of two alternative events for the same observation, we add them together (remembering to subtract the joint probability if they are not exclusive!).

When we have two or more different observations, and we want the probability for the events that happen in compound, we multiply them.

Finally, for completeness' sake, we can give alternative, more formal definitions of the concepts of exclusivity and independence based on the arithmetic we have discussed.

> Events are **independent** if and only if $P(A \wedge B) = P(A) \times P(B)$, which implies that neither event (A or B) influences the probability of the other event.

Saying events A and B are *mutually exclusive* is saying: $P(A \wedge B) = 0$, which is also equivalent to saying $P(A \vee B) = P(A) + P(B)$. Saying events A and B are *independent* is the same as saying: $P(A \wedge B) = P(A) \times P(B)$.

# 10.2 Conditional probability

> The **conditional probability** of event A, given that event B occurs, is defined as
>
> $$P(A|B) = \frac{P(A \wedge B)}{P(B)}.$$

We have seen above that probability is essentially an attempt to measure uncertainty. However, the aim of biology, like the rest of science, is to help us reduce uncertainty. Some people may catch a potentially fatal disease; the probability of them recovering from the disease could be expressed as a simple probability. However, we would expect the probability to change if the person were treated in hospital, because of some relationship between treatment and recovery. We believe (or at least hope) that the probability of recovery is greater *if you are in hospital*. This is an example of how probabilities are usually used in biology: a **conditional probability**.

A conditional probability is written as P(A|B), and can be read as the probability of event A occurring *given that*, or *knowing that*, event B has occurred. If we recall that the simple probability P(A) is equal to the proportion of *all outcomes* that correspond to event A, we can see another way of thinking of conditional probability: P(A|B) is the proportion of *outcomes that correspond to event B* for which A also occurs. In other words, P(A|B) equals P(A ∧ B) as a proportion of P(B).

$$P(A|B) = \frac{P(A \wedge B)}{P(B)}. \tag{EQ10.5}$$

Let us return to our example of a recessive genetic trait, for which 20 % of the genes in the population are the recessive allele. We saw that the chance of selecting an individual at random who is of recessive phenotype would be 4 %, and that 32 % of the population are heterozygous, thus 36 % of the population have at least one recessive allele in their genotype. What if we selected an individual who shows the dominant phenotype: what would be the probability that this individual possesses a recessive gene?

We know that 96 % of the population show the dominant phenotype, and that 32 % of the population are heterozygous. So, the 96 % of individuals who show the dominant phenotype is made up of the 32 % of the population who have a recessive gene, and the 64 % who are homozygous dominant. Thus the probability of an individual with the dominant phenotype having a recessive gene = 32/(32 + 64) = 0.333, or one third.

However, we can get all the answers we have obtained so far simply and quickly by applying the equations for probability introduced in Section 10.1.

If 4 % of the population show a recessive phenotype, what is the probability that someone with a dominant phenotype has a recessive gene?

Let P(R) = probability of a gene being recessive.

P(homozygous recessive) = $P(R)^2$ = 0.04.
Therefore P(R) = 0.2.

Thus 20 % of genes are recessive allele, and 80 % are dominant.

P(heterozygous) $= 1 -$ P(homozygous recessive)

$\qquad\qquad\qquad - $ P(homozygous dominant)

$\qquad\qquad\qquad = 1 - (0.2)^2 - (0.8)^2 = 0.32.$

Thus 32 % of individuals are heterozygous.

P(heterozygous|dominant phenotype)

$$= \frac{\text{P(heterozygous} \wedge \text{dominant phenotype)}}{\text{P(dominant phenotype)}}$$

$$= \frac{\text{P(heterozygous)}}{\text{P(dominant phenotype)}} = \frac{0.32}{0.96} = \frac{1}{3}.$$

There are several things about conditional probability that follow from its definition.

- Because

$$P(A|B) = \frac{P(A \wedge B)}{P(B)},$$

  it follows that the joint probability of both A and B can be found from P(B) multiplied by P(A|B).

  $$P(A \wedge B) = P(A|B) \times P(B). \qquad\qquad\qquad (EQ10.6)$$

This makes sense if we think of a simple probability tree like that in Figure 10.5, where we choose B or $\sim$B, then A or $\sim$A. For the joint probability we need to first go down the branch for B and then along the branch for A, given B has occurred, so the value of $P(A \wedge B)$ will be $P(B) \times P(A|B)$.

- If A and B are mutually exclusive, $P(A \wedge B) = 0$, then

  $$P(A|B) = \frac{P(A \wedge B)}{P(B)} = 0.$$

  The conditional probability of any event given a *mutually exclusive* event is zero.

This agrees with what we know about mutually exclusive alternatives: given one has happened, the other is impossible.

- If A and B are independent, $P(A \wedge B) = P(A) \times P(B)$ but $P(A \wedge B) = P(A|B) \times P(B)$, therefore $P(A|B) = P(A)$. The conditional probability of *independent* events is the same as their simple probability. The conditional probability $P(A|B)$ is equal to the simple probability $P(A)$, if, and only if, A and B are independent events. This agrees with what we know about independent events: knowing that event B has occurred should not change the probability of event A, if they are independent events.

When we talk about conditional probability, there are three rather similar concepts that are easy to confuse with one another. As well as the conditional probability of A given that B occurs, P(A|B), there is the reverse conditional probability P(B|A) and the joint probability P(A ∧ B). Although these may often seem quite similar, they can have very different values, especially when either A or B is very likely or very unlikely!

To illustrate, we can look at a simple example from the biology of everyday life: sneezing. When people are infected with one of the rhinoviruses that causes them to have a common cold, they will probably sneeze. So if someone is sneezing in winter, we might decide that they probably have a cold. In terms of conditional probability notation:

- P(S|C), the probability that someone with a cold will sneeze, is high.
- P(C|S), the probability that someone sneezing has a cold, is high.

Because both of these are high, we might make the mistake of believing that these two probabilities are saying more or less the same thing, but they are not. The difference in meaning becomes more obvious if we think about the summer months. In summer, when the pollen count is high, many people will sneeze due to hay fever, but rather few will have a cold. So if someone sneezes in summer, it is not very likely they have a cold, but if they *do* have a cold, it is still likely they will sneeze:

- P(S|C), the probability that someone with a cold will sneeze, is just as high.
- P(C|S), the probability that someone sneezing has a cold, is low.

## Bayes' equation

The relationship between a conditional probability and its reverse conditional probability is quite simply derived, by combining EQ10.5   P(A|B) = P(A ∧ B)/P(B)   with   a   version   of   EQ10.6, P(A ∧ B) = P(B|A) × P(A). This gives the simplest statement of what is known as Bayes' theorem, and so is known as *Bayes' equation*:

$$P(A|B) = P(B|A) \times \frac{P(A)}{P(B)}.$$
(EQ10.7)

This shows us that we cannot convert a conditional probability to a reverse conditional probability unless we know the ratio of the simple probabilities P(A) and P(B).

Intuition leads many people to assume that because P(B|A) is large, then P(A|B) must also be large, if not the same value. This is not generally correct. As we can see from EQ10.7, for any value of P(B|A) the value of P(A|B) depends upon the probabilities P(A) and P(B). In particular, as we decrease P(A) or increase P(B) then the value of P(A|B) will decrease. This can be illustrated by an important example: diagnosing a rare disease.

Imagine there is a rare disease that affects 1 in 10 000 people. The disease is difficult to detect, but our test is 99 % accurate: 99 % of people who have the disease test positive, and 99 % of people who do not have the disease test negative. If a randomly selected person tests positive, how likely are they to have the disease? The answer seems obvious – most people would say that the chances are 99 % – but this answer is wrong, because it ignores the last two points we have made above.

To do it the long way, we could consider what we would expect if we tested 1 000 000 people at random. Obviously, almost all (999 900) of them will probably not have the disease, but we would expect 1 % of them to test positive anyway – we expect 9999 positive tests. We would also expect around 100 people to have the disease, and around 99 of these to test positive. So, out of our 1 000 000 population, we would expect there to be $9999 + 99 = 10\ 098$ positive tests. Given only 99 of these are from people who have the disease, the probability of someone who tests positive having the disease is $= 99/10\ 098 = 0.98$ %.

Again, we can derive this answer using simple algebra and the formula for Bayes' equation.

Let D be the event 'having the disease' and T be the event 'testing positive'. $P(D) = 1$ in $10\ 000 = 0.0001$. $P(T|D) = 0.99$ as the test is 99 % accurate for those with disease, whereas $P(T|{\sim}D) = 0.01$ as the test is also 99 % accurate for those without. To use Bayes' equation (EQ10.7), we also need $P(T)$. There are two exclusive alternatives that can give a positive test: either the person has, or does not have, the disease. We add the probability for each source of a positive test.

P(no disease *and* positive test):

$$P({\sim}D \wedge T) = P(T|{\sim}D) \times P({\sim}D) = 1\ \% \times 0.9999 = 0.009999.$$

P(disease *and* a positive test):

$$P(D \wedge T) = P(T|D) \times P(D) = 99\ \% \times 0.0001 = 0.000099.$$

$$P(T) = P({\sim}D \wedge T) + P(D \wedge T) = 0.0101\ \text{(three significant figures).}$$

Thus, from Bayes' equation:

$$P(D|T) = P(T|D) \times \frac{P(D)}{P(T)} = 0.99 \times \frac{0.0001}{0.0101} = 0.98\ \%.$$

There is a 0.98 % chance that an individual who tests positive has the disease.

Suppose that the disease affected as many as 1 % of the population? In this case

$$P(D|T) = \frac{P(T|D) \times P(D)}{P(T|D) \times P(D) + P(T|\sim D) \times P(\sim D)}$$

$$= \frac{0.99 \times 0.01}{0.99 \times 0.01 + 0.01 \times 0.99} = 0.5.$$

The chance that an individual who tests positive has the disease would now be 50 %: this increase reflects the fact that the disease is much more common.

## 10.3 Probability distributions

So far, we have been thinking about the probability of individual events, such as the offspring of heterozygous parents being homozygous recessive, homozygous dominant, or heterozygous. Using tree diagrams and thinking about the probability of each outcome individually is fine as long as the number of possible outcomes is small. This approach is obviously unworkable or impossible if the number of possible alternative outcomes is very large or unlimited (as with, say, a measurement that could take on any number in a given range). So we need a different way to write and think about probability.

We do this by describing all alternative possibilities in a single description, which we can call a **probability system** (mathematicians call it a probability space, but we will use the simpler term). This way of thinking about probability is the basis of all the mathematics of probability, including that which we have already met. The probability trees we saw at the start of this chapter are one way of drawing the probability system describing 'alternative possible genotypes'. As we saw with our tree diagrams, a probability system combines two things. First, a set of possible alternative outcomes, which must be *exhaustive* and *exclusive*: that is, include every possibility exactly once. Second, the set of corresponding probabilities of each outcome, which must sum up to one. An actual observation of the outcome is called a *sample* from the system.

The simplest way to think about what a probability system describes is probably the boring example of the six-sided dice used in many games. Imagine we have two dice: one is an ordinary, fair die, and the other is from a crooked casino, and is loaded so that it rolls a '2' more often than a '5'. There are two types of question we might ask about dice rolls. First, we could ask, 'what might happen if we roll this die?' Second, we could ask 'what number was rolled?' The answer to the first question is a *probability system* (possible values and their corresponding probabilities, which are different for the two dice). The answer to the second question is a *sample* (a single number).

Our probability system is a mathematical description of what might happen when we roll each die. So what exactly should a probability system consist of? It must tell us the *possible numbers* the die can roll, as well as the *probabilities* of each possible roll. Knowing the probability system description of the dice, or of genotype selection, means that we know the probability of any given event occurring when we sample (that is, the probability of a description of the sample turning out to be true).

As we shall see, this approach is very useful, and allows us to answer questions that seem to be impossible at first glance. Nonetheless, the concepts involved are not very familiar, so we must go slowly! Luckily, probability system descriptions can easily be applied to the examples of biological probability we have met so far. We will start with one of these examples to illustrate the key concepts of probability systems, before showing how these concepts can be used to solve novel problems in biology. As our example, we will consider the set of alternative genotypes that may occur in the offspring of heterozygous plants. We saw above how to calculate the probability of each of three possible outcomes: P(homozygous dominant) = 25 %, P(heterozygous) = 50 %, P(homozygous recessive) = 25 %.

## Discrete probability systems and probability mass functions

We know that genotype can take one of three different values, and that there is a probability associated with each. We can describe genotype selection using a probability system, which we will call $G$. We know exactly what $G$ must be: the set of values {'homozygous recessive, heterozygous, homozygous dominant'}, and the corresponding properties {'0.25, 0.5, 0.25'}.

We can therefore draw the probability system $G$ as a graph showing the probability for each possible value. Because there are a limited number of discrete values, $G$ will look like a simple column graph, as shown in Figure 10.8.

The graph is called the *probability mass function* (or simply the probability function) of $G$. It is a 'function' in the sense that it relates each possible value of the observed genotype to another value, namely the probability of that value occurring. We must include all the possible values (outcomes) on our graph *exactly once*. That way, if we sum up the probabilities for each value, the total is 100 % (the sum of probabilities of an exclusive, exhaustive set).

This graph therefore allows us to describe the whole set of alternatives and their probabilities at once: it is a picture of $G$, which we call a *discrete probability system*. A discrete probability system can be used to describe any variable quantity or concept (for example something that we can observe in an experiment) that can take on any one of a set of possible values. It describes the possible values the measurement could take, *along with* the set of probabilities of it taking on each one.

So we know the probability mass function for $G$. What does this tell us? When we breed two heterozygous plants, then we don't know exactly what our first offspring will be. However, because we know the probability system $G$, we know what we *might* get, and *how likely we are* to get it. For example, the probability of the value 'heterozygous' being sampled from $G$ is 0.5. We can write this as

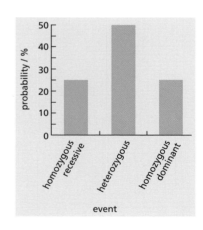

**Figure 10.8**
Probability mass function showing the probability of each genotype for offspring of heterozygous parents.

$P(G = \text{heterozygous}) = 0.5.$

We should be careful though – the equals sign inside the parentheses can cause confusion. The thing that is equal to 'heterozygous' is a *possible single sample* from G, rather than G itself. We are saying that there is a 50 % chance of sampling 'heterozygous' from G. We are not saying that there is a 50 % chance that the probability system G will somehow equal 'heterozygous'.

This is an important concept, so it is worth going over it carefully. We know from Chapter 1 that we can use a variable to 'stand in for' something. Imagine we want to describe just one sampled event, or measurement (say, the genotype of the first offspring). We might use the variable $g$ to refer to it. In this case '$g = \text{heterozygous}$' means that the first offspring was a heterozygous plant. In this case, $g$ is a variable standing in for a single observation: the genotype of *one plant*.

Our probability system, G, refers to the set of alternatives from which an observation is sampled. Thus $g$ refers to the value *sampled* from our probability system G. This is just like the difference between knowing the value rolled by a die, and knowing the characteristics of the dice that rolled the value.

If we take many observations, in different experiments, we might end up with a large set of different $g$ values: around 25 % of which would be homozygous recessive, 50 % heterozygous, and 25 % homozygous dominant. This set of observations, and these proportions, are another way of thinking about what our probability system describes: 'the set of values we would expect if we took a very large number of observations'.

So, to summarize thus far:

- A discrete probability system is a set of *alternative outcomes*, and a set of *corresponding probabilities* (a *probability function*).
- Given what we know about probability, every probability function must give a non-negative number for each alternative (i.e. possible outcome), and the sum of all these numbers must be one (the alternatives are exhaustive and exclusive).
- This is just another way of writing what we already know about probability. If we know the probabilities of some alternative events A, B, C . . . are P(A), P(B), P(C) . . . , then we can define a probability system $X$ to summarize all of this. $X$ can be described as a probability function, or graph, $f(x)$ such that $f(A) = P(X = A) = P(A)$.

## Random variables

All our discussion of probability to this point has been in terms of outcomes that differ in *type* (for example genotype). However, frequently in biology we are interested in probability systems for observations that may differ numerically (that is, in number or amount). A probability system describing a set of probabilities where every outcome is a number is called a **random variable**. Random variables are useful because, as we will see below, they allow us to summarize the set of possible outcome values using properties (mean and variance), which are equivalent to the descriptive statistics discussed in Chapter 9.

You might think that the phrase 'random variable' seems confusing, but the word 'random' here does *not* imply that every outcome is equally likely. Why not just say 'variable quantity'? Well, the 'random variable' terminology is something you may come across whenever probability theory is used to model biological systems, so you need to know it and use it correctly. Mathematicians like to be clear, so they use the term random variable to describe a *probability system* (for example a description of a loaded die), whereas they use the term 'variable' to refer to a single *value* (for example a description of a die roll). Note that we tend to use upper case for random variables, and lower case for ordinary variables.

Let us take a simple example. A scientist is collecting samples of flies. In the population of flies, each fly has one, two, or three spots. Fifty per cent of flies are one-spot flies, 30 % are two-spot flies, and 20 % have three spots.

We can define a random variable, $S$, to describe the possible number of spots, $s$, on a randomly selected fly. $S$ is a probability system where every outcome is a number. The possible numbers are 1, 2, and 3. The probability mass function $f(x)$ for $S$ gives the probability of sampling each value, $x$, that is $P(S = x)$. In this case, P('number of spots = 1') = 0.5, so $f(1) = 0.5$. Similarly, $f(2) = 0.3$ and $f(3) = 0.2$; $f(x) = 0$ for all other values of $x$.

A final point worth making about random variables is that more than one random variable might be used to describe the same underlying probability system. We have just considered the probability function that a scientist would find a fly that has one, two, or three spots. However, imagine the scientist was paid 10¢ ($0.1) for each spot on each fly he collected. What do we know about the amounts he could be paid for the next fly? We know enough to define a random variable $M$ describing *the amount of money he might make* (in dollars) from the next capture. Let us construct the probability mass function for this variable $f_M$. Remember that $f_M(x)$ must tell us $P(M = x)$, the probability that he will be paid $x$ dollars for the next capture. It should be clear that this function is therefore: $f_M(0.1) = 0.5$, $f_M(0.2) = 0.3$, $f_M(0.3) = 0.2$, and $f_M(x) = 0$ for all other values of $x$. You should now be able to work out for yourself a full description of a random variable describing the possible 'number of spots' observed on a single roll of a standard six-sided die.

At this stage, you might well be forgiven for asking 'what is so useful about thinking in terms of random variables?' As stated above, one main advantage is that we can use random variables to summarize variable quantities in terms of one or two properties, just as we did for samples. It should be obvious that we can turn any probability system describing the world into a random variable by assigning a number to each outcome (by means of some measurement scale, such as we saw in Chapter 9). Therefore, as long as we use a sensible measurement scale for converting possible outcomes into numbers, using random variables allows us to summarize *possibilities* in the same way that we can summarize *data*.

We will look at how we calculate these summary properties of random variables in Section 10.4. First, we need to consider the more common situation in biological measurement: when the set of possible outcomes is all the numbers on a continuous scale.

## Continuous random variables and probability density functions

Let us return to the genotype example from earlier in the chapter, but this time we consider a gene that influences the height of a plant. For a certain species this gene has two co-dominant alleles that control the height of a plant. Homozygous 'small' plants tend to be around 75 cm tall, heterozygous 'medium' plants to be around 1.5 m tall, and homozygous 'tall' plants around 2.25 m tall.

Above, we introduced a probability system that described the possible genotypes for offspring of heterozygous parents. Obviously, we could use a similar probability system to describe the possible phenotypes: this would be the set of values {'short', 'medium', 'tall'} with the corresponding probabilities {0.25, 0.5, 0.25}. If we want to use a random variable (with the outcomes being *numbers*) we must give each possible outcome a number, such as typical height values in centimeters. This gives us a random variable with a mass function $P(75) = 0.25$, $P(150) = 0.5$, $P(225) = 0.25$, and $P(x) = 0$ for all other values.

However, this would not be a very plausible biological model, as the real measured heights are of course not going to be exactly 75, 150, or 225 cm. What we need to describe possible heights accurately is a random variable that gives the *actual* (rather than the 'approximate') possible heights of the offspring, along with the probability of measurement of each value. However, this is a little tricky, because height could actually take on any value in a given range. In other words, height is not a discrete variable.

Taking this one step at a time, we will assume for now that we can measure height to the nearest multiple of 5 cm. We could then have a better random variable describing all possible height measurements for the offspring of two heterozygous parent plants to the nearest 5 cm. For our particular species, the probability mass function turns out to look something like Figure 10.9.

**Figure 10.9**
Probability mass function showing the probability of different 'measurement intervals' for measured heights of offspring. We would need a slightly different function if we were measuring height with different accuracy.

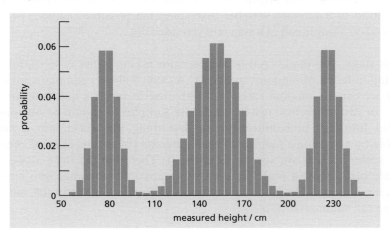

This graph is quite similar to the probability function we saw in Figure 10.8, although as the $x$ axis corresponds to a continuous scale, it could be regarded as a histogram. We know that there is a 25 % probability that a plant would be 'short' (therefore around 75 cm tall). This function tells us more: for example we can see that a plant is more likely to be measured as 75 cm tall than as 60 cm tall.

Of course, we might want to know height more accurately than just 'to the nearest 5 cm': we might want to know about heights to the nearest centimeter, or the nearest 2 mm. To understand each of these possible measurement circumstances exactly, we would need a histogram like Figure 10.9 with the appropriate column widths. So can we just draw another probability function showing the probability for each possible *exact* value of height? The answer is no: as we measure more and more accurately, the probability of a value becomes smaller and smaller. The probability of a plant being exactly 70.00001 cm tall, rather than, say, 70.00002 cm, would be very small indeed. In fact, taken to its extreme limit, the probability would be zero for each (infinitely accurate) different measurement. So when we are describing continuous measurements (where any value is possible), a probability mass function is not useful.

What we need is therefore a graph, similar in shape to Figure 10.9, but with columns 'as wide or narrow as we want'. This sounds impossible, but actually we can do it quite easily. The way we do it is to draw a continuous graph which looks like a smoother version of the histograms we want – do not worry about how we get this graph for now – and this curve is scaled such that the *area under the whole curve is one*. Figure 10.10 shows the curve we need. This graph is called the probability density graph, or **probability density function**, $f_{pdf}(x)$ of the random variable that describes actual height values.

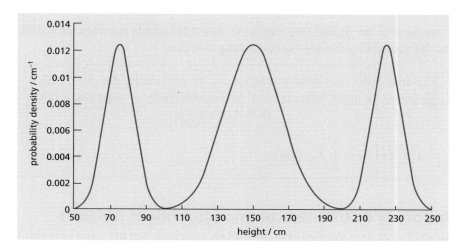

**Figure 10.10**
A probability density function for measured height, which could be used with any set of measurement intervals. Probability density functions are like smoothed histograms for continuous variables.

So how do we use a probability density function? The graph is scaled such that the area under the total curve is equal to one; because of this, the area under the curve *in any interval* gives the probability of the variable taking a value in that interval. For example, we can see in Figure 10.11 a shaded region corresponding to the interval 70–90. From the definition of the probability density function, we know that the area of this shaded region is equal to the probability of getting a measurement within the 70–90 range. Clearly, by varying the width of the shaded area, we could establish the probability of a measurement falling into any range we liked.

**Figure 10.11**
The area under a probability density curve gives the probability of sampling a value within a given range.

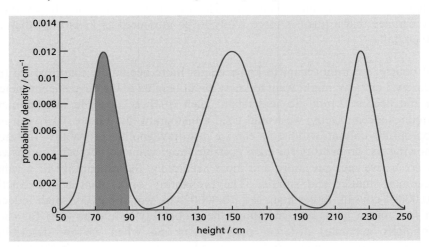

We can use this kind of function to create random variables that describe the possible alternatives in the most common measurement situation; that is, a situation where any value within a particular range is possible. Such a description is called a **continuous random variable**: a range of values along with a probability density function (which allows us to calculate the probability of sampling within any interval we choose to consider).

The area under the probability density function (the *definite integral*) over a given range is equal to the probability of sampling a value in that range. The total area under the curve must be one, and the function can never be negative.

If we wish to be formal, we can write this relationship between probability and the probability density function using calculus:

> The area under (the *definite integral* of) the probability density function $f_{pdf}(x)$ over a given range is equal to the probability of sampling a value in that range.
>
> $$P(a < x \leqslant b) = \int_a^b f_{pdf}(x)\mathrm{d}x. \qquad \text{(EQ10.8)}$$

## Uniformly distributed continuous random variables

We shall start with the simplest case: when a variable has the same probability density over the whole range, known as a *uniform distribution*. A random variable with a uniform distribution has the same probability (or probability density) for all alternative outcomes.

For example, suppose a prey hides from a predator beneath a shelf of rock 2 m long by selecting a position at random along the shelf and remaining stationary. We could describe the set of possible locations at which the prey is hiding using a random variable, $D$, that describes the distance of the prey from the northern end of the rock. What is the probability density function, $f_{pdf}(x)$, that describes the possible locations?

The probability density function of $D$ is shown in Figure 10.12. This is just a graphical way of showing that the prey animal 'is equally likely to be anywhere under the shelf': the probability density has the same constant value $f_{pdf}(x) = k$, throughout the possible range $0 < x \leqslant 2$ m, and it is zero everywhere else. What is the value of $k$?

The distance from the northern end of the rock is somewhere between 0 m and 2 m. The total area under the density curve is given by the area of a rectangle that is $k$ high and 2 m wide. This area under the curve is 1, therefore the height, $k$, of the density function must be 0.5 m$^{-1}$:

$$f_{pdf}(x) = \frac{1}{2 \text{ m}} = 0.5 \text{ m}^{-1} \text{ (for } 0 < x \leqslant 2 \text{ m; } f_{pdf}(x) = 0 \text{ otherwise).}$$

Now imagine that the predator starts to search along the shelf from the northern end at a constant speed of 0.005 m s$^{-1}$, until it finds the prey. We could define another continuous random variable $T$ that describes the possible durations of this search. The time taken for a search distance of 0 m would be 0 s, and the time taken for a search distance of 2 m would be 2 m/0.005 m s$^{-1}$ = 400 s. So the range of possible values for a duration $t$ sampled from $T$ is: 0 s $< t \leqslant$ 400 s.

We know that the probability density function of $D$ (distance to search required) is flat. Because the duration of search is simply determined by the distance searched (distance divided by speed), this means that the probability density function of $T$ must also be flat. As before, the total area under the curve must be 1, so the height of the curve must be

$$f_{pdf}(t) = \frac{1}{400 \text{ s}} = 0.0025 \text{ s}^{-1} \text{ (for } 0 \text{ s} < t \leqslant 400 \text{ s; } f_{pdf}(x) = 0 \text{ otherwise).}$$

As we can see from these examples, if the probability density function is constant across the whole possible range, the value of this constant must be the reciprocal of the width of this possible range (because the total area under the density function is one). We will return to this example as we go on, to show how understanding these random variables can give answers that (even for such a simple model) we do not have any other way of obtaining.

Before continuing, try answering the following question for yourself. If a disturbance that is sufficient to make the predator flee (abandoning any search) occurs after 2 minutes, what is the probability of the prey not being found?

To answer this question the easiest thing to do is to work out the probability that the prey *has* been found inside the interval. The probability that the prey has been found after 120 s is given by the area under $f_{pdf}(t)$ between $t = 0$ s and $t = 120$ s, i.e. 120 s $\times$ 0.0025 s$^{-1}$ = 30 %. The probability that it has not been found is 70 %.

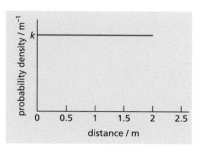

**Figure 10.12**
A uniform probability density describing the possible locations of prey hiding under a rock.

## Cumulative probability functions

Figures 10.9 and 10.10 show two different probability functions: one for a discrete random variable, and one for a continuous random variable. However, despite being the same shape, and explaining the same situation, the vertical scale on which they are plotted (probability and probability density) are different. Can we find some way of representing probability that will work for both discrete and continuous random variables?

We can plot the probability of sampling a value from a random variable $H$ (describing possible heights) that is no larger than a specific value of $x$. This value is known as the **cumulative probability** of the value $x$, and is written $P(H \leq x)$.

> The **cumulative distribution function $F_{cdf}(x)$** of a random variable is a function that describes the probability of sampling a value from the random variable that is no larger than $x$.
>
> $$F_{cdf}(x) = P(X \leq x). \hspace{2cm} \text{(EQ10.9)}$$

Note that $F_{cdf}(x)$ never decreases with increasing $x$, and that $F_{cdf}(-\infty) = 0$, $F_{cdf}(+\infty) = 1$.

For a discrete random variable, the cumulative distribution function represents the *sum of the heights of all bars up to and including $x$* as we move along the $x$ axis; for a continuous random variable, the cumulative distribution function represents the *area under the probability density function to the left of the point* along the $x$ axis. The cumulative distribution functions for both the discrete and continuous forms of the random variable $H$ are shown in Figure 10.13.

The cumulative distribution function is very simple to work out for a discrete random variable. If $x_0$ is the lowest possible value, then $F_{cdf}(x_0) = P(x_0)$. If $x_1$ is the next lowest, and so on, we simply add each probability onto the cumulative probability for the previous item, namely

**Figure 10.13**
Cumulative distribution functions (right) can represent discrete or continuous random variables. They show the probability of sampling a value less than or equal to the $x$ value, so they always increase from zero at the minimum value to one at the maximum value.

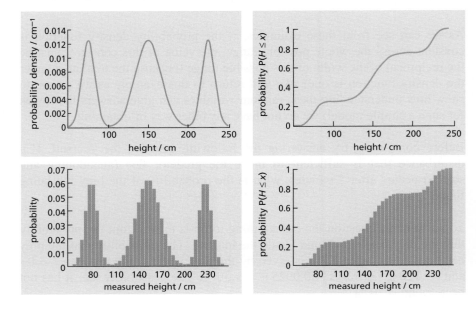

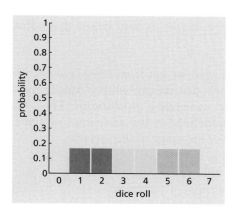
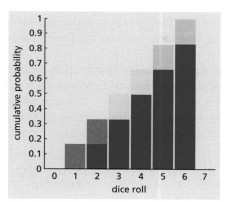

**Figure 10.14**
Constructing a cumulative distribution function (cdf). The left panel shows a discrete probability mass function describing the roll of a die. The right panel shows the corresponding cdf. The colors are used to illustrate that the cdf can be thought of as a running total of the probability values: adding the corresponding bar of the probability mass function onto the previous value.

$$F_{\mathrm{cdf}}(x_1) = \mathrm{P}(x_1) + F_{\mathrm{cdf}}(x_0),$$
$$F_{\mathrm{cdf}}(x_2) = \mathrm{P}(x_2) + F_{\mathrm{cdf}}(x_1) \dots.$$

To see how this works for discrete random variables, look at Figure 10.14. This shows the probability mass function and cumulative probability function for a discrete random variable describing possible dice rolls. Each roll value (1–6) has the same probability $\mathrm{P}(x) = 1/6$, thus all of the six bars are the same height. The bars are shaded different colors to show how the cumulative probability function is obtained: to find the probability that we sample 'at least $x$', just add up the exclusive probabilities for all values up to and including $x$, thus $F_{\mathrm{cdf}}(x) = \mathrm{P}(x) + F_{\mathrm{cdf}}(x-1)$.

For example, returning to the 'spotted fly' example, the probability mass function for the random variable $S$ (describing the number of spots on a fly) was:

$$f(1) = 0.5,$$
$$f(2) = 0.3,$$
$$f(3) = 0.2.$$

Therefore the cumulative distribution function for $S$ is:

$$F_{\mathrm{cdf}}(1) = \mathrm{P}(1) = 0.5,$$
$$F_{\mathrm{cdf}}(2) = \mathrm{P}(2) + F_{\mathrm{cdf}}(1) = 0.3 + 0.5 = 0.8,$$
$$F_{\mathrm{cdf}}(3) = \mathrm{P}(3) + F_{\mathrm{cdf}}(2) = 0.2 + 0.8 = 1.$$

It is not quite so simple for a *continuous* variable. However, if we think back to the calculus methods in Chapter 6, the way to obtain a cumulative probability function from a probability density function becomes obvious. The cumulative probability of sampling a value less than $x$ from a random variable $X$ is the summed area beneath the probability density function to the left of that point. In calculus terms, therefore, the cumulative distribution function is the *definite integral* of the density function from the lowest possible value up to $x$,

$$F_{\mathrm{cdf}}(x) = \mathrm{P}(X \leq x) = \int_{-\infty}^{x} f_{\mathrm{pdf}}(x)\, \mathrm{d}x. \qquad \text{(EQ10.10)}$$

We have used $-\infty$ as the lower boundary in this equation, as it is then always the correct formula. If there is a particular 'lowest possible value', then this

The probability density function, $f_{pdf}(x)$, is the slope (derivative) of the cumulative distribution function $F_{cdf}(x)$ of a continuous random variable.

implies that the area under the probability density is zero below that point, i.e. between $-\infty$ and the lowest value.

If the cumulative distribution function is what we get from *integration* of the probability density function, then it follows that the probability density must be what we get by *differentiating the cumulative probability*. Thus the probability density function $f_{pdf}(x)$ can be defined as the *derivative*, or slope, of the cumulative distribution function over the range of the variable.

$$f_{pdf}(x) = \frac{d}{dx} F_{cdf}(x) = \frac{d}{dx} P(X \leqslant x) \qquad \text{(EQ10.11)}$$

Thus what the height of a probability density function tells us is how likely we are to sample a value *near to* the point in question: the higher the density, the higher the rate of increase in probability of sampling in an interval as we increase the width of the interval.

What we already know about probability tells us a variety of things about the nature of the cumulative distribution function. We know that it must be zero from $-\infty$ to the lowest possible value, it must only increase or stay flat with increasing $x$, and it must be equal to one from the highest possible value to $+\infty$.

Cumulative distribution functions are often the most useful form of probability function, certainly when we wish to calculate probabilities for continuous random variables. Once we know the cumulative distribution function of a continuous variable, we can calculate the probability of sampling from any interval as the *difference between* two cumulative probabilities.

$$P(a < X \leqslant b) = P(X \leqslant b) - P(X \leqslant a) = F_{cdf}(b) - F_{cdf}(a). \qquad \text{(EQ10.12)}$$

If you cannot see why this is true, look back at Figure 10.14. Consider the probability of rolling a value in the range $2 < X \leqslant 4$, namely the probability of rolling a 3 or a 4. These are exclusive, so the probability of *either* is given by the sum of $P(X = 3)$ and $P(X = 4)$. So $P(2 < X \leqslant 4)$ is given by the sum of the heights of the two pink bars in the middle of the probability mass function. Looking at the cumulative distribution function, it should be obvious, therefore, that $P(2 < X \leqslant 4) = F_{cdf}(4) - F_{cdf}(2)$.

We can see how we can use EQ10.12 in our simple 'search' example for a continuous random variable. Recall that a prey hides from a predator beneath a shelf of rock and that the predator searches along the shelf until it finds the prey after $t$ seconds, where $0 \text{ s} \leqslant t < 400 \text{ s}$. The random variable $T$ describing possible values of $t$ has the probability density function:

$$f_{pdf}(t) = \frac{1}{possible\ range} = \frac{1}{400\ \text{s}} = 0.0025\ \text{s}^{-1}.$$

The cumulative distribution function at some time $t$ is given by the area under the curve from zero to $t$, that is, the *integral* of this function from the lowest possible value up to $t$.

$$F_{cdf}(t) = \int_0^t (f_{pdf}(t))\,dt = \int_0^t (0.0025\ \text{s}^{-1})\,dt = 0.0025\ \text{s}^{-1} \times t.$$

Knowing the cumulative distribution function means we can find the probability of the search duration in any given interval by the difference in $F_{cdf}$ between the two times. For example, the probability of a duration between 75 s and 125 s is given by

$$F_{cdf}(125) - F_{cdf}(75) = 0.0025 \times (125 - 75) = 0.125.$$

Of course, we do not need to start with a probability distribution to find the cumulative distribution function. We could have derived the cumulative distribution function directly by arguing as follows.

The cumulative probability for $T$ at time $t$ is the probability of sampling a duration less than or equal to $t$ seconds. To get a duration value less than $t$, the prey must have been found before that time. So we need the probability of 'having found' the prey within $t$ seconds. The prey has selected a position at random, so the probability of finding the prey within a given search duration $t$ seconds is equal to the proportion of the shelf that could be searched in that time. The cumulative distribution function is thus simply the proportion that could be searched in time $t$, namely:

$$F_{cdf}(t) = t \times 0.005\ \text{m s}^{-1}/2\ \text{m} = t \times 0.0025\ \text{s}^{-1}.$$

If we used this approach to find the cumulative distribution function, we could obtain the probability density function by differentiating the cumulative distribution function to give:

$$f_{pdf}(t) = \frac{d}{dt} F_{cdf}(t) = 0.0025\ \text{s}^{-1}.$$

Reassuringly, these answers are the same as the ones we saw above!

We can see here that the cumulative distribution function for a uniform variable is just as simple as the probability density: a straight line between zero (at the lowest possible value) and one (at the highest possible value). This makes sense: the probability density is equal to the slope of the cumulative distribution function, and is uniform throughout the possible range.

Given that cumulative probability functions are so useful, why do we bother with probability density functions at all? Well, when we define a random variable it is often easier to work out what the density function is than the cumulative distribution function. Also, when we want to visualize or describe the probability functions of a random variable, the shape of the density function shows which ranges or intervals are likely to be sampled much more clearly than does the cumulative function.

We have now covered all the theory we need to use probability theory to model real biological situations. Let us take stock of what we know:

- A random variable is a probability system where each possible outcome is a number. It has a set (or range) of possible values, along with a function that gives a probability associated with each value, or range of values.
- If the random variable is discrete, the probability of each possible value is given directly by a probability mass function. For a continuous random variable, because there are infinitely many values within any interval, the probability of an *exact* value is zero, and we calculate the probability of being within a certain range or interval.
- The total probability (the area under the probability density curve, or the total of the discrete probability values) is one.
- Any random variable (discrete or continuous) has a cumulative probability distribution.
- For a discrete random variable, the cumulative probability at a particular value is the sum of the heights of the bars up to and including the value, from the graph of the probability mass function.
- For a continuous random variable, the cumulative probability at a particular value is the area under the probability density curve up to and including the value.

## 10.4 Describing sampling distributions

Most biological measurements are well described by random variables, in that they do not have a single value: when we talk about the 'number of leaves on a tree in this forest', we are referring to many possible values (one value for each possible tree), each with an associated probability of actually being observed, or sampled. As we have seen in Section 10.3, we could therefore describe the measurement in terms of a random variable, whose probability function tells us how likely we are to sample any value (or range of values) if a single tree is picked at random.

We would say that such a measurement had a *sampling distribution*: certain values are likely, others less so. This is, of course, exactly the same thing as the *probability function* of the random variable that we might use to describe the set of possible alternative measurements. So, as we have already said, probability functions are useful in biology because they describe the sampling distribution of experimental measurements.

However, even when we can generate a random variable as a model of a situation, we might not want to express every piece of biological theory in terms of a graph of a sampling distribution for a measurement value. Rather, we might want to summarize a *typical* value for the possible measurements, and how *varied* the possible measurements are. One of the most useful things that knowledge of sampling distributions can give us is the ability to calculate this kind of summary in a formal, precise way.

The description of the typical measurement is called the *mean*, or expected value, of the random variable, and the description of the variability is called the *variance*. These *parameters* are essentially the same as the *statistics* we used to describe samples in Chapter 9.

## Mean, standard deviation, and variance

The mean of a random variable, just like the sample mean, tells us how large, on average, values sampled from a random variable are. It is formally known as the **expected value**, or **expectation**, of a random variable $X$, and is generally written $E[X]$. The expected value of a random variable is what we would expect to get if we took a very large number of samples from the random variable, and calculated their *sample* mean. Because the sample mean of a data set is used a great deal in data analysis, the mathematical analysis of expected values is key to many of the statistical techniques we will see in Chapter 11.

The expected value of a discrete random variable is easy to calculate. If we take a very large number ($N$) of samples, the proportion of them that we expect to have value $x$ is, of course, $P(x)$. So, for each value of $x$, we expect to have $N \times P(x)$ lots of $x$. So the contribution to the sum of all values in our large number of samples is $N \times P(x) \times x$. The expected value, or average, is simply this sum divided by $N$, so each value of $x$ contributes $x \times P(x)$ to the expected value:

$$E[X] = \sum x P(x). \tag{EQ10.13}$$

Returning to our 'spotted fly' example, the probability mass function for the random variable $S$ was: $f(1) = 0.5, f(2) = 0.3, f(3) = 0.2$. The expected value of $S$ can be found by using EQ10.13:

$$E[S] = \sum s P(s) = 1 \times 0.5 + 2 \times 0.3 + 3 \times 0.2 = 1.7.$$

The mean of a probability distribution is usually written $\mu$, and is simply another symbol for the expected value of the sampling distribution (the long-term average of sampled values). Thus the expected value of $S$ of 1.7 corresponds to a statement that the mean number of spots on flies in the population described by $S$ is 1.7.

$$\mu_X \equiv E[X]. \tag{EQ10.14}$$

It is not quite so simple to calculate the expected value for a continuous variable, but we essentially do the same thing using calculus. The probability of a value in each (infinitesimal) range is given by the area under the curve, namely the definite integral, so we multiply this by $x$ at each point:

> The mean or expected value of a probability distribution is the 'long-term average' of repeated samples from the distribution.

$$E[X] = \int_{-\infty}^{\infty} x f(x)\, dx. \tag{EQ10.15}$$

As an example, we can return to the random variable $T$ from Section 10.3, which described the possible durations of a predator's search along a 2 m shelf, at a search speed of 0.005 m s$^{-1}$. We showed above that if the target of the search (the prey) selected a place along the shelf at random, the probability density function for the search was $f_{\mathrm{pdf}}(t) = 0.0025$ s$^{-1}$.

Knowing this, we can now calculate the mean of the possible search durations by plugging this formula for $f_{\mathrm{pdf}}$ into EQ10.15 to obtain the expected value (mean) of the random variable:

$$\mu_T = E[T] = \int_{-\infty}^{\infty} t \times f_{\mathrm{pdf}}(t)\mathrm{d}t = \int_{0}^{400\ \mathrm{s}} t \times f_{\mathrm{pdf}}(t)\,\mathrm{d}t$$

$$= \int_{0}^{400\ \mathrm{s}} (t \times 0.0025\ \mathrm{s}^{-1})\,\mathrm{d}t$$

$$= \left[\frac{t^2}{2} \times 0.0025\ \mathrm{s}^{-1}\right]_{0}^{400\ \mathrm{s}}$$

$$= \left[\frac{(400\ \mathrm{s})^2}{2} \times 0.0025\ \mathrm{s}^{-1}\right] - [0]$$

$$= 200\ \mathrm{s}.$$

So, the mean duration of possible searches is 200 s.

Figure 10.15 shows three sampling distributions (probability density functions) for the heights from different populations of plants. In fact, the three populations are the three genotypes that make up the possibilities shown in Figures 10.9–10.11. The vertical scale has changed so that the area under *each* of these curves is one. All three distributions have the same overall shape, which is extremely common in real life, and in statistics. It is called the *normal distribution*, and we will discuss it in more detail in Section 10.5.

**Figure 10.15**
Three normally distributed populations (of heights) with different means and variances. Each has a similar distribution shape, and is scaled so that the area underneath **each** curve is 1.

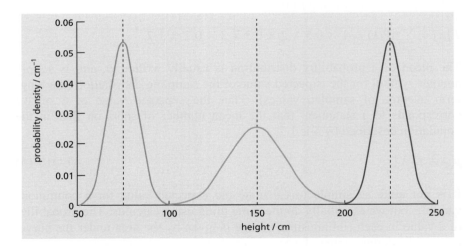

The expected values for these three sampling distributions are 75, 150, and 225 cm. We can see that a change in the expected value corresponds to a change in the 'location' of the sampling distribution on the *x* axis. However, it is also clear from this figure that knowing the mean of a distribution does not tell us everything. The middle of the three curves is more spread out than the others. We saw in Chapter 9 that the sample mean tells us only one property of a sample of data, and that another important property is the spread or dispersion.

This spread is usually measured in terms of the *variance* of the distribution, just like the sample variance we met in Chapter 9. The variance of a probability distribution is usually written $\sigma^2$, and defined as the expected value (average) of the squared difference between a sampled value and the mean,

$$\sigma^2 = E[(X - \mu)^2].$$  (EQ10.16)

A bit of algebra gives us an alternative, often more useful, way of writing this. It can be proved that this algebra 'works' for expected values, but we will not go into that here. Because $E[\mu] = \mu = E[X]$, and $E[\mu^2] = \mu^2$, we can derive the following:

$$\sigma^2 = E[(X - \mu)^2]$$

$$= E[X^2 - 2\mu X + \mu^2]$$

$$= E[X^2] - 2\mu E[X] + \mu^2,$$

$$\sigma^2 = E[X^2] - \mu^2.$$  (EQ10.17)

Here, $E[X^2]$ means the average of 'values sampled from $X$, then squared'. We can obtain this by substituting $x^2$ for $x$ in our equation for expected value. Thus for a discrete variable, $E[X^2] = \sum x^2 P(x)$, while for a continuous variable $E[X^2] = \int_{-\infty}^{\infty} x^2 f(x) \, dx$.

Because the variance is in squared units, we commonly use its square root, which is called the *standard deviation* ($\sigma$) of the probability distribution:

$$\sigma = \sqrt{E[(X - \mu)^2]}.$$  (EQ10.18)

The standard deviations for the three sampling distributions in Figure 10.15 are 8, 16, and 8 cm indicating that the heights are *more spread out* in the medium-sized population.

Let us return to our two examples of discrete and continuous random variables to show how we calculate the variance in practice.

The discrete random variable $S$, describing the possible number of spots on a fly capture, has probability mass function $f(1) = 0.5$, $f(2) = 0.3$, $f(3) = 0.2$, and an expected value is 1.7. The variance, from EQ10.17, is given by

$$\sigma_S^2 = E[S^2] - \mu^2 = \sum s^2 P(s) - \mu^2$$

$$= 1^2 \times 0.5 + 2^2 \times 0.3 + 3^2 \times 0.2 - 1.7^2$$

$$= 0.5 + 1.2 + 1.8 - 2.89 = 0.61.$$

The standard deviation is the square root of this value,

$$\sigma_S = \sqrt{\sigma_S^2} = \sqrt{0.61} = 0.781 \text{ (three significant figures)}.$$

The random variable $T$, describing possible search durations, has the probability density function $f_{pdf}(t) = 0.0025 \text{ s}^{-1}$ and expected value of 200 s. We can use EQ10.17 to find the variance as

$$\sigma_T^2 = E[T^2] - \mu^2 = \int_0^{400\text{ s}} t^2 \times f_{\text{pdf}}(t)\,dt - \mu^2$$

$$= \int_0^{400\text{ s}} (t^2 \times 0.0025\text{ s}^{-1})\,dt - \mu^2$$

$$= \left[\frac{t^3}{3} \times 0.0025\text{ s}^{-1}\right]_0^{400\text{ s}} - \mu^2$$

$$= \left[\frac{(400\text{ s})^3}{3} \times 0.0025\text{ s}^{-1}\right] - [0] - (200\text{ s})^2$$

$$= 13333.\dot{3}\text{ s}^2$$

and hence calculate the standard deviation to be

$$\sigma_T = \sqrt{\sigma_T^2} = \sqrt{13333.\dot{3}\text{ s}^2} = 115.47\text{ s} = 115\text{ s (three significant figures)}.$$

So, now we can describe a probability distribution in terms of two main parameters, the mean and standard deviation (or variance). What do these values tell us about the random variable? We can see what these values tell us by relating the three distributions shown in Figure 10.15 to their corresponding parameter values (given above):

- For a symmetrical probability distribution, the mean represents the center of the distribution. For the normal distribution (see below) it is also the *most likely* value to be sampled.
- The standard deviation reflects the spread of the distribution, so it gives us an indication of how *far from the mean* a value is likely to be.

However, as a final example, let us consider a woman who is pregnant with her first child. We can easily define a discrete random variable $M$ that describes the number of male children she will have. Obviously, the possible values are 0 and 1, and the probability mass function for this variable is $P(M = 0) = 0.5$, $P(M = 1) = 0.5$.

If we calculate the expected value of the random variable, using EQ10.13, we will obviously get a value of 0.5. However, a child halfway between male and female, or half a male child, is definitely not what the proud parents are 'expecting'!

This illustrates an important point about the expected value: it does not simply mean the 'thing we expect'. In fact, for a discrete random variable (such as sex of a fetus), it is often a value that is *impossible* to sample. Even when the variable is continuous such that any value is possible, calculating the expected value is not always the same thing as working out the 'most likely value'. This is only true when the peak of the probability density function happens to be at the expected value, which is true for the normal distribution.

It may help you to remember that the 'expected value' for a lottery ticket payout in the UK lottery is around 45 pence. Of course, if anyone really *expected* to win 45 pence every time, they would never pay £1 to enter!

# 10.5 The normal distribution

The three sampling distributions in Figure 10.15 all have the same shape. As mentioned above, this distribution shape is called the *normal distribution*. Anything that is described by a random variable with a sampling distribution of this shape is said to be *normally distributed*.

The probability density functions of normally distributed variables are symmetrical and bell-shaped. A particular distribution is completely described by the two parameters we met above, namely the mean and variance (where the middle of the distribution is, and how spread out the distribution is around its center).

Generally, what we are interested in is the probability associated with sampling from a normally distributed variable within a given range. To do this, we need to know the cumulative probability function, that is the area underneath our bell-shaped probability density curve. The probability density and cumulative distribution for a normal distribution are shown in Figure 10.16.

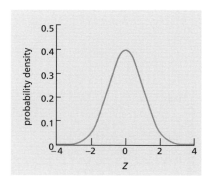

Unfortunately, the general formulae for the normal probability distribution functions are rather complicated. For example, the probability density function for a variable with mean $\mu$ and standard deviation $\sigma$ is:

$$f_{\text{pdf}}(x) = \frac{e^{-\left((x-\mu)^2/2\sigma^2\right)}}{\sigma\sqrt{2\pi}}. \qquad \text{(EQ10.19)}$$

We really do not want to use this formula! So, when we are interested in a normally distributed variable, the first thing we do is to simplify everything by rescaling, or *standardizing*, our variable. To do this we scale it so that it has a mean of zero, and a standard deviation of one.

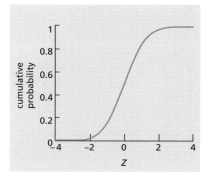

Scaling a variable to have a mean of zero and a standard deviation of one is actually very simple, once we know the mean and standard deviation of some random variable $X$. For each value, we subtract the expected value (so it is expressed as a deviation from the mean), and then divide by the standard deviation. A standardized value is usually written as (lower case) $z$, and known as a '$z$ statistic':

$$z_x = \frac{x - \mu_X}{\sigma_X}. \qquad \text{(EQ10.20)}$$

**Figure 10.16**
The probability density and cumulative distribution functions for the standard normal distribution. The cumulative distribution function is S-shaped, and the density function is bell-shaped and symmetrical.

The normal distribution with these parameters is called the *standard normal distribution*, and is usually denoted with the letter $Z$.

Why does this rescaling help us? Well, for one thing, using the standard normal curve greatly simplifies the equations for the probability density and cumulative distribution functions for the distribution. The probability functions of the standardized normal distribution, $Z$, are:

$$f_{\text{pdf}}(z) = \frac{1}{\sqrt{2\pi}} e^{-\left(z^2/2\right)}, \tag{EQ10.21}$$

$$F_{\text{cdf}}(z) = \Phi(z) = \frac{1}{\sqrt{2\pi}} \int_{-\infty}^{z} e^{-(x^2/2)} \mathrm{d}x. \tag{EQ10.22}$$

Although still not the kind of thing we want to spend too much time with, these equations are at least a little bit simpler. However, the real benefit of using the standard normal distribution is that other people have already worked out the associated probability functions for us! In fact, these are the functions shown in Figure 10.16.

If a random variable $X$ is normally distributed with a standard deviation $\sigma$ and mean $\mu$, any value $x$ for that variable will be associated with the same cumulative probability (and probability density) as the corresponding $z$ statistic. So all we need to answer questions about normally distributed variables is the cumulative distribution function of $Z$. This distribution is so useful, it even has its own symbol, $\Phi(z)$. The cumulative distribution function is given in Appendix 4, and can be obtained from a computer. In Microsoft Excel®, the worksheet function '=NORMSDIST(z)' returns the cumulative probability $\Phi(z)$.

To see how this works, let us tackle a probability problem based on our plant distributions. The heights of homozygous 'tall' plants are normally distributed with a mean of 225 cm, and a standard deviation of 8 cm. The heights of heterozygous 'medium' plants are normally distributed with a mean of 150 cm, and a standard deviation of 16 cm. What is the probability that a plant from each population will be greater than 2 m tall?

The first thing we need to do is to convert 2 m to a $z$ value for each population. Heterozygous plants have a mean of 150 cm, and standard deviation of 16 cm. Using EQ10.20, we see that a value of 200 cm is equivalent to a $z$ value of

$$z = \frac{x - \mu}{\sigma} = \frac{200 - 150}{16} = \frac{50}{16} = 3.12.$$

From Appendix 4 (or by using our computer), we find that the cumulative probability associated with this $z$ value is 0.9994. Thus 99.94 % of heterozygous plants are *less than or equal to* 2 m tall. This is the converse probability of the one we need, so there is 0.06 % probability of a heterozygous plant growing to more than 2 m tall.

Similarly, to calculate the probability that a homozygous 'tall' plant will be less than 2 m tall, we must first convert 2 m into a $z$ value for the homozygous tall plants. These have a mean of 225 cm, and standard deviation of 8 cm; therefore the $z$ value we need is

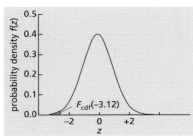

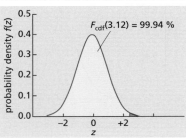

**Figure 10.17**
Using the cumulative normal distribution function. The area below a $z$ value of −3.12 corresponds to the area above a $z$ value of +3.12, which can be obtained from a table. This type of sketch is always useful when using the cumulative normal distribution.

$$z = \frac{x - \mu}{\sigma} = \frac{200 - 225}{8} = \frac{-25}{8} = -3.12.$$

Appendix 4 does not give the cumulative probability associated with negative $z$ values. However, the cumulative probability for $z = +3.12$ is 0.9994. We know that the normal distribution is symmetrical, so if 99.94 % of $z$ values are *less than or equal to* 3.12, then 99.94 % of $z$ values must be *greater than or equal to* $-3.12$. So there is a 99.94 % probability of a homozygous 'tall' plant growing to more than 2 m tall. This relationship is most obvious graphically, as shown in Figure 10.17.

As we can see here, it is not always obvious what is meant by the cumulative probability associated with each $z$ value. We suggest that you briefly sketch the sampling distribution when you are using the cumulative normal distribution to obtain a probability (for example using Appendix 4), as this makes it much easier to visualize how to use the cumulative probability associated with a $z$ value.

A final word of caution about how to use the standard normal distribution. It is perfectly reasonable (and common) to standardize the values from any random variable if we know the mean and standard deviation of the variable. Standardized values, namely values expressed as 'standard deviations above the mean', are generally written as $z$ values. However, it is important to remember that standardizing the random variable does not change the shape of its probability function: it does not make the variable any more 'normally distributed'. The probabilities associated with $z$ values are only useful for variables we know to follow the normal distribution.

## Box 10.3 The binomial distribution

There are many situations in which we may be interested in the proportion of times that an event happens: the proportion of individuals within a population who catch a certain disease, or the number of times a rat chooses the correct lever, or the number of offspring who show a recessive trait. These situations can all be modeled mathematically in a similar way: a series of *independent trials* that each has the same probability $\pi$ of being counted as '*successful*' (an outcome for which the event in question is true).

This simple model (called a Bernoulli model, after the 17th Century Swiss mathematician Jacob Bernoulli) allows us to calculate the probability mass function for a random variable $B(n, \pi)$ that describes the possibilities for the 'total number of successes' we could see after $n$ trials, if each has a probability $\pi$ of success. (The probability of success on one trial is usually represented by the Greek letter $\pi$, rather than $p$, because $p$ is used to denote an observed proportion. Clearly, $\pi$ in this context has nothing to do

with the ratio of the circumference and diameter of a circle.)

So how do we work out this random variable? The set of possible values is obvious: we cannot get fewer than zero, or more than $n$, successes. The probability function can also be calculated quite simply for a given value of $n$ and $\pi$. If every trial has the probability $\pi$ of being a success, it follows that each trial (each day, individual, or offspring) also has the probability $(1 - \pi)$ of not being counted as a success. You should now be able to work out for yourself that the probability of a sequence of $x$ successes (each with probability $\pi$) followed by $n - x$ failures (with probability $(1 - \pi)$) is given by: $\pi^x \times (1 - \pi)^{n-x}$.

Because the trials are independent, the probability of *any* sequence containing the same number of successes is the same. In other words, the probability of the sequence {'success', 'success', 'failure'} is the same as the probability of {'success', 'failure', 'success'} and {'failure', 'success',

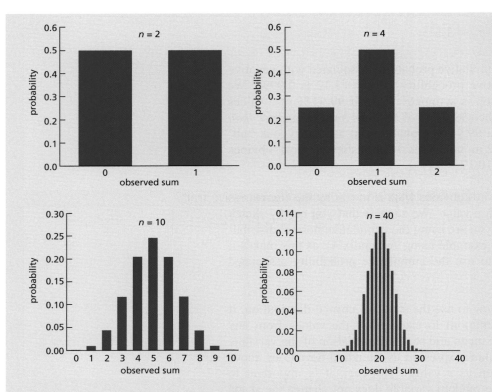

**Figure 1**
Examples of the binomial distribution for various values of *n*, with $\pi = 0.5$.

'success'}. So to complete the probability function we just need to know how many of the possible sequences of *n* trials will give us *x* successes. This number, known as the binomial coefficient, is given by:

$$\frac{n!}{x!(n-x)!}$$

*n*! ('*n* factorial') means $n \times (n-1) \times (n-2) \times \ldots \times 2 \times 1$; thus $5! = 5 \times 4 \times 3 \times 2 \times 1 = 120$. We also define $0! = 1$.

The binomial coefficient tells us how many sequences give *x* successes. As the sequences are exclusive alternatives, we can simply multiply this number by the probability associated with each one to get the total probability of *x*:

$$P(x) = \frac{n!}{x!(n-x)!} \times \pi^x \times (1-\pi)^{n-x} \quad 0 \leqslant x \leqslant n$$

$P(x) = 0$ otherwise.

This probability function is called the binomial distribution. It has a mean of $n \times \pi$ and a variance of $n \times (\pi - \pi^2)$. Some examples of binomial distributions for $\pi = 0.5$ are shown in Figure 1. These graphs show the possible numbers of successes you would expect when running various numbers of trials (for example, possible numbers of heads when tossing a coin). From the $n = 40$ graph, we see that we are likely to get around 20 heads in total, but very unlikely to get more than about 30 heads in total.

Binomial probabilities can be obtained from a computer package such as Microsoft Excel®. The worksheet function '=BINOMDIST(*x*, *n*, $\pi$, 0)' gives the probability P(*x*) for exactly *x* successes in *n* trials with probability $\pi$ of success on each. Similarly, '=BINOMDIST(*x*, *n*, $\pi$, 1)' gives the corresponding cumulative probability; that is, the probability of '*x* or fewer' successes.

Looking at Figure 1, we can see that as we increase the number of trials, the binomial distribution graphs get closer to the shape of the normal curve. In fact, when we look in detail at the binomial distributions for a variety of values of *n* and $\pi$, we always find the shape is very close to the normal curve, *as long as both $n \times \pi$ and $n \times (1 - \pi)$* are above five. Because we know the mean and variance of the binomial distribution, we can express a possible number of successes as a standardized normal variable, or *z* value, and use this value to obtain a quick approximation of the cumulative binomial probability for our value of *x*.

# Presenting Your Work

## QUESTION A

In a large population, 49 % of the organisms express a phenotype known to be recessive; 51 % show the phenotype of the dominant allele. A mother, who expresses the recessive trait, has mated with a male who expresses the dominant trait. Assuming that no other alleles of this gene exist in the population, and that breeding is random, answer the following questions:

(a) What proportion of the population is likely to be heterozygous?

(b) What is the probability that the male is heterozygous?

(c) What is the probability that their first offspring will express the recessive trait?

(d) If the first offspring does not express the trait, what is the probability that the second offspring does?

(a) Let $R$ = proportion of genes recessive.

$R^2 = 49\% \Rightarrow P(R) = 70\%$

$P(\text{homozygous dominant}) = (1 - R)^2 = 0.09$

$P(\text{heterozygous}) = 100\% - 49\% - 9\% = 42\%$

(b) $P(\text{heterozygous} \mid \text{dominant}) = P(\text{heterozygous} \wedge \text{dominant})/P(\text{dominant})$

$= 0.42/0.51 = 82.35\%$

(c) $P(\text{recessive}) = P(\text{paternal gamete recessive}) = 0.5 \times P(\text{heterozygous}) = 41.17\%$

(d) Father could be homozygous dominant, or heterozygous.

Need to know P(father heterozygous | 1st dominant) so can use Bayes' rule

P(father heterozygous | 1st dominant)

= P(father heterozygous) × P(1st dominant | father heterozygous)/P(1st dominant)

P(father heterozygous) = 0.8235

P(1st dominant | father heterozygous) = 0.5

P(1st dominant) = P(1st dominant ∧ father heterozygous) + P(1st dominant ∧ father homozygous)

= 0.5 × 0.8235 + 1.0 × 0.1765 = 0.5882

P(father heterozygous | 1st dominant)

= P(father heterozygous) × P(1st dominant | father heterozygous)/P(1st dominant)

= 0.8235 × 0.5/0.5882 = 70%

## QUESTION B

A species of fly lays eggs in domestic fruit bowls, which leads to fruit being damaged. A bowl contains three items of fruit, and each piece has an independent 10 % chance of being damaged in this way. What are the possible numbers of damaged fruit, and what are the probabilities of each. What is the expected value for the number of damaged fruit?

*Range of possible values*

Any value between 0 and 3 damaged fruit is possible. Possible values are 0, 1, 2, or 3.

*Probabilities of each value*

Let events A, B, C be damage to each of the three fruits. $P(A) = P(B) = P(C) = 0.1$

Probabilities are independent, we simply multiply them together.

$P(3) = P(A)^3 = (0.1)^3 = 0.001$

$P(0) = P(\sim A)^3 = (0.9)^3 = 0.729$

Three exclusive alternatives for 1 damaged (exclusive $\Rightarrow$ add probabilities):

$P(A \wedge \sim B \wedge \sim C) = 0.1 \times 0.9 \times 0.9 = 0.081$

$P(\sim A \wedge B \wedge \sim C) = 0.9 \times 0.1 \times 0.9 = 0.081$

$P(A \wedge \sim B \wedge \sim C) = 0.9 \times 0.9 \times 0.1 = 0.081$

$\therefore P(1) = 3 \times 0.081 = 0.243$

$P(2) = 1 - (P(0) + P(1) + P(3)) = 1 - 0.973 = 0.027$

*Expected value*

$E[\# \text{ damaged}] = \Sigma x P(x) = 0 + 1 \times 0.243 + 2 \times 0.027 + 3 \times 0.001 = 0.3$

## QUESTION C

The height of a species of plant grown from seed in the laboratory is normally distributed with a mean of 136 cm, and a standard deviation of 12 cm.

(a) What proportion of these plants will be greater than 150 cm tall?

(b) What is the probability of a randomly sampled plant being more than 120 cm tall?

The seedlings are thinned, such that all the plants less than 120 cm tall are now removed, and all the others remain.

(c) What is the probability that one of the remaining plants, selected at random, will be more than 150 cm tall?

(a) We need to convert plant heights from their distribution to the standardized normal distribution (that is, to z values). 150 cm represents a z value of $\frac{150-136}{12} = 1\frac{2}{12} = 1.1\dot{6}$.

From Appendix 4, we find that the proportion of the normal distribution below $z = 1.17$ is 0.879.

Thus the proportion above 1.17 is 0.121, namely 12.1% of the plants will be greater than 150 cm.

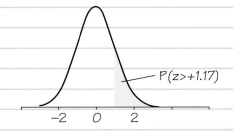

P(z>+1.17)

(b) Similarly, 120 cm corresponds to a z value of $-16/12 = -1.33$. Appendix 4 shows that the proportion of the standardized normal distribution lower than +1.33 is 0.908. We know that the standardized normal curve is symmetrical around zero, so this is equal to the proportion *greater* than $-1.33$. Thus the probability of a plant being more than 120 cm tall is 0.908.

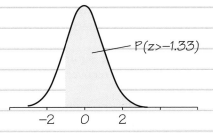

P(z>-1.33)

(c) The probability of a remaining plant being greater than 150 cm is essentially the conditional probability of a plant being over 150 cm tall *given that it is* more than 120 cm tall. From the definition of conditional probability, P(> 150 cm | > 120 cm) = P(> 150 cm and > 120 cm)/P(> 120 cm). In this case, P(> 150 cm and > 120 cm) is the same as P(> 150 cm).

The probability of a surviving plant being taller than 150 cm is 0.879/0.908 = 0.968.

## End of Chapter Questions

(Answers to questions can be found at the end of the book)

### Basic

**1.** A computer picks a whole number, at random, between 1 and 9 (inclusive). What is the probability it picks an even number?

**2.** A box contains 4 red counters and 1 white counter. Four counters, chosen at random, are removed. What is the probability that one of the counters removed is white?

**3.** A regular deck of 52 cards contains 4 aces and 4 kings. What is the probability of drawing a card from a shuffled deck that is *not* an ace or a king?

**4.** A card is drawn from a pack at random. Consider the events 'the card is an ace', 'the card is a king'. Are these events (a) exhaustive, (b) independent, (c) exclusive?

**5.** Construct a probability tree to describe the possible outcomes of three coin tosses.

**6.** In a game, I toss a coin up to three times, stopping as soon as I get a 'head'. Draw a probability tree to describe the possible sequences.

**7.** Half the cards in a deck are red, including two of the kings. What is the probability of drawing a card that is *either* a king or a red card (or both)?

**8.** Draw the probability graph describing the possible outcomes of 'suit' when drawing from a regular pack of cards.

**9.** Write the probability function of a random variable describing the possible 'number of spots' observed on a single roll of a standard six-sided die.

**10.** What are the mean and variance of the possible rolls of a fair six-sided die?

### Intermediate

**11.** The mass of a species of shrew is approximately normally distributed with a mean of 10 g and a standard deviation of 5 g.
(a) What is the probability that a random selection of 9 shrews has a total mass of more than 100 g?
(b) What is the probability that a single $n = 10$ sample has a mean greater than 12 g?

**12.** IQ scores are standardized to a normal distribution with a mean of 100 and a standard deviation of 15. What proportion of the population have an IQ of 100 or higher?

**13.** A male chimpanzee makes sexual advances to 12 females. If he has an (independent) 10 % probability of mating on each occasion, what is the probability that he mates at least once?

**14.** The sampling distribution of length in a particular species of insect is normally distributed with a mean of 12 cm and a standard deviation of 3 cm.
(a) What is the probability of an insect being less than 11 cm long?
(b) What is the probability of an insect being between 10 cm and 12 cm long?
(c) What is the probability of an insect that is between 10 cm and 12 cm long being less than 11 cm long?

**15.** A wild species of plant has four different types: 40 % have yellow flowers, 32 % have red flowers, 18 % have blue flowers, and the rest have white flowers.
(a) What is the probability that the flowers on a randomly selected plant will be blue or red?
(b) What is the probability that a colored (non-white) flower is blue?

**16.** A squirrel caches nuts in two locations by a certain tree. He has a 56 % chance of remembering the location of the first cache, and a 25 % chance of recalling the location of the second.
(a) What is the probability he will recover both caches?
(b) What is the probability that the squirrel will recover at least one cache?
(c) What is the probability that the squirrel will have recovered the first cache, if he recovers only one of the two caches?
(d) What is the expected value for the number of caches the squirrel will recover?

**17.** A rat is placed in a 'radial arm maze', which has eight alternative 'arms' leading out from the center. One of these arms contains food. On each trial, the rat is placed in the center and allowed to explore one arm only. Once the rat finds the food, or has completed four trials unsuccessfully, the experiment is over. The rat has a good memory and never explores the same arm twice.

(a) Using a probability tree diagram, calculate the probability that the rat will find the food on the first trial.
(b) What are the probabilities of the rat finding food on the second, third, and fourth trials?
(c) What is the probability of the rat failing to find the food after four attempts?

**18.** A rat with a very poor memory is placed in an eight-arm maze for a series of up to four trials, as described in the previous question. Owing to its lack of memory, the rat is equally likely to select any arm on each trial (whether or not it was previously explored). Calculate the probabilities of the rat finding the food after each possible number of trials, or failing altogether.

**19.** What are the expected values for the number of unsuccessful trials for rats in the eight-arm maze task described in (a) question 17, (b) question 18?

## Advanced

**20.** A random variable describes the time to the next event in radioactive decay. The probability function of this variable (known as an exponential probability distribution) is described by a function of the form $f(t) = Ae^{-3t}$, where $t \geq 0$. Determine the value of $A$.

**21.** A continuous random variable $X$ has a cumulative distribution function $F_{cdf}(X \leq x) = x^2$ over the interval $x = 0$ to $x = 1$.
(a) What is the probability density function for the variable over this range?
(b) Which value has the highest probability density?
(c) What are the mean and variance of the probability distribution?
(d) What is the probability of sampling a value above the mean?
(e) What is the probability of sampling a value more than one standard deviation away from the mean?

**22.** A female fish searches the bed of a river for food. It takes her 15 minutes to cover an area of 20 m$^2$. Assume that there are two pieces of food (A and B) independently randomly positioned in this area, and that she eats the first piece found without noticeably changing the rate of search.
(a) What is the cumulative probability that she will have found food A after $t$ seconds?
(b) What is the expected time to find piece A?
(c) What is the cumulative probability that she will have found food B after $t$ seconds?

(d) What is the cumulative probability that she will have found *both* pieces of food after $t$ seconds?
(e) What is the expected time to find both pieces?
(f) What is the cumulative probability that she will have found *either* piece of food after $t$ seconds?
(g) What is the expected time to find either piece?

**23.** A fruit falling from an apple tree may roll a little when it hits the ground. The cumulative probability of each apple coming to rest within a distance $x$ meters of the edge of the trunk is approximately given by: $P(x) = 0.5x^{0.5}$. According to this model:
(a) What is the furthest distance an apple can roll?
(b) What is the probability density function describing the sampling distribution of possible distances?
(c) What is the average distance rolled?

**24.** A species of water lily has circular leaves, whose radii follow a log-normal distribution, such that a logarithmic transformation of the radius data (for example taking the natural logarithm of each radius divided by 1 cm), produces a normal distribution.

The mean of the transformed distribution corresponds to a lily radius of 4 cm. A value one standard deviation above the mean of the distribution corresponds to a radius of 6 cm.
(a) Show that the distribution of log-transformed radii has mean $\mu_x = 1.39$, and variance $\sigma_x^2 = 0.164$, to three significant figures.
(b) Show that the distribution of log-transformed *areas* is normal, with mean $\mu_x = 3.92$, and variance $\sigma_x^2 = 0.658$, to three significant figures.
(c) Show that around 81 % of leaves have an area of less than 25 cm$^2$.
(d) What is the distribution of differences in logarithm of leaf area for two randomly selected leaves?
(e) Using the answer to (d), show that the proportion of leaf pairs for which the larger leaf is more than twice the area of the smaller is around 55 %.

# CHAPTER 11

# Statistical Inference

When we perform a biology experiment, we are attempting to measure things and make theories about the living world.

A plant scientist may be interested in whether fertilizer treatments influence resistance to disease in a species of crop plant, a neuroscientist in the way human brains respond to threatening stimuli, and a pharmacologist in the degree to which a certain drug binds to a particular type of receptor protein. The type of experiment each scientist would run is, of course, very different. Perhaps the plant scientist will study growing plants using a variety of fertilizer treatments; the neuroscientist will present pictures of threatening faces to volunteers in a functional MRI scanner; whereas the pharmacologist will carefully measure drug binding at a variety of doses in the laboratory.

**Figure 11.2**
A pharmacologist carefully preparing a variety of doses of a drug.

**Figure 11.1**
A lettuce field in Arizona. A plant scientist may study growing plants using a variety of fertilizer treatments. Image courtesy of Cronkite News Service. Photo Hailey Gindlesperger.

Although these experiments are very different in lots of ways, they have this important thing in common: it is impossible for us to measure everything (every plant, every person, or every dose). Instead, we take a sample of possible measurements, and then try to draw general conclusions from this sample. This chapter describes how to decide when we may be justified in drawing conclusions from these data, and how confident we should be in these conclusions.

# 11.1 Inferential statistics

One of the most well-known quotes about statistics is a phrase made famous by Mark Twain (although he attributed it to British Prime Minister Benjamin Disraeli): 'There are three types of lies: lies, damn lies, and statistics'. Of course, correctly calculated statistics are not really untrue: it is the conclusions that these statistics seem to support which may be lies.

What we will discuss in this chapter is how to draw conclusions that are not lies, or at least not likely to be lies, based on experimental data.

We will look first at how we try to draw conclusions that go beyond our data: that is, generalize from our results to the wider world. In Chapter 9 we looked at describing and summarizing a set of data. These techniques allow us to tell people what results we get when we run an experiment. Description is not the only thing we want to do with our experimental results. It might be nice to presume that anything that is true of our summarized sample data will be true in general, but we know from the probability theory discussed in Chapter 10 that this will not generally be the case. Therefore describing results accurately is only the first small step in a statistical analysis.

Imagine that we are testing the effects of a new compound, and wish to know the heart rate of those who have taken the compound (perhaps the compound is designed to influence the effect of the hormone adrenalin, which increases heart rate by acting on $\beta$-receptors). We take a sample of 16 participants who have taken the new drug and, by electrocardiography, we measure the heart rate of each individual.

We know how to describe a sample of data in terms of descriptive statistics (the sample mean, sample variance, etc.). So we know how to calculate these and say exactly what average heart rate was sampled. It turns out that the heart rates in our sample have a range 60–90 beats per minute (b.p.m.), with a sample mean of 75 b.p.m., and a sample standard deviation of 5.0 b.p.m.

Now, those reading the research (a drug company, or a doctor) are not only interested in the results of *what heart rates* we observed in our participants. They will also want to know what heart rates they should expect to observe from *people in general* who take the drug. Of course, we do not know this, as we have only measured the heart rates of a sample of 16 people! However, we could use a combination of our experimental data and probability theory to improve any guess about what would happen to 'people in general'. This is really what science, and statistics, is for: trying to predict what will happen based on what we have found in an experiment.

It should be obvious that we cannot simply conclude that any individual who takes the drug will have a heart rate of exactly 75 b.p.m. (there was a range of 60–90 b.p.m. in our sample). However, what can we conclude? That *on average* the heart rate will be 75 b.p.m.? That *everyone* will have a heart rate within the range 60–90 b.p.m. after taking the drug?

These conclusions are also not justified, because we know that individuals, and groups of individuals, are not all exactly alike. We saw in Chapter 9 that the properties of different samples tend to be a little different, because of sampling variation. A better conclusion might be to say that the average heart rate of people in general is more likely to be in the range of 70–80 b.p.m. than it is to be in another range such as 50–60 b.p.m. This conclusion seems more justified, but how confident should we be that the average heart rate does fall in that range? Alternatively, what range of values should we quote in order to be 95 % confident that we have a range that contains the true mean?

Answering this kind of question requires us to combine the techniques that we have met in the last two chapters, to produce what are known as **inferential statistics**. These techniques give us a basis for drawing conclusions, or inferences, from our data, and for measuring how confident we should be in these conclusions.

We saw in Chapter 10 that most things we can measure will not give us exactly the same answer every time, and thus our measurements are variable quantities. Mathematically, we can describe a measurement process as sampling from a random variable: if our measurements are not the same value each time, there must be a set of possible values we could measure, along with a corresponding set of probabilities. It is common to refer to this random variable (the set of all possible measurements) as the **underlying population** of the sample. The mathematical notion of an underlying population of possible measurements is thus similar to the idea of a population of organisms.

When we run an experiment we obtain a sample, that is, a randomly selected subset of the possible measurements. In Chapter 9 we saw how to describe the numbers in a sample. What we would like to do now is to use this sample to draw conclusions about the population of 'all possible measurements'. The aim of inferential statistics is therefore to draw conclusions about the possible values of the *population* properties (*parameters*) that might plausibly have given us our *sample* properties (*statistics*).

The techniques that help us to make decisions based upon estimates of the parameters of the underlying population are called **parametric** techniques. There are other techniques, known as **non-parametric** techniques, that try to help us make inferences when we cannot estimate these parameters (for example, if we cannot treat our data as *interval data*, and thus cannot calculate a sample mean).

Before we go on, it is worth stopping to make a point about notation. There is a simple convention that is followed throughout almost all disciplines in science and mathematics: population parameters (that we are estimating, or drawing conclusions about) are written as Greek characters (such as $\sigma^2$ for a variance, and $\pi$ for a proportion). Sample statistics (that we calculate directly from experimental data) are written as roman (ordinary alphabet) characters (such as $s^2$ for a sample variance, and $p$ for a proportion).

## Box 11.1 Notation for sample and population statistics

We need to distinguish between the descriptive statistics that characterize our sample (which we know), and the corresponding parameters that characterize the population (to which we wish to generalize). To help keep this distinction clear, we use Greek characters to denote population parameters, as opposed to normal (or roman) characters for sample statistics.

Thus the mean of a population or variable is denoted $\mu$ (the Greek letter mu), the standard deviation of the population $\sigma$ (the Greek letter sigma), and the variance of a population $\sigma^2$.

The corresponding properties of a sample are: $\bar{x}$, $s_x$, $s_x^2$ (the sample mean, sample standard deviation, and sample variance, respectively).

Sometimes, we might use the sample statistics as our 'best guess' or point estimate of a population parameter. In such a case, we place a 'hat' symbol above the symbol for the parameter we are estimating.

**Table 1 Comparison of symbols used for the properties of an underlying population, and for statistics calculated from data**

| population parameter | symbol | sample statistic | symbol |
|---|---|---|---|
| mean | $\mu$ | sample mean | $\bar{x}$ |
| variance | $\sigma^2$ | sample variance | $s_x^2$ |
| standard deviation | $\sigma$ | sample standard deviation | $s_x$ |
| median | $\nu$ | sample median | $\tilde{x}$ |
| covariance | $\sigma_{xy}$ | sample covariance | $s_{xy}$ |
| correlation | $\rho_{xy}$ | sample correlation | $r_{xy}$ |
| proportion | $\pi$ | proportion in sample | $p$ |

For example, if we write $\hat{\mu}_X = \bar{x}$ this tells us that the unknown parameter ($\mu_X$) of a population is being estimated by the mean of a sample ($\bar{x}$).

When it is the population that we are really interested in (which is almost always the case), the sample statistics we choose are those that tell us something about the properties of the population. We mentioned in Chapter 9 that certain descriptive statistics – the sample mean and sample variance – can give us a good estimate of the corresponding properties of the underlying population of possible measurements.

# 11.2 Interval estimates and confidence intervals

We know that the sample mean and sample variance are good estimates of the underlying population parameters. However, although the value of the sample mean is a good estimate, it will in general not be *exactly* equal to the population mean; likewise the sample variance will not be exactly equal to the population variance. Thus although these statistics give us an *estimated value* of the population parameter, this type of estimate does not allow us to draw quantitative *conclusions* about such parameters. It is obvious that we are not justified in stating that 'we conclude that the population mean is exactly 75 b.p.m.' from our experiment on heart rate. Furthermore, recalling the nature of a probability distribution function for a continuous variable (see Chapter 10), we cannot improve the situation by giving a probability $p$ such that $p = $ 'the probability that the population mean is exactly 75 b.p.m.'.

There is something useful we could say, however: 'we can be confident that the population mean lies between 70 and 80'. This is known as an **interval estimate**. By comparison, a precise estimated value such as the sample mean is a **point estimate**. We obtain an interval estimate by *working backward* from what we know about sampling.

## Confidence intervals when the variance is known

Let us take a very general case: we have a normally distributed population, $X$, for which we know the mean ($\mu$) and standard deviation ($\sigma$). We know from the cumulative probability function of the normal distribution that only 2.5 % of normally distributed values are more than 1.96 standard deviations above the mean (that is, only 2.5 % correspond to $z$ statistics larger than +1.96). The symmetry of the normal curve tells us that only 2.5 % of normally distributed values are lower than 1.96 standard deviations *below* the mean.

Thus when we sample a value, $x$, from this normal distribution, there is a 95 % probability that it is not more than 1.96 standard deviations away from the mean. Putting this mathematically, there is a 95 % probability that:

$$\mu - 1.96\sigma \leqslant x \leqslant \mu + 1.96\sigma \tag{EQ11.1}$$

or (equivalently)

$$x - 1.96\sigma \leqslant \mu \leqslant x + 1.96\sigma. \tag{EQ11.2}$$

There is a 95 % probability that a particular sampled value of $x$ (sampled from $X$) will fall within $1.96\sigma$ of $\mu$.

This is where we 'work backward' to use the theory to generate interval estimates. We can re-express this relationship in the following form:

$$\mu = x \pm 1.96\sigma. \tag{EQ11.3}$$

This expression (it is not strictly an equation, as the right-hand side is expressing a range of values) is used as shorthand for the double inequality in EQ11.2. The right-hand side describes the upper and lower limits of the interval estimate of $\mu$ (based on the point estimate, $x$), and the equals sign is taken as meaning 'lies between'.

How would we use this expression? Often, we do not know the value of the population parameter $\mu$. Instead, we have a single estimate of the value given by $x$. If we know the variance of possible $x$ values, we can generate an interval estimate for the parameter $\mu$. This estimate is called the **95 % confidence interval** for $\mu$. The definition of a 95 % confidence interval is that if we calculated such an interval for each of a large number of randomly sampled $x$ values, then in the long run we would find that 95 % of these intervals would contain the true parameter $\mu$. See Figure 11.3 for a graphical explanation of this technique.

**Figure 11.3**
Single-value samples have been taken repeatedly from a population to estimate $\mu$, the population mean. For each observation, a 95 % confidence interval has been calculated. One of the observations, shown in red, lies outside the 95 % region either side of $\mu$; the corresponding 95 % confidence interval does not therefore contain $\mu$.

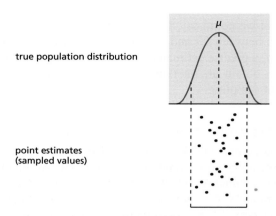

true population distribution

point estimates
(sampled values)

if 95 % of point estimates fall in a range this wide...

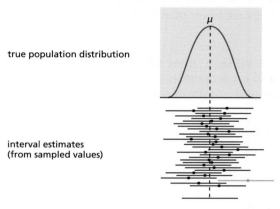

true population distribution

interval estimates
(from sampled values)

... then 95 % of interval estimates this wide contain the true value.

In practice, if we have a single $x$ value (the results of an experiment), and we know (from theory) the standard deviation of possible $x$ values, we can use EQ11.3 to obtain an interval that we are 95 % confident will contain the population parameter. We are '95 % confident' because 95 % of all possible interval estimates that we could calculate in this way from different $x$ values (i.e. from different experiments) will include the true value of $\mu$. At this point, we should note two things that limit the usefulness of EQ11.3. The cut-off values of '1.96 standard deviations' are only suitable when we know that the estimates of a parameter will be normally distributed. Furthermore, the confidence interval for a value can only be calculated if we know the standard deviation of the population of values.

One sampling distribution that we know from theory will generally be very close to normal in shape is the distribution of possible *sample means* from a population. We also know that the standard deviation of this distribution of sample means (called the standard error of the mean) can be calculated from the standard deviation of the population of individual values. It is important for what follows that we know the difference between the standard error of the mean, $\sigma_{\bar{x}}$, and the population standard deviation of individual values, $\sigma_x$. Look back at the last section of Chapter 9 if you do not recall the difference between these two values!

Because sample means are (approximately) normally distributed around the population mean, then if we know the standard error of the mean, we can use

EQ11.3 to give us a confidence interval estimate for the population mean. Our single estimate is now the sample mean $\bar{x}$, and the standard deviation of this estimate is $\sigma_{\bar{x}}$:

$$\mu = \bar{x} \pm 1.96\sigma_{\bar{x}} = \bar{x} \pm 1.96\frac{\sigma_x}{\sqrt{n}}. \qquad \text{(EQ11.4)}$$

Let us review how EQ11.4 is obtained. The distribution of possible sample means is approximately normally distributed, has a mean equal to the true population mean ($\mu$), and a standard deviation equal to the standard error of the mean ($\sigma_{\bar{x}}$). Because 95 % of sampled values from a normal population lie within 1.96 standard deviations of the mean of that population, then we can be 95 % confident that $\mu$ (the true population mean) is no more than $1.96 \times \sigma_{\bar{x}}$ away from the value of $\bar{x}$ (our obtained sample mean).

We can look at a simple example to see how this works in practice. Recall that in our simple experiment, we had a sample of $n = 16$ volunteers, in which the sample mean of the resting heart rate was found to be 75 b.p.m. after taking a new drug. If we happen to know that the distribution of resting heart rate measurements is approximately normal, with a standard deviation of 5 b.p.m., what is the 95 % confidence interval for the true mean heart rate after drug consumption?

Using EQ11.4, the 95 % confidence interval for the population mean (based on the known population variance) is given by

$$\mu = \bar{x} \pm 1.96\sigma_{\bar{x}} = \bar{x} \pm \frac{1.96\sigma_x}{\sqrt{n}} = 75 \pm \frac{1.96 \times 5}{\sqrt{16}} = 75 \pm 2.45.$$

Thus the 95 % confidence interval based on this sample mean is the range from 72.55 b.p.m. to 77.45 b.p.m.

You should now attempt question 1 in the End of Chapter Questions before proceeding to the next section.

## Interpreting confidence intervals

When talking about 95 % confidence intervals, we have been very careful not to say 'there is a 95 % probability that the true mean lies within the confidence interval'. There is a reason for that, which becomes obvious when we notice that two 95 % confidence intervals, calculated as estimates of the same value, sometimes do not overlap. For example, in Figure 11.3 the line marked in red does not overlap the line above it.

Imagine that two scientists run identical experiments and obtain confidence intervals that do not overlap. The first scientist decides that there is a 95 % probability that the population mean is inside the confidence interval she calculated, that is between 10 and 20 units. This means that there is at most a 5 % chance that it lies in any range that is outside this interval. The second scientist, on the other hand, calculates his 95 % confidence interval and finds that it is between 25 and 35, all of which is 'outside' the range predicted by the first scientist. He therefore concludes that there is a 95 % probability that the true value lies in that range. These conclusions seem to contradict one another: we know that the probability cannot be both 5 % and 95 %.

This contradiction is why we say '95 % confidence' instead of '95 % probability', and is resolved when we realize that experiments are *not* repeatedly sampling different values of the population parameter: the 'truth' is fixed. What the two scientists were sampling are *interval estimates*. It might be easiest to understand this in terms of a simple analogy.

Imagine that there are 20 cards, each of which has a single digit on the face. Suppose that 19 of the cards have the same digit ($v$) printed on the face, and the other one has a different digit ($z$). We do not know the values of $v$ and $z$. All the cards are face down, and we have to guess what $v$ is. We turn over a single card (at random) and observe the number $n$ on the card. Suppose that we find the card has the digit 5 on its face. Can we write down a probability for $v = 5$? The answer seems obvious – 'our confidence that $v = 5$ is 95 %, because 95 % of cards contain the right answer'.

Now, we turn over a second card, and observe the digit 3 on it. As a result, we are now a lot less certain that $v = 5$. We do not need to do much calculation to know the probabilities: there are only two possible values $v$ could have (3 or 5) and either is equally likely. Thus our confidence that $v = 5$ would now be 50 % (as it should be for $v = 3$). This change makes sense, because increasing our knowledge obviously *changes our confidence* that a statement is true. Finally, we turn over the other cards and observe they all have the digit 3 on them. Now our confidence that $v = 5$ is zero.

How does all of this fit with the theory of probability we gave in Chapter 10, where probabilities were described by a sampling distribution? Can we generate a sampling distribution for the different possible values of $v$ during our card sampling game?

No, we cannot give a sampling distribution for $v$, because we are *not* repeatedly sampling different values of $v$. Instead, $v$ is a fixed value and we are actually sampling $n$ (the number on the card), which is our *estimate* of $v$. Of course, there *is* a sampling distribution for $n$ in this game, but we do not know what it is, until we know the values of $v$ and $z$.

The situation with non-overlapping confidence intervals is equivalent to the card game. We know that 95 % of confidence intervals are good estimates, just like 95 % of cards have the true value of $v$ on them. If one confidence interval is all each scientist has, then each should have 95 % certainty that the true value does indeed lie in that interval. However, once they acquire other evidence – for example from the other experiment – their confidence should change. In essence, they are getting more information about *how good their first estimate probably was*. Just as with the cards, when we obtain more evidence (another experiment, with different results) the probability that the estimate lies inside the confidence interval will change, which is why it is not a *fixed* probability of 95 %.

Another thing we have been careful not to say about confidence intervals (which you might see elsewhere – it is a very common mistake) is that they tell us the range in which we have a 95 % chance of finding the results of another experiment conducted in the same way, known as a *replication*. It can be seen in Figure 11.3 why this is not the case: the top of the figure shows the interval that has a 95 % probability of containing a replication, and several confidence intervals are shown in the lower part of the figure. The confidence intervals are the same width as the true '95 % capture interval', but of course, they are not generally in the right place! On average, therefore, a very much lower proportion than 95 % of replications will fall into the 95 % confidence interval from a previous experiment.

In general, the most sensible statement that can be made about confidence intervals is based on the original definition: '95 % of confidence intervals contain the true value'. If you stick to this statement when presenting or interpreting confidence intervals, you will not go wrong!

## Confidence intervals for proportions

In most experimental situations, when we do not know the value of a population parameter, we also do not know the population standard deviation of the possible estimates. However, there is one occasion when the variance of an underlying distribution of estimates can be calculated from the sample itself: estimating *proportions*. The formula for a 95 % confidence interval estimate of a population proportion ($\pi$), based on a single observed sample proportion ($p$), from a sample of size $n$, is given in EQ11.7. Before using this expression, however, we will show how it is derived from what we know about sample means.

Calculating proportions does not seem at first glance to have anything to do with taking a sample mean, but actually it is directly equivalent. Imagine we are interested in the proportion of a population of lizards that are green, based on a sample. Obviously, 'color' is a categorical variable, not a numeric one, so we cannot use a 'sample mean' for color. Instead, we define a numeric quantity $x$, for each observation in our sample. We set $x = 1$ for green individuals, and $x = 0$ for non-green individuals. Thus $x$ is 'the number of green individuals in an observation'. Statisticians call this a 'dummy variable'.

Now, we look at a sample of size $n$, and count the number, $g$, that are green. The sample mean of $x$ (our dummy variable) is given by the standard formula (EQ9.1). Because there are $g$ green lizards, which have $x = 1$ (and the remaining lizards have $x = 0$), the sample mean of $x$ is simply the number of green individuals in the sample, divided by the total number in the sample. However, of course, this formula also gives us the *proportion*, $p$, of green individuals in the sample:

$$\bar{x} = \frac{\sum x}{n} = \frac{g}{n} = p. \tag{EQ11.5}$$

In this way, a sample proportion can *always* be thought of as the sample mean of a suitable dummy variable. Why is this important? Well, if we want to estimate the proportion of lizards in the population (written $\pi$) that are green, then we can apply what we know about estimating population means from a sample. Because the sampled proportion is a form of sample mean, the possible sample proportions are generally approximately normally distributed, and the sample proportion ($p$) gives us the best point estimate of the population proportion ($\pi$).

To get an *interval* estimate, we also need to know the standard deviation of the sampled proportions. The variance of sampled proportions is relatively easy to calculate from the variance of the random variable, $X$, describing our dummy variable, $x$, based on what we covered in Chapter 10.

From EQ10.13, we note that $E[X] = \sum x P(x) = \pi$, and $E[X^2] = \pi$ (because $x = x^2$ for both of the possible values of our dummy variable). Putting these into EQ10.17 gives $\sigma_X^2 = E[X^2] - (E[X])^2 = \pi - \pi^2 = \pi(1 - \pi)$.

The variance of individual $x$ values (1s or 0s) is therefore $\pi(1 - \pi)$. A sample proportion is simply the mean of $n$ such observations (as long as each observation is independent, which is true if we are sampling from a large population). Thus we can use the formula for the standard error of the mean (EQ9.5) to give the standard deviation of possible sample proportions:

$$\sigma_p = \sqrt{\frac{\pi(1 - \pi)}{n}}.$$  (EQ11.6)

Using $p$ as an estimate of $\pi$, we substitute these values into EQ11.3 to obtain an interval estimate (95 % confidence interval) of the population proportion ($\pi$) based on a sample proportion ($p$) from a sample of $n$ items.

$$\pi = p \pm 1.96 \times \sigma_p = p \pm 1.96\sqrt{\frac{p - p^2}{n}}.$$  (EQ11.7)

This formula gives a good estimate of the 95 % confidence interval for a population proportion, providing the distribution of sample proportions is roughly normal in shape. This will be the case as long as the proportion is not too large, nor too small (in general, at least five items in the sample must be included and at least five excluded, for the approximation to hold).

A confidence interval for a population proportion is used in just the same way as a confidence interval for a population mean. Let us look at a simple example.

In a random, $n = 25$, sample of volunteers, the resting heart rate was found to increase in 18 individuals in the 30 minutes after administration of a new drug. What is the 95 % confidence interval for the proportion of a population who will show increased heart rate after a similar dose?

We have a *point estimate* of the proportion who will show an increase from the sample proportion, $p$, which is $18/25 = 72\%$. Using EQ11.7, we can obtain a 95 % confidence interval estimate for the population proportion:

$$\pi = p \pm 1.96\sqrt{\frac{p - p^2}{n}} = 0.72 \pm 1.96\sqrt{\frac{0.72 - 0.518}{25}}$$

$$= 0.72 \pm 1.96 \times 0.09 = 0.72 \pm 0.176.$$

Thus the 95 % confidence interval for the population proportion is the range from 54.4 % to 89.6 %.

## Box 11.2 Adjusted confidence intervals

The formula given in EQ11.7 provides a fairly good estimate of the 95 % confidence interval for a population proportion, providing that the distribution of sample proportions is roughly normal in shape. However, this is not the case if the population proportion is extreme (close to 1 or 0), or if the sample is small. In these situations, less than 95 % of possible samples would produce a confidence interval containing the true population proportion.

There is a slight adjustment to the procedure for a 95 % confidence interval on proportions that gives a better interval estimate, which works well even for smaller samples or more extreme proportions. This adjustment procedure is so simple that, in fact, some statisticians advise using it for *all* samples.

To make the adjustment, we simply add two extra observations of each type (for example, 2 'green' and 2 'non-green' items) to our sample, *then recalculate n and p and* proceed with EQ11.7. To see how this works, let us apply the adjustment for the example in the text.

In the original experiment, heart rate increase was observed in 18 out of $n = 25$ individuals. We do not adjust the *point estimate* of the proportion of the underlying population who would show an increase: this is still the sample proportion: $18/25 = 72$ %.

To get the adjusted *interval estimate*, we recalculate $p$ and $n$ by adding two observations of 'increase' and two observations of 'decrease' to our data. This gives $n = 29$ observations, of which 20 were increases. Thus the adjusted $p = 20/29 = 69$ %.

Using EQ11.7, we can obtain a revised 95 % confidence interval estimate for the population proportion:

$$\pi = p \pm 1.96\sqrt{\frac{p - p^2}{n}}$$

$$= 0.69 \pm 1.96\sqrt{\frac{0.69 - 0.48}{29}}$$

$$= 0.69 \pm 1.96 \times 0.086 = 0.69 \pm 0.168.$$

Thus our best estimate of the underlying proportion is 72 %, and our 95 % confidence interval for the population proportion is the range from 52.1 % to 85.8 %. This interval is very similar to the unadjusted interval seen in the main text, although slightly closer to 50 %.

This adjustment procedure should only be used when estimating 95 % confidence intervals for proportions.

A confidence interval is most easily understood as an *estimate* of the population value. However, there is another interpretation that is equally important, in terms of what is called a *hypothesis test*. Hypothesis tests have played a key role in almost all areas of biological sciences, and it is important that you understand the principles and terminology, so we will cover this material next, before showing what they have in common with confidence interval estimates.

# 11.3 Hypothesis testing

There are times when conducting research when, as well as knowing the range of values that represents our estimate of the parameter (for example the true mean of the values), we need to decide whether a particular possible value of the parameter can be 'ruled out' on the basis of our data. For example, we might want to know if it is plausible that our data showing a change in blood pressure could have arisen if there were *no systematic change* in heart rate (that is, any apparent change was just due to sampling error). Alternatively, it might be that average heart rate after taking the drug would have to be below 85 b.p.m. for the drug to be considered safe.

Obviously, any such conclusion will be uncertain, because we do not know whether our observations are misleading. Sampling error implies that they may be atypical, for example if we happen to pick 25 people with very high heart rate as our sample. To determine which of our uncertain conclusions we should trust, we can use an approach that is called a *hypothesis test*.

A **hypothesis** is a statement about the world, which may or may not be true. A hypothesis test is based on determining whether we have evidence from our data to indicate that a particular hypothesis is false. The hypothesis that is tested is generally called the **null hypothesis ($H_0$)** – this will usually be something like 'there is no effect of the drug on heart rate', or 'there is no difference between the two population means'.

Now, the null hypothesis seems pretty dull – we are usually interested in something else, often called the **research**, or **alternative**, **hypothesis $H_1$** (for example 'there is an effect of this drug on heart rate'). The most important thing to remember about significance tests is that the research hypothesis *is not the thing we are testing*. It may seem odd that we test the null hypothesis, rather than the hypothesis we are really interested in. The reason we test a null hypothesis is that when it is precisely specified, we can calculate the conditional probability of observing data like those we obtained.

The first thing we do in a hypothesis test is to determine the null hypothesis ($H_0$) that we wish to test. We then calculate the probability ($p$) that we would obtain data 'as weird as our sample' if $H_0$ were true (and we ran a similar experiment). Obviously, if this $p$ value is small, then this would lead us to suspect that the null hypothesis was not a good description of the real world.

Let us discuss a simple example to illustrate the logic. Imagine we know that the fish ponds in a certain area contain only two species of fish, one blue and the other red. We can capture one fish at a time, determine its color, and put it back. Our first experiment involves ten separate fish-spotting events. If we found that all ten fish have the same color (say, blue), we would obviously conclude that there are 'probably more blue fish than red fish in this pond'. However, how confident should we be of this conclusion?

As we know from Chapter 10, the probability of seeing ten blue fish in a row depends upon the proportion of blue fish. Because we do not know the *true* proportion, we cannot calculate the true probability of sampling ten blue fish.

The $p$ value is the probability that, if the null hypothesis were true, we would sample data 'at least as unusual' as what we observed. Technically, $p$ is the proportion of all possible samples (if $H_0$ is true) that are as, or more, extreme than our sample.

As our aim is to find out whether the two proportions are equal, we work out what the probability of catching ten blue fish in a row would be *if there were equal proportions of each*.

The exact probability of observing ten blue fish, if there were equal numbers of each color, is 1/1024. If there were, in fact, equal numbers of each species, then the chance of seeing ten blue fish in a row in this experiment is a little less than 1 in 1000. Of course, if there were equal numbers of each species then there would be just the same probability of all ten fish being red. These are exclusive probabilities and can be added to calculate the probability of observing *ten fish of just one color*, given equal numbers of each color. This probability is about 1 in 500.

In terms of a hypothesis test, what we have just done is calculate a $p$ value based on the null hypothesis ($H_0$) that the proportion of blue fish is 50 %. The $p$ value is the conditional probability of our data if $H_0$ was true, P(ten fish of one color $\mid H_0$) = 0.002.

Now, imagine that we repeat the experiment in a second pond, from which we observe seven blue and three red fish. It turns out that the probability of seven or more out of ten fish being blue is 176/1024, if there are equal proportions of red or blue fish. Again, there would be the same probability of sampling seven or more red fish, so the $p$ value for the second experiment is 352/1024 = 0.344. In other words, if $H_0$ was true, then we would expect to see samples as 'lopsided' as our sample (7 blue : 3 red) in more than a third of experiments. This does not make it seem that unlikely that there are equal numbers of each color in the second pond.

Note that when we calculated our $p$ values in these two cases we calculated not only the probability of the exact sample observed, but rather the total probability of all possibilities that 'are at least as unusual'. The reason for this is simple: if we got excited about seven out of ten fish being blue (that is, we felt it was evidence against the null hypothesis) then we would get even more excited about eight, nine, or ten fish being blue. Similarly, we would be just as excited about as high a proportion of red fish.

We want the $p$ value to be the probability of 'getting excited' if the null hypothesis is true. In the first example, above, we include the probability of 'exactly ten red fish', and for the second we calculate the probability of 'seven or more fish of either color'. This may seem slightly strange, but all we are really doing is finding the answer to the question 'assuming the null hypothesis *is* true, what proportion of possible samples would be *at least* this unusual?'.

So what conclusion would we want to draw for the two situations above? The experiment in the first pond gave us results we would only expect to see 1 time in 500 (if there were equal numbers). The second pond gave us a sample no more lopsided than we would see 1 time in 3 (if there were equal numbers). From this, it seems unlikely that the first pond contains equal proportions, whereas the second pond could quite easily contain equal proportions. In other words, we feel that a decision that the null hypothesis is *not* true is more justified when the $p$ value is lower.

**Figure 11.4**
A pond contains only two species of fish: one blue and the other red. A fish is removed, its color noted, and the fish replaced. This process is repeated to generate ten observations. In all, seven blues and three reds were recorded. The diagram shows a repeated simulation of this experiment under the assumption (null hypothesis) that the proportions of both species in the pond are equal (50 %). It shows that finding seven or more fish of one color would be quite likely, so the actual observation does not provide sufficient evidence to reject the null hypothesis.

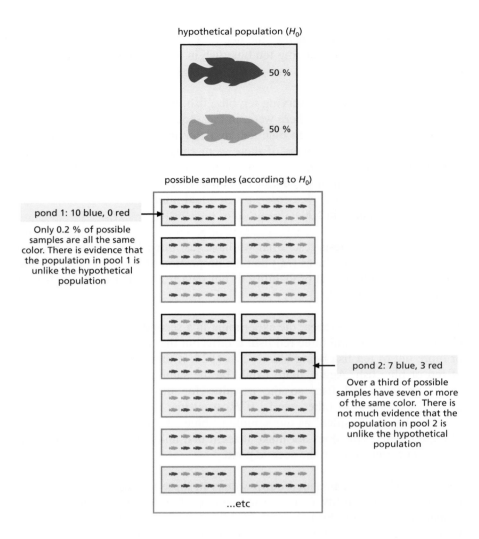

hypothetical population ($H_0$)

50 %

50 %

possible samples (according to $H_0$)

pond 1: 10 blue, 0 red

Only 0.2 % of possible samples are all the same color. There is evidence that the population in pool 1 is unlike the hypothetical population

pond 2: 7 blue, 3 red

Over a third of possible samples have seven or more of the same color. There is not much evidence that the population in pool 2 is unlike the hypothetical population

...etc

If we make a decision that the null hypothesis is false on the basis of whether the $p$ value is low, we must first adopt some criterion for 'low enough'. If our $p$ value is less than or equal to this criterion we say that the data are *statistically significant* evidence against the null hypothesis; when the $p$ value is larger than our criterion, we say that the data are *non-significant*, and do not provide evidence against the null hypothesis. The decision criterion (how low a $p$ value has to be in order to be considered 'evidence') is called the significance level, and written $\alpha$. The usual criterion that is adopted in most cases is $\alpha = 0.05$.

This example conveys pretty much all there is to hypothesis tests. We will go through this again, to make sure that it is clear.

In the first experiment, ten blue fish in a row corresponds to a $p$ value (probability given the null hypothesis) of 0.002, so we decide that this is statistically significant (unlikely enough to be taken as) evidence against $H_0$. We therefore have evidence to believe that there are more blue fish than red fish in the first pond.

**Statistical significance**
Data are said to be statistically significant evidence against $H_0$ if the $p$ value obtained from them is less than or equal to a criterion value, called the significance level, $\alpha$.

In the second experiment, we found seven out of ten fish were blue, which gives a $p$ value of 0.344. This is not statistically significant (no reliable evidence against the null hypothesis) and we have no evidence to make a decision about equal numbers (there may or may not be equal numbers of fish).

Let us take a moment to review the terminology, as this is often a little confusing until you get used to it. To understand the terminology, all we really need to remember are the principles of conditional probability: hypothesis tests are all based upon conditional probabilities.

The null hypothesis, $H_0$, is a description of the world from which we calculate our $p$ value. Usually, $H_0$ can be thought of as simply 'no effect'. The exact null hypothesis will depend upon the test we use: for example, in the $t$ test we will look at later, it may be that the means of two underlying populations are equal; in the example above the null hypothesis was that there are equal proportions of blue and red fish in the pond.

The $p$ value is the conditional probability, given the null hypothesis, that we would (in a similar experiment) obtain sample properties at least as unlikely as those we found in our sample. 'At least as unlikely' usually means 'as large or larger' or 'as small or smaller'. In our example, it is the probability of obtaining at least as many fish of one color, if there were truly equal proportions of each.

We make a decision about whether the data are evidence against $H_0$ for each test we perform. If the data would be unlikely to be observed if $H_0$ were true, then we conclude that the data are evidence against $H_0$. For consistency, we adopt a standard criterion for how 'unlikely' our data have to be before we take them as evidence against $H_0$. This is known as a significance level, $\alpha$.

If $p$ is less than or equal to $\alpha$, then the effect in the data is said to be statistically significant at the $\alpha$ level; otherwise, the effect is said to be non-significant. You can therefore always read 'statistically significant' as shorthand for 'unlikely enough given the null hypothesis'. Strictly, 'statistically significant' does not mean anything without a level, but the 5 % level ($\alpha = 0.05$) is used so commonly, it is sometimes taken as read, unless another level is given, so 'the effect is significant' is common shorthand for 'we have evidence against $H_0$ because $p < 0.05$'.

The lower the $\alpha$ value, the greater the significance level is said to be. Observations that are more unlikely given the null hypothesis are therefore more significant. This can be confusing: talking about data with high significance means data with a low $p$ value.

A lower $p$ value is, rather confusingly, sometimes said to be 'more highly significant', as it is 'more unlikely given the null hypothesis'.

There are two types of error we can make when we draw a conclusion from a hypothesis test: we can 'tell lies' (draw a false conclusion) by deciding we have evidence to reject a true $H_0$ – this is called a **type I error**. Or, we could 'fail to tell the truth' (miss a real effect) by deciding we have no evidence about $H_0$ when it is false – a **type II error**. Because the type II error does not involve a specific false conclusion, most statistical analysis focuses upon reducing the type I error rate.

## Box 11.3 Significance and errors

A very common misunderstanding about hypothesis tests is to believe that keeping $\alpha = 0.05$ means that 5 % of our significant results will be 'errors'. However, $\alpha$ does not tell us what proportion of significant (or non-significant) results will be errors. To see this, imagine two scientists, Anne and Bob, who are both running lots of hypothesis tests. Anne is detecting unicorns (running tests on data for which the null hypothesis is always true); Bob is researching whether we need oxygen to breathe (running a series of tests for which the null hypothesis is false). It does not matter what value of $\alpha$ Anne and Bob use: all of Bob's significant conclusions are correct rejections of $H_0$ (because we do need oxygen); whereas all of Anne's significant results are type I errors (there are no unicorns).

Remember that $p$ is not a simple probability: it is the conditional probability of type I errors when the null hypothesis is true. So, when Anne does a significance test for which the null hypothesis is true, she has a chance of $\alpha$ of making a type I error for each experiment. So, if we stick to our conventional level for $\alpha = 0.05$, although all of her significant *conclusions* are type I errors, we know that only around 5 % of her *experiments* will give rise to these 'falsely significant' results. In turn, Bob will not make any type I errors, as the null hypothesis will be false for all of his experiments.

In general, the criterion of $\alpha = 0.05$ means that somewhere between zero and 5 % of *experiments* should give rise to type I errors. For this reason, $\alpha$ is also referred to as the *maximum type I error rate*.

It might seem that this upper limit of 5 % is an alarmingly high number of false conclusions for Anne to be making, so why do we not simply use a lower value of $\alpha$? The answer is that the probability of a type II error gets very much higher for lower $\alpha$ values. If Bob is too skeptical, he may decide there is no evidence that we need oxygen!

The process of hypothesis testing is summarized in Table 1 (a version of which can be found in just about every textbook that mentions significance tests).

The lower the value of $\alpha$, the more skeptical the test is, so the less likely we are to be persuaded by a given set of data, and the less likely we are to go along the reject $H_0$ row of Table 1. As we can see from the table, this will increase $\beta$, the chance of making a type II error, when $H_0$ is false. In other words, a low $\alpha$ is good when the null hypothesis *is* true (it makes us less likely to tell lies). When $H_0$ is false on the other hand, choosing a lower $\alpha$ value means we make more type II errors (we fail to 'tell the whole truth').

### Table 1 The possible outcomes of a hypothesis test

| | | truth about the world | |
| --- | --- | --- | --- |
| | | $H_0$ is true (no effect) | $H_0$ is false (effect) |
| **decision** | reject $H_0$ (conclusion) probability | type I error (an incorrect conclusion) $\alpha$ | correct conclusion that there is an 'effect' $(1 - \beta) =$ 'power' |
| | do not reject $H_0$ (no conclusion) probability | 'correct' lack of conclusion (there is 'no effect') $(1 - \alpha)$ | type II error (missing a correct conclusion) $\beta$ |

Type I errors are only possible when the null hypothesis being tested is true, in which they occur with a probability equal to the significance level. Decreasing $\alpha$ means we are more likely to be on the lower row of this table: it improves our performance when $H_0$ is true, but our tests have less power when $H_0$ is false.

Similarly, there are two ways we could get it right, either by 'telling the truth' (deciding that we have evidence that $H_0$ is false when it *is* false), or by 'saying nothing when there is nothing to say' (deciding that there is no evidence to reject $H_0$ when it is in fact true). The probability of correctly deciding that there is no evidence when the null hypothesis is true is, by definition, $1 - \alpha$ (the probability of not making a type I error). The probability of detecting a particular effect (finding evidence to reject $H_0$ when it is indeed false) is called the **power** of the test to detect that particular effect (the power of a test is greater for larger effects).

Because we only draw a conclusion when $p < \alpha$, it may seem at first glance as if only type I errors really matter: after all, if we do not draw a conclusion from an experiment with non-significant results, then how can this be an error? Well, imagine if your experiment was investigating whether a new type of coffee machine caused cancer. If it does cause cancer, and you failed to notice this (a type II error), this might have terrible long-term consequences. On the other hand, if you decide that it did cause cancer when it does not (a type I error), you can cause a health scare, but this is arguably much less dangerous to the public (except to those who manufacture coffee machines!).

In fact, one of the problems many people seem to have when learning about statistical tests is confusing 'statistical significance' with 'practical importance'. Remember that *statistically significant* means only 'unlikely to occur given the null hypothesis', which can be very different from 'important'. If we ran a very, very large study, we might find that eating vitamin pills has a statistically significant effect of raising the life expectancy of humans by an average of 20 minutes. Although this effect might be statistically significant, it is also so small as to be of no practical importance.

Our advice is to try to avoid the phrase 'significant' on its own: only use the full term 'statistically significant' when discussing $p$ values, and try not to say 'significant' when you mean 'important'. This will help you, and those with whom you are communicating, to avoid this type of misunderstanding.

> Do not say 'significant' if you mean important.
> Do not assume that statistically significant *implies* 'important'.

## 11.4 Hypothesis testing using test statistics

In the red fish, blue fish hypothesis test above, we calculated $p$ directly. Although the logic is the same in any null hypothesis statistical test, in complex real-life situations a direct calculation of $p$ might well be impossible, or at least very tedious. What we usually do when performing a hypothesis test is to calculate a **test statistic** from our data. The way in which the test statistic calculations are designed means that we know the sampling distribution we would expect if $H_0$ is true.

Why is this useful? Well, if we know the sampling distribution of the statistic when $H_0$ is true, then we can work out the probability of obtaining a value of the statistic as extreme as the value obtained. This probability is our $p$ value. The statistic is chosen such that when there *is* an effect, the test statistic will follow a different distribution, and we will be more likely to get values that correspond to low $p$ values (that is, values *unlikely* to be observed if the null hypothesis is true).

There is a whole range of techniques for different types of hypothesis, but all follow this same logic:

- Decide the null hypothesis ($H_0$) that we wish to test (for example 'no difference' if we are testing for a difference).
- Calculate a statistic from the data. The distribution of the statistic is known when $H_0$ is true.
- The $p$ value is the probability we would have of sampling a value of the statistic as extreme as the value found, if $H_0$ were true.
- If $p \leq \alpha$, we have evidence that $H_0$ is untenable. We may therefore conclude that an alternative hypothesis $H_1$ is more tenable.
- If $p > \alpha$, we do not have enough evidence to reject $H_0$. We do not draw any conclusions.

**Figure 11.5**
How hypothesis testing works. A null hypothesis $H_0$ is used to calculate critical values. If the real situation corresponds to an alternative hypothesis $H_1$ that is not too far removed from $H_0$, then it is quite likely that an observed statistic would fall between the critical values and that we would not be able to reject the null hypothesis. If, however, $H_1$ is far from $H_0$, then it is very likely that we would reject the null hypothesis.

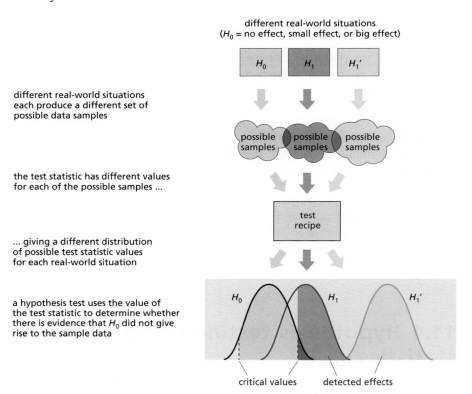

The process is shown schematically in Figure 11.5 for three different situations.

- In green, we see the situation when the world is exactly as described by the null hypothesis. In this situation, the test statistic will vary in a way that we can predict exactly. Certain *critical values* of the test statistic (shown with the vertical lines) can be drawn so that only a small proportion of possible samples ($\alpha = 5\%$) give a test statistic more extreme than these values.
- An alternative to the null hypothesis situation is shown in pink. Here there is a very large effect, and the world is very different to the null hypothesis. In this case, the statistic will still vary between samples, but it will stay

outside the range we would predict from the null hypothesis, namely outside the critical values. The hypothesis test will detect the effect as statistically significant in all experiments in this situation.

- The final alternative, shown in blue, is that there may be a small effect. In this case, the statistic will *sometimes* be inside the range we would predict from the null hypothesis, and sometimes it will be outside the range. The hypothesis test will detect the effect as statistically significant for *some* of the experiments in this situation.

To decide whether an effect is statistically significant (that is, whether $p \leqslant \alpha$), we only need to know the value of the test statistic that corresponds to a $p$ value of exactly $\alpha$, known as the *critical value*. If the statistic is as extreme, or more extreme, than this critical value then $p \leqslant \alpha$. If the statistic is less extreme than the critical value, then $p > \alpha$. These critical values are also often useful for generating confidence intervals, as we shall see.

For example, we know that 95 % of all normally distributed values are within 1.96 standard deviations of the mean. This means that any value sampled from a standard normal distribution (a $z$ statistic) has a probability of 0.05 of being more extreme than 1.96. Thus, if the null hypothesis implied that a calculated test statistic, $z$, followed the standard normal curve, we would simply have to decide if the magnitude of $z$ was larger or smaller than 1.96 to perform a significance test. Any value less extreme than 1.96 is non-significant, and a value greater than 1.96 is statistically significant; 1.96 is thus said to be the critical value of the normal distribution for $\alpha = 0.05$.

Each test can be simplified to a simple 'recipe' of instructions. The simplest, and most generally useful tests, are based around theory we have already encountered. We will look at six of the most common test recipes, which together enable us to test hypotheses about single values, test for differences between means, and test hypotheses about observed frequency counts.

## Hypothesis tests for a normally distributed variable, with known variance

This test can be used to assess whether the unknown mean of a normally distributed population is equal to a particular value, based on a single sampled value, as long as we know the variance of the distribution.

Step 1: specify a null hypothesis to be tested. This describes the mean $\mu_{H_0}$ of a normal distribution whose variance, $\sigma_x^2$, is known, and from which our experimental value, $x$, may have been sampled.

Step 2: convert the experimental value into a $z$ statistic, based on the null hypothesis:

$$z = \frac{x - \mu_{H_0}}{\sigma_x}.$$

(EQ11.8)

Step 3: determine the probability of sampling a $z$ statistic at least as extreme as the obtained $z$ statistic (that is, twice the probability of sampling a $z$ statistic larger than the absolute value of $z$). This is the $p$ value.

Step 4: if the $p$ value is lower than a significance level $\alpha$, then there is evidence at this significance level against the null hypothesis. In other words, there is evidence that the experimental value was not sampled from a population with the given distribution.

We will return to the heart rate example from the start of the chapter to illustrate how this works in practice. In a random $n = 16$ sample of volunteers from a population of adults, the sample mean of the resting heart rate was found to be 75 b.p.m. after taking a new drug. The distribution of resting heart rates in the population is known to be normal, with a mean of 72 b.p.m. and a standard deviation of 5 b.p.m. Is this statistically significant evidence that the drug influences mean heart rate?

To answer this, we need to test the null hypothesis that the population mean of possible sample means (that is, the results obtained from possible experiments just like this one) is 72 b.p.m. We know that the set of possible sample means is normally distributed, with a mean equal to the population mean, and a standard deviation equal to the standard error of the mean. If we assume that the drug does not influence the variability of heart rates, we can calculate the standard deviation (standard error) of this distribution:

$$\sigma_{\bar{x}} = \frac{\sigma_x}{\sqrt{n}} = \frac{5 \text{ b.p.m.}}{\sqrt{16}} = 1.25 \text{ b.p.m.}$$

Using EQ11.8, we can convert our observed sample mean into a test statistic, $z$. If the null hypothesis is true, this value would vary (across samples) according to the standard normal probability distribution.

$$z = \frac{\bar{x} - \mu_{H_0}}{\sigma_{\bar{x}}} = \frac{(75 - 72) \text{ b.p.m.}}{1.25 \text{ b.p.m.}} = 2.4.$$

From the table of the normal distribution in Appendix 4, we find that the probability of sampling a value greater than $z = +2.4$ is 0.008. If the null hypothesis were true, there would be the same chance of sampling a statistic of less than $z = -2.4$, so the total $p$ value (the proportion of possible samples this extreme, given the null hypothesis) is 0.016.

This is lower than the conventional significance level of 0.05, so we conclude that this experiment provides statistically significant evidence that the drug increases the mean heart rate. If we did not need to calculate the $p$ value exactly (although we should do this, when we can), we could have drawn the conclusion that this was statistically significant at the 0.05 level by observing that the value of $z = 2.4$ is greater than the 0.05 critical value ($z = 1.96$) for the normal distribution.

You should now attempt question 3 from the End of Chapter Questions before proceeding to the next section.

## Combining hypothesis tests and confidence intervals

Now, it so happens that we chose a null hypothesis mean of 72 for our significance test in the last example above. However, of course, we could have

| $H_0$ mean | 70 | 71 | 72 | 73 | 74 | 75 | 76 | 77 | 78 | 79 | 80 |
|---|---|---|---|---|---|---|---|---|---|---|---|
| **Table 11.1 The outcome of a hypothesis test varies with the exact hypothesis tested** | | | | | | | | | | | |
| difference | 5 | 4 | 3 | 2 | 1 | 0 | −1 | −2 | −3 | −4 | −5 |
| $z$ score | **4** | **3.2** | **2.4** | 1.6 | 0.8 | 0 | −0.8 | −1.6 | **−2.4** | **−3.2** | **−4.0** |
| $p$ value | **< 0.001** | **< 0.001** | **0.016** | 0.110 | 0.424 | 1.00 | 0.424 | 0.110 | **0.016** | **< 0.001** | **< 0.001** |

(if we wished) compared our sample mean with some other possible value, say 74. In fact, we could run exactly the same test for a whole range of different possible null hypothesis means between 70 and 80, as shown in Table 11.1.

Unsurprisingly, we find that some of these null hypothesis means give a $z$ value corresponding to a statistically significant $p$ value (bold), while other possible values do not give a statistically significant $p$ value. So, our data are statistically significantly different from some values, and not statistically significantly different from others.

As the difference between our null hypothesis value and the sample mean increases, the $z$ statistic gets further from zero; that is, large magnitude differences give rise to large magnitude $z$ statistics. Thus values in the range from 73 to 77 give values of $z$ within 1.96 (our critical value) of zero and thus our data are *not statistically significantly different from values in this range*.

This range corresponds to one we have already calculated at the start of this chapter: it is the 95 % confidence interval for the population mean. This property of a confidence interval can be used as an alternative (although less precise) definition. The 95 % confidence interval is the range of possible parameter values from which our sample data are not statistically significantly different at the 5 % significance level.

This shows that the logic of calculating a confidence interval is essentially the same as that for hypothesis testing. The 95 % confidence interval is calculated such that only 5 % of values are this far (or further) away from the true value. Only 5 % of samples will give a sample mean so far away from the population parameter that their 95 % confidence interval does not include the true value.

The limits of the 95 % confidence interval will be the values of the null hypothesis mean that would give a $z$ value of 1.96, the critical value for a hypothesis test using $\alpha = 0.05$. This in turn means that the $p$ value when testing a hypothesis for a population mean *outside* the corresponding confidence interval will always be less than 5 %.

Although we have been focusing on the 'conventional' level of significance so far (5 % significance level, and the 95 % confidence interval), exactly the same logic applies for other significance levels. For example, the value of $z = 2.58$ is the critical value for the 1 % significance level, and can be used to give 99 % confidence intervals (when used in place of 1.96 in the formulae for 95 % confidence intervals). This is much less common than calculating a 95 % confidence interval. The relative size of the two critical values suggests that the 99 % confidence interval will always be wider than the 95 % confidence interval. This makes sense, as we will always need a *wider* range to be *more* confident that our interval estimate contains the true value.

If confidence intervals seem equivalent to hypothesis tests, why do we need both? Well, the two forms of analysis give complementary information about the data, as follows:

- The *p* value allows us to decide exactly how implausible a certain value (often zero) is, and whether this value can be rejected. It allows us to test a specific hypothesis.
- The confidence interval allows us to decide which range of values is plausible, and which ranges are not plausible. It is a way of generating an *interval estimate* of a parameter.

If we were interested only in whether a certain hypothetical value can be rejected (and not in the 'true' value beyond this), we would only need to consider the *p* value. On the other hand, if we wanted only to estimate the true value, and had no particular desire to test whether a certain hypothetical value was plausible, we would only be interested in a confidence interval. When we perform experiments we are usually interested in both of these things, so we should report the *p* value and the confidence interval.

## Hypothesis tests when the population variance is estimated (*t* tests)

We saw above that we can use the normal distribution to test hypotheses about, and generate confidence intervals from, a sample mean when we know the variance or standard deviation of the underlying population. In the real world, it is rare for us to be sampling from a population whose variance we already know, so these formulae cannot be used. However, we can use a very similar method that is based on using the sample variance to estimate the population variance.

It transpires that when we estimate the population variance using the sample variance, we expect a slightly different distribution of our statistic. The statistic follows a distribution known as *t*, which is slightly different to the standard normal curve. Sampled values, written *t* instead of *z*, are more likely to be extreme values.

The exact nature of the distribution is determined by 'how good' an estimate of the variance we have, known as the **degrees of freedom** (*df*) of the *t* statistic or distribution. Figure 11.6 shows what happens to the *t* distribution, as the *df* increases from 1 to 1000; with high *df*, the *t* distribution becomes the same shape as the normal curve.

Because the shape of the *t* distribution is influenced by the *df*, when we perform a hypothesis test or construct a confidence interval, we need to know the *df* of our *t* statistic. As a result, for each of the techniques using *t* statistics, we will therefore explain how to calculate the *df*, as well as how to calculate the *t* statistic.

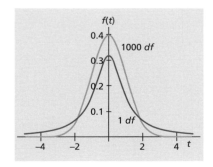

**Figure 11.6**
Plot of the *t* distributions with 1 degree of freedom (*df*) and with 1000 *df*.

Most students (and teachers) find the concept of *df* slightly confusing. Luckily, you do not need to understand the mathematics behind the *df* fully to use *t* statistics. We can think of the *df* of a *t* statistic as corresponding to the number of 'independently sampled differences' that have been averaged in an estimate of a standard error, or variance. You may remember from Chapter 9 that a sample variance is based upon an average of squared differences from the sample mean. Larger samples contain a larger number of differences to be averaged, and therefore provide a better estimate of the population variance.

> The shape of the *t* distribution, and the critical values, depend upon the number of degrees of freedom.

When using a computer, we can obtain a precise probability associated with each *t* value (for example, by using the Microsoft Excel® worksheet function '=TDIST(*value, df*)'). However, if we wish to convert a *t* value to a probability using a table, such as we do for the standard normal curve, we have a problem: because the distribution of *t* is different for each value of the *df*, we would need a different table for each *df*, showing the cumulative probability of each value of *t*. This would clearly require a very large appendix!

What we usually do, therefore, is to tabulate only the *critical values* of *t* for each value of *df*. This table can be used to perform hypothesis tests at the corresponding significance levels: if the calculated *t* statistic is more extreme than a given critical value, the *p* value must be less than the significance level. The critical values are also needed to generate confidence intervals. For example, the 95 % confidence interval for the population mean can be obtained using the critical value of *t* for $\alpha = 5$ %.

The recipes for performing a *t* test are the same as for a test that uses the normal distribution (a *z* test), but we use the sample variance in place of the known population variance (that is, we *estimate* the population variance from our data).

## Hypothesis tests and confidence intervals for the mean of a single sample

This technique allows us to test hypotheses about the mean of a population, based upon the properties (sample mean and variance) of a single random sample. It enables us to decide whether we have evidence that the population mean is different from a particular value. It also allows us to give an interval estimate for the value of the population mean. The procedure works as follows.

Step 1: decide on the null hypothesis to be tested (for example the 'expected' value of the population mean). This describes a possible mean, $\mu_{H_0}$, for the population from which our sample was taken.

Step 2: calculate a *t* statistic based on the sample mean, $\bar{x}$, and the estimated standard error of the mean (calculated from the sample standard deviation, $s_x$). This statistic has $(n-1)$ *df*.

$$t_{n-1} = \frac{\bar{x} - \mu_{H_0}}{s_x/\sqrt{n}}.$$

(EQ11.9)

Step 3: if the magnitude (ignoring the sign) of the obtained $t$ value is larger than the critical value corresponding to $(n - 1)$ $df$ for the significance level $\alpha$, then there is evidence at this significance level against the null hypothesis. We declare that the difference between the observed mean and the null hypothesis mean is statistically significant: the data can be taken as evidence that the data were not sampled from a population with the mean specified in $H_0$.

Step 4: a confidence interval for the population mean can be calculated with limits

$$\mu = \bar{x} \pm t_{\text{critical}} \times \frac{s_x}{\sqrt{n}}. \tag{EQ11.10}$$

Let us see how this works in practice. Imagine that we have a sample of nine measurements, whose mean is 7, and standard deviation is 12. Is this significant evidence (at the 5 % level) against a hypothesis that the underlying population has a mean of zero? What is the 95 % confidence interval for the mean of the population?

We can follow the recipe step by step. Our null hypothesis is that the mean of sample means is $\mu_{H_0} = 0$. The estimated standard error of the sample means is $s_{\bar{x}} = s_x/\sqrt{n} = 4$.

Using EQ11.9, we get a $t$ statistic of

$$t = \frac{\bar{x} - \mu_{H_0}}{s_x/\sqrt{n}} = \frac{7 - 0}{12/3} = 7/4 = 1.75,$$

which has $(n - 1) = 8$ $df$.

Appendix 5 shows that the critical value for $t$ for $\alpha = 5$ % with 8 $df$ is 2.306. Our test statistic does not exceed this critical value, and thus we do not have statistically significant evidence against the null hypothesis. We report this by saying that our observed mean is *not* statistically significantly different from zero.

Remember that a non-significant result does not mean that we should conclude that the true population mean *is* zero. If we want to draw a conclusion about the value of the population mean, we can use a confidence interval. The 95 % confidence interval for the population mean is given by EQ11.10:

$$\mu = \bar{x} \pm t_{\text{critical}} \times \frac{s_x}{\sqrt{n}} = 7 \pm 2.306 \times 4 = 7 \pm 9.224.$$

The 95 % confidence interval for the population mean is therefore the interval:

$$-2.224 \leqslant \mu \leqslant +16.224.$$

We can see once again that the confidence interval and the hypothesis test are equivalent: the null hypothesis value is inside the confidence interval, thus the $t$ statistic is smaller than the critical value and therefore statistically non-significant. If our null hypothesis value had been *outside* the confidence interval, then the $t$ statistic would have been larger than the critical value, and therefore would have been declared statistically significant.

You should now attempt question 4 from the End of Chapter Questions before proceeding to the next section.

## Hypothesis tests for differences between two means

The most common research questions are about differences between samples, asking whether one set of numbers is bigger than another set. We often want to know if one measurement is larger (in general) than another. Do male ferrets have longer or shorter tails than females? Do patients treated with gene therapy produce more or fewer antibodies?

We attempt to answer this type of question by measuring two different samples (male and female ferrets; patients with and without gene therapy). We can calculate the sample means, which may suggest to us that the population means are different. What we need to do for statistical inference is to ask, 'how unlikely is it that we would sample a difference as big as this, assuming that there was no difference between the means of the two under-lying populations?'.

The first question we must address, however, is whether our data can be considered in pairs, such that one value in the first sample is matched to one value in the second sample. Usually this occurs because the same person or animal was measured in two conditions (for example before and after gene therapy). Other reasons why data might be thought of as paired include male and female siblings, mating pairs, or subjects divided into two groups as matched pairs on some measurement. For example, we might decide that one of the tallest two individuals should be assigned to each group, then repeat for the next tallest pair, etc. In all of these cases it is possible that the sampling of the two groups is correlated, meaning that if we sampled an 'extreme' individual in one group, we may be more (or less) likely to sample an extreme measurement for the corresponding individual in the other group.

It is important to know whether data are from related samples. If samples are related, then there is a possibility that the samples are correlated, so that when we sample an extreme value in one sample we tend to get an extreme value for the matching item in the other sample. Of course if samples are not related, then this cannot be the case (as measurements are independent and there is no matching item in the other sample). Luckily, it is very easy to remember which technique to use.

We simply need to decide whether we can re-express the data sensibly as a single set of 'difference' or 'improvement' scores. This is the first stage of analyzing related samples, as we will see below. If we can calculate difference scores it must be because we have got paired data, so we should calculate these scores, and analyze these data as related samples. If we cannot calculate difference scores sensibly (because the data points do not come in matching pairs), then we must perform an independent samples test.

## The paired (related samples) *t* test

We will consider the technique for paired data first. This test is very simple, and in fact we already know how to do it! We saw above how to test whether or not the mean of a *single* sample is zero, by using a one-sample test. We can extend this to comparing two related samples by turning these two samples into a single sample of 'difference scores'.

If we take the null hypothesis as the statement that the means of the two (matched) populations are equal, then this is just the same as saying that the population mean of the possible difference scores is zero. The only assumption we need to make to perform this test is that the distribution of *sample means of difference scores* will be normal in shape.

> The paired *t* test does not assume that samples are normal in shape, nor that they have equal variance. It only makes assumptions about the distribution of difference scores.

We want to decide whether there is a difference between the populations. That is, we need to look for evidence that the means of the populations sampled are not equal. So the null hypothesis to be tested is that the mean of possible difference scores is zero. It is also possible to test whether the mean difference score is some value other than zero, but it is rarely necessary to do this.

Step 1: for each pair of data, subtract the value in the second sample from the matching value in the first sample, to give a difference score. The sign (positive or negative) of the difference is important, and must be included. Calculate the mean $\bar{d}$ and standard deviation $s_d$ of the **difference scores**. From now on we are now doing a one-sample *t* test, using these difference scores as our single sample.

Step 2: calculate a *t* score based on the mean of the sampled difference scores, and the estimated standard error of this mean (calculated from the standard deviation of difference scores). The *t* statistic has $(n - 1)$ *df*, where *n* is the number of pairs of data.

$$t_{n-1} = \frac{\bar{d}}{s_d/\sqrt{n}}. \tag{EQ11.11}$$

Step 3: if the magnitude (ignoring the sign) of the obtained *t* value is larger than the critical value corresponding to $(n - 1)$ *df*, for the significance level $\alpha$, then there is evidence at this significance level against the null hypothesis. We declare that the mean difference score is statistically significantly different from zero, which can be taken as evidence that the data were not sampled from populations with equal means.

Step 4: a confidence interval for the mean difference score (that is, the difference in sample means) can be calculated from

$$\mu_d = \bar{d} \pm t_{\text{critical}} \times \frac{s_d}{\sqrt{n}}. \tag{EQ11.12}$$

Let us look at an example of how this might work in practice.

In an experiment, two skin samples were taken from each of 16 frogs. One sample was exposed to sunlight, and the other kept in darkness, for 2 hours. The amounts of a certain hormone released by the samples during the following 12 hours are shown in Table 11.2. Is this evidence for sunlight changing the amount of hormone released?

**Table 11.2 The amount of a certain hormone released by skin samples taken from frogs**

| frog no. | 1 | 2 | 3 | 4 | 5 | 6 | 7 | 8 | 9 | 10 | 11 | 12 | 13 | 14 | 15 | 16 |
|---|---|---|---|---|---|---|---|---|---|---|---|---|---|---|---|---|
| sunlight | 11 | 12 | 13 | 13 | 13 | 15 | 16 | 20 | 21 | 24 | 25 | 26 | 26 | 27 | 29 | 32 |
| dark | 7 | 15 | 11 | 15 | 15 | 12 | 17 | 15 | 24 | 20 | 19 | 17 | 12 | 16 | 13 | 15 |

The first stage in the analysis is to decide which test to use. In this case, we wish to compare the means of these *related* data sets, so the paired *t* test is appropriate. These are related data because each of the values in the 'sunlight' condition can be matched to the value from the same frog in the 'dark' condition. We can therefore calculate a difference score for each frog, which measures the 'effect of sunlight'.

**Table 11.3 The difference score for each frog measures the 'effect of sunlight' on the skin tissues**

| frog no. | 1 | 2 | 3 | 4 | 5 | 6 | 7 | 8 | 9 | 10 | 11 | 12 | 13 | 14 | 15 | 16 |
|---|---|---|---|---|---|---|---|---|---|---|---|---|---|---|---|---|
| effect (difference) | 4 | −3 | 2 | −2 | −2 | 3 | −1 | 5 | −3 | 4 | 6 | 9 | 14 | 11 | 16 | 17 |

We now proceed with the recipe given above, to compare the sample mean of these difference scores to zero. The sample mean of the difference scores is $\bar{d} = +5$. The sample standard deviation of difference scores is $s_d = 6.713$. Now we can calculate a *t* statistic by putting these values into EQ11.11:

$$t_{15} = \frac{\bar{d}}{s_d/\sqrt{n}} = \frac{5}{6.713/4} = 2.98.$$

This statistic has $(n - 1) = 15$ *df*. Remember that it is the number of *difference scores* that we count here.

To determine whether this value is statistically significant, we compare it with the critical value for 15 *df*. It exceeds the critical value for 0.05 (which Appendix 5 tells us is 2.131), meaning that the result is statistically significant at the 0.05 level. We can also generate a 95 % confidence interval for the difference in mean between the sunlight and dark samples by using EQ11.12:

$$\mu_d = \bar{d} \pm t_{\text{critical}} \times \frac{s_d}{\sqrt{n}} = 5 \pm 2.131 \times \frac{6.713}{4} = 5 \pm 3.576.$$

So we can conclude that the difference is statistically significant at the 0.05 level, and that the 95 % confidence interval for the true mean difference is:

$$1.42 \leqslant \mu_d \leqslant 8.58.$$

However, the critical value for a $t$ test at the 0.01 level with 15 $df$ is 2.947, which is less than the $t$ value calculated above. Thus the $p$ value is lower than 0.01, indicating that there is less than a 0.01 chance of sampling a mean difference score this far from zero, if the population mean of difference scores is truly zero. The effect is therefore statistically significant at both the 0.05 *and* the 0.01 levels. We only need to report the more highly significant $\alpha$, so we state that the difference is statistically significant at the 1 % significance level.

This example also shows that a result can be statistically significant at more than one level. This follows on from the definitions of significance levels and $p$ values. If $p < 0.01$, then it must also be true that $p < 0.05$! Although the test was significant at the 1 % level in this case, it is still appropriate to report the 'standard' 95 % confidence interval as our estimate of the true mean difference (just as we would report the 95 % confidence interval if the result was non-significant).

## The independent (or unrelated samples) $t$ test

When our samples cannot be expressed as a single sample of difference scores, we have to use a different strategy. If we can estimate the sampling distribution of the mean of each sample, then we can also estimate the sampling distribution of the difference between two means: that is, how likely we would be to find a difference as big as this between two sampled means, given that the two population means are equal.

If the two underlying populations do have the same mean (the null hypothesis is true), then the mean of 'possible differences between sample means' is zero. It will be normally distributed assuming that it is the difference between two normally distributed sample means. So we can perform a $t$ test if we have some way to estimate the variance of the difference between two means. Remember the variance sum law we saw in Section 9.5? This tells us that the variance of the difference between independent sample means is the sum of the variances of the two sample means. So we now have all we need.

The test procedure is quite simple when the numbers in each sample are equal, and when we assume that the variances of the two populations are similar (homogenous), so this is the only case we will cover here. It is perfectly possible to extend the same logic to situations where sample sizes are unequal, or variances are dissimilar, and appropriate tests are available on all computer packages. However, these situations make the formulae for the $t$ statistic a little more complicated, so we will not go into the detail of the analyses here.

## Box 11.4 Assumption of homogeneity in the *t* test

All statistical software packages (including Microsoft Excel®) provide two alternative forms of the independent samples *t* test. The simplest version (as presented in this chapter) assumes that the variance is equal in the two populations from which our samples are drawn, known as the assumption of 'homogenous variance'. This assumption may seem strange, but it is a completely sensible assumption in many experimental situations that assign samples or organisms at random to each group, and then give the two groups different treatments. The underlying population is the same (and therefore has equal variance) for both groups *unless* the treatment has some effect on the measurement.

It is possible to conduct the *t* test without making this assumption, although the resulting test is somewhat less powerful. The standard *t* test (assuming homogenous variances) is known to be very robust, providing the two samples are of similar sizes. Some older textbooks may recommend running some form of statistical test for equality of variance to determine which test to use, but this approach has recently been shown to be completely unhelpful.

The 'unequal variances' version should be used whenever we suspect that the variances are very different (generally if one variance is more than twice the other), or when the sample sizes are unequal (if one sample is more than twice the size of the other).

We want to decide whether there is a difference between two population means. So we test the null hypothesis that the means of the populations sampled are equal. If this is true, then the mean of possible differences between sample means must also be zero.

Step 1: for each sample of data, calculate the mean and standard deviation.

Step 2: calculate a *t* statistic based on the difference in sample means, and the estimated standard error of this difference (calculated from the sample standard deviations). The *t* statistic has $2(n-1)$ degrees of freedom, where $n$ is the number of data points in each sample.

$$t_{2(n-1)} = \frac{\bar{x}_1 - \bar{x}_2}{\sqrt{\dfrac{s_1^2 + s_2^2}{n}}}.$$

(EQ11.13)

Step 3: if the magnitude (ignoring the sign) of the obtained *t* value is larger than the critical value corresponding to $2(n-1)$ *df* for the significance level $\alpha$, then there is evidence at this significance level against the null hypothesis. We declare that the difference between the sample means is statistically significantly different from zero, which can be taken as evidence that the data were not sampled from populations of equal mean.

Step 4: a confidence interval for the difference between the means of the two populations can be calculated from

$$\mu_d = (\bar{x}_1 - \bar{x}_2) \pm t_{\text{critical}} \times \sqrt{\dfrac{s_1^2 + s_2^2}{n}}.$$

(EQ11.14)

Again, it is easiest to see how this works by example.

In a certain species of South American underground shrew, the females are nocturnal and the males diurnal. Twelve female shrews and 12 male shrews were independently selected and tested on how fast they learned to find their way around a new tunnel system. The times in seconds for each group were as follows:

Females:
342, 475, 423, 435, 398, 455, 351, 426, 338, 547, 438, 355.

Males:
331, 417, 522, 312, 400, 338, 460, 369, 414, 460, 456, 326.

To determine whether there is evidence of a difference in performance between males and females on this task, we have to perform a suitable test. We wish to compare 'average performance', so we will test the sample means. We do not *know* the underlying population variances, so we will have to use a *t* test. Finally, we have two samples that are not related (since we are told that they are independently selected). This means that these data are not 'matched' in any sense, so we cannot calculate difference scores. The suitable test is therefore the independent samples *t* test.

The null hypothesis is that the difference in population means is zero (i.e. the population means are equal).

Our sample means are males = 400.42, females = 415.25. Our sample standard deviations are males = 66.393, females = 62.508. For simplicity, we analyze the numerical values of our data, i.e. time measured in seconds.

From EQ11.13, we obtain a *t* statistic of

$$t_{22} = \frac{\bar{x}_1 - \bar{x}_2}{\sqrt{\dfrac{s_1^2 + s_2^2}{n}}} = \frac{415.25 - 400.42}{\sqrt{\dfrac{(66.393)^2 + (62.508)^2}{12}}} = 0.563.$$

The critical value for *t* with 22 *df* is 2.074. Our value is less than this, and thus the observed difference between sample means is not statistically significant (i.e. we have no evidence against the hypothesis that the sexes take equal time). We can estimate the difference in means between the sexes using a 95 % confidence interval, obtained from EQ11.14:

$$\mu_d = (\bar{x}_1 - \bar{x}_2) \pm t_{\text{critical}} \times \sqrt{\frac{s_1^2 + s_2^2}{n}} = 14.83 \pm (2.074 \times 26.32).$$

So the 95 % confidence interval for the difference in time between the male and female populations is between 69.43 seconds (with males faster) and −39.76 seconds (meaning females are faster).

## One-tailed and two-tailed tests

In all the examples so far, we have been interested in the probability of sampling a statistic (or data) at least as extreme as that found, if the null hypothesis were true. This is called the two-tailed $p$ value, and when testing against a significance level, we could state that the difference in our experiment was significant, or non-significant, *two-tailed*.

In some circumstances, using a normally distributed ($z$) test statistic we might want to calculate the probability of sampling a difference *in the direction observed* that is at least as extreme as we found. This probability is called the one-tailed $p$ value because we use only one tail, or end, of the distribution. It should be obvious, that for a symmetrical distribution such as the normal, or $t$ distribution, the one-tailed probability is always half that of the two-tailed probability, as shown in Figure 11.7. We have a different critical value for a one-tailed probability of 5 %.

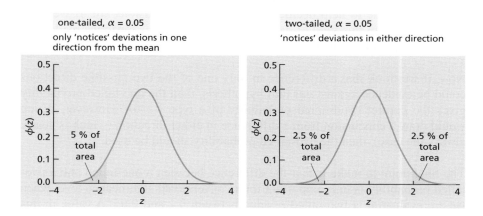

**Figure 11.7**
One-tailed versus two-tailed probability. A one-tailed probability for a test statistic, *z*, is the probability of sampling a value from the normal curve that is as (or more) extreme as the *z* value in the same 'end' or tail of the distribution. This probability is always half of the corresponding two-tailed probability.

To illustrate what a one-tailed probability means, we can think back to our 'fish counting' example, where we wished to decide whether there were equal numbers of red and blue fish. Imagine we took a sample of ten fish, and nine of them were red. The two types of $p$ value can be understood as follows.

- One-tailed $p$ value: 'the probability that we would see at least *9 red fish* if there were equal numbers'.
- Two-tailed $p$ value: 'the probability that we would see at least *9 fish of one color* if there were equal numbers'.

For a $t$ test, a larger than expected value for a sample mean in one of the samples will give a positive $t$ score, whereas a smaller than expected value for that mean will give a negative $t$ score. If we are 'interested' only in whether the mean is *larger* than expected, we should use the *one-tailed* probability associated with the $t$ score. In this context, 'interested' means that only one kind of result could be regarded as evidence against the null hypothesis.

The occasions on which only one 'tail' might be evidence are rare, but they might include replications of a previous experiment. If you did exactly the same experiment again, but got the opposite result on the second occasion, you could not decide that the results of the second experiment were 'true', and ignore the first set. Alternatively, the prediction of a theory may suggest a certain difference in one direction, and so you can only find evidence for this prediction if the data show a difference of the right sign!

One of the most common questions asked by students about hypothesis tests is 'when are we allowed to use a one-tailed test?'. This might be a common question, but it is not a very sensible one, as statistical theory does not give you 'permission' to use a technique. The more sensible version of the question is: when *should* we use a one-tailed test, and why?

The answer to this question is clear when we think about testing in terms of type I errors, or 'false positive' conclusions that can be made when we decide that we have found evidence against a null hypothesis that is, in fact, true. We want our $p$ value to help decide how 'safe' our conclusions are. If we draw conclusions from this kind of data, how often would this mean we made type I errors if the null hypothesis were really true?

Now, if we think that a difference in only one of the two possible directions could be evidence against our null hypothesis, then the one-tailed probability gives the correct conditional probability of a type I error. However, if you would draw a conclusion from a difference in either direction (which is nearly always the case), then the two-tailed probability should be used.

The final point to make is that the difference between a one-tailed and a two-tailed probability is only important for a conventional significance level when we are evaluating a two-tailed $p$ value between 0.1 and 0.05. A value in this range is statistically significant if we take the one-tailed probability, but not if we take the two-tailed probability. A $p$ value larger than this range is not statistically significant one-tailed *or* two-tailed, whereas a $p$ value that is smaller than this range is statistically significant both one-tailed and two-tailed.

If you remember this, you will avoid a mistake that people often make with one-tailed and two-tailed tests. Some textbooks call the one-tailed test a *directional test*, and the two-tailed test a *non-directional* test. These terms can be misleading, as they make students think that the conclusions are different in the two cases. For example, a common mistake is to think that a statistically significant result for a two-tailed test means we decide that 'the means are different', but not *which* is larger, whereas we can only conclude that 'one is larger than the other' from a one-tailed test. This cannot be true, because a result that is significant with the two-tailed test will *also* be significant if we used the one-tailed test.

For either test, we simply conclude from a statistically significant result that we have evidence against $H_0$: that is, evidence that there *is* a difference, in the direction observed.

# 11.5 Hypothesis tests for categorical data

So far, we have looked only at hypothesis tests for comparing measurements of amount; that is, measurements on interval or ratio scales that represent the magnitude of some quality (height, weight, blood cell count, and so on).

The final techniques we will introduce are those for assessing measurements on a purely nominal scale. Here each measurement is of the type of a given observation (hair color, sex, blood group, and so on), rather than its magnitude. When we have this kind of measurement, we end up with data that tell us how many observations are of each type, so we know the frequency, or proportion, of a sample that falls within each category.

There are research hypotheses that we might like to test for such data, including the very first example of a hypothesis test in this chapter 'are red fish more common than blue fish?'. Although we were able to work out a $p$ value for this type of question directly, it is much less obvious how to do this for a hypothesis with more than two categories, for example 'do the proportions of offspring from crossbreeding between heterozygous parents occur in the predicted genotype ratio of 1:2:1?'.

It turns out that there is a very simple way to test this type of hypothesis. The test statistic most commonly used to determine the fit to a simple model is written $\chi^2$ (**chi-square**), because this is the name of the distribution it will follow if the null hypothesis is true.

The $\chi^2$ statistic has two main uses. The first use is based on comparing data to a simple model of proportions. The second use compares data to a slightly more complex model based on the relationship between two categorical variables. The principle is the same for both types of test.

First, we generate an expected frequency ($E$) for each possible type, or category, of observation. This is the long-term average number of observations of that particular type we would expect to see in a given size of sample if the null hypothesis were true. The $\chi^2$ statistic is calculated from the difference between the expected number ($E$) and the actual number of observations ($O$) for each category.

$$\chi^2 = \sum \frac{(O-E)^2}{E}.$$

(EQ11.15)

The total here is summed across all possible categories. Looking at the formula, it should be obvious that the chi-square statistic is always positive, and measures how much the data deviate from the expected model. If there is a big difference between the observed and expected values, the value of chi-square will be large. By contrast, when there is little difference, the chi-square value will be small. Thus a chi-square value that is *large enough* to be unlikely (that is, statistically significant) can be taken as evidence that the proposed model is incorrect.

## Pearson's chi-square test for goodness-of-fit for categorical data

We can use this procedure to decide if a sample of observations do not 'fit' with what we would expect if a particular situation, or null hypothesis, was true: if the fit is bad enough (i.e. the data are not what we would expect), then we will find statistically significant evidence against the hypothesis. The procedure for testing whether categorical data are likely to be observed from a particular model of proportions is as follows.

Step 1: start with a null hypothesis, which describes an expected proportion of observations for each category.

Step 2: work out the expected count ($E$) for each possible category; that is, the number of observations from a sample of this size that we would expect to fall into each category, on average, if $H_0$ were true. This is the total number of observations multiplied by the expected proportion for that category. We must record the expected number of observations, not a proportion or percentage.

Step 3: using the observed count ($O$) for each category, calculate the contribution to the total chi-square statistic, according to EQ11.15. The total of the contributions from each category is our test statistic (chi-square value).

Step 4: as with the *t* test, the chi-square statistic has different critical values depending upon the degrees of freedom. When testing a simple model in this way, it has 'categories − 1' degrees of freedom.

Step 5: compare the value of the chi-square test statistic with the critical value for the appropriate *df* to assess whether it is statistically significant.

If the value is statistically significant, then it provides evidence that the null hypothesis model is *not* correct.

Let us look at an example to see how this works. During an experiment studying rat behavior, 50 female rats are tested in a T-shaped maze. Each rat is placed in the bottom of the T, and allowed to explore, until it enters a 'goal area' at the end of one of the arms. The rat is then scored as having chosen that arm. One arm is treated with a solution of rat pheromone, and the other is untreated. Twelve of the rats choose the treated arm. Do these female rats show a statistically significant preference for one of the two arms?

Here, we are interested in the research question 'is there a preference?'. So our null hypothesis must be 'no preference'. If there is no preference, we would expect that, on average, equal proportions (50 %) of female rats will choose each arm.

> A goodness-of-fit chi-square test is used when we have a single 'class' of categories. The degrees of freedom are one less than the number of categories.

We need an expected count ($E$) for each category, and the number of observations ($O$) that fall into each category; remember that we need the number of observations, not percentages. There are 50 female rats, so the expected number ($E$) according to $H_0$ is 25 choosing each arm. We know 12 female rats chose the treated arm, which means that 38 chose the untreated arm. From Table 11.4 we can calculate

$$\chi^2 = \sum \frac{(O - E)^2}{E} = 13.52.$$

**Table 11.4 Calculations for a goodness-of-fit chi-square test on 50 female rats in a T-shaped maze**

| choice | observed | expected | $(O - E)^2/E$ |
|---|---|---|---|
| treated arm | 12 | 25 | 6.76 |
| untreated arm | 38 | 25 | 6.76 |

One arm of the T-shaped maze has been treated with a solution of rat pheromone. The null hypothesis is that rats do not show any preference for either arm, so that we would expect 25 rats to choose each arm.

The degrees of freedom $df = $ (categories $- 1$) $= 1$. In case you are wondering, the reason the degrees of freedom is one less than the number of possible categories is that we know the proportion in the last category, as soon as we know the proportion in all the others, so it is not free to vary. This is easy to see here: once we know what proportion turned down one arm – we know the proportion that selected the other arm, so we have only one 'variable' measurement.

Appendix 6 shows critical values for $\chi^2$. Our value for chi-square is larger than the critical value for $df = 1$ at the 0.05 level, which is 3.84. Thus, the difference between the observed data pattern and that predicted by the null hypothesis model is statistically significant at the 5 % level. Our data therefore provide evidence against the null hypothesis (that there is no preference), and we conclude we have evidence for a preference for the untreated arm in female rats.

You should now attempt question 7 from the End of Chapter Questions before proceeding to the next section.

The procedure above allows us to test whether the observed set of frequency counts is consistent with a particular null hypothesis set of population proportions. This test can easily be extended to answer a much more common research question: are the proportions seen in different categorizations related?

An example of this type of research question would be, 'do a higher proportion of arts students smoke than medical students?' or 'is scent preference related to the sex of an animal?'.

Questions of this sort investigate the *contingency* between two categorizations; that is, whether the probability of being in a particular category on one measurement scale (for example smoker or non-smoker) depends upon some other measurement scale (such as subject studied).

The null hypothesis we need to test is that there is *no such relationship* in the population, for example that the proportion of arts students who smoke is equal to the proportion of medical students who smoke.

If we find evidence against this null hypothesis, we have evidence that there *is* a relationship (for example the probability of an individual student being a smoker depends upon the subject that they study).

> A contingency test is used when we have two 'classes' of categories. It tests whether the category that an observation is in on one scale influences which category it is likely to be in on the other scale.

## Pearson's chi-square test for contingency (relationship between categorical measures)

The procedure for testing whether two categorical measures are related is as follows.

Step 1: the null hypothesis states that the probability of falling into a specific category on one measurement is unrelated to the category of an observation on another measurement. Draw up a table with the possible values of one measurement as the columns, and those of another measurement as the rows.

Step 2: each cell of the table represents a possible observation. Note the number observed ($O$) in each cell of the table, and the total observed in each row and column (and the overall total number of observations). The table must contain the actual number of observations, not proportions or percentages.

Step 3: work out the expected frequency ($E$) for each possible cell using the following formula:

$$E = \frac{\text{row total} \times \text{column total}}{\text{overall total}}.$$   (EQ11.16)

Step 4: calculate the contribution to the chi-square statistic, according to EQ11.15, for each cell. The total of the contributions from each cell is our chi-square value.

Step 5: the degrees of freedom of the statistic is given by

$$df = (\text{rows} - 1) \times (\text{columns} - 1).$$

Step 6: compare the value of our chi-square test statistic with the critical value for that $df$ to assess its statistical significance.

This may all seem a bit difficult to follow, so let us return to the rat example to see how it works.

During the experiment described above on rat behavior, 50 male rats were tested in the same way, and 20 of these rats chose the treated arm. We know that 12 out of 50 female rats also chose the treated arm, so is this evidence that the two sexes differ in their choice behavior?

To answer this we must determine whether preference for which arm is chosen is related to the sex of the animal. Another way of asking this is: 'is the proportion of choices of the treated arm the same for both sexes of rat?'.

This requires a contingency test. We need to test the null hypothesis that choice is *the same* for the two sexes. We start by constructing a table of the categories into which observations could fall, with one variable (e.g. choice) along the rows of the table, and the other (sex) making up the columns. We could equally well choose to have 'choice' as the columns, and 'sex' as the rows, as it makes no difference to the calculations.

First, we fill in the number of observations ($O$) that fall into each category.

**Table 11.5  Observed frequencies for 50 female rats and 50 male rats in a T-shaped maze**

|  | male | | female | | row total |
|---|---|---|---|---|---|
|  | *O* | *E* | *O* | *E* |  |
| treated arm | 20 |  | 12 |  | 32 |
| untreated arm | 30 |  | 38 |  | 68 |
| column total | 50 |  | 50 |  |  |

One arm of the T-shaped maze has been treated with a solution of rat pheromone.

We need the total number of observations in each row and column, in order to calculate the expected value ($E$) for each category. Our null hypothesis is that the bias is the same for each sex. So, the number of males choosing the treated arm should be the total number of males × the overall proportion of rats who choose the treated arm.

The simplest way to calculate this is by using EQ11.16. So, for males choosing the treated arm, we have:

$$E = \frac{\text{row total} \times \text{column total}}{\text{overall total}} = \frac{32 \times 50}{100} = 16.$$

After calculating all the expected frequencies in this way, we can complete our table. As it happens, all the expected values are integers in this case. Often expected values are not whole numbers – in such cases we *do not* round them, but use exact expected values in the table and calculations.

**Table 11.6  Contingency calculations for a chi-square test on 50 female rats and 50 male rats in a T-shaped maze**

|  | male | | female | | row total |
|---|---|---|---|---|---|
|  | *O* | *E* | *O* | *E* |  |
| treated arm | 20 | *16* | 12 | *16* | 32 |
| untreated arm | 30 | *34* | 38 | *34* | 68 |
| column total | 50 |  | 50 |  |  |

One arm of the T-shaped maze has been treated with a solution of rat pheromone. The null hypothesis is that gender does not alter any preference for the choice of arm, so we would expect the same proportion of male rats as female rats to choose the treated arm (and the same proportion of male rats as female rats to choose the untreated arm).

Next, we calculate our chi-square test statistic, adding the contribution from each of the four cells.

$$\chi^2 = \sum \frac{(O-E)^2}{E} = \frac{(20-16)^2}{16} + \frac{(30-34)^2}{34} + \frac{(12-16)^2}{16} + \frac{(38-34)^2}{34}$$
$$= 2.94.$$

In this case, $df = (\text{rows} - 1) \times (\text{columns} - 1) = 1$.

This value is smaller than the critical value for the 5 % level (3.84, from Appendix 6), and therefore is not statistically significant at this level. Our data do not provide evidence against the null hypothesis that the preference is the same in the two sexes. So, we conclude that we have no evidence for a difference in preference for the treated arm between the two sexes.

This example was chosen to help illustrate two things that often confuse students with chi-square tests.

First, it seems odd that we have only one degree of freedom when we have four categories. This is a result of the way in which we calculated the expected values. Because the row and column totals of expected values are equal to the observed totals, there is actually only one 'amount' of variation between $O$ and $E$ possible – if one expected value is known, the other three values follow automatically. Thus the value of $(O-E)^2$ is the same (16) for all categories.

Second, the conclusion from this analysis shows why we need a contingency test. You might wonder why we cannot simply test for a preference in the male and female rats separately. We have already seen that the females show a statistically significant preference. If we analyze the preference in male rats only, using a two-category goodness-of-fit test as we did for females we get a chi-square value of 2.0, with 1 $df$. You should check that you can obtain this value. This value is not significant; thus we must conclude that we have no evidence that male rats prefer either arm.

This result seems inconsistent with the contingency test. If males showed no statistically significant preference, but females did show a preference, how can we not have evidence that the sexes differ? The apparent paradox shows that, with statistical significance, we cannot reason by extension in this way. Although we have evidence for a preference in females, we have *no evidence* about males (which is not the same thing as having evidence that they are indifferent). Evidence for a difference in one group, but not in another group, does not imply that we have found evidence for a difference between the groups. This mistake is commonly made in the interpretation of $t$ tests, as well as chi-square tests.

If it helps you to understand why this is the case, imagine that we decided that a 'difference in height of 2 feet' between individuals was big enough to be 'noticeable'. If John is 6 feet 6 inches, and James is 6 feet 2 inches, then John is noticeably taller than Kate (4 feet 4 inches), but James is not. However, it would be a mistake to conclude from this that John is *noticeably taller than James*, even though John is noticeably taller than Kate but James is not. The same logic would apply if we use the phrase 'statistically significantly' in place of the word 'noticeably'.

All chi-square tests are similar to one or other of the examples described above. The difference between the two procedures is the way in which the expected values are calculated, which leads to a difference in the formula for the degrees of freedom. The easiest way to remember which of the two forms of the test to use is to see whether your data fit into a *table* and you are interested in a *relationship*. If so, then you should be performing a contingency test. If your data just fit into a row of categories, and you already know the proportions you are testing, then you need a goodness-of-fit test.

In both cases, the null hypothesis describes a set of probabilities, or proportions. The null hypothesis tells us the probability that any one observation would fall into each category. A statistically significant result means that this model is unlikely to be correct (i.e. the data obtained would be unlikely if the null hypothesis were true). In a goodness-of-fit test this is easily forgotten: a statistically significant chi-square value is evidence to *reject*, not to *confirm*, the model being tested.

Finally, the chi-square test generally detects differences in multiple directions, so it does not have the one-tailed/two-tailed distinction. With the *t* test, we get a large positive value for one extreme possibility, and a large negative value for the other. The chi-square value, on the other hand, is always positive, and increases as a result of *any* difference from the expected values. So it therefore does not make sense to talk about 'two-tailed' or 'one-tailed' probability when doing a chi-square test. We are always measuring 'total variation from the model', so we are only interested in large values of chi-square.

> When one effect is significant and another is not, this is not evidence that the effects are of a different size! The significance of any difference must be tested directly.

## 11.6 Assumptions and validity conditions

A set of assumptions are made by any statistical test in order to obtain a *p* value. For example, for an independent *t* test we may assume that the variance of two populations is the same (the '*homogeneity of variance*' assumption), and that all our data points are independent (the value of one measurement will not influence another). The tests are designed to give good estimates of *p* if all such assumptions are met. However, what if they are not? A test is said to be: **robust** if it gives a *good* estimate of *p* even when we violate its assumptions; **liberal** if it gives an *underestimate* of *p* when we violate its assumptions; or **conservative** if it gives an *overestimate* of *p* when we violate its assumptions.

> All statistical tests make some assumptions, usually about the way in which data are sampled, or the properties of the underlying population. When the assumptions are not true, the test may be unaffected (robust), may give too many type I errors (liberal), or may give too many type II errors (conservative).

Obviously, we would like our tests to be robust. A liberal test is 'worse' than a conservative test if we are worried about type I errors, but we should be wary of using a conservative test if we are very keen to detect any effect (that is, if type II errors might be costly).

When we design an experiment, or decide how to conduct a field survey, it is very important to be aware of the assumptions on which the various statistical tests are based, as well as the null hypotheses that they test.

If we collect data that do not meet the assumptions of the statistical tests we plan to perform, we may be in trouble. It is better to think about how you will analyze your data, and what assumptions they must meet, before you start the experiment or survey. It may be tedious, but it would be more frustrating to find that your experiment cannot be analyzed as you hoped because of a mistake in your design.

What are the common assumptions made by the statistical tests we have met?

## Random sampling

Statistical tests typically assume that data are a random sample of the population from which they are drawn. So when we are collecting data, we need to be sure that we are taking measurements truly at random. Every member of the original population (every point in an area, or every moment in a period of time) should have an equal chance of being in the sample.

The best way to achieve this is usually to select locations, times, or individuals by using a random number table, or by generating a series of random numbers using a calculator or computer (or even drawing numbered pieces of paper from a hat). We should not ignore these numbers because they are inconvenient, or make the sample seem a bit unevenly distributed, otherwise the final sample is not truly random.

> A confounding factor is something that may influence measurements in an experiment, whose effect cannot be distinguished from a variable of interest owing to non-random sampling.

Random selection of locations, times, or individuals is important because there may be **confounding factors** that vary between them. These are factors that can affect the measurements we are taking in ways that we wish to ignore.

For instance, suppose we have a set of 60 plants that have been grown in the same conditions on a bench in a greenhouse for 2 weeks, and we now want to subject them to four different fertilizer treatments. If we put all the biggest plants in one treatment, when we come to measure the plants a few weeks later, we will have biased the results in favor of finding bigger plants in the treatment with that particular fertilizer. The initial size of the plant, in other words, is a potential confounding factor in this experiment.

By assigning the plants randomly to the various treatments, though perhaps ensuring an equal number of plants receive each of the four fertilizers, we can avoid such systematic biases. We may be confident that we would never be so foolish as to do something like assigning all the big plants to one treatment, but random allocation is still a good idea because there may be lots of other variables influencing our measurements that we have not considered.

For example, suppose we simply take the 15 plants from the front of the bench and assign them to the first treatment, then take the 15 plants in the next row back and assign them to the second treatment, and so on. This might seem random enough, but the plants at the front might have had better watering over the past two weeks, and those at the back may be directly over heating pipes or against the window. Randomization is the best way to guard against such confounding factors that we are not aware of. Other factors that need to be considered are the time of measurement, and who collects the data (we must avoid doing all the measurements for one group at a certain time, and other measurements at a different time, in case there is regular variability over time that we are not aware of).

## Independent measurements

Most statistical tests also assume that the measurements in your sample are independent, which means that the value of any one measurement does not influence the likely value of other measurements.

When we collect data, it is therefore important to ensure that our measurements are independent, and to avoid **pseudo-replication**, which occurs when we take several measurements that are not truly independent. Pseudo-replication is dangerous because it leads us to draw stronger conclusions than are justified, by inflating the apparent sample size.

Suppose, for instance, we wish to determine whether female mice negotiate a maze more quickly than male mice. You know that a larger sample size gives you a better chance of detecting any difference in speed between the sexes: comparing one male and one female mouse is of little use; comparing ten males and ten females is much more likely to reveal any difference between the sexes; 100 males and females is even better!

It would be tempting to inflate our sample size artificially by running ten males and ten females through the maze ten times each. This might seem to give us 100 maze times for comparison, but analyzing it in this way is an example of pseudo-replication. Because we have only examined ten mice of each sex, we have not gained as much extra information about the distribution of speeds for males and females as if we had tested 100 of each sex. Instead, we have measured ten mice more carefully. Because the ten speeds recorded for each female mouse are not independent, you cannot use them as separate pieces of evidence about whether females *in general* negotiate the maze more quickly than males. In this situation, we would perform our statistical test on the ten average speeds (one for each mouse). Even when measuring different individuals, it is possible for measurements to be non-independent. If you put ten males in the maze all at once (and later did the same with ten females), the speeds recorded would not be independent. Mice with poor navigation might be able to follow a companion to leave the maze quickly; alternatively, one slow individual might hold up the rest.

This issue is also important in the distinction between the paired and unpaired *t* tests. When data are paired in some way, they *must* be analyzed with the paired *t* test, as the values may correlate between samples. The formula for the unpaired *t* test is based upon the variance sum law for independent variables. The independent *t* test is highly conservative if this correlation is positive, and becomes very liberal if the correlation is negative.

## Normally distributed data

Many statistical tests, including the *t* tests we introduced in this chapter, involve estimating the properties of the population(s) from which our data are drawn, assuming that these populations are normally distributed. Such tests are described as parametric tests because they try to estimate population parameters. Alternative tests that do not try to estimate these parameters are available, and are known as non-parametric tests.

---

### Box 11.5 Non-parametric approaches: permutation and resampling

The *t* test provides a parametric approach to statistics, as it estimates the parameters of the population from which we have sampled. Other approaches are possible, which are not based on these estimations, and thus do not make the same assumptions about the distribution of the data. For example a Mann–Whitney–Wilcoxon test is a non-parametric test that may be used to compare the central tendencies of two samples, just like a *t* test. It is important to note that these tests do make *other* assumptions about the data, so they are not simply always a 'better idea'; they are also generally less powerful than parametric tests.

We will not discuss how to perform these tests in detail, as they are usually performed using dedicated computer software. The interpretation of *p* values, or confidence intervals, is generally the same as we have seen for the simple parametric tests in this chapter.

Most modern non-parametric tests work by generating a large set of 'pseudo-samples' based on the original sample. The *p* value for any difference between two samples is given by the proportion of these pseudo-samples that show as large a difference as that seen in the data.

According to the null hypothesis, scores in different groups are equivalent, and thus could be 'swapped'. Permutation tests can be used with small data sets. These tests look at all of the possible rearrangements of the original data, based on swapping data points that are equivalent according to the null hypothesis.

Resampling techniques do not 'swap' data points: rather, they generate many thousands of pseudo-samples at random by picking data points from the original sample. Each of the data points in a pseudo-sample is chosen at random from the set of equivalent points (allowing a given data point to appear more than once in a given pseudo-sample).

The commonest modern approach is a form of resampling known as 'bootstrapping', which can be used to test hypotheses, or calculate confidence intervals on statistics (such as the mode or median) for which other tests are too complicated, or non-robust. This technique works best with relatively large samples.

---

So what should we do if, when we plot a histogram of our samples, the values they contain appear to deviate markedly from a normal distribution? Under these circumstances, we may suspect that the data are not drawn from normally distributed populations. Consequently, it might not be appropriate to evaluate hypotheses about these populations using parametric tests such as a *t* test. If the assumptions of the tests are violated, we may expect them not to give reliable results.

Statisticians have shown that *t* tests are fairly robust even when the data are not normally distributed, and sample sizes are relatively small. A slight deviation of our data from normality should not trouble us unduly; but a marked deviation is potentially a problem. The test becomes more robust with larger samples, as the distribution of possible sample means will always become increasingly normal in shape as the sample size increases. However, how large do our samples need to be before the mean is approximately normally distributed? These questions do not have strict answers. Some people prefer always to have at least eight observations, whereas other researchers might stipulate ten or twelve observations as the absolute minimum.

However, whatever rule of thumb we use, there will be times when our data look too odd for us to wish to analyze them with a *t* test. For instance, suppose we wish to compare the numbers of moths caught by two different kinds of trap. We set out several traps of each type at randomly chosen locations, leave them overnight, and return the following morning to count the number of moths in each one. However, when we plot histograms of the number of moths caught for each trap type, we find that the data are markedly skewed to the right (see Section 9.4). This suggests that the population of 'possible measurements' is not normally distributed.

How can we compare the mean number of moths caught in the two trap types? A *t* test would assume that each sample was drawn from an approximately normally distributed population, which does not appear to be the case here. Recall from Chapter 9 that we can transform data values, for example by calculating the square root of each value. We might find that the square roots of our data are approximately normally distributed, meaning that we could therefore compare the sample mean of the square roots of the number of moths caught in each trap type using a *t* test. Because taking the square root of our values does not change their rank order, the hypothesis that the number of moths caught by trap type A is typically greater than that caught by trap type B may be evaluated by assessing whether the mean square root of the number of moths caught by trap type A is typically greater than the mean square root of the number caught by trap type B.

## Assumptions of chi-square procedures

When we run a chi-square test, the null hypothesis assumes that all of the observations are independent, and have the same probabilities of falling into each category. This is often forgotten. If some of our data points are related in some way, for example observations from the same animal, we cannot mix them with another set of related data.

The expected proportions must add up to 1. They represent the probability of any one observation falling into a particular category. This is very easy to get wrong: we must analyze all the observations, for example the number of failures as well as the number of successes. This is generally only a problem with contingency tests. In the T-shaped maze example above, we examined the frequency counts from two sexes 'male and female' that chose the two types of arm 'treated and untreated'. We might have been tempted to compare the number of males that chose the treated arm (20) with the number of females that chose the treated arm (12). This would be a mistake, and could produce a very misleading form of analysis in some situations: imagine that we saw the same numbers of each sex choosing the treated arm (20 and 12) after we had run 5000 males and 50 females. Obviously, we would not want to draw the same conclusions about bias!

> There are several important assumptions about data made by the chi-square test: observations should be (equally) independent, all observations should be included, and all expected frequencies should be at least 5.

Finally, the expected frequencies should not be too small (i.e. less than 5). The test relies on an approximation to the normal curve, which only works when all the expected counts are 5 or above, just as with the confidence intervals on a proportion. You can overcome this problem by combining similar categories together. For example, if you observed four colors of lizard, 'red', 'green', 'blue', 'orange', you might combine these as 'red or orange', 'green', 'blue' if your expected frequencies for orange turned out to be very low.

# Presenting Your Work

## QUESTION A

In a study of smoking behavior in the workplace, 100 staff (50 staff of each gender) are randomly selected from a large workforce, and their smoking status noted: 23 females and 13 males are smokers, the remaining staff in the sample all being non-smokers.

(a) Provide point and interval estimates of the proportion of each gender who smoke.

(b) Is there evidence for a difference in smoking behavior between the genders?

---

(a) For male sample, $n = 50$, proportion of smokers $p = 13/50 = 0.26$. Point estimate of male smokers is 26 %.

Interval estimate:

$$\pi = p \pm 1.96\sqrt{\frac{p - p^2}{n}} = 0.26 \pm 1.96\sqrt{\frac{0.26 - 0.07}{50}} = 0.26 \pm 1.96 \times 0.062$$

$$= 0.26 \pm 0.122 = 14\text{ % to } 38\text{ %}.$$

For female sample, $n = 50$, proportion of smokers $p = 23/50 = 0.46$. Point estimate of female smokers is 46 %.

Interval estimate:

$$\pi = p \pm 1.96\sqrt{\frac{p - p^2}{n}} = 0.46 \pm 1.96\sqrt{\frac{0.46 - 0.22}{50}} = 0.46 \pm 1.96 \times 0.070$$

$$= 0.46 \pm 0.138 = 32\text{ % to } 60\text{ %}.$$

(b) To test for difference, we need a chi-square contingency test.

Observed values:

|        | smokers | non | total |
|--------|---------|-----|-------|
| female | 23      | 27  | 50    |
| male   | 13      | 37  | 50    |
| total  | 36      | 64  | 100   |

The null hypothesis states that gender and smoking are independent, so we would expect equal proportions across rows/columns.

Expected values (row total × column total / grand total):

|        | smokers | non | total |
|--------|---------|-----|-------|
| female | 18      | 32  | 50    |
| male   | 18      | 32  | 50    |
| total  | 36      | 64  | 100   |

Chi-square statistic for testing this hypothesis has $df = 1 = (\text{rows} - 1) \times (\text{columns} - 1)$

$$\chi_1^2 = \sum \frac{(O - E)^2}{E} = \frac{(23 - 18)^2}{18} + \frac{(13 - 18)^2}{18} + \frac{(27 - 32)^2}{32} + \frac{(37 - 32)^2}{32} = 4.34.$$

This value is above the critical value of 3.84 for $\alpha = 0.05$ significance level, so there is evidence against $H_0$ at this significance level.

There is significant evidence that the genders differ in the proportion of smokers, $p < 0.05$.

## QUESTION B

Thirty-two farms are involved in trials of different possible crops for biodiesel production. Half of the farms are randomly allocated to grow and harvest each of two possible crop species. The yields (in kilograms per square meter) for the farms are given below:

Species A  4.0  1.7  5.1  3.6  4.1  4.9  5.5  4.5  4.4  3.7  5.4  3.2  5.8  3.4  3.1  3.7

Species B  2.9  5.5  4.2  5.2  2.4  2.6  3.9  3.6  2.7  3.4  2.5  2.6  2.4  4.5  2.9  3.6

(a) Provide a confidence interval estimate of the true mean yield for each species.

(b) Does either species give a yield significantly above the commercial target of 3 kg m$^{-2}$?

(c) Is there evidence for a difference in yield between the species?

---

(a) For species A, $n = 16$, mean yield $= 4.13$ kg m$^{-2}$, standard deviation yield $s = 1.07$ kg m$^{-2}$. For 95 % confidence intervals, we need 5 % critical value for $t$ with $df = n - 1 = 15$; $t = 2.13$.

95 % confidence interval for mean yield:

$$\mu = \bar{x} \pm t_{n-1} \times \frac{s}{\sqrt{n}} = 4.13 \text{ kg m}^{-2} \pm 2.13 \times \frac{1.07 \text{ kg m}^{-2}}{\sqrt{16}} = (4.13 \pm 0.57) \text{ kg m}^{-2}.$$

For species B, $n = 16$, mean yield $= 3.43$ kg m$^{-2}$, standard deviation yield $s = 0.996$ kg m$^{-2}$. 95 % confidence interval for mean yield:

$$\mu = \bar{x} \pm t_{n-1} \times \frac{s}{\sqrt{n}} = 3.43 \text{ kg m}^{-2} \pm 2.13 \times \frac{0.996 \text{ kg m}^{-2}}{\sqrt{16}} = (3.43 \pm 0.53) \text{ kg m}^{-2}.$$

(b) 3 kg m$^{-2}$ is below the 95 % confidence interval for species A, so this species produces significantly more yield than the target value.

3 kg m$^{-2}$ is within the 95 % confidence interval for species B, so this species does not produce significantly more yield than the target value.

(c) To test for difference in yield we need an *independent samples t test*.

Null hypothesis is that population mean yields are equal for two species.

$$t_{30} = \frac{\bar{x}_A - \bar{x}_B}{\sqrt{\dfrac{s_A^2 + s_B^2}{n}}} = \frac{(4.13 - 3.43)}{\sqrt{\dfrac{(1.07)^2 + (0.996)^2}{16}}} = \frac{0.7}{0.36} = 1.92.$$

Independent $t$ test, $df = n_1 + n_2 - 2$.

The critical value of $t$ with 30 $df = 2.04$ for $\alpha = 0.05$ significance level. Obtained $t$ is below this value, so there is no evidence against $H_0$ at this significance level.

There is no significant evidence that species differ in yield, $p > 0.05$.

## Basic

### 1. Confidence intervals when population variance is known

(a) A particular normally distributed variable has a standard deviation of 1, and a mean of 3. Using EQ11.1, determine the interval in which 95 % of observations of this variable will fall.

(b) A normally distributed population, $X$, has a standard deviation of 1, with an unknown mean, $\mu$. A single value, $x$, is sampled from this population. Given that $x = 4$, use EQ11.3 to obtain a 95 % confidence interval estimate of $\mu$.

(c) A normally distributed population, $X$, has a standard deviation of 10, with an unknown mean, $\mu$. A random sample of $n = 16$ items is taken from this population. Given that the mean of this sample is 4, use EQ11.4 to obtain a 95 % confidence interval estimate of $\mu$.

(d) A random sample of $n = 16$ items is drawn from a normally distributed population with a standard deviation of 12. The mean of the sample is 5. Provide a 95 % confidence interval estimate for the population mean.

(e) A random sample of $n = 25$ items is drawn from a normally distributed population with a variance of 100. The mean of the sample is 17. Provide a 95 % confidence interval estimate for the population mean.

### 2. Confidence intervals for proportions

(a) The proportion, $\pi$, of males in a large population is 50 %. Using EQ11.6, work out the standard deviation of possible sample proportions, $p$, for the possible random samples of size $n = 20$, drawn from the population.

(b) Using EQ11.1, determine the interval in which the 'proportion of males' will fall for 95 % of possible $n = 20$ samples.

(c) A single random sample, $n = 20$, is drawn from a population for which the proportion of males is *unknown*. Eight of the individuals in the sample are males, with 12 females. Using EQ11.7, provide a 95 % confidence interval for the proportion of males in the whole population.

(d) In a random sample of 30 observations of a certain species of grass in a large field, 12 plants tested positive for a particular gene. Provide a 95 % confidence interval for the proportion of plants of that species in the field that carry the gene.

(e) In a random sample of 100 coin tosses from a 'trick' coin, 75 were heads and 25 were tails. Provide a 95 % confidence interval for the probability of a head on each flip.

### 3. Hypothesis tests using a normal distribution

(a) A particular normally distributed variable has a standard deviation of 1, with an unknown mean, $\mu$. A single value, $x$, is sampled from this population to test the null hypothesis that $\mu = 2$. Given that $x = 4$, use EQ11.8 to obtain a $z$ statistic to test this null hypothesis. Is this $z$ value statistically significant at the conventional level (that is, more extreme than 1.96)?

(b) A normally distributed population, $X$, has a standard deviation of 10, with an unknown mean, $\mu$. A random sample of $n = 16$ items is taken from this population to test the null hypothesis that $\mu = 3$. Given that the mean of this sample is 4, use EQ11.8 to obtain a $z$ statistic to test this null hypothesis, and determine whether it is statistically significant. (Note: you will need to calculate the *standard error* of sample means.)

(c) A random sample of $n = 16$ items is drawn from a normally distributed population with a standard deviation of 12. The mean of the sample is 5. Is this statistically significant evidence against the hypothesis that the mean of the population is 7?

(d) A random sample of $n = 25$ items is drawn from a normally distributed population with a variance of 100. The mean of the sample is 17. Is this statistically significant evidence against the hypothesis that the mean of the population is 7?

(e) Compare the hypothesis test questions 3(b)–3(d) with the confidence intervals calculated in questions 1(c)–1(e). Can you see a relationship between the confidence intervals and the statistical significance of the test of each null hypothesis?

### 4. Hypothesis tests when the population variance is estimated (*t* tests)

(a) A particular normally distributed variable has an unknown mean, $\mu$, and standard deviation, $\sigma$. A random sample of $n = 16$ items is taken from this population to test the null hypothesis that $\mu = 3$. Given that the mean of this sample is 4, and the standard deviation is 20, use EQ11.9 to obtain a $t$ statistic to test this null hypothesis.

(b) A random sample of $n = 32$ items is drawn from a normally distributed population. The sample mean is 5, and the sample standard deviation is 12. Is this statistically significant evidence against the hypothesis that the mean of the population is 7?

(c) A random sample of $n = 40$ items is drawn from a normally distributed population. The mean of the sample is 12, and the variance is 120. Provide a confidence interval for the population mean. Does the sample provide evidence against the hypothesis that the mean of the population is 7?

(d) The following random sample was drawn from a normally distributed population:

7, 13, 7, 10, 9, 17, 6, 3, 10, 2, 12, 7, 23, 16

Provide a 95 % confidence interval for the population mean based upon this sample.

(e) What are the 95 % and 99 % confidence intervals for the mean of the following sample of numbers?

88.5, 92.4, 94.3, 79.8, 94.3, 75.9, 96.2, 99.8

Is the mean of these numbers statistically significantly different from a hypothetical mean of 100 at the 5 % and 1 % levels?

## 5.  The paired (related samples) *t* test

(a) Two measurements were taken from each of a sample of 30 organisms, one before and one after a treatment. The increase from the first to second measurement was recorded. The mean of these $n = 30$ increases was 5 units, and the sample standard deviation of the 30 increases was 10 units. Using EQ11.11, calculate a *t* statistic to determine whether the observed increase in measurement was statistically significantly greater than zero.

(b) Using EQ11.12, calculate a confidence interval on the population mean of possible increases for the data in question 5(a).

(c) A random sample of 7 items was taken, and measured at two time points. The following measurements were made (matched items are arranged in vertical pairs):

Sample 1:  14.5   12.3   7.9   16.8   5.1    9.2   9.7
Sample 2:  14.3   12.7   9.0   16.7   5.4   12.1   9.9

By calculating difference scores for each individual, perform a *t* test to determine if the change in sample mean between sample 1 and sample 2 is statistically significant.

(d) Provide a confidence interval for the mean increase in measurement between the two time points for the data in question 5(c).

(e) A sample of individuals was tested twice. The following measurements were taken (matched items are arranged in vertical pairs):

Test 1:  127  163  149  101  137  125  141  142  133
Test 2:  135  170  181  111  151  120  138  153  140

Provide a confidence interval on the mean increase in measurement between the two tests. Is the observed change in mean statistically significantly greater than zero?

## 6.  The independent *t* test

(a) Two populations are being compared by sampling. The sample from population A has $n = 10$, a sample mean of 7, and a sample standard deviation of 12. The sample from population B has $n = 10$, a sample mean of 9, and a sample standard deviation of 10. Using EQ11.13, calculate a *t* statistic to determine whether the difference between the sample means is statistically significant.

(b) Using EQ11.14, calculate a confidence interval on the true difference between population means for the data in question 6(a).

(c) The following random samples were taken independently from different normally distributed populations. Use a *t* test to determine whether they provide evidence for a difference in means between the populations from which the samples are taken, and provide a confidence interval for the difference between the means.

A:  10    7  10  14  13    6  14  12    4
B:  12  14  19  14  18    7  15  13  19

(d) The following random samples were taken independently from different normally distributed populations. Use a *t* test to determine whether they provide evidence for a difference in means between the populations from which the samples are taken, and provide a confidence interval for the difference between the means.

Sample 1:
114    113    111    111    124    114    119    114
Sample 2:
116    112    102    125    121    109    121    118

(e) The following random samples were taken independently from different normally distributed populations. Use a $t$ test to determine whether they provide evidence for a difference in means between the populations from which the samples are taken, and provide a confidence interval for the difference between the means.

A: 75 70 72 68 72 69 63 57 69 74 66 59
B: 62 63 59 66 62 63 66 56 65 66 64 68

**7. Pearson's chi-square test for goodness-of-fit for categorical data**

(a) In a random sample of 100 organisms from a population, 75 are male and 25 female. Using EQ11.15, calculate a test statistic to determine whether the sample provides statistically significant evidence that there are unequal numbers of each sex in the population.

(b) In a random sample of 100 trees in a forest, 19 are inhabited by squirrels. Is this evidence for more than 10 % of the trees in the forest being inhabited?

(c) In a random sample of 60 male lizards, 12 have green spots, 28 have yellow spots, and 20 have no spots. Is this evidence against the population having equal numbers of each spot type?

**8. Pearson's chi-square test for contingency**

(a) In a study of 200 undergraduate students at a large university, 100 students of each gender are asked about smoking behavior: 40 are female smokers, 60 are female non-smokers, 25 are male smokers, and 75 are male non-smokers. Is this evidence for a relationship between gender and smoking behavior?

(b) In an experiment on mice, 40 out of 45 mice bred without a certain gene develop diabetes, whereas only 7 of 50 with normal genotype develop the disease. Is this evidence that the absence of the gene makes the disease more likely?

(c) In a random sample of 60 male lizards, 12 have green spots, 28 have yellow spots, and 20 have no spots. Of these lizards, 8 of the 'green spot', 17 'yellow spot', and 5 'no spot' are observed to mate with 'no spot' females in a laboratory study. Is this evidence for spot color influencing mating preference?

**9.** In the general population, 10 % of adults is left-handed. Of 215 students starting a biology degree in one year, 34 were left-handed. Is this statistically significantly more left-handers than we would expect?

**10.** In the examples below, the data show the values obtained in two different tests of the same organism (the data are arranged in matched pairs, the members of each pair being shown one above the other). For each example, determine whether the average difference in measurement between test 1 and test 2 is statistically significant.

(a) Test 1: 4.5 2.3 7.9 6.8 5.3   6.2   5.7
    Test 2: 4.3 2.7 9.0 6.7 5.6 10.1   6.9

(b) Test 1: 127 163 149 101 137 125 141 142 133
    Test 2: 135 170 181 111 151 120 138 153 140

(c) Test 1: 5 3 7 11 9   4 3 2
    Test 2: 7 4 6 12 6 10 9 3

**11.** The following random samples were taken independently from two different normally distributed populations. Use a $t$ test to determine whether they provide evidence for a difference in means between the populations from which the samples are taken.

A: 43 70 51 35 60 77 48 62 57
B: 90 45 73 64 86 59 88 72 89

**12.** A craps die from a casino is rolled 330 times. The various sides appear with the following frequencies.

1 2 3 4 5 6
75 49 53 36 60 57

Is there evidence against the null hypothesis that the die produces the expected proportion of 1/6 for each number?

**13.** The following ten measurements were made of the light intensity at a field site at midday using a light meter. Within what interval can we infer that the true mean (that is, population mean, or mean of a very large number of such measurements) lies, with a 95 % confidence of being right?

4.32 5.07 4.29 6.02 5.11 4.93 3.98 4.83 5.50 6.10

**14.** Fifty male pigeons and 50 female pigeons are put in a bird maze. Twenty of the males escape from the maze and find food whereas 30 get lost. Only 8 female pigeons get out of the maze while 42 become lost. Is this good evidence that female pigeons get lost in bird mazes more than males?

## Intermediate

**15.** In an experiment to measure the speed of a new type of mechanical switch, the following 12 closure times (in milliseconds) were recorded:

605  460  752  321  550  612  700  680  800  491
523  594

(a) Provide a point estimate of the true (population) mean closure time, to the nearest millisecond.
(b) Provide interval estimates of the true mean closure time corresponding to 99 % and 95 % confidence levels.
(c) To be useful, the mean closing time for the switch must be less than 750 ms. Does this study suggest that the new switch will be useful?

**16.** Reaction times to a visual stimulus were measured for randomly selected students specializing in arts (A) and science (S) subjects. The table below gives each student's mean reaction time (in milliseconds) with his subject group:

A   A   S   A   S   S   S   S   A   S   A   A
58  73  71  69  82  59  64  68  77  60  63  52

S   A   A   S   S   S   A   S   A   A
66  74  65  70  72  61  79  74  70  62

Is there a statistically significant difference between the reaction times of arts and science students?

**17.** In a large forest, a random sample of 150 trees was examined. Eighty per cent of the trees were native species. Sixty per cent of the native-species trees contained nests for at least one species of bird, compared with 50 % of non-native-species trees.
(a) Provide a 95 % confidence interval estimate for the proportion of trees in the forest that were native species.
(b) Calculate a table showing the number of trees containing nests, for native and non-native species.
(c) Is there a statistically significant difference in the proportion of trees containing nests between native and non-native species?

**18.** Reviewing a series of studies, a scientist observes the following estimates of the true proportion of a particular strain of mouse that will develop diabetes.

25 %, 46 %, 35 %, 42 %, 38 %, 28 %, 45 %

When asked for an estimate of the true population proportion of mice that develop diabetes, the scientist concludes that we should be confident that the true proportion is between 46 % and 25 % (the range of observed values).
What is the 95 % confidence interval for the true proportion?
Should we be more, or less, than 95 % confident that the mean lies in the range reported by the scientist?

**19.** The swimming speed of 12 marine organisms was measured twice (once in each of two different artificial brine solutions). A paired $t$ test gives a $t$ value of 2.85 for the difference in swimming speed.
(a) How many degrees of freedom would this $t$ statistic have?
(b) Is the observed difference statistically significant?
(c) Why might the statistic be reported as significant at $\alpha = 0.01$?

**20.** A field study produced the following table of results when measuring the favorite scent out of three popular perfume brands:

| | scent preference | | |
|---|---|---|---|
| **age of female** | A | B | C |
| under 14 | 3 | 8 | 5 |
| 14–30 | 9 | 7 | 4 |
| over 30 | 6 | 11 | 16 |

Is scent preference statistically significantly different for different age groups?

**21.** Twelve volunteers are engaged in 'experimental conversation'. In the 'positive' condition they are 'reinforced' by an approving 'uh-huh' from the experimenter whenever they use the personal pronoun 'I'. In the 'negative' condition they are 'punished' by a disapproving 'huh' when they say 'I'. The experimenter expects that the personal pronoun will be used more in the positive condition. The rates of 'I' utterances in the experiment (10 minute intervals) are as follows:

| participant | A | B | C | D | E | F |
|---|---|---|---|---|---|---|
| positive condition | 17 | 62 | 20 | 11 | 31 | 25 |
| negative condition | 14 | 68 | 19 | 3 | 27 | 26 |

| participant | G | H | I | J | K | L |
|---|---|---|---|---|---|---|
| positive condition | 15 | 38 | 47 | 22 | 26 | 8 |
| negative condition | 9 | 22 | 40 | 19 | 20 | 11 |

Is there evidence for the feedback having any effect?

**22.** A sample of 20 regions from the same large field of cabbages was randomly assigned to one of two fertilization treatments: treatment A, artificially synthesized fertilizer; and treatment B, 'natural' fertilizer. The fertilizers (artificial or natural) were applied to each region over a 1 month period, and then the mean weight of the cabbages in each region was measured, giving the following data:

cabbage weight (in grams)
treatment A: 205  318  170  236  283  198  244
treatment B: 224  165  145  320  223  162  263

cabbage weight (in grams)
treatment A: 345  277  301
treatment B: 187  156  263

Is there a statistically significant difference in mean weights?
What range of differences between the mean weights of cabbages grown in the two types of fertilizer are plausible given these data, at the 95 % confidence level?

## Advanced

**23.** Identify what is wrong with the following reported conclusions.
(a) The sample mean of observed durations was 25 s (95 % confidence interval estimate for population mean = 20 s to 30 s; 99 % confidence interval = 22 s to 28 s).
(b) The two groups were significantly different, $t_7 = 1.99, p < 0.05$, two-tailed.
(c) The following data were obtained from 8 males and 8 females.

|  | males | female |
|---|---|---|
| enjoy public life | 6 | 4 |
| enjoy private life | 3 | 6 |

A chi-square analysis shows that there is a statistically significant association between sex and public/private life enjoyment.

**24.** Fifteen students from two colleges are tested on statistics.

college 1:    67  72  48  33  24  87  33  55
college 2:    56  74  46  65  70  57  43  25

college 1:    47  56  54  35  15  58  64
college 2:    24  53  86  45  53  78  25

(a) Do the colleges statistically significantly differ in mean mark?
(b) What is the 95 % confidence interval for the difference between the mean score between the colleges?
(c) Assuming a pass mark of 50, what can we conclude about the relationship between college and pass rate?

**25.** A biologist records the number of eggs laid by several spawning pairs of cichlid fish, obtaining the values shown below:

9  4  5  51  5  17  8  13  91  9  31  11  11  5  8
24  1  56  14  10  15  5  13  18  12

(a) Use a histogram to determine the direction of skew.
(b) Calculate the mean of these data, and generate a 95 % confidence interval estimate for the population mean.
(c) Re-express these data as $\log_{10}$ (number of eggs). Calculate the sample mean, and a 95 % confidence interval estimate for the population mean of these log egg numbers.
(d) Express the sample mean and the confidence limits of log egg number, in terms of egg numbers. Which approach (analysis of egg number or log egg number) gives the more informative answer?

**26.** The scores of 30 students in an examination are given below, along with the gender (M for male, F for female) of each student.

| M | F | M | F | M | F | F | F | M | M |
|---|---|---|---|---|---|---|---|---|---|
| 40 | 44 | 44 | 50 | 50 | 58 | 58 | 62 | 62 | 63 |

| M | M | M | M | F | M | M | F | M | M |
|---|---|---|---|---|---|---|---|---|---|
| 63 | 64 | 65 | 66 | 67 | 67 | 68 | 69 | 69 | 69 |

| F | F | F | F | F | F | F | M | F | M |
|---|---|---|---|---|---|---|---|---|---|
| 70 | 72 | 73 | 73 | 74 | 74 | 79 | 81 | 87 | 87 |

(a) Determine the mean and 95 % confidence interval for the scores of each gender.
The marks are classified as follows: <40 fail, 40−49 third, 50−59 lower second, 60−69 upper second, 70−79 first class, 80+ distinction.
Determine if these data provide evidence for the following research questions.
(b) Does either gender tend to gain more marks, on average, than the other on the examination?

(c)  Are students of each sex equally likely to achieve at least a first-class mark?

(d)  The university guidelines suggest first-class degrees should be awarded to around 25 % of students overall, and at least 65 % should achieve at least an upper second. Does the marking of these students differ from these guidelines?

# CHAPTER 12

# Biological Modeling

This chapter illustrates how biologists actually use the mathematical techniques that we have discussed in the previous chapters. In general, these techniques allow us to generate biological models; that is, mathematical descriptions of biological systems.

However accurate our biological theories are, we know from Chapters 9–11 that our data will not be *exactly* as predicted. We have seen how probability theory can be used to analyze variation from a particular theoretical description, or model, to estimate how likely it is that our data would be sampled if that model were correct. In the first half of this chapter, we will see how to find the parameters, or indeed the theoretical model, that give us the *best* description of our data by determining which curve gets closest to the data points when plotted on a graph.

Biological theories frequently make predictions in terms of the rates of change of certain values, rather than about the values themselves. For example, we may predict that the number of individuals in a population that fall prey to predators every day (the *rate* of predation) is proportional to the sizes of both the population of prey and the population of predators. Similarly, a theory of enzyme kinetics might predict that the *rate of progress* of a biochemical reaction will depend upon the concentrations of the reactants and the availability of enzyme-binding sites. In Chapter 5, we saw how to use differentiation to determine the rate of change (or *derivative*) of a function that describes a biological process.

**Figure 12.1**
Simple biological models allow us to analyze the relationships observed in the wild between population levels of predators and prey.

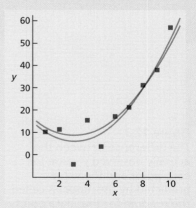

**Figure 12.2**
Biological models also allow us to understand and predict the relationships observed in experiments: the blue points represent observed data; the two curves are alternative models to describe the experimental data.

In the second part of this chapter, we will see how to construct *differential equations* (equations based on rates of change) to model growth of populations and the progress of chemical reactions. As you might predict, converting these differential equations into statements about quantities, rather than rates of change, usually relies on *integration*, as we saw in Chapter 6.

# 12.1 Fitting a model to experimental data: linear models

The simplest type of biological model is based on the concept of a correlation. For example, we would predict that the amount of food consumed by a population of apes will increase in proportion to the number of apes in the population. If such a model were true, then we would predict a positive linear correlation between a measurement of the number of apes, and a measurement of the amount of food they consumed.

In Chapter 9, we saw how, if there is a correlation between two variables, the data points will tend to cluster around a line on a scatterplot. When we obtain data of this form, they might suggest to us that a linear relationship (or correlation) is a good model for our data. However, if we wish to ensure that an increasing ape population has enough to eat, we also need to know *how much* the food consumption increases with population.

If we think of this in terms of a scatterplot of apes and food consumption, what we need is a technique that tells us which line best describes our observed data. The technique we use is called linear regression, which is the simplest example of a general technique known as least squares estimation for fitting a curve to observed data.

For any given straight line, we know that we can write an equation using two parameters: the slope ($B$) and the intercept ($A$):

$$y = A + Bx. \tag{EQ12.1}$$

If we take a pair of values for $A$ and $B$, then this gives us the equation for a possible line. Using this line (or equation) for each $(x, y)$ point, we can calculate a predicted value of $y$ based on the particular $x$ value, $\hat{y} = A + Bx$;

**Figure 12.3**
The difference between the actually observed *y*-value for each point and the value predicted by the model is called the *residual* for that data point.

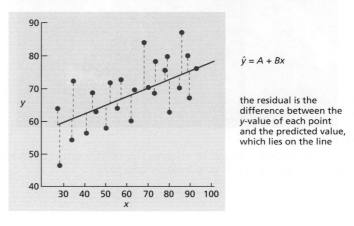

$\hat{y} = A + Bx$

the residual is the difference between the *y*-value of each point and the predicted value, which lies on the line

the little 'hat' on the $y$ indicates that the equation gives the 'predicted value of $y$', rather than an *actual* measured value of $y$.

Our overall aim is to find a line on a scatterplot that goes as close to all the points as possible. To do this, we obviously need to define 'close to' in some way that can be measured. The measurement of 'closeness' that we use is the same as that used for the measure of dispersion in a sample variance (and in the variance of a probability distribution); that is, the *square* of the residual.

## Box 12.1 Why use *squared* residuals?

When we perform a regression of one variable $Y$ upon another variable $X$, we find the line that, for our data, best predicts each value of $y$ that we observed. A fit is better if the line is closer to the points, and the total distance is measured as the sum of the squared differences between the observed value of $y$, and the value of $\hat{y}$ calculated for the corresponding observed $x$ value.

It is reasonable to ask why regression uses the squared vertical distance from each point to the line. The use of *vertical* distance means that we end up with the best prediction of $y$ (generally, we would not get the same answer when predicting $x$ from $y$ values). The use of *squared distance* means that we prefer a line that passes through the 'middle' of the set of points to one that goes closer to some points, but further away from others.

Why not just use something similar, such as the *total* vertical distance from the line? Consider Figure 1. It is

obvious that line A seems to give the best fit of the three lines to the six data points. The residuals for two of the points are shown, $a$ and $b$.

The value of residual $a$ is negative, and the value of $b$ is positive. Taking all six data points together, the positive residuals and negative residuals for line A are equal, and thus the *total* of all the residuals around line A is zero. So why do we not use this as our measure of which line is best? Well, it turns out that the same is true for line C, which also has equal positive and negative residuals!

It can easily be demonstrated that for *any* slope we choose, we can always find an intercept value such that all of the positive and negative residuals will cancel out. For example, if we choose a slope of zero, we find that an intercept equal to the sample mean of the $y$-values will give equal positive and negative residuals. Thus choosing simply the 'total' of the residuals will give us no way of deciding which possible slope is *best*.

A similar problem arises if we try to use *absolute* distance (ignoring the positive or negative sign). Now, if we move line A down (toward line B), we will increase the positive residuals (for example $b$), and *decrease* the negative residuals (for example $a$). For this reason, the total *absolute* distance from point 1 and point 2 is the same for all three lines shown in Figure 1: in fact, all three lines have the same total *absolute* distance from all the data points!

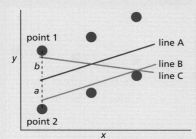

**Figure 1**
Three lines are shown with six data points. The residuals are shown for two of the points for line A ('$a$' and '$b$'). The residual '$a$' is negative because it is below the line. Adding all the residuals gives a total of zero for lines A and C. Adding all the residuals (ignoring sign) gives the same *total absolute distance* for all three lines. It is only when we consider the total *squared* residuals that line A counts as 'closer' to the six data points than lines B or C.

The problem is solved if we measure *squared* distance: this measure 'prefers' a line to go in the middle of two vertically separated points (for example point 1 and point 2), because if the total of $a + b$ is kept constant, the sum of $a^2 + b^2$ is smallest when $a = b$. It is only when we measure squared distance that the line through the 'middle' is regarded as the closest to the data points.

The total distance from a given line to our set of data is measured as the sum of squared differences between each data point and the corresponding point on the line. These differences are illustrated in Figure 12.3.

The difference between the *observed y-value* (the value in our data) and the *predicted y-value* (the point on the line of our equation using $A$ and $B$ values) is called the residual, or residual error, of that data point. If we move the line by changing the values of $A$ and $B$, we will change the residuals for our data points. Thus each line has a different value of total distance (total squared residuals) from the data points. We choose the one with the lowest total squared distance as the *best-fitting* line.

One field that requires this type of curve fitting is enzyme kinetics. Imagine we run a study to look at how the initial rate of an enzyme-catalyzed reaction varies with the concentration of a substrate. A fixed amount of enzyme is mixed with solutions of various concentrations of the substrate, and the rate of product formation is closely monitored. The data we get are shown in Table 12.1.

| Table 12.1  The results of an experiment measuring the initial velocity, or rate of a reaction, as the concentration of a substrate is varied | | | | | | | | | |
|---|---|---|---|---|---|---|---|---|---|
| concentration / units ($x$) | 1.00 | 2.00 | 3.00 | 4.00 | 5.00 | 6.00 | 7.00 | 8.00 | 9.00 |
| rate / units ($y$) | 24.45 | 47.72 | 54.67 | 62.47 | 69.05 | 67.40 | 71.98 | 73.99 | 76.96 |

We have plotted these data in Figure 12.4, along with two possible lines that one might have drawn 'by eye' to describe the relationship. One has a slope of $B = 12$, and intercept $A = 0$, the other has a slope of $B = 4$, and an intercept $A = 40$.

**Figure 12.4**
A graph of the data from Table 12.1, along with two possible straight-line 'fits' to the data that could have been drawn 'by eye'.

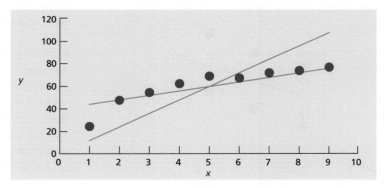

From looking at the figure, we would probably judge that the less-steep line ($A = 40$, $B = 4$) appears to give a better fit. However, rather than relying on a visual judgment, we can measure 'how much' this is true in terms of the total of squared residuals: we calculate the vertical difference between the line and the data point for each item, square it, and add them all up.

The calculations (shown in Table 12.2) give a larger total squared error (the sum of squared errors, $SSe$) for the steeper line ($y = 12x$) than for the other line ($y = 40 + 4x$), confirming our visual impression that the second line provides a *better fit* to the data.

**Table 12.2  Predicted values for the data in Table 12.1 using the linear models $A = 0$, $B = 12$ (table A) and $A = 40$, $B = 4$ (table B), along with the differences (residuals) and squared differences**

table A

| concentration/units ($x$) | 1.00 | 2.00 | 3.00 | 4.00 | 5.00 | 6.00 | 7.00 | 8.00 | 9.00 |
|---|---|---|---|---|---|---|---|---|---|
| rate/units ($y$) | 24.45 | 47.72 | 54.67 | 62.47 | 69.05 | 67.40 | 71.98 | 73.99 | 76.96 |
| predicted rate ($\hat{y} = 0 + 12x$) | 12.00 | 24.00 | 36.00 | 48.00 | 60.00 | 72.00 | 84.00 | 96.00 | 108.00 |
| difference ($y - \hat{y}$) | 12.45 | 23.72 | 18.67 | 14.47 | 9.05 | −4.60 | −12.02 | −22.01 | −31.04 |
| squared difference | 155.00 | 562.79 | 348.59 | 209.50 | 81.86 | 21.16 | 144.37 | 484.62 | 963.53 |

Total of squared differences $SSe = 2971.06$

table B

| concentration/units ($x$) | 1.00 | 2.00 | 3.00 | 4.00 | 5.00 | 6.00 | 7.00 | 8.00 | 9.00 |
|---|---|---|---|---|---|---|---|---|---|
| rate/units ($y$) | 24.45 | 47.72 | 54.67 | 62.47 | 69.05 | 67.40 | 71.98 | 73.99 | 76.96 |
| predicted rate ($\hat{y} = 40 + 4x$) | 44.00 | 48.00 | 52.00 | 56.00 | 60.00 | 64.00 | 68.00 | 72.00 | 76.00 |
| difference ($y - \hat{y}$) | −19.55 | −0.28 | 2.67 | 6.47 | 9.05 | 3.40 | 3.98 | 1.99 | 0.96 |
| squared difference | 382.20 | 0.08 | 7.13 | 41.91 | 81.86 | 11.56 | 15.88 | 3.94 | 0.92 |

Total of squared differences $SSe = 545.46$

This idea of 'total squared error' ($SSe$) gives us the ability to decide which of two (or more) different lines gives the best fit to our data. We could do this for any line (that is, any pair of values for $A$ and $B$ in EQ12.1), and determine what pair of values give the lowest $SSe$ out of those we have selected. However, unless we happen to choose the *best possible* values of $A$ and $B$, we would only find the best answer of those we try, and could not be sure to obtain the *best possible* answer.

To make sure we get the best possible line, we can use some techniques from earlier chapters. The formal approach to this problem involves deriving a formula for the total $SSe$ as a function of our intercept $A$ and slope $B$, and then using calculus to determine which values of $A$ and $B$ give the minimum value of this function. This method reveals that the values of $A$ and $B$ that give the lowest possible value for $SSe$ can be expressed in terms of the sample means ($\bar{x}$ and $\bar{y}$), the sample standard deviations ($s_x$ and $s_y$), and the *correlation coefficient r*, that were introduced in Chapter 9:

$$B = \left( r \frac{s_y}{s_x} \right), \qquad\qquad\qquad\qquad\qquad \text{(EQ12.2)}$$

$$A = (\bar{y} - B\bar{x}). \qquad\qquad\qquad\qquad\qquad \text{(EQ12.3)}$$

We will not look in detail at the derivation of these equations, but will take an informal approach to show that these answers are what we would expect from a model in which $Y$ varies in proportion with changes in $X$.

When $X$ has a typical value, we expect $Y$ to have a typical value. In other words, when $X$ is at its mean, we would predict that $Y$ is also at its mean value; therefore our line should go through ($\bar{x}$, $\bar{y}$).

In Chapter 9, we gave an interpretation of the correlation coefficient, which is that the value of $Y$ tends to increase by $r$ standard deviations ($s_y$) for every standard deviation ($s_x$) of increase in $X$.

Putting these two statements together gives the following:

$$\hat{y} = \bar{y} + \frac{r(x - \bar{x})s_y}{s_x} = \left(\bar{y} - \bar{x}r\frac{s_y}{s_x}\right) + \left(r\frac{s_y}{s_x}\right)x. \qquad \text{(EQ12.4)}$$

Looking more carefully at this equation, we should be reassured to see that it is a restatement of EQ12.1–EQ12.3. The slope of this line is the same as the expression for $B$ given in EQ12.2, and (using the $B$ term for simplicity), the intercept is the same as that given in EQ12.3.

These equations enable us to solve the problem of finding the *best possible* straight line to fit the data in Table 12.1. Calculating the required statistics for the data in Table 12.1 (see Chapter 9) gives: $s_x = 2.74$; $s_y = 16.60$; $r = 0.90$; $\bar{x} = 5$; $\bar{y} = 60.97$. From these values, we can obtain our values of the $A$ and $B$ coefficients (using EQ12.2 and EQ12.3):

$$B = \left(r\frac{s_y}{s_x}\right) = \left(0.90 \times \frac{16.60}{2.74}\right) = 5.47, \qquad \text{(EQ12.5)}$$

$$A = (\bar{y} - B\bar{x}) = 60.97 - 5.47 \times 5 = 33.60. \qquad \text{(EQ12.6)}$$

And we can substitute these values into EQ12.1 to obtain the formula for the best possible line, which is shown in Figure 12.5,

$$\hat{y} = A + Bx = 33.60 + 5.47x. \qquad \text{(EQ12.7)}$$

**Figure 12.5**
A graph of the data from Table 12.1, with the best-fitting straight line.

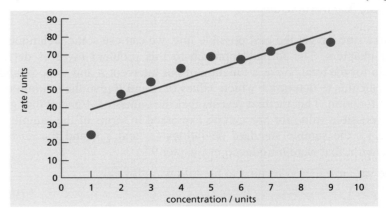

The best straight line for predicting $Y$ from $X$ can always be obtained by these formulae. The values of $A$ and $B$ are often called the **regression coefficients**, and the technique for finding the coefficients is called the **linear regression** of $Y$ on $X$. From looking at the equations, it should be clear that if we swapped all the $Y$ values for $X$ values we could obtain equations for predicting $X$ from $Y$. However, if we were to do this, we would get the equation for a different line. The line that gives the smallest vertical distance to the points is not the same as the line that gives the smallest horizontal distance to the points, except for the unusual case where all the points lie on a straight line.

Unlike correlation, the technique of linear regression is asymmetrical, so the line used when predicting $X$ from $Y$ is not the same as the line for predicting $Y$ from $X$.

Although we can obtain the slope and intercept of a best-fitting line, this does not give us all the information we need to assess our model. Figure 12.6 shows two data sets that have the same best-fitting line. However, it is clear that the fit for the data set on the left is much better than is the case with the set on the right. We need some way to describe how well our data fit the model, as well as simply describing the model itself.

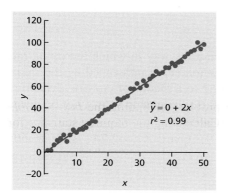

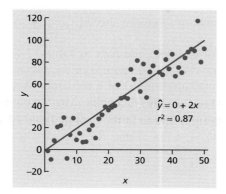

**Figure 12.6**
Two data sets which have exactly the same regression (best-fitting) line. Although the model is the same in each case, the fit to the model is much better in the left-hand panel than in the right-hand panel, which is reflected in the values of the coefficient of determination, $r^2$.

We have measured a model's fit in terms of the $SSe$, or total squared error from the regression line. For the graph in Figure 12.5, $SSe = 406.8$ units$^2$, which is of course lower than the values for the other lines in Table 12.2. However, as this value will depend upon many factors (including, for example, the number of data points), we usually express the fit in one of two ways.

The *average* squared distance from the line (**mean squared error**, or $MSe$) is obtained by dividing $SSe$ by $(n - 2)$, where $n$ is number of data points. In this case $MSe = 58.1$ units$^2$. The square root of this value, which is in the same units as the $y$ values, is called the **standard error of the estimate**. It is often used as a measure of how close items are to a fitted line. In this case it is 7.62 units.

The second way of expressing the fit is in terms of the square of the correlation coefficient, $r^2$, which is called the **coefficient of determination**. This statistic describes the relationship between the value of $MSe$ and the sample variance in $y$, according to the following formula:

$$MSe = (1 - r^2)s_y^2 \frac{(n - 1)}{(n - 2)}.$$  (EQ12.8)

This formula can be derived using algebra from the equation for the best-fitting line (we do not need to worry about exactly how). To see what this formula implies, we can make a simplifying approximation that $(n - 1) \approx (n - 2)$, which gives:

$$MSe \approx (1 - r^2)s_y^2.$$  (EQ12.9)

This approximation shows us that the $MSe$ is less than the sample variance. The difference, as a proportion of the sample variance, is $r^2$. Thus $r^2$ is the proportion of the *total variation in $y$* that is explained by the linear relationship.

> The regression coefficients (slope $B$, and intercept $A$) describe the best-fitting model. The coefficient of determination tells us how good a fit to the data the model is.

# 12.2 Fitting nonlinear models

The linear regression approach we saw above is simple, and always gives the best answer if we wish to model a relationship as a straight line. However, many biological theories will predict that the observed data should follow functions that are curved.

For example, the data in Table 12.1 concern the way in which the initial rate (or velocity) of an enzyme-catalyzed reaction ($v$) is influenced by the initial concentration ($s$) of a substrate. According to a theory of enzyme dynamics, known as the Michaelis–Menten model, this relationship should not be linear, but will be governed by the following equation:

$$v = \frac{V_{max}s}{K_M + s}.$$
(EQ12.10)

A graph showing the predicted relationship of the reaction rate to the concentration of substrate is shown in Figure 12.7.

**Figure 12.7**
The rate of an enzyme-catalyzed reaction as a function of substrate concentration, according to the Michaelis–Menten model.

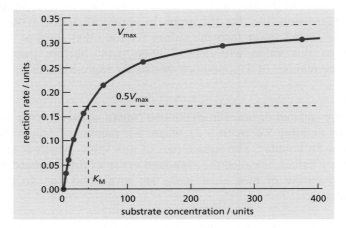

It is clear from the graph that the predicted relationship is not linear. We can understand the behavior predicted by the Michaelis–Menten mechanism by considering three cases.

First, if $s$ is much larger than $K_M$, the denominator of the rate equation $(K_M + s) \approx s$ so the velocity $v \approx V_{max}$. This represents a case where there is enough substrate to saturate the enzyme (effectively, all of the enzyme present will be catalyzing the reaction), and the reaction proceeds as rapidly as possible.

Second, we can see from the equation that when $s = K_M$, the initial rate $v = 0.5V_{max}$. Thus the Michaelis constant $K_M$ can be defined as the concentration of substrate at which the rate is half of its maximal value.

Finally, if $s$ is much smaller than $K_M$, the bottom half of the equation $(K_M + s) \approx K_M$. In this case, there is plenty of enzyme to go around, and the rate of formation of product is now linearly dependent on the concentration of the substrate:

$$v \approx \frac{V_{max}s}{K_M}. \qquad\qquad (EQ12.11)$$

We used linear regression to fit a straight line to some data for an enzyme-catalyzed reaction in Figure 12.5. However, these data seem to show more of a curved pattern than a straight line. This suggests that we might get a better fit to the data using the Michaelis–Menten model. To do this, we need to extend our linear regression approach to one that fits a **curved** line to the data.

We will not be fitting a straight line, so we cannot use EQ12.2 and EQ12.3 to obtain regression coefficients. In fact, we cannot easily derive the best possible fit using theory at all. We must therefore use trial and error to find the curve that gives us the lowest sum of squared residuals. We can compare the fit of two curves with different values of $K_M$ and $V_{max}$ (just as we did for comparing straight lines in Table 12.2) and choose the values that give the better fit.

The general approach will be:

- Start with some possible values for our model parameters (in this case, $K_M$ and $V_{max}$); that is, the unknown values in the equation.
- From these parameters, we calculate the predicted initial rate ($y$ value) for each initial concentration ($x$ value) for each data point.
- Measure the fit of the parameters by calculating the squared error (difference between observed and predicted $y$ values) for each point.
- Try to adjust the starting values in a manner that should reduce the total squared error, and start again.

The trick here is to know what adjustments to make in the final step. Luckily, we can use specialized computer programs (such as the 'Solver' tool that is built into Microsoft Excel®) to try this for us. This type of program enables us to follow the steps above quite rapidly, allowing trial-and-error fitting of different shapes of curve to our data.

To perform this task in Microsoft Excel®, we first need to set up our worksheet, as shown in Figure 12.8.

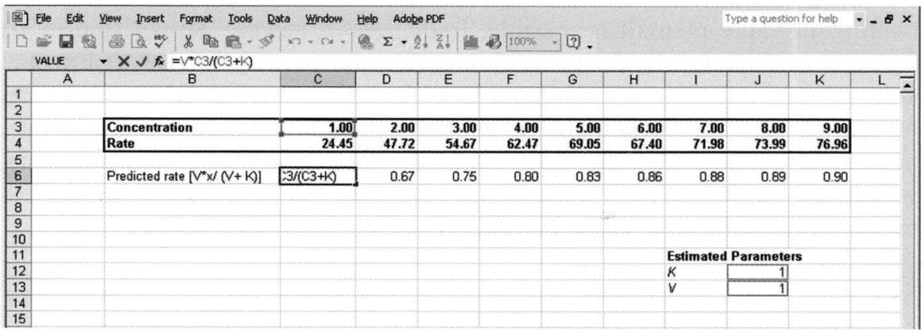

**Figure 12.8**
The data from Table 12.1 entered into a worksheet in Microsoft Excel®. The cells below the data contain a set of predicted values of rate, based on the concentration data and estimated parameters for the Michaelis–Menten model.

Step 1: we start by copying our data into a table of an Excel® worksheet. This can be seen at the top of Figure 12.8, in bold.

Step 2: we assign two cells of the worksheet to contain our estimated values for $K_M$ and $V_{max}$. These cells are shown bottom right of Figure 12.8. These cells must be given some starting values; in this case we have chosen a value of 1 for both of these parameters.

Step 3: we produce a row of cells that contain the predicted values of rate (based on the estimates of $K_M$ and $V_{max}$). These cells contain a calculation formula that predicts the rate based on the Michaelis–Menten formula (EQ 12.10). In Figure 12.8, we can see that the predicted rate for column C is given by the worksheet formula '= V*C3/(C3+K)'. Here, C3 is the cell containing the concentration for which we are predicting the rate, and K and V are names that have been defined to refer to the two cells containing our estimated values for $K_M$ and $V_{max}$. If you do not define names, it is possible to use cell references '$J$12' and '$J$13' in place of K and V.

After completing the row of equivalent formulae, one for each data point, we now have a set of *predicted rates* in our worksheet, which will change when we change our estimates of $K_M$ and $V_{max}$.

Step 4: we add two additional rows to the table, containing the residual and squared residual for each data point. For each column (that is, data point), we calculate the difference between the predicted value and the observed value, along with the squared difference. The sum of these squared values is *SSe* for the model based on the current values of our estimated parameters.

We now have a worksheet that is similar to Figure 12.9.

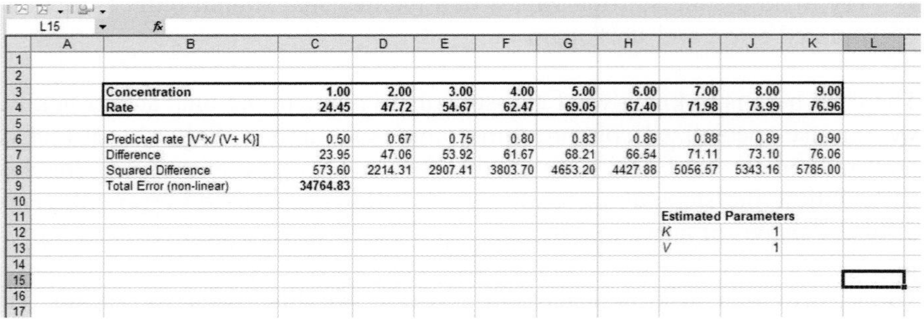

**Figure 12.9**
The worksheet from Figure 12.8, after the differences between predicted and observed reaction rates have been calculated, along with the squares of the differences, and the total squared difference, or error.

| | A | B | C | D | E | F | G | H | I | J | K | L |
|---|---|---|---|---|---|---|---|---|---|---|---|---|
| 1 | | | | | | | | | | | | |
| 2 | | | | | | | | | | | | |
| 3 | | Concentration | 1.00 | 2.00 | 3.00 | 4.00 | 5.00 | 6.00 | 7.00 | 8.00 | 9.00 | |
| 4 | | Rate | 24.45 | 47.72 | 54.67 | 62.47 | 69.05 | 67.40 | 71.98 | 73.99 | 76.96 | |
| 5 | | | | | | | | | | | | |
| 6 | | Predicted rate [V*x/ (V+ K)] | 0.50 | 0.67 | 0.75 | 0.80 | 0.83 | 0.86 | 0.88 | 0.89 | 0.90 | |
| 7 | | Difference | 23.95 | 47.06 | 53.92 | 61.67 | 68.21 | 66.54 | 71.11 | 73.10 | 76.06 | |
| 8 | | Squared Difference | 573.60 | 2214.31 | 2907.41 | 3803.70 | 4653.20 | 4427.88 | 5056.57 | 5343.16 | 5785.00 | |
| 9 | | Total Error (non-linear) | 34764.83 | | | | | | | | | |
| 10 | | | | | | | | | | | | |
| 11 | | | | | | | | | Estimated Parameters | | | |
| 12 | | | | | | | | | K | 1 | | |
| 13 | | | | | | | | | V | 1 | | |
| 14 | | | | | | | | | | | | |
| 15 | | | | | | | | | | | | |
| 16 | | | | | | | | | | | | |
| 17 | | | | | | | | | | | | |

Step 5: now that we have a measure of *SSe* (total squared error) in this worksheet, we can ask Microsoft Excel® to do the 'trial-and-error' process of making the value as small as possible. To do this, we run the 'Solver' add-in. At the time of writing, the Solver add-in is distributed with Microsoft Excel®, although typically it is not installed. Once installed, the Solver command can be started from the Tools menu. If the Solver command is not visible, it may need to be installed from the Add-ins menu. Using the dialog box that appears, we can ask the Solver to vary the values in the cells named V and K, in order to make the *SSe* value as small as possible. This step is shown in Figure 12.10.

Step 6: when the Solver has completed, it will report the values it has found which give the lowest *SSe* value. In this case the values found are $K_M = 2.33$ and $V_{max} = 96.9$, which give an *SSe* value of 48.0 units$^2$. This is much lower than the value of 406.9 units$^2$ obtained with a linear fit. The final values and a graph of the fitted model are shown in Figure 12.11.

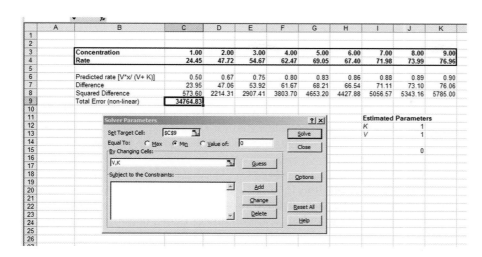

**Figure 12.10**
The worksheet from Figure 12.9, along with the Microsoft Excel® Solver dialog box, in which the values to be adjusted are chosen. See text for details.

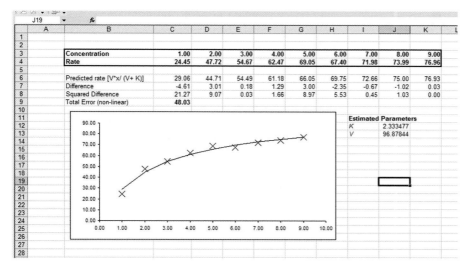

**Figure 12.11**
The worksheet from Figure 12.9, after the Solver has found the solution, along with a graph showing the best-fitting model.

This approach can be used to search for the best possible fit for any model we can express in a worksheet formula. However, it is not *guaranteed* to find the best answer for every set of starting values. As such, it is often worth starting with several different values of your parameters, and checking which answer found is the best (in terms of the lowest *SSe*).

## Box 12.2 Linearizing: fitting linear models to nonlinear data

The Michaelis–Menten model has been around since 1913, long before computers (and Microsoft Excel®) were around. So how did enzymologists go about estimating values of $K_M$ and $V_{max}$?

We saw in Section 12.1 that fitting a straight line to our data can *always* be done by theory, and thus does not need the help of specialized software such as the Solver. Because this 'general solution' applies to linear functions (i.e. straight lines), we can use this approach if we can find

a way to re-express our nonlinear model as a linear relationship between two variables. This is actually very easy to do in the case of the Michaelis–Menten equation.

If we take reciprocals of both sides of the Michaelis–Menten equation (EQ12.10), we find that it gives a linear relationship between $1/v$ and $1/s$:

$$\frac{1}{v} = \frac{(K_M + s)}{V_{max}s} = \frac{1}{V_{max}} + \frac{K_M}{V_{max}}\left(\frac{1}{s}\right).$$

This shows us that the relationship between $1/v$ and $1/s$ should be a straight line, with a slope ($B$) and intercept ($A$) given by:

$$A = \frac{1}{V_{max}},$$

$$B = \frac{K_M}{V_{max}}.$$

This way of expressing the Michaelis–Menten equation is known as the 'Lineweaver–Burk' equation, and is illustrated in Figure 1.

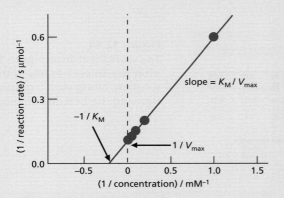

**Figure 1**
The Lineweaver–Burk plot of the Michaelis–Menten model plots the reciprocal of the velocity against the reciprocal of the concentration for each data point. This shows the values of the model parameters directly, in terms of the slope and the intercept of the line.

Until the computer-based approach used in Section 12.2 was available, fitting a straight line to the reciprocals was used to try to find the best values of $K_M$ and $V_{max}$ from a data sample like the one we analyzed above.

Unfortunately, despite its simplicity, this type of approach does not give answers as good as those from the nonlinear

fitting approach we saw above. This is easy to see in Figure 2, which shows the data from Table 12.1 in both 'raw' format (right) and on a reciprocal–reciprocal 'Lineweaver–Burk' plot (left).

If we generate values for $K_M$ and $V_{max}$ using linear regression, we obtain a model shown on both plots by the black lines, whereas the values we obtained with the Solver are shown by the red lines. It should be obvious that, although the black line does a good job of fitting the reciprocals (especially the one on the extreme right of the left-hand panel), it does not provide as good a fit to the raw data as the one we obtained using the Solver; the curve based on values of $K_M$ and $V_{max}$ from the linearized graph (the black line) has a standard error of the estimate of 4.34 units predicting the raw data, compared with 2.62 units for the red curve.

This might seem strange, but the reason can be seen if we look carefully at Figure 2. The difference between the Solver model and the data for the smallest values of concentration and rate have a large impact in the Lineweaver–Burk plot (left), because these points are 'spread out'; in effect, the straight line approach predicts these points more exactly, but at the expense of being able to predict other values quite as well. Another common linear plot is the Hanes–Woolf plot, in which the concentration divided by the velocity ($s/v$) is plotted against the concentration ($s$). This approach is less influenced by extreme reciprocals than the Lineweaver–Burk plot. However, we should not try to fit our data with a straight line just because we can: the approach of nonlinear fitting is preferred because it treats all data points equally.

Although the linear versions of the equations are not the best way of *fitting* the models, they remain a common (and therefore important) way of summarizing data because they show the values of $K_M$ and $V_{max}$ directly.

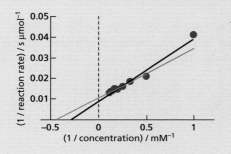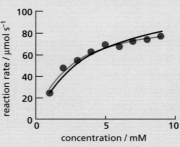

**Figure 2**
The data from Table 12.1 plotted as raw values (right) and as a Lineweaver–Burk plot of the reciprocals of each value (left). The black lines show the model obtained from fitting a straight line on the left-hand plot. The red lines show the model fit obtained using the nonlinear approach.

# 12.3 Differential equations: modeling changing systems

The previous two sections showed how to assess how closely our data fit a particular biological model, and how to work out the best parameters for such a model. The final sections of this chapter will demonstrate where these models come from. In general, biological models represent a way of writing down our theories in a precise mathematical way. In the example above, we started by fitting a model stating that 'the initial rate of reaction is proportional to the concentration of the reagent'. To assess this model, we need to measure the velocity of the reaction in terms of the rate of change in reactant concentration. However, what should we do if we can only measure the total amount of some process that occurs? The remainder of this chapter describes how to use *differential equations* to describe models for the *amount* of something that is measured, if the theoretical statements are based purely upon the *rate of change* of that amount.

We will start with a very simple example. We observe the leaf fall over a year in a forest, and find that dead leaves accumulate on the ground in an area of forest at a rate of 3 g cm$^{-2}$ of ground area per year. Furthermore, we estimate that in this environment leaves decompose at a rate of just over 0.2 % per day, which equates to 75 % per year. In other words the *current* rate of decay is a rate that would destroy 75 % of the current leaf mold mass if decay continued at the same exact rate for a whole year.

**Figure 12.12**
The rate of leaf fall and decomposition is one example where we can model a biological system in terms of rates of change.

This is a model based on rates of change, and we can use it to make predictions. For example, if nothing (wind, animals, and so on) disturbed the leaves, what would be the stable or 'steady-state' density (in terms of mass per unit area) of rotting leaves on the forest floor?

Two different factors affect the rate of change in the mass of dead leaves per square centimeter on the forest floor: the rate at which leaves *fall* and the rate at which they *decay*. The amount of leaf fall is a constant 3 g cm$^{-2}$ yr$^{-1}$. By contrast, the rate of decrease depends on the quantity of leaves present. We estimate that rate of decay at *any given instant* is such that, if it continued at that exact rate for a year, it would remove 75 % of the current amount of leaf mold.

Note that this simple model does not predict that 75 % of the mold would decay in a year if no more leaves fell, because the rate of decay is not constant. You may remember from our discussion of radioactive decay in Section 8.3 that when the rate of decay in something is proportional to the amount present, we will observe exponential decay. We will discuss such exponential growth and decay in the next section of the chapter.

Based on our knowledge of the two processes, we can build a simple model of the net rate of change in mass of leaf mold on the forest floor. The overall rate of change will be the rate of leaf falling (increasing the mass) minus the rate of decay (decreasing the mass). As we have simple expressions for each of these, we can therefore write an equation for the rate of change in the density of leaf mold mass, which we denote *mass*,

$$\text{rate of change} = 3 \text{ g cm}^{-2} \text{ year}^{-1} - 0.75 \text{ year}^{-1} \times mass$$
$$= -0.75 \text{ year}^{-1} \times \left(mass - 4 \text{ g cm}^{-2}\right). \tag{EQ12.12}$$

To simplify things a little bit, we can define some quantities that take care of the units. If we use the symbol $M$ to denote the current density of leaf mold mass *measured in grams per square centimeter* and we define $t$ as the time from some starting point *measured in years*, we can rewrite this expression without the units.

Where does $t$ come in? Recall from Chapter 5 that the rate of change of $M$ over time can also be called the *derivative of $M$* with respect to time $t$, so:

$$\frac{dM}{dt} = -0.75(M - 4). \tag{EQ12.13}$$

What we have written here is an equation for the *derivative* of $M$, called a *differential equation*, which can be used to answer questions relating to the rate of change in $M$. For example, we asked what the steady-state level of leaves would be, that is the rate at which the rate of decay matches the rate of leaf fall. To answer this, we need to find the value of $M$ for which the rate of change will be zero. This represents one type of solution to our differential equation, often (incorrectly) known as the *equilibrium solution* but more correctly described as the *steady-state solution* of the equation.

When the system is in a steady state, the processes of leaf fall and decay are balanced and the net rate of change is zero. This means that $-0.75(M - 4) = 0$, which can only be true when $M = 4$. The steady-state mass density of leaves on the forest floor must therefore be 4 g cm$^{-2}$.

## Classifying differential equations

Many differential equations can look a little more complicated than the one for the accumulation and decay of leaves that we have just constructed

(EQ 12.13). However, equations of the same type can usually be solved by similar approaches, so it is useful to learn how to distinguish between different types of differential equation.

First, if all of the unknown functions in the equation are functions of just one variable (such as $x$ or $t$), then the derivatives are known as ordinary derivatives, and the equation is known as an ordinary differential equation. By contrast, when the unknown functions depend on more than one variable, then the derivatives that occur will be called partial derivatives, and the equation will be a partial differential equation.

Second, the order of a differential equation is just the order of the highest-order derivative in the equation. Ordinary differential equations may involve an unknown function $y$ and an independent variable $x$ in expressions that include $x$, $y$, $dx/dy$, $d^2x/dy^2$, and so on. If $d^nx/dy^n$ is the highest derivative, then the equation is said to be of $n$th order.

For example $dy/dx = 4$ is a first-order differential equation, whereas $d^2y/dx^2 = 4$ is second order.

For example $dy/dx + 4y = 1$ is first order, whereas $d^2y/dx^2 - y = 2x$ is second order.

Third, a linear differential equation is one in which the function $y$, and all the derivative terms of $y$, appear as multiples, but not as powers or products.

For example $d^2y/dx^2 + 2dy/dx + 4y = 0$ is linear and second order.

For example $d^3y/dx^3 + (dy/dx)^2 + y = 4x$ is nonlinear and third order.

Fourth, and finally, if $y$ and its derivatives are set to zero, a differential equation is said to be homogeneous if the equation is still satisfied. Otherwise, it is said to be inhomogeneous.

For example $d^2y/dx^2 - y = 2x$ is second order, linear, and inhomogeneous.

For example $dy/dx + y/x = 0$ is first order, linear, and homogeneous.

We can now classify the equation we derived for the mass of leaves on a forest floor in terms of order, linearity, and homogeneity.

$$\frac{dM}{dt} = -0.75(M - 4). \hspace{4cm} \text{(EQ12.13, repeat)}$$

It is a first-order differential equation ($dM/dt$ is the only derivative).

It is an inhomogeneous differential equation (it does not hold when $M = 0$ and $dM/dt = 0$).

It is a linear differential equation ($dM/dt$ and $M$ are not raised to a power).

## Solving differential equations

In general we might want to know about the amount of something, not just about its rate of change. In our leaf fall example, we might be investigating the recovery of an area of forest after the leaves were cleared for some reason. Let us say that we know $M$ is currently 2 (that is, the mass of leaf mold is currently 2 g cm$^{-2}$), and we want to predict how $M$ will change over the next year.

To approach this type of question, we need to get an equation in terms of $M$. An equation for $M$ that satisfies the differential equation for d$M$/d$t$ is called a *solution* to the differential equation. If a differential equation that describes the change of some quantity, $y$, with respect to $x$ can be written in the form:

$$\frac{dy}{dx} = f(x), \qquad \text{(EQ12.14)}$$

where $f(x)$ is any function of $x$ (including $f(x)$ being zero or any other constant), then the way to solve the equation is obvious. If the derivative of the quantity $y$ is equal to $f(x)$, then the formula for the quantity as a function of $x$, $y(x)$, must be the integral of $f(x)$.

In other words, to solve the differential equation we integrate $f(x)$, add a constant of integration – and that is it! This type of solution is called **direct integration**. Returning to the leaf mold example, imagine we are only interested in the *total* leaf fall ($T$, when measured in grams per square centimeter) that occurs in an area of forest over a period of $t$ years. We know that the rate of fall of leaf mass is simply:

$$\frac{dT}{dt} = 3. \qquad \text{(EQ12.15)}$$

We need to find an equation for $T$ (as a function of $t$) that is a solution to this equation. We can easily obtain this by integrating both sides using the methods described in Section 6.2:

$$T = \int \frac{dT}{dt} \, dt = \int 3 \, dt = 3t + c. \qquad \text{(EQ12.16)}$$

The relationship $T = 3t + c$ is called the *general solution* to the differential equation, as it contains an unknown constant, $c$. This is the familiar constant of integration, and is required because an infinite number of solutions satisfy the differential equation (see Section 6.1). Essentially the solution to a first-order differential equation tells us how much one variable ($T$) changes when another variable ($t$) changes. However, because $T$ could have any value at the point that $t$ starts to change, there are many possible answers for $T(t)$.

If we want to end up with a *particular solution* of the equation, which describes a specific state of affairs, then we will need *additional information* to distinguish the single correct solution from all of the others that are possible. This extra information is called a *boundary condition*.

For example, we could ask how much total leaf fall we would have after $t$ years, if we started collecting with no leaves at all (i.e. when $t = 0$, $T(t) = 0$). This is an equation of the form $T(t) = 3t + c$ whose graph passes through the origin; that is, it satisfies the equation $T(0) = 0$. This tells us the constant $c$ must be zero, so the particular solution is $T = 3t$.

It is also possible to solve some simple second-order differential equations using the same method of direct integration. For example, if

$$\frac{d^2 y}{dx^2} = 2x, \quad \text{then} \quad \frac{dy}{dx} = \int \frac{d^2 y}{dx^2} \, dx = \int 2x \, dx = x^2 + c. \qquad \text{(EQ12.17)}$$

We can integrate a second time to show that

$$y = \int \frac{dy}{dx} \, dx = \int (x^2 + c) \, dx = \frac{x^3}{3} + cx + k. \qquad \text{(EQ12.18)}$$

Because we integrated twice while attempting to solve this differential equation, we end up with *two* undetermined constants in the general solution. We therefore need two boundary conditions to create a particular solution. This is true for all second-order differential equations.

> Generally, to obtain the particular solution to a differential equation problem we need as many boundary conditions as there are unknown values in the general solution.

## Separation of variables

The other common method for solving a first-order differential equation is known as **separating the variables**. This method is useful when the rate of change can be written in the form of a product (or quotient) of two functions,

$$\frac{dy}{dx} = f(x)g(y), \qquad \text{(EQ12.19)}$$

where $f(x)$ is a function of $x$ alone (possibly a constant) and $g(y)$ is a function of the dependent variable $y$ alone. This equation can be rewritten as follows:

$$f(x) = \frac{1}{g(y)} \frac{dy}{dx}. \qquad \text{(EQ12.20)}$$

What does this achieve? Well, we have 'separated' the variables in the equation, so that the function of $x$ appears on one side of the equals sign and the functions of $y$ and the derivative appear on the other. It is now possible to integrate both sides of this expression with respect to $x$:

$$\int f(x) \, dx = \int \frac{1}{g(y)} \frac{dy}{dx} \, dx. \qquad \text{(EQ12.21)}$$

Here comes the neat trick: we use the rule for changing the variable (see Section 7.6) of an integral on the right-hand side . . .

$$\int f(x) \, dx = \int \frac{1}{g(y)} \frac{dy}{dx} \, dx = \int \frac{1}{g(y)} \, dy. \qquad \text{(EQ12.22)}$$

Now the problem has been reduced to two separate integrals, one with respect to $x$ and the other with respect to $y$. If we can solve each of these, then we have a solution to the original differential equation.

Let us apply this to the leaf mold example. We saw that the differential equation that describes the current mass $M(t)$ of leaves (measured in grams per square centimeter) on the floor of the forest as a function of time $t$ (measured in years) is:

$$\frac{\mathrm{d}M(t)}{\mathrm{d}t} = -0.75(M - 4). \tag{EQ12.13, repeat}$$

What is the equation for the mass of leaves over time, if we start with no leaves at all? First, we need to find the general solution. We can separate the variables, by rearranging the equation to put all terms containing $M$ or derivatives of $M$ on the same side:

$$\frac{1}{M - 4} \times \frac{\mathrm{d}M(t)}{\mathrm{d}t} = -0.75. \tag{EQ12.23}$$

Next, we integrate both sides with respect to $t$:

$$\int \frac{1}{M - 4} \times \frac{\mathrm{d}M}{\mathrm{d}t} \, \mathrm{d}t = -\int 0.75 \, \mathrm{d}t. \tag{EQ12.24}$$

Further working using the change of variable rule (EQ7.72) and integrating the reciprocal of a function (EQ8.12) gives:

$$\int \frac{1}{M - 4} \, \mathrm{d}M = \ln(M - 4) = -0.75t + c, \tag{EQ12.25}$$

where $c$ is a constant of integration. Taking exponentials of both sides of this equation gives:

$$e^{\ln(M-4)} = (M - 4) = e^{-0.75t+c} = Ce^{-0.75t}, \tag{EQ12.26}$$

where $C = e^c$. Rearranging this equation gives our general solution:

$$M(t) = 4 - Ce^{-0.75t}. \tag{EQ12.27}$$

What about the particular solution? To find the particular solution we need to determine a value for the constant $C$. We can substitute $M(0) = 0$ and $t = 0$ into EQ12.27, to model build up of leaf mold after all existing mold was cleared. This gives $0 = 4 - C$, therefore $C = 4$. So the particular solution is

$$M(t) = 4 - 4e^{-0.75t}. \tag{EQ12.28}$$

We could have used different initial values for the starting amount of leaves. What effect would different values have on the long-term value of $M$? As $t \to +\infty$, we can see that $e^{-0.75t}$ will tend toward zero, thus even for the *general solution* (EQ12.27) $M(+\infty) = 4$. In other words, the initial mass of leaves has no effect on the long-term value of $M$, which will approach 4 from any initial value. This confirms what we had already deduced about the steady-state solution above.

## 12.4 Using differential equations I: population growth and decline

In Section 8.3 we saw that the exponential function $e^x$ has the unique property of being equal to its derivative. In other words, the rate of change (or slope) of this function is equal to its value. This is important in biology, because there are many situations where we might expect the rate of increase or decrease in some quantity to be proportional to its current amount. In such cases we would expect the exponential function to be a good mathematical model of the situation.

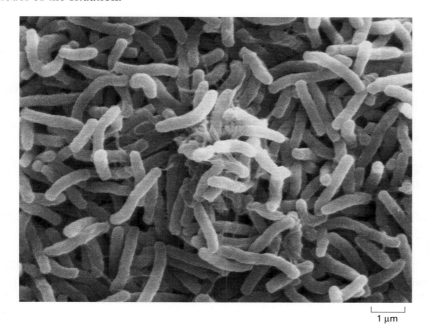

1 μm

**Figure 12.13**
Scanning electron microscope image of *Vibrio cholerae* bacteria, which infect the digestive system. If each bacterium in a colony reproduces at a constant rate, the size of the colony will grow in an exponential pattern.

For a simple example, we can construct a mathematical model for the growth of a population, if net growth in population is proportional to population size. Mathematically, our theory simply states that the rate of increase in population at any given time is proportional to the size of the population, $N$:

$$\frac{dN}{dt} = rN. \qquad \text{(EQ12.29)}$$

In this expression the constant of proportionality, $r$, is called the *growth rate*. If $r$ is positive then the population increases with time, but if $r$ is negative the population declines. If we want a function to describe the population as a function of time, we need a solution to this differential equation.

The solution can be found by separating out the variables and integrating both sides of the equation with respect to $t$:

$$\frac{1}{N}\frac{dN}{dt} = r, \qquad (EQ12.30)$$

$$\int \frac{1}{N}\frac{dN}{dt}\,dt = \int r\,dt. \qquad (EQ12.31)$$

Tackling the left-hand side first, we get:

$$\int \frac{1}{N}\frac{dN}{dt}\,dt = \int \frac{1}{N}\,dN = \ln N + c', \qquad (EQ12.32)$$

where $c'$ is a constant of integration. Because $r$ is a constant, the right-hand side is easier to deal with:

$$\int r\,dt = rt + c'', \qquad (EQ12.33)$$

where $c''$ is another constant of integration. We can therefore write that

$$\ln(N) = rt + c, \qquad (EQ12.34)$$

where $c'$ and $c''$ have been merged into a single constant ($c = c'' - c'$). Taking the exponential of both sides (to get rid of the natural logarithm) gives us:

$$N = e^{rt+c}. \qquad (EQ12.35)$$

If $N_0$ denotes the size of the population when $t = 0$, we can see that $e^c = N_0$, so the solution may be written:

$$N(t) = N_0 e^{rt}. \qquad (EQ12.36)$$

As we expected, the solution is a form of exponential function. When $r$ is positive, the model is said to represent exponential growth of a population; when $r$ is negative, the model represents exponential decay or decline. Examples of different exponential functions with different values of $r$ are shown in Figure 12.14.

We already saw in Section 8.3 that exponential decay or decline can be equivalently written in terms of the time it takes for the population or amount to reduce by one half, known as the *half-life* of the decay. We can do the exact same thing with exponential growth:

$$N = N_0 e^{rt} = N_0 (e^{\ln(2)})^{rt/\ln(2)} = N_0 2^{\frac{t}{\tau}}. \qquad (EQ12.37)$$

When $r$ is positive, the value $\tau = \ln(2)/r$ is known as the *doubling time*, and reflects the time taken for the size of the quantity to double from its current level at any point. Thus we can quickly characterize any exponential growth as a series of regular 'doublings' in size, and vice versa.

Consider the case of folding a piece of paper 0.1 mm thick. If we fold it in half, the resulting doubled piece is 0.2 mm thick. What is the equation for thickness $T$ as a function of number of folds $f$?

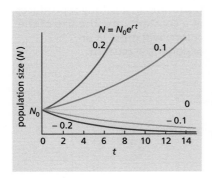

**Figure 12.14**
Exponential functions showing the effect of the value of $r$, the growth rate. When $r$ is more positive, the population grows rapidly. When $r$ is negative, the population size decays.

Every fold is doubling the thickness, so thickness will grow exponentially with the number of folds. Because it doubles after every fold, the equivalent of the doubling time in the equation is one fold. Thus if $T_0$ is the thickness after zero folds (0.1 mm $= 10^{-4}$ m), the equation for $T(f)$ is simply:

$$T = T_0 \times 2^f = 2^f \times 10^{-4} \text{ m}. \tag{EQ12.38}$$

If it were possible to fold the paper repeatedly, how thick would it be after 10 or 50 folds? If you folded it 10 times, it would be $2^{10} \times 10^{-4}$ m $= 1024 \times 10^{-4}$ m $= 0.1024$ m (about 10 cm). If you folded it 50 times, it would be $2^{50} \times 10^{-4}$ m $= 112\,589\,990\,684.2624$ m (about 112 million kilometers; you had better stop folding now, as the sun is only 150 million kilometers from the earth!).

## Box 12.3 Models and experiments

The simple models discussed in this chapter are based upon specific values of parameters such as leaf fall and decay rates. A more *general* approach to modeling biological processes would be to proceed without knowing or assuming these parameter values. We might use such a general model as the basis for experimental evaluation (does the model provide a good account of the real world?), or to conduct experiments based on the model to determine parameter values.

In the first simple model we discussed, we assumed that two processes affect the rate of change in the amount of leaf mold: the rate at which leaves *fall* and the rate at which the leaf mold *decays*. The rate of increase is assumed constant at a rate $F$, whereas decay depends on the quantity of leaf mold present. Our model states that the rate of decay is simply proportional to the amount of leaf mold present; that is, rate of decay $= k_D \times M$ where $M$ is the amount of leaf mold present at any instant and $k_D$ is a rate constant for decay.

The equation for net rate of change in mass of leaf mold is easy to write:

$$\text{rate of change in } M = \frac{dM}{dt} = F - k_D M.$$

Note that we have written this as a *physical value* equation, rather than using numerical values (in this equation $M$ means 'mass density' not 'mass density measured in grams per square centimeter'). Let us consider the steady-state solution for this value. At steady state, the rate of change is zero, so $dM/dt = 0$, which requires that $M = F/k_D$. Thus the rate constant can be expressed as a ratio of the steady-state value of mass $M_S$, and leaf fall rate $F$: $k_D = F/M_S$.

This gives us a means of establishing the rate constant via experimental measurement. For instance if we measured the rate of leaf fall as 3 g cm$^{-2}$ yr$^{-1}$ and the steady-state amount as 4 g cm$^{-2}$, we could obtain an estimate of $k_D$:

$$k_D = \frac{F}{M_S} = \frac{3 \text{ g cm}^{-2} \text{ year}^{-1}}{4 \text{ g cm}^{-2}} = 0.75 \text{ year}^{-1}.$$

We can also solve our physical value differential equation to obtain a general equation to describe the predicted mass $M(t)$ of leaves on the floor of the forest as a function of time $t$:

$$\frac{dM}{dt} = F - k_D M.$$

We can separate the variables, by rearranging the equation to put all terms containing $M$ or derivatives of $M$ on the same side:

$$\frac{dM}{dt} = -k_D \left( M - \frac{F}{k_D} \right)$$

$$\frac{1}{M - F/k_D} \frac{dM}{dt} = -k_D.$$

Integrate both sides with respect to $t$:

$$\int \frac{1}{M - F/k_D} \frac{dM}{dt} dt = -\int k_D dt.$$

Integrating each side in turn:

$$\int \frac{1}{M - F/k_D} dM = \ln(M - F/k_D) + c_0.$$

$$-\int k_D dt = -k_D t + c_1.$$

This gives:

$$\ln(M - F/k_D) + c_0 = -k_D t + c_1.$$

We have a problem with the left-hand side, as the contents of the bracket are not a number (they are a physical quantity, but we can only take logarithms of numbers; see Section 3.7). However, we can dodge this with a bit of nifty algebra. Combining constants $(c = c_1 - c_0)$ gives:

$$\ln(M - F/k_D) - c = -k_D t.$$

If we define a new constant $M_U$ such that $\ln(M_U) = c$, we have

$$\ln(M - F/k_D) - \ln(M_U) = -k_D t$$

$$\ln\left(\frac{M - F/k_D}{M_U}\right) = -k_D t.$$

Now, if we understand the new constant $M_U$ to represent some unknown *density* of leaf mold, we now have a ratio of densities, namely a number, for our logarithm. Taking exponentials of both sides of this equation gives:

$$\exp\left\{\ln\left(\frac{M - F/k_D}{M_U}\right)\right\} = e^{-k_D t}.$$

Rearranging this equation gives our general solution:

$$\frac{M - F/k_D}{M_U} = e^{-k_D t}$$

$$M = M_U e^{-k_D t} + F/k_D.$$

What about the particular solution? To find the particular solution we need to determine a value for $M_U$. Substituting $M = M_0$ at $t = 0$ gives

$$M_0 = M_U + F/k_D$$

$$M_U = M_0 - F/k_D$$

and the particular solution is

$$M(t) = M_0 e^{-k_D t} + (1 - e^{-k_D t})F/k_D.$$

How can we use this model experimentally? First, we could consider what happens to leaf mold if we prevent any more leaf fall ($F = 0$). In this case, the model predicts exponential decay of leaf mold with a half-life of $t_{1/2} = \ln(2)/k_D$. (You should be able to derive this result for yourself from the equation above.) This provides an alternative experimental method of determining $k_D$, as well as an experimental test of the model. A demonstration that the level of leaf mold follows an exponential decay would be a major confirmation of the assumptions we made.

Instead, we might perform an experiment in which we clear away all of the mold, and watch to see how it accumulates. The model predicts that the mold will build up toward the steady-state level $M_S$, and that the *difference* between the current level and $M_S$ will be given by:

$$M_S - M(t) = (F/k_D)e^{-k_D t}.$$

Again, you should be able to obtain this from the particular solution. This result shows that the difference between $M_S$ and $M(t)$ decays exponentially with time.

In this example, we started with plausible statements (although they are, obviously, simplifications) about the processes occurring: there is a constant rate of leaf fall, $F$, and the rate of decay of leaf mold is proportional, with constant $k_D$, to the amount of leaf mold present. Using algebra and calculus, we converted these into statements about the amount of leaf mold present as a function of time. Experiments can be used both to test the validity of the model *and* to determine values for $F$ and $k_D$.

## Logistic growth

The folding paper example is a clear illustration of why the simple exponential growth model described in the previous section is not realistic for sustained growth. Common sense dictates that a natural population cannot grow indefinitely. Instead, the population will eventually be restricted by the constraints of the environment in which it lives.

The most common way to model this is in terms of what is called a *logistic growth model*, which is illustrated in Figure 12.15. The logistic model (red line) starts with growth very similar to an exponential growth function, but then levels off.

The logistic model assumes that there are two components that influence the rate of growth. First, we assume that growth of the population depends upon the number in the population, as we did for exponential growth. This factor dominates for small populations, thus the two curves shown in Figure 12.15 are very similar for low population sizes.

Second, we assume that the environment can support a maximum population of size $K$, known as the environmental *carrying capacity*. Growth is therefore limited by the degree to which the environment can support further growth, that is by the 'spare capacity', $(K - N)$. Thus as the population approaches its maximum value, $K$, the rate of growth should approach zero, as we see in Figure 12.15.

According to such a model, the rate of change in the population is based on a product of two factors: the number in the population ($N$) and the degree to which it can still grow ($K - N$). Based on these assumptions, the differential equation is given by

$$\frac{dN}{dt} = cN(K - N),$$ (EQ12.39)

where $c$ is a combined constant of proportionality. You may well see this written in a slightly different form, known as the Verhulst equation, using a different constant of proportionality $r = Kc$:

$$\frac{dN}{dt} = cN(K - N) = \frac{r}{K}N(K - N) = rN\left(1 - \frac{N}{K}\right).$$ (EQ12.40)

To avoid units here, we will define $t$ as time measured in, say, hours. Otherwise, we would give $r$ and $c$ units of reciprocal time, e.g. $h^{-1}$.

The general solution to EQ12.40 is given by:

$$N(t) = \frac{KN_0}{N_0 + (K - N_0)\,e^{-rt}}$$ (EQ12.41)

where $N_0$ is the value of $N$ at time $t = 0$.

Examples of this curve for different values of $r$, $K$, and $N_0$ are shown in Figure 12.16. We see that the graphs all have a characteristic 'S-shape', which is called a sigmoidal curve. When $N$ is low, growth starts out like exponential growth with growth rate $r$. Growth is roughly linear around $N = K/2$, and then slows as the size approaches the asymptote of $N = K$, with the gap to the asymptote, $K - N$, following approximately exponential decay.

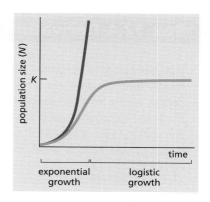

**Figure 12.15**
A comparison of an exponential (blue line) and logistic (red line) model of a population growth. The logistic model predicts initial growth very similar to the exponential model for small populations, but the population size quickly levels out and approaches an asymptote of a maximum value ($K$) known as the carrying capacity.

## Box 12.4 Derivation of the logistic growth function

This derivation is presented simply to show that the logistic function does not arrive 'by magic', but can be derived purely from the mathematics we have covered in the book, although it takes several stages. Although you are unlikely ever to need to *know* this derivation, it is useful to see how it arises from a simple differential equation.

The theoretical background of the model is the assumption that if growth rate is limited by some constant carrying capacity, $K$, then at any instant, growth is proportional to the size of the population, $N$, and to the 'spare capacity for growth' $(K - N)$. We start by writing this down as a simple differential equation:

$$\frac{dN}{dt} = cN(K - N).$$

The value $c$ is a combined constant of proportionality. This differential equation is separable because we can treat $c$ as a function of $t$.

$$c = \frac{1}{N(K - N)} \times \frac{dN}{dt}.$$

Integrating both sides of this equation with respect to $t$ gives:

$$\int c \, dt = \int \frac{1}{N(K - N)} \times \frac{dN}{dt} \, dt = \int \frac{1}{N(K - N)} \, dN.$$

The right-hand side needs a bit of algebraic manipulation before we can integrate it:

$$\frac{1}{N(K - N)} = \frac{K}{KN(K - N)}$$

$$= \frac{K - N}{KN(K - N)} + \frac{N}{KN(K - N)} = \frac{1}{K}\left(\frac{1}{N} + \frac{1}{(K - N)}\right).$$

We can therefore rewrite the differential equation as follows:

$$\int c \, dt = \frac{1}{K} \int \left(\frac{1}{N} + \frac{1}{(K - N)}\right) dN.$$

Therefore:

$$ct + Q_0 = \frac{1}{K} \int \left(\frac{1}{N} + \frac{1}{K - N}\right) dN = \frac{1}{K}(\ln N - \ln(K - N) + Q_1)$$

where $Q_0$ and $Q_1$ are constants of integration. Combining the constant terms ($Q = Q_0 - Q_1/K$), then

$$ct + Q = \frac{1}{K}\ln\left(\frac{N}{K - N}\right).$$

Now, to get a value for $Q$, we need a boundary condition. If $N = N_0$ when $t = 0$, then

$$Q = \frac{1}{K}\ln\left(\frac{N_0}{K - N_0}\right) - c \times 0.$$

Substituting this into our solution, gives

$$ct + \frac{1}{K}\ln\left(\frac{N_0}{K - N_0}\right) = \frac{1}{K}\ln\left(\frac{N}{K - N}\right).$$

This last equation can be rewritten in the form

$$Kct = \ln\left(\frac{N}{K - N}\right) - \ln\left(\frac{N_0}{K - N_0}\right) = \ln\left(\frac{N(K - N_0)}{N_0(K - N)}\right).$$

Taking exponentials of both sides:

$$e^{Kct} = \frac{N(K - N_0)}{N_0(K - N)}.$$

Rearranging this equation to make $N$ the subject gives, at last, the logistic function:

$$N = \frac{KN_0}{N_0 + (K - N_0)\,e^{-Kct}}.$$

Let us look at an example of how the logistic function can be used. A colony of bacteria is grown in a dish containing a medium that is capable of supporting a maximum population of only 600. One hundred bacteria are introduced initially, and after an hour 120 are counted. Assuming logistic growth, find the equation to describe the population of bacteria over time, measured in hours.

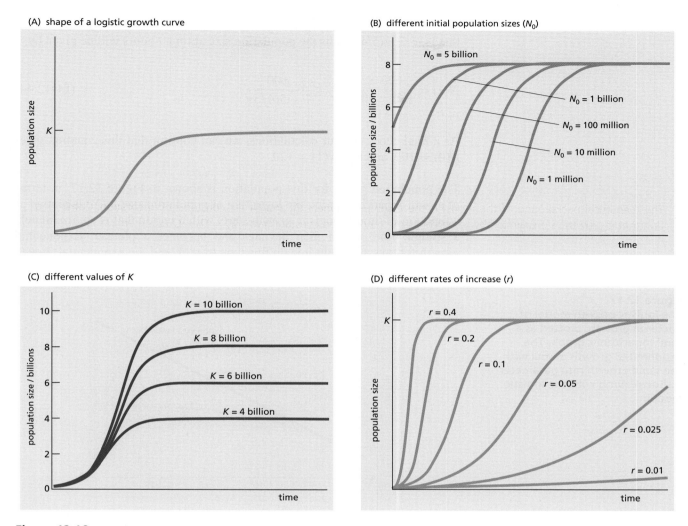

**Figure 12.16**
The effect of different parameters upon the logistic growth model.

The formula for logistic growth is:

$$N(t) = \frac{KN_0}{N_0 + (K - N_0)\,e^{-rt}}.$$

$K = 600$, $N_0 = 100$, and $N(1) = 120$, so we can solve this equation for $r$:

$$120 = \frac{600(100)}{100 + (600 - 100)e^{-r \times 1}} = \frac{600}{1 + 5e^{-r}}. \qquad \text{(EQ12.42)}$$

Therefore

$$e^{-r} = \frac{1}{5}\left(\frac{600}{120} - 1\right) = \frac{4}{5}, \qquad \text{(EQ12.43)}$$

that is, $r = \ln(5/4)$. Thus the population size at time $t$ hours will be given by:

$$N(t) = \frac{600 \times 100}{100 + 500e^{-rt}} = \frac{600}{1 + 5(4/5)^t} . \qquad \text{(EQ12.44)}$$

As a final check for our calculations, we can confirm that this equation gives $N(0) = 100$, and that $N(1) = 120$.

The population graph for this population is shown in Figure 12.17, in terms of a semi-logarithmic plot: the logarithm of population size $\ln(N)$ is plotted as a function of time $t$. Logistic growth starts with a region that is approximately a straight line, which then deviates and becomes a plateau. Exponential growth follows a straight line on this type of plot, and the exponential growth curve with growth rate $r = \ln(5/4)$ is shown in the figure for comparison.

**Figure 12.17**
The logistic growth model of bacterial growth, plotted as a semi-logarithmic graph. The exponential growth model with the same growth rate parameter is shown along with the logistic model.

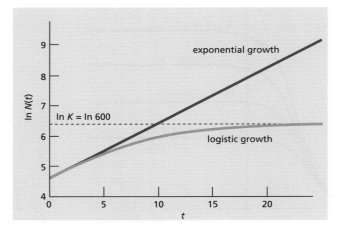

Finally, we will look at the logistic formula written in another way, which may help us to see why the curve has the shape it does:

$$N(t) = \frac{KN_0}{N_0 + (K - N_0)\,e^{-rt}} = \frac{KN_0 e^{rt}}{N_0 e^{rt} + (K - N_0)} = \frac{N_0 e^{rt}}{1 + \dfrac{N_0}{K}(e^{rt} - 1)} .$$

$$\text{(EQ12.45)}$$

- The top half of this function is the term $N_0 e^{rt}$, which is simply the equation for an exponential increase with growth rate $r$.
- The bottom half of the fraction is the term $1 + N_0/K(e^{rt} - 1)$, which is always one or more, so the logistic curve will never go above an exponential curve with growth rate $r$. This confirms what we see in Figures 12.15 and 12.17.
- When $t$ is close to zero, the bottom half of the fraction is close to one, so the logistic curve will start out being almost the same as an exponential growth curve. As $t$ is increased, the bottom of the fraction becomes larger than one, so the logistic curve falls ever further below the exponential growth function.

- When $t$ is large, the terms that contain $e^{rt}$ become very much larger than everything else. Ignoring all other terms in the fraction shows us that the asymptote will be $N = K$.

The logistic growth model is common in ecology. The reproductive behavior of species is often described on a spectrum that is labeled according to terms in this equation: '$r$ strategist' species produce many offspring (enabling rapid growth from small populations in unstable environments), whereas '$K$ strategist' species invest heavily in a smaller number of offspring (aiding their survival in populations that are close to a stable environmental carrying capacity).

# 12.5 Using differential equations II: biochemical reactions

Differential equations also provide useful models for many biochemical applications. As an example, we can work out the progress of a first-order chemical reaction in which substance A converts into substance B:

$A \rightarrow B$.

The law of mass action states that if the conversion of reactant A into product B represents an elementary reaction step, then the rate of the reaction (that is, the rate of formation of product) will be proportional to the concentration of A. In other words

$$\text{rate of formation of product} = \frac{db(t)}{dt} = ka(t), \qquad \text{(EQ12.46)}$$

where $a(t)$ is the concentration of the reactant A at time $t$, $b(t)$ is the concentration of the product B, and $k$ is the rate constant for the reaction. Again, to avoid units, we will define $a$, $b$, and $t$ as numerical values corresponding to concentrations or times measured on an appropriate scale. The rate of formation of the product must be equal to the rate of consumption of the starting material, as we cannot have product appearing from nowhere. Thus:

$$\frac{db(t)}{dt} = -\frac{da(t)}{dt} = ka(t), \qquad \text{(EQ12.47)}$$

or, more simply:

$$\frac{da}{dt} = -ka. \qquad \text{(EQ12.48)}$$

This type of differential equation is called the *rate equation* of the chemical reaction.

To get an equation that describes *how much* product is formed in a given amount of time, we must get rid of the derivative. To do this, we use integration to solve the differential equation as before. This is very similar to the differential equation for exponential growth we discussed in the previous section, and can be solved in the same way.

First, we separate the variables:

$$\frac{1}{a}\frac{da}{dt} = -k. \tag{EQ12.49}$$

Integrating both sides of this expression with respect to *t* we get:

$$\int \frac{1}{a}\frac{da}{dt}\, dt = -k \int 1\, dt. \tag{EQ12.50}$$

Solving each integration in turn, and merging the two constants, we get:

$$\ln a = -kt + c. \tag{EQ12.51}$$

Taking the exponential of both sides gives us:

$$a(t) = e^{-kt+c} = A_0\, e^{-kt} \tag{EQ12.52}$$

where $A_0 = e^c$.

This is confirmation of what we stated in Section 8.4 in the discussion of radioactive decay, namely that this equation is the solution of the simple differential equation that we started with. The time dependence of $a(t)$ is an exponential decay from a finite value at the start ($A_0$) to a value of zero at infinite time. The form of the curve is identical to the decay of a radioactive isotope, and the concentration of the reactant, A, even possesses a half-life.

So far, we have been considering how the concentration of the *reactant* in this chemical reaction changes over time. However, what about the concentration of the *product*? When we wrote down the rate equation, we stated that

$$\text{rate of formation of product} = \frac{db(t)}{dt} = -\frac{da(t)}{dt} = ka(t). \tag{EQ12.53}$$

Now that we know how *a* varies as a function of time, we should be able to work out the time dependence of the concentration of the product, *b*. This will give us a new differential equation to solve:

$$\frac{db(t)}{dt} = ka(t) = kA_0 e^{-kt}. \tag{EQ12.54}$$

This equation is telling us that the rate of increase in the concentration of the product B will *decrease with time*, as more of the reactant A is consumed and the term $e^{-kt}$ approaches zero.

If we want a formula for the amount of product, then we need to solve this equation. As before, we can begin in a straightforward way by integrating both sides of the equation with respect to $t$.

Integrating the left-hand side gives:

$$\int \frac{db(t)}{dt} \, dt = b(t), \tag{EQ12.55}$$

while integrating the right-hand side gives:

$$\int kA_0 \, e^{-kt} \, dt = kA_0 \times \frac{e^{-kt}}{-k} + c = -A_0 \, e^{-kt} + c, \tag{EQ12.56}$$

where $c$ is a constant of integration. Thus:

$$b(t) = -A_0 e^{-kt} + c. \tag{EQ12.57}$$

If there is no product present when we start monitoring the reaction, then $b(0) = 0$, which means that $c = A_0$. Therefore,

$$b(t) = -A_0 e^{-kt} + A_0 = A_0(1 - e^{-kt}). \tag{EQ12.58}$$

This time the concentration is initially zero and grows rapidly, but then gradually approaches a maximum value of $A_0$ at infinite time. The equation for $b(t)$ looks like that for exponential decay, although it has a $+A_0$ term. This tells us that the *gap between b and $A_0$* decays exponentially. Plots of both of these functions for $k = 2$ are shown in Figure 12.18.

## Multi-stage processes and steady-state models

Some biochemical reactions proceed through the formation of an intermediate, as in the pair of first-order reactions shown here:

$$A \xrightarrow{k} B \xrightarrow{k'} C.$$

Figure 12.19 shows functions $a(t)$, $b(t)$, and $c(t)$ that describe the variation of the concentrations of the reactant, intermediate, and product as a function of time, for two different values of $k$ and $k'$.

In the upper panel, the initial step of the reaction sequence is fast and the second is slow, $k = 10k'$. This leads to a build up in the concentration of the intermediate B before any significant amounts of the end-product C are formed. Under these circumstances, one could say that the conversion of B to C is the 'rate-determining step' of the reaction sequence.

This is not the case in the lower panel of Figure 12.19, which shows the case where the rate of the second stage of the reaction is much larger than the first, with $k = k'/10$. If $k'$ is large compared with $k$, so that the intermediate is much more highly reactive than the starting material, then the conversion of A to B will now be the 'bottleneck' in the process. As a result, almost as soon as intermediate B is formed it is converted rapidly into the end-product, C. It

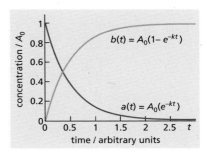

**Figure 12.18**
The progress of a first-order chemical reaction can be modeled with exponential functions.

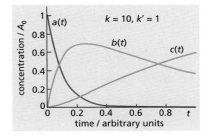

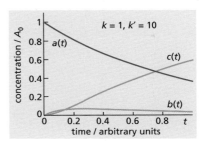

**Figure 12.19**
Plots of the functions $a(t)$, $b(t)$, and $c(t)$ which describe the theoretical concentrations of the reagent, intermediate, and product of a two-stage chemical reaction. The top panel shows a situation where the creation of the intermediate proceeds more rapidly than the conversion of the intermediate into the final product. The lower panel shows a situation in which the conversion of the intermediate is more rapid than its creation.

follows that the amount of the intermediate B being consumed will be approximately the same as the amount being produced, over a small time period.

In this kind of situation, we can model even quite complicated processes by using what is called a 'steady-state' approximation: this is an assumption that the concentration of one component will change at a rate that is much smaller than the rates of change of other components.

Let us see how this approximation might be used in this reaction. We will use $a(t)$, $b(t)$, and $c(t)$ to represent the changing numerical values of the concentrations of A, B, and C, respectively, over time.

A is converted to B at a rate $ka(t)$, so that:

$$\frac{\mathrm{d}a(t)}{\mathrm{d}t} = -ka \ . \tag{EQ12.59}$$

If only pure A is present at the beginning of the experiment, then as we have already seen, the concentration of A will decay exponentially over time:

$$a(t) = A_0 e^{-kt}. \tag{EQ12.60}$$

Similarly, C is formed by the conversion of B, and so:

$$\frac{\mathrm{d}c(t)}{\mathrm{d}t} = +k'b. \tag{EQ12.61}$$

The concentration of the intermediate B *increases* because of the consumption of A, and *decreases* by being converted into the end-product C, so its rate equation is written:

$$\frac{\mathrm{d}b(t)}{\mathrm{d}t} = +ka - k'b. \tag{EQ12.62}$$

For the situation shown in the lower panel of Figure 12.19, we could therefore make the approximation that $\mathrm{d}b(t)/\mathrm{d}t \approx 0$, or, in other words, assume that the concentration of intermediate B is maintained at a relatively *steady state*. Combining and solving all the above equations is simplified greatly once we assume that B is consumed at approximately the same rate that it is produced. If $\mathrm{d}b(t)/\mathrm{d}t \approx 0$ it follows that $ka \approx k'b$. Rewriting with $b$ as the subject gives:

$$b(t) \approx \frac{ka}{k'} = \frac{k}{k'} A_0 e^{-kt}. \tag{EQ12.63}$$

Note that using this approximation does not actually result in a *constant* concentration for B, which seems odd when describing this as a *steady-state* approximation. What we are actually assuming is that the rate of change in the concentration of B is *much smaller* than the changes in A and C.

Following on from this:

$$\frac{dc(t)}{dt} = k'b \approx kA_0 e^{-kt}, \quad \text{so} \quad c(t) \approx \int kA_0 e^{-kt} \, dt = -A_0 e^{-kt} + R$$

$$\text{(EQ12.64)}$$

where $R$ is a constant of integration. However, if we start with no C at all, we know that $c(0) = 0$, so $R = A_0$, which gives $c(t) = A_0(1 - e^{-kt})$. These equations indicate that formation of C is approximated by a first-order reaction from A.

How good an approximation does this approach provide? A comparison of this solution with one that does not use the steady-state approximation is given in Figure 12.20. Comparison of these two panels shows that the approximation gives a very good fit to the exact solution.

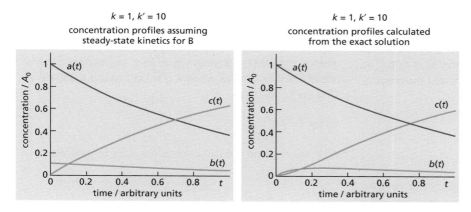

**Figure 12.20**
A comparison of the concentration functions $a(t)$, $b(t)$, and $c(t)$ produced when modeling a two-stage reaction using steady-state approximation (left panel) and using the exact solution (right panel).

A steady-state approximation can be very useful for simplifying the rate equations of complex reaction systems, and is an example of the way that special knowledge about the biology or chemistry of a particular system can be used to make a difficult mathematical problem easier.

## Box 12.5 The Michaelis and Menten model

In 1913, Michaelis and Menten proposed a reaction mechanism for the mode of action of an enzyme, from which the equations used in the text may be derived. This derivation is presented here simply to show how the tools that we have met in this book are capable of turning complex biological theories into simple equations.

We will label the enzyme E and the substance it acts on, the substrate, S. The overall reaction converts the substrate S into a product P with no net change in the concentration of the enzyme E:

$$E + S \rightarrow P + E.$$

Experiments show that the rate of formation of the product depends on the concentration of the enzyme, and so although the net reaction is just $S \rightarrow P$, this must reflect an underlying mechanism with steps that involve the enzyme. One simple mechanism is the following:

$$E + S \underset{k_{-1}}{\overset{k_1}{\rightleftharpoons}} ES_C \overset{k_2}{\longrightarrow} P + E.$$

In this scheme, $ES_C$ represents a bound, active complex formed between the enzyme and the substrate, which may decay with first-order kinetics, either producing the products or regenerating the original materials. This scheme assumes that there is no back reaction or inhibition of the enzymic reaction once products are formed. We are interested in the rate of formation of the product, P, a quantity that enzymologists call the 'velocity' of the reaction, $v$.

We will choose $e(t)$, $s(t)$, $c(t)$, and $p(t)$ to denote the time-dependent concentrations of the enzyme E, the substrate S, the enzyme–substrate complex $ES_C$, and the product P, respectively. Thus,

$$v = \frac{dp(t)}{dt} = k_2 c(t).$$

To deal with this equation it is necessary to know the concentration $c$ of the enzyme–substrate complex $ES_C$. We therefore need to set up the rate equation for $ES_C$.

$$\frac{dc(t)}{dt} = \underbrace{k_1 e(t)s(t)}_{\text{formation from E and S}} - \underbrace{k_{-1}c(t)}_{\text{decay into E and S}} - \underbrace{k_2 c(t)}_{\text{decay into product}}.$$

We can assume that the $ES_C$ complex is formed at approximately the same rate as it decays, i.e. using a 'steady-state' assumption $dc/dt \approx 0$.

$$k_1 e(t)s(t) - k_{-1}c(t) - k_2 c(t) \approx 0,$$

so that, using $K_M = (k_{-1} + k_2)/k_1$:

$$c(t) \approx \frac{k_1 e(t)s(t)}{k_{-1} + k_2} = \frac{e(t)s(t)}{K_M}.$$

The quantities $e(t)$ and $s(t)$ are the concentrations of the *free* enzyme and the *free* substrate: in other words they do not include the enzyme and substrate molecules that are tied up in the enzyme–substrate complex ($ES_C$). However, we know that the total amount of enzyme present, $E_0$, will remain constant throughout the reaction sequence, whether the enzyme is free in solution or is present in the complex when it is bound to the substrate. Thus we know that:

$$e(t) = E_0 - c(t).$$

With this in mind we can rewrite our approximate equation for $c(t)$:

$$c(t) \approx \frac{[E_0 - c(t)]s(t)}{K_M}.$$

Rearranging this expression to make $c$ the subject of the formula gives:

$$c(t) \approx \frac{E_0 s(t)}{K_M + s(t)}.$$

Therefore the rate of formation of the product is:

$$v = \frac{dp(t)}{dt} = k_2 c(t) \approx \frac{k_2 E_0 s(t)}{K_M + s(t)}.$$

If $s(t)$ is very large (much greater than $K_M$), then $K_M + s(t) \approx s(t)$ and $v \approx k_2 E_0$. Also, when $s(t)$ is very large, the enzyme will become 'saturated' with substrate, so that all of the enzyme present is in the $ES_C$ complex. At this point, the velocity is as large as it can be. Therefore, $V_{max} = k_2 E_0$. Thus we can rewrite our equation for $v$ as:

$$v \approx \frac{V_{max} s(t)}{K_M + s(t)}.$$

## Box 12.6 Two-stage reactions *not* assuming steady states

Earlier in the text we modeled a two-stage process by which A spontaneously converts to C through an intermediate B.

$$A \xrightarrow{k} B \xrightarrow{k'} C$$

We saw how a close approximation to this two-stage reaction could be modeled quite easily, providing that $k'$ is much greater than $k$, indicating that it is difficult to convert A into B, whereas B is highly reactive. This allows us to assume a *steady-state hypothesis* for B.

How can we proceed if we cannot make this assumption, for example if $k$ and $k'$ are very similar?

We showed that the following relationships hold:

$$a(t) = A_0 e^{-kt}; \quad \frac{dc(t)}{dt} = +k'b(t); \quad \frac{db(t)}{dt} = +ka(t) - k'b(t).$$

If we do not make the steady-state assumption, the differential equation for intermediate B is:

$$\frac{db(t)}{dt} = kA_0 e^{-kt} - k'b(t).$$

This can be solved, but requires a method beyond the scope of this book. However, you should be able to confirm, by differentiating $b(t)$, that the following is the general solution

$$b(t) = \frac{kA_0}{k' - k}(e^{-kt} - e^{-k't}).$$

This equation shows that the concentration of B is governed by both positive and negative exponential decay terms. If the positive term dominates, then the concentration of the intermediate will increase, but if the negative term is dominant, the intermediate will be consumed.

Finally, the rate of production of the end-product C depends on the concentration of B:

$$\frac{dc(t)}{dt} = k'b(t) = \frac{kk'A_0}{k' - k}(e^{-kt} - e^{-k't}).$$

Using the boundary conditions that at time $t = 0$, $a(0) = A_0$, $b(0) = 0$, and $c(0) = 0$, this differential equation can be solved by the direct integration method to give:

$$c(t) = A_0\left\{1 + \left(\frac{1}{k - k'}\right)(k'e^{-kt} - ke^{-k't})\right\}.$$

## Presenting Your Work

### QUESTION A

A model states that a foraging population of size $x$ will find an amount of food $y(x)$ in proportion to the number of individuals searching, and in inverse proportion to the amount of food already found. The differential equation derived to model this relationship is:

$$\frac{dy}{dx} = \frac{x}{y}.$$

Find the general solution, and determine the particular solution of the equation that satisfies the boundary condition that $y = 10$ when $x = 5$.

---

Separating the variables of the equation

$dy/dx = x/y$ therefore $y \times dy/dx = x$.

Then integrate both sides with respect to $x$:

$$\int y \times \frac{dy}{dx}\, dx = \int x\, dx \text{ and thus } \int y\, dy = \int x\, dx.$$

Therefore $y^2/2 + c' = x^2/2 + c''$. Let $c = 2(c'' - c')$, and the general solution is:

$$y^2 = x^2 + c.$$

If $y = 10$ when $x = 5$, then $c = 100 - 25 = 75$, so the particular solution is

$$y^2 = x^2 + 75.$$

---

**QUESTION B**

Confirm that

$$P(t) = \frac{kP_0 e^{rt}}{k + P_0(e^{rt} - 1)}$$

is a solution to the Verhulst formula

$$\frac{dP}{dt} = rP\left(1 - \frac{P}{k}\right),$$

with boundary condition $P = P_0$ when $t = 0$.

---

$$P(t) = \frac{kP_0 e^{rt}}{k + P_0(e^{rt} - 1)}$$

Let $u = kP_0 e^{rt}$

Let $v = k + P_0(e^{rt} - 1)$

$$\frac{du}{dt} = rkP_0 e^{rt} = ru$$

$$\frac{dv}{dt} = rP_0 e^{rt} = \frac{r}{k}u$$

$$\frac{dP}{dt} = \frac{d}{dt}\left(\frac{u}{v}\right) = \frac{v\dfrac{du}{dt} - u\dfrac{dv}{dt}}{v^2}$$

$$\frac{dP}{dt} = \frac{v(ru)}{v^2} - \frac{u\left(\dfrac{r}{k}u\right)}{v^2}$$

$$\frac{dP}{dt} = \frac{ru}{v} - \frac{ru^2}{kv^2}$$

But $P = \dfrac{u}{v}$, therefore

$$\frac{dP}{dt} = rP - \frac{rP^2}{k}$$

$$\frac{dP}{dt} = rP\left(1 - \frac{P}{k}\right)$$

Thus $P(t) = \dfrac{kP_0 e^{rt}}{k + P_0(e^{rt} - 1)}$ is a solution to $\dfrac{dP}{dt} = rP\left(1 - \dfrac{P}{k}\right)$.

When $t = 0$, $P(0) = \dfrac{kP_0 e^0}{k + P_0(e^0 - 1)} = \dfrac{kP_0}{k + P_0(1 - 1)} = \dfrac{kP_0}{k} = P_0$.

Thus $P(t) = \dfrac{kP_0 e^{rt}}{k + P_0(e^{rt} - 1)}$ is the correct solution for boundary condition $P = P_0$ when $t = 0$.

**QUESTION C**

The number of cars using a new car park exactly doubles every day from the park opening, until it reaches capacity. This happens for the first time on the fourteenth day of opening.

(a) How many days did it take to become more than one quarter full?

(b) If $C(d)$ is a function describing the number of cars on day $d$ of the park being open, write the equation for the increase in cars as a function of the number of cars on the previous day.

(c) Write an equation for $C(d)$, in terms of $C_1 =$ the number of cars on the first day. Over what range of days is this function an accurate description?

(d) What is the smallest number of parking spaces the park could have?

---

(a) If the car park was full on the fourteenth day, the first time it was at least half full was the thirteenth day. This means that it must have been at least one quarter full (for the first time) on the twelfth day.

(b) $C(d) = 2\,C(d-1)$, so change $= C(d) - C(d-1) = 2C(d-1) - C(d-1) = C(d-1)$.

(c) If the number of cars on the first day is $C_1$, then the total number of cars in the park is $C(d) = C_1 \times 2^{(d-1)}$, where $d$ is the number of days since opening.

This function is accurate for days 1–13, namely $1 \leqslant d \leqslant 13$.

(d) The smallest possible value for $C_1$ is 1 car (if $C_1 = 0$, the car park would still be empty on day 14) so the smallest value for $C(13) = 2^{12} = 4096$. The car park was full on day 14, but not day 13, so the number of spaces must be between 4097 and 8192; the smallest possible number of spaces is 4097.

## End of Chapter Questions

(Answers to questions can be found at the end of the book)

### Basic

**1.** What would be the particular solution for the differential equation $dM/dt = 3$, if the boundary condition stated that $M = 4$ when $t = 0$?

**2.** A colony of bacteria is grown in a dish. Sixty bacteria are introduced initially, and after an hour 120 are counted. Assuming exponential growth, find the equation to describe the population of bacteria $N$ after $t$ hours.

**3.** A colony of bacteria is grown in a dish; 100 bacteria are introduced initially, and after an hour 190 are counted. Assuming logistic growth, what would be the equation to describe the population of bacteria over time, if the dish is capable of sustaining a constant population of 1900?

**4.**

| X | 6 | 7 | 8 | 9 | 10 | 11 | 12 |
|---|---|---|---|---|---|---|---|
| Y | 276.19 | 308.79 | 353.79 | 389.69 | 429.29 | 468.19 | 518.09 |

The best-fitting line for predicting $Y$ from $X$ for these data is $Y = 32 + 40X$. What is the total squared error in the prediction for this line?

**5.** Classify the following ordinary differential equations according to order, linearity, and homogeneity:

(a) $\dfrac{dy}{dx} = x^{-4/3}$

(b) $\dfrac{d^2y}{dx^2} + 3\dfrac{dy}{dx} + y = 0$

(c) $\dfrac{dN}{dt} = KN\left(1 - \dfrac{N}{K}\right)$

(d) $\dfrac{dy}{dx} = \left(\dfrac{d^2y}{dx^2} + x\right)^{0.5}$

(e) $\dfrac{dy}{dx} - x^3y = e^x$

**6.** Obtain general solutions to the following differential equations:

(a) $\dfrac{dy}{dx} = x^{-4/3}$

(b) $\dfrac{d^2y}{dx^2} = 2x - 10$

(c) $\dfrac{d^2y}{dx^2} = e^{2x}$

**7.** Substance S converts to product P by means of a simple chemical reaction S → P. The concentration of S at time $t$ is given by $s(t) = S_0e^{-kt}$ where $S_0$ is the concentration of S at time $t = 0$. If $s = 0.25S_0$ after 30 minutes, determine the value of $k$.

**8.** Verify that the function $y = e^{-4t}$ satisfies the differential equation $dy/dx + 4y = 0$.

### Intermediate

**9.** The rate of growth of the population $P(t)$ of frogs (in thousands) after $t$ years is given by the equation

$$\frac{d}{dt}P(t) = (4 + 0.2t)^{3/2}.$$

(a) Find the general solution for $P(t)$ for this differential equation.

(b) Find the particular solution for the population growth over the first years of the millennium, if the population of frogs was 100 000 at the beginning of the year 2000.

(c) What is the population at the beginning of the year 2015?

**10.** Obtain general solutions to the following differential equations:

(a) $\dfrac{dy}{dx} = \dfrac{2x}{x^2 + 4}$

(b) $\dfrac{dy}{dx} - 4x^3y = 0$

(c) $\dfrac{dy}{dx} + \dfrac{y^2}{x + 2} = 0$

**11.** After administration, the amount of cocaine in human blood plasma decays exponentially with a half-life of 1.5 hours. If initially there is 0.2 mg of cocaine in the bloodstream, determine:

(a) the formula for the amount of cocaine in the bloodstream at any time, $t$;

(b) the time at which the amount drops below 0.01 mg;

(c) the percentage that will remain after 1 day.

**12.** Between 1950 and 1990 the human world population increased approximately exponentially at a rate of 1.86 % per year. Estimate the doubling time of the world population, according to this approximation.

**13.** Solve the following differential equation with the boundary condition that $y = 0$ when $x = 0$:

$$\frac{dy}{dx} = \frac{1 - y}{1 - x}.$$

**14.** A chemical reaction involving chemicals $X$ and $Y$ (with concentrations $x$ and $y$, respectively) is described by the differential equations $X$: $dx/dt = y$ and $Y$: $dy/dt = k^2 x$ (where $k$ is a rate constant).
(a) By differentiating one of the equations, obtain second-order equations involving $x$ or $y$ alone.
(b) Show that $x = (e^{kt} - e^{-kt})/k$ and $y = (e^{kt} + e^{-kt})$ are solutions to these equations.

**15.** The following values of $Y$ were found in an experiment varying some quantity $X$.

| $X$ | 1 | 2 | 3 | 4 | 5 | 6 | 7 | 8 | 9 | 10 |
|---|---|---|---|---|---|---|---|---|---|---|
| $Y$ | 6.9 | 7.2 | 9.6 | 12.0 | 11.4 | 14.2 | 16.1 | 16.5 | 18.5 | 20.1 |

The sample means of $X$ and $Y$ are 5.5 and 13.25, respectively; the sample standard deviations are 3.03 and 4.59; and the correlation $r$ is +0.990.
Determine the equation of the straight line that best predicts $Y$ from $X$.

**16.** According to theory, $Y$ varies with some quantity $X$ according to the hyperbolic function $Y = 100/(1 + kX)$. The following values of $Y$ were found in an experiment varying $X$.

| $X$ | 1 | 2 | 3 | 4 | 5 | 6 | 7 | 8 | 9 | 10 |
|---|---|---|---|---|---|---|---|---|---|---|
| $Y$ | 46 | 33 | 23 | 22 | 20 | 16 | 11 | 10 | 11 | 10 |

Determine which of the following values of $k$ gives the best account of the data, by calculating the total squared error in prediction for each value:
(a) 0.95
(b) 1.0
(c) 1.05
(d) Using Microsoft Excel®'s Solver tool, establish the value of $k$ that gives the lowest $SSe$.

**Advanced**

**17.** After a contaminated reservoir is treated with a bactericide, the rate of change of harmful bacteria $t$ days after treatment is given by:

$$\frac{dN}{dt} = -\frac{3000t}{1 + t^2}$$

where $N(t)$ is the number of bacteria in 1 ml of water.
(a) State with a reason whether the count of bacteria increases or decreases during the period $0 \leqslant t \leqslant 10$.
(b) Find the greatest rate of change in population over this period, and the time at which it occurs.
(c) If the initial count of bacteria is $N_0$, derive an expression for $N(t)$.
(d) Given that $N_0$ is 10 000 bacteria per milliliter, determine the count of bacteria after 10 days.
(e) After how many days (to one decimal place) does the bacteria count reach 500 bacteria per milliliter?

**18.** The power used by an electrical booster varies as a function of duration of use, $t$ (in minutes $0 \leqslant t \leqslant 10$) as follows:

$$\text{power} = \frac{dE}{dt} = \frac{3}{2} + 5\exp\left(\frac{t}{2} - 5\right).$$

An approximation for this function is proposed: $dE/dt = 0.3t + 1$.
(a) Derive the general solution for the energy consumption $E(t)$ according to the first model.
(b) Derive the general solution for the energy consumption $E(t)$ according to the approximate model.
(c) Give the particular solutions for each model, given that $E(0) = 0$.
(d) Taking the integer values from $t = 0$ to $t = 10$, determine how large the deviation (in terms of squared error) is between the true and approximate values of energy consumption.
(e) Does the linear approximation $E = t \times 2.049$ provide a better fit to these data points?

**19.** A newborn baby requires a complete blood transfusion. New blood is pumped in at a fixed rate $Q$ liters per second. A mixture of old and new blood is pumped out at the same rate. If $V$ is the volume of blood in the baby measured in liters and $c(t)$ is the proportion of new blood at time $t$ after pumping begins, then the following differential equation might model $c(t)$:

$$\frac{dc}{dt} = \frac{Q}{V} \times (1 - c).$$

(a) Explain why this equation is a reasonable model of the transfusion process.

(b) What assumption has been made in formulating the model?

(c) Suppose that $V = 0.5$ liters and $Q = 10^{-3}$ liters per second. How long does it take before there is only 1 % of the old blood left?

(d) How much new blood will have been used by this time?

**20.** Drug A binds to a receptor with association constant $K_a = N_a/(N_f \times [a])$, where $N_f$ is the number of free receptor sites (no molecule bound), $N_a$ is the number of sites bound to drug A, and $[a]$ is the concentration of the drug. The total number of receptor sites (bound and unbound) $N_T = N_f + N_a$.

(a) Show that the amount bound is

$$N_a = \frac{N_T \times K_a \times [a]}{1 + K_a \times [a]}.$$

(b) The results of an experiment to measure $K_a$ are shown below:

| $[a] / \mu M$ | $N_a / \text{pmol}$ |
|---|---|
| 10 | 38.0 |
| 30 | 71.0 |
| 100 | 132.0 |
| 200 | 161.0 |
| 300 | 165.2 |
| 400 | 179.2 |
| 600 | 191.8 |

Using a spreadsheet program, such as Microsoft Excel®, generate a table to compare these obtained values with predicted values based on estimated values of $K_a = 0.05 \ \mu M^{-1}$ and $N_T = 175$ pmol, and to calculate the sum of squared error in the predictions.

(c) If you have access to the Solver tool in Microsoft Excel®, use it to estimate the values of $K_a$ and $N_T$ that minimize the sum squared errors in prediction.

(d) Plot a graph of the results of the experiment, and a curve showing the fit of your model.

## Chapter 1

**Answer 1**

$12.2 \text{ mm s}^{-1}$

**Answer 2**

(a) 0.45

(b) 0.02

(c) 0.0000374

(d) 2.37

**Answer 3**

(a) $67\%$

(b) $0.000024\%$

(c) $20\%$

(d) $567.1\%$

**Answer 4**

(a) $68\,°F$, $293.15\,K$

(b) $37\,°C$, $310.15\,K$

(c) $-269.15\,°C$, $-442.47\,°F$

**Answer 5**

3 343 000 000

**Answer 6**

$56 \text{ mmol kg}^{-1}$

**Answer 7**

$8.55 \text{ mmol l}^{-1}$ assuming that the density of this dilute solution is $1 \text{ kg l}^{-1}$

**Answer 8**

(a) $67.9\%$ (w/w)

(b) $6.17 \text{ mol kg}^{-1}$

**Answer 9**

$$1.25 \frac{\text{pmol}}{\text{cm}^2 \text{ s}} \times \frac{1 \text{ mol}}{10^{12} \text{ pmol}} \times \left(\frac{100 \text{ cm}}{1 \text{ m}}\right)^2$$
$$= 1.25 \times 10^{-8} \text{ mol m}^{-2} \text{ s}^{-1}$$

**Answer 10**

Put the starting solution with concentration $C_1$ in the first tube and 990 µl of the solvent in each of the successive tubes. Transfer 10 µl of the original solution into the second tube. Mix. The concentration in this tube is now $C_2 = C_1 \times 10/(10 + 990) = C_1/100$. Take 10 µl of this solution and add it to the third tube, the concentration in this tube is now $C_3 = C_2 \times 10/(10 + 990) = C_2/100 = C_1/10\,000$. Continue for as long as necessary.

**Answer 11**

(a) Doubled

(b) Halved. The number of reactions per unit volume will go down four-fold, but there will be twice as much volume.

(c) Doubled

(d) Quadrupled

**Answer 12**

$2 \times 10^3 \text{ M}^{-1} \text{ s}^{-1} \times (10^{-6} \text{ M})^2 \times 0.1 \text{ l} = 2 \times 10^{-10} \text{ mol s}^{-1}$

**Answer 13**

Relative molecular mass $= 2$, so mass of $A = 6.023 \times 10^{23}$ molecules is $2 \times 10^{-3}$ kg. Therefore the mass of one molecule is $2 \times 10^{-3} \text{ kg}/6.023 \times 10^{23} = 3.32 \times 10^{-27}$ kg.

**Answer 14**

$6.023 \times 10^{23} \text{ion/mol} \times 1.6 \times 10^{-19} \text{C/ion} = 9.64 \times 10^4 \text{ C}$

**Answer 15**

(a) (Mass of CsCl) $= 15 \text{ g} \times 0.2$
$\qquad\qquad\qquad = \text{(mass of stock solution)} \times 0.6$
(mass of stock solution) $= 5$ g

(b) Mass of CsCl accompanying 1 kg water is
$5 \text{ mol} \times 0.16837 \text{ kg mol}^{-1} = 0.84185$ kg
so mass fraction of CsCl is
$0.84185/(1.84185) = 0.457 = 45.7\%$
Mass of CsCl $= 15 \text{ g} \times 0.457$
$\qquad\qquad\qquad = \text{(mass of stock solution)} \times 0.6$
Mass of stock solution $= 11.4$ g to which must be added 3.6 g water to make up the 15 g

(c) Mass of CsCl in 10 ml is
$0.01 \text{ l} \times 3 \text{ mol l}^{-1} \times 168.37 \text{ g} = 5.05$ g
Mass of 10 ml solution is
$1.385 \times (1 \text{ g}/1 \text{ ml}) \times 10 \text{ ml} = 13.85$ g
Mass of stock containing 5.05 g CsCl is
$5.05 \text{ g}/0.6 = 8.42$ g to which must be added
$12.85 \text{ g} - 8.42 \text{ g} = 4.43$ g water

**Answer 16**

The initial stock concentrations and volumes must be high enough that the solutes are easily weighed in the laboratory but the concentrations must remain below the solubility limits. Large volumes can be inconvenient and wasteful. The volumes to be pipetted must also be convenient, ideally all the same so that the steps in the procedure are easy to remember.

Choosing 10 ml as the volume to pipette, the stock concentrations should be NaCl 3.5 M, KCl 100 mM, $MgCl_2$ 12.5 mM, $CaCl_2$ 37.5 mM, Hepes 250 mM.

## Chapter 2

Answer to question in text after EQ2.64: $\quad x = 7/19$

Answer to question in text after EQ2.69: $\quad x = \dfrac{b_1 c_2 - b_2 c_1}{b_1 a_2 - b_2 a_1}$

Note that this can be written down 'by inspection' just by replacing $a$ with $b$ and $b$ with $a$ each time they occur in the solution for $y$.

**Answer 1**

(a) 51

(b) 47

(c) 7

(d) 1

(e) 190

**Answer 2**

(a) 9/14

(b) 5/14

(c) 1/14

(d) 7/2

(e) 63/8

**Answer 3**

(a) 1.375

(b) 0.13

(c) 6.25 %

**Answer 4**

(a) $4.38 \times 10^6, 4.4 \times 10^6$

(b) $2.35 \times 10^3, 2.3 \times 10^3$

(c) $3.37 \times 10^{-5}, 3.4 \times 10^{-5}$

**Answer 5**

300 %, 75 %

**Answer 6**

6-fold increase, 8-fold decrease

**Answer 7**

(a) $\pm 0.02$

(b) $\pm 0.2$

(c) $\pm 0.06$

**Answer 8**

$2.236 \ (2.236^2 = 4.999696 < 5 < 5.004169 = 2.237^2)$

**Answer 9**

(a) $13x - 2$

(b) $4x^2 + 5x + 4$

(c) $x^4 + 8x^3 + 24x^2 + 32x + 16$

(d) $3x^4 y^{12}$

(e) $x^4 + 4ax^3 + 6a^2x^2 + 4a^3x + a^4$

**Answer 10**

(a) $\dfrac{2(x - 3)}{x(x - 2)} = 2(x - 3)/(x(x - 2))$

(b) $x/(x + 1)$

(c) $(x + 1)/(2x + 1)$

(d) $(x + 2)/(x - 3)$

**Answer 11**

$a^2 + 2ab + b^2$

$a^3 + 3a^2b + 3ab^2 + b^3$

$a^4 + 4a^3b + 6a^2b^2 + 4ab^3 + b^4$

$a^5 + 5a^4b + 10a^3b^2 + 10a^2b^3 + 5ab^4 + b^5$

**Answer 12**

(a) $x = W/2 - 500$

(b) $t = (2 - 1/N)/3, \ (N \neq 0)$

(c) $z = Tb/(a - Tc), \ T \neq a/c$

**Answer 13**

$x = 2.62$

**Answer 14**

$x = -2$ and $x = -6$

**Answer 15**

$x = -1.76$ and $x = -6.24$

**Answer 15a**

$x = -1.17$

**Answer 16**

$x = 13/5, \ y = -3, \ z = -2/5$

**Answer 17**

$$\frac{1}{a \times b} \times (a \times b) \equiv 1$$

$$\frac{1}{a \times b} \times a \times b \equiv 1$$

Multiply both sides of the second form by $1/b$ and use that $b \times (1/b) \equiv 1$

$$\frac{1}{a \times b} \times a \times b \times \frac{1}{b} \equiv \frac{1}{b}$$

$$\frac{1}{a \times b} \times a \equiv \frac{1}{b}.$$

Now multiply both sides by $1/a$ and use that $a \times (1/a) \equiv 1$

$$\frac{1}{a \times b} \equiv \frac{1}{b} \times \frac{1}{a}.$$

But because $a \times b \equiv b \times a$ we can also write

$$\frac{1}{a \times b} \times b \times a \equiv 1$$

and after multiplying first by $1/a$ and then by $1/b$ we have that

$$\frac{1}{a \times b} \equiv \frac{1}{a} \times \frac{1}{b} \equiv \frac{1}{b} \times \frac{1}{a}.$$

**Answer 18**

$$\frac{1}{\dfrac{1}{a} + \dfrac{1}{b}} = \frac{1}{\dfrac{1}{a}\left(\dfrac{b}{b}\right) + \dfrac{1}{b}\left(\dfrac{a}{a}\right)} = \frac{1}{\dfrac{b}{ab} + \dfrac{a}{ab}} = \frac{1}{\dfrac{a + b}{ab}} = \frac{ab}{a + b}$$

**Answer 19**

$x^2 + dx + e = 0$ where $d = b/a$ and $e = c/a$.

$$x = -\frac{d}{2} \pm \frac{\sqrt{(d^2 - 4e)}}{2}$$

**Answer 20**

$ax^2 + bx + c = 0$

and $b^2 - 4ac = 0$;

the second equation means $c = b^2/4a$

so the first can be rewritten

$ax^2 + bx + b^2/4a = 0$

divide by $a$

$x^2 + (b/a)x + (b/2a)^2 = 0$

$(x + b/2a)^2 = 0$.

**Answer 21**

$a = 1, b = 0, c = x^2 - 1$

$y = \pm\sqrt{(1 - x^2)}$

Real solutions exist for $-1 \leqslant x \leqslant 1$.

**Answer 22**

$x = 0, y = 1$ and $x = 1, y = 0$

**Answer 23**

$x = 2.89$ and $y = -2.77$, $x = -0.485$ and $y = 3.97$

## Chapter 3

**Answer 1**

$x = 1$ with $y = 0$ and $x = 0$ with $y = 1$

**Answer 2**

(a) 2.6214

(b) $-4.3786$

(c) 1.2428

(d) $-3.6214$

(e) $-0.1893$

**Answer 3**

108320

**Answer 4**

2.577

**Answer 5**

(a) 9.974

(b) 0.01382

(c) $\ln(231) = 5.442$

(d) $112.4 \times \ln(23.9/15) = 52.36$

**Answer 6**

10 units

**Answer 7**

pH 5.5

**Answer 8**

$20 \log_{10}(5 \times 10^{-3}/(2 \times 10^{-5}))$ dB $= 48$ dB

**Answer 9**

$B = 300, \tau = t_{\text{double}}/\ln(2) = 36$ minutes

**Answer 10**

(a) $t = \tau \ln(N_0/N)$

(b) $2.37 \ln(18/5) = 3.036$

**Answer 11**

The semi-log plot; $200 \exp(0.3466t)$; 204800. When the population gets too big, food will become scarcer, and this will limit population growth.

**Answer 12**

$f(u) = f(a^v) = vf(a) = v = \log_a(u)$

## Chapter 4

**Answer 1**

(a) $37.4° \times (\pi \text{ rad} / 180°) = 0.2078 \, \pi \text{ rad}$

(b) $0.2078 \, \pi \text{ rad} = 0.6528$ rad

Note: the unit rad is optional

**Answer 2**

(a) 0.7944

(b) 0.5878

(c) $-1.664$

(d) 0.2500

**Answer 3**

(a) 52°, 128°

(b) 44°, 136°, 224°, 316°

(c) 73°, 253°

(d) 262°, 278°

(e) No values

**Answer 4**

$69\,020$ m$^2 \approx 6.90 \times 10^4$ m$^2 = 6.90$ ha

**Answer 5**

$8825$ µm$^2 = 8.8 \times 10^3$ µm$^2 = 8.8 \times 10^{-9}$ m$^2$

**Answer 6**

$9.2$ mm$^3 = 9.2 \times 10^{-9}$ m$^3$

**Answer 7**

0°, 67°

**Answer 8**

$400 + 60 \cos(\pi t)$

**Answer 9**

665 mm

**Answer 10**

$4.8 \times 10^4$

**Answer 11**

$7.32$ m, $23^2 \times 75/12\pi$ m$^3 = 1052$ m$^3$

$= 1.05 \times 10^3$ m$^3$, $4.7 \times 10^5$ kg (470 tonnes)

**Answer 12**

34.1 mm, 51°, 26°

**Answer 13**

Take $B = A\cos(\varphi)$ and $C = A\sin(\varphi)$

**Answer 14**

To prove EQ4.24 refer to Figure 4.12

$$x = c\cos(\alpha)$$
$$c^2 = h^2 + x^2$$
$$a^2 = h^2 + (b-x)^2 = h^2 + b^2 - 2bx + x^2$$
$$= h^2 + b^2 - 2bc\cos(\alpha) + c^2 - h^2$$
$$= b^2 + c^2 - 2bc\cos(\alpha)$$

To prove $\dfrac{\sin(\alpha)}{a} = \dfrac{\sin(\gamma)}{c}$ refer to Figure 4.12

$$\sin(\alpha) = \frac{h}{c}$$
$$\sin(\gamma) = \frac{h}{a}$$
$$\sin(\alpha) \times c = h = \sin(\gamma) \times a$$
$$\frac{\sin(\alpha)}{a} = \frac{\sin(\gamma)}{c}$$

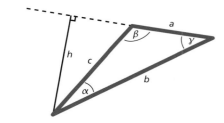

To prove $\dfrac{\sin(\beta)}{b} = \dfrac{\sin(\gamma)}{c}$ refer to the figure in this answer

$$\sin(180° - \beta) = \sin(\beta) = \frac{h}{c}$$
$$\sin(\gamma) = \frac{h}{b}$$
$$\sin(\beta) \times c = \sin(\gamma) \times b$$
$$\frac{\sin(\beta)}{b} = \frac{\sin(\gamma)}{c}$$

# Chapter 5

**Answer 1**

(a) (i) Average rate of change between weeks 8 and 12

$$= \frac{1530 - 1249}{12\text{ week} - 8\text{ week}} = 70.25\text{ week}^{-1}$$

(ii) Average rate of change between weeks 10 and 14

$$= \frac{1598 - 1501}{14\text{ week} - 10\text{ week}} = 24.25\text{ week}^{-1}$$

(iii) Average rate of change between weeks 12 and 16

$$= \frac{1710 - 1530}{16\text{ week} - 12\text{ week}} = 45\text{ week}^{-1}$$

(b) Instantaneous rate of change at week 12 is approximately

$$\frac{1620 - 1440}{15.6\text{ week} - 8.8\text{ week}} = 26.5\text{ week}^{-1}$$

**Answer 2**

(a) $2x + 1$

(b) $3x^2$

**Answer 3**

(a) $I = 0$

(b) $I = B$

**Answer 4**

(a) $\dfrac{dy}{dx} = 4 \times 8 \times x^{8-1} = 32x^7$

(b) $\dfrac{dy}{dx} = -1 \times 12 \times x^{12-1} = -12x^{11}$

(c) $\dfrac{dy}{dx} = \dfrac{1}{\pi} \times 1 \times x^{1-1} = \dfrac{x^0}{\pi} = \dfrac{1}{\pi}$

(d) $\dfrac{dy}{dx} = -\dfrac{1}{3} \times x^{-1/3-1} = -\dfrac{1}{3} \times x^{-4/3}$

(e) $\dfrac{dy}{dx} = 2x^{-1/2} = 2/\sqrt{x}$

**Answer 5**

$$\frac{\Delta P \pi r^3}{2\eta L}$$

**Answer 6**

(a) $\dfrac{dy}{dx} = 2 \times x^{2-1} + 3 \times 1 \times x^{1-1} - 5 \times 0 = 2x + 3$

(b) $\dfrac{dy}{dx} = -\dfrac{1}{3} \times x^{-1/3-1} - 1 \times 1 \times x^{1-1} - 6 \times 0$

$$= -\frac{1}{3}x^{-4/3} - 1$$

(c) $\dfrac{dy}{dx} = 1 \times x^{1-1} + (-2) \times x^{-2-1} = 1 - 2x^{-3}$

(d) $\dfrac{dy}{dx} = 2 \times (-1) \times x^{-1-1} - 3 \times \dfrac{1}{2} \times x^{1/2-1}$

$$= -2x^{-2} - \frac{3}{2}x^{-1/2}$$

(e) $\dfrac{dy}{dx} = 5 \times 4 \times x^{4-1} - (-3) \times x^{-3-1} = 20x^3 + 3x^{-4}$

**Answer 7**

(a) The rate at which the growth rate increases; m year$^{-2}$

(b) Jerk, that is rate of increase of acceleration; mm ms$^{-3}$

## Answer 8

The answers can be seen by reference to the graph below. Maxima occur at points B, F, and J; minima at D and H; points of inflexion at A, E, I, C, G, and K.

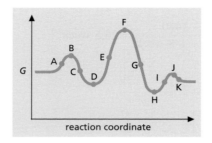

## Answer 9

$$\frac{dV}{dr} = \frac{24\varepsilon}{r}\left[\left(\frac{\sigma}{r}\right)^6 - 2\left(\frac{\sigma}{r}\right)^{12}\right]$$

## Answer 10

(a) $-1600\ \mathrm{Nm^{-5}}$

(b) $V = \dfrac{8 \times 10^{-5}\ \mathrm{Nm}}{P}$

$$\frac{dV}{dP} = 8 \times 10^{-5}\ \mathrm{Nm} \times (-1) \times P^{-2} = -\frac{8 \times 10^{-5}\ \mathrm{Nm}}{P^2}$$

(c) $-1580\ \mathrm{Nm^{-5}}$

## Answer 11

(a) The population curve (shown in brown) displays restricted growth from an initial population of 100 to a maximum population of about 1000. The population has nearly reached its maximum value by year 11 and in future would grow only very slightly.

(b) $\Delta N/\Delta t$ (red curve) goes through a maximum at $t \sim 4.5$ years, indicating that the population curve goes through a point of inflexion. The rate of change of population should be at its greatest at this point. $\Delta N/\Delta t$ remains positive throughout, so the population continues increasing.

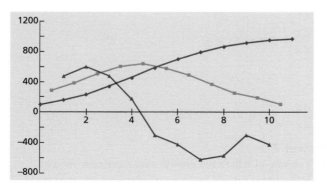

(c) The $\Delta^2 N/\Delta t^2$ plot (blue curve) cuts the $t$ axis at $t \sim 4.5$ years, confirming that this is where the population curve passes through a point of inflexion. This curve looks rather noisy because we have multiplied the signal by a factor of 25!

## Answer 12

$x = 12, 5.4\ ^\circ\mathrm{C}$

## Answer 13

$8\ \mathrm{mg\ l^{-1}}$

## Answer 14

$C/2$

## Answer 15

$c = 0.5t$

## Answer 16

(a) $y = 3x + 1$

(b) $3y = -x + 1$

(c) $(-0.2, 0.4)$

## Answer 17

$(-2, 4)$ and $(+2, 4)$

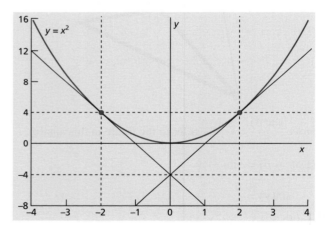

## Answer 18

Set $R = Ar^{-4}$, show that $dR/dr = -4Ar^{-5}$, and use this to demonstrate that $\delta R/R = -4\delta r/r$.

## Answer 19

Trout:

Distance,
$x = 300t^{1.6} = 300 \times (0.1)^{1.6} = 7.54$ cm.

Velocity,

$$\frac{dx}{dt} = 300 \times 1.6t^{0.6} = 480 \times (0.1)^{0.6} = 120.57\ \mathrm{cm\ s^{-1}}.$$

Acceleration,

$$\frac{d^2x}{dt^2} = 480 \times 0.6t^{-0.4} = 288 \times (0.1)^{-0.4} = 723.42\ \mathrm{cm\ s^{-2}}.$$

If length $= 14.4$ cm, velocity $= 8.4$ lengths per second.

Sunfish:
Distance,

$x = 210t^{1.71} = 210 \times (0.1)^{1.71} = 4.09$ cm.

Velocity,

$\dfrac{dx}{dt} = 210 \times 1.71t^{0.71} = 359.1 \times (0.1)^{0.71} = 70.02$ cm s$^{-1}$.

Acceleration,

$\dfrac{d^2x}{dt^2} = 359.1 \times 0.71t^{-0.29} = 254.961 \times (0.1)^{-0.29}$

$\qquad = 497.13$ cm s$^{-2}$.

If length $= 8.0$ cm, velocity $= 8.8$ lengths per second.

Comment: the larger trout is moving faster in real terms, but when normalized for size the velocities of the two fish are rather similar.

## Answer 20
(a) The populations of lions and zebra are stable when both $dL/dt$ and $dZ/dt$ are zero.
(b) The zebra will become extinct when $Z = 0$.
(c) Both $(Z, L) = (0, 0)$ and $(Z, L) = (5000, 50)$ correspond to stable populations. Both species will survive if $(Z, L) = (5000, 50)$.

## Answer 21
5 %

# Chapter 6
## Answer 1
(a) $\displaystyle\int 7x^0 \, dx = \dfrac{7x^{0+1}}{0+1} + c = 7x + c$

(b) $\dfrac{4x^{6+1}}{6+1} + c = \dfrac{4x^7}{7} + c$

(c) $\dfrac{x^{-2+1}}{-2+1} + c = \dfrac{x^{-1}}{-1} + c = -\dfrac{1}{x} + c$

(d) $x^2 + x + c$
(e) $15x^{7/5} + c$
(f) $x^2 - 2x^{3/2} + 3x^{-1/3} + c$

## Answer 2
(a) $\displaystyle\int (x^2 + 6x + 9) \, dx = \dfrac{x^3}{3} + 3x^2 + 9x + c$

(b) $\displaystyle\int (x - x^{-2}) \, dx = \dfrac{x^2}{2} + \dfrac{1}{x} + c$

(c) $y = \displaystyle\int (x^{2/3} - x^{3/2}) \, dx = \dfrac{x^{5/3}}{5/3} - \dfrac{x^{5/2}}{5/2} + c$

$\qquad = \dfrac{3x^{5/3}}{5} - \dfrac{2x^{5/2}}{5} + c$

## Answer 3
(a) 64
(b) $\dfrac{15}{128} = 0.117$
(c) 17.36
(d) 8

## Answer 4
(a) 28
(b) $-1000/3$
(c) $1 - \dfrac{\pi}{2} = -0.571$

## Answer 5
(a) 0.5
(b) 0.667
(c) 0.8
(d) 1

## Answer 6
(a) $\displaystyle\int_0^3 (t^2 - 2t + 3) \, dt = 9$

(b) $\dfrac{155}{3} = 51.67$

(c) 69

(d) $\dfrac{1}{3-0} \displaystyle\int_0^3 (t^2 - 2t + 3) \, dt = \dfrac{9}{3} = 3$

## Answer 7
(a) $100 + \int_0^{15} R(t) \, dt$ represents the total bee population at the end of 15 weeks.
(b) kg
(c) $\int_5^{10} G(t) \, dt$ represents the increase in the mass of the child between its fifth and tenth years.

## Answer 8
$\dfrac{14}{9} = 1.556$

## Answer 9
$y(t) = \dfrac{2kt^{3/2}}{3}$

## Answer 10
$p = \dfrac{2}{3}t^{1/2} + 30$

## Answer 11

$$y = \frac{x^3}{3} + \frac{7}{3}$$

## Answer 12

The area of the wound after 10 days is 0.02 cm$^2$.

## Answer 13

(a) $p(1) = (1)^2 + (1)^3 + 100 = 102$
(b) $p(10) = 100 + 1000 + 100 = 1200$

## Answer 14

$v(t) = v_0 - gt$; $s(t) = s_0 + v_0 t - \frac{1}{2}gt^2$; 0.31 s; 0.46 m

## Answer 15

1/3

## Answer 16

4.5

# Chapter 7

## Answer 1

(a) $1 + \cos(x)$
(b) $-0.5\sin(x) + 2x$
(c) $3\cos(x) - 4\sin(x)$

## Answer 2

(a) $36(3x - 1)^{11} + \dfrac{5}{x^2}$

(b) $\dfrac{1}{2}\cos\left(\dfrac{x}{2}\right)$

(c) $\sin(x) + x\cos(x)$
(d) $1/(x + 1)^2$

(e) $\dfrac{2\sin(x)\cos(x) - 2\sin^2(x)}{x^3}$

(f) $24x(2 + 3x^2)^3$

## Answer 3

(a) $(22x - 19)(2x + 1)^9$

(b) $\dfrac{2x^3 + 3x^2 p - p^3}{(x + p)^2} = 2x - p$

(c) $\dfrac{-4x}{(x^2 - 1)^2}$

(d) $-\dfrac{2\pi}{T}\sin\left(\dfrac{2\pi x}{T} + \phi\right)$

(e) $\dfrac{\cos(x^3) + 3x^3\sin(x^3)}{\cos^2(x^3)}$

(f) $q$

## Answer 4

(a) $\sin(x) + \cos(x) + c$

(b) $4x + 2\sin\left(\dfrac{x}{2}\right) + c$

(c) $y_0 x - \dfrac{T}{2\pi}\cos\left(\dfrac{2\pi x}{T} + \phi\right) + c$

(d) $-\dfrac{1}{2}\cos(x^2) + c$

## Answer 5

(a) $-\dfrac{1}{3\sin^3(x)} + c$

(b) $2\sqrt{2 + \sin(x)} + c$

(c) $\dfrac{x^2}{2} + px + c$

(d) $\dfrac{(3x - 1)^{12}}{36} + \dfrac{5}{x} + c$

## Answer 6

(a) $I = 45.3$
(b) $I = 12.7$
(c) $I = 1/8$
(d) $I = 10.8$

## Answer 7

$$\frac{dl}{d\alpha} = \frac{r\sin(\alpha)}{\sqrt{2(1 - \cos(\alpha))}}$$

## Answer 8

(a)

| $\delta x$ | 0.01 | 0.1 | 0.5 |
|---|---|---|---|
| Approximation | 1.00333 | 1.03333 | 1.16667 |
| Correct result | 1.00332 | 1.03228 | 1.14471 |
| Error (%) | 0.001 | 0.102 | 1.918 |

(b)

| $\delta x$ | 0.01 | 0.1 | 0.5 |
|---|---|---|---|
| Approximation | 1.00000 | 1.00000 | 1.00000 |
| Correct result | 0.99995 | 0.99500 | 0.87758 |
| Error (%) | 0.005 | 0.503 | 13.950 |

(c)

| $\delta x$ | 0.01 | 0.1 | 0.5 |
|---|---|---|---|
| Approximation | 0.99000 | 0.90000 | 0.50000 |
| Correct result | 0.99010 | 0.90909 | 0.66667 |
| Error (%) | 0.010 | 1.000 | 25.000 |

## Answer 9

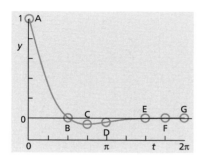

A = (0, +1); B = (π/2, 0); C = (3π/4, −0.067); D = (π, −0.043); E = (3π/2, 0); F = (7π/4, +0.003); G = (2π, +0.002). C is a local minimum. F is a local maximum. A, D, and G are points of inflexion. No asymptotes.

## Answer 10

Rate of recovery, $\dfrac{\mathrm{d}r}{\mathrm{d}t} = \dfrac{100}{(1+t)^2}$

## Answer 11

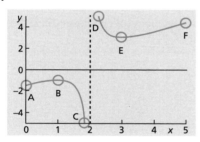

A = (−2, −1.5); B = (+1, −1) local maximum; C = (+1.83, −5); D = (+2.27, +5); E = (+3, +3) local minimum; F = (+5, +4.33). Vertical asymptote at $x = 2$. Note that this graph also possesses an asymptote around the line $y = x − 1$.

## Answer 12

17.2 cm

## Answer 13

2/π for $0 \leqslant t \leqslant π/2$, and 0 for $0 \leqslant t \leqslant 2π$

## Answer 14

100 175 (to the nearest whole frog)

## Answer 15

(a) Amplitude is 5 mm;
frequency = 15/2π = 2.39 s⁻¹;
period = 1/2.39 = 0.42 s

(b)

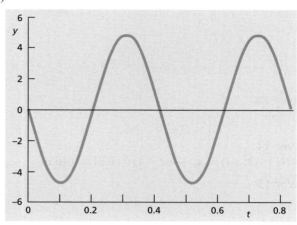

(c) π/15, 3π/15
(d) 67.5 kg s⁻²

## Answer 16

$$\frac{\mathrm{d}}{\mathrm{d}x}(\sin(x)) = 1 - \frac{3x^2}{3 \times 2 \times 1} + \frac{5x^4}{5 \times 4 \times 3 \times 2 \times 1}$$

$$- \frac{7x^6}{7 \times 6 \times 5 \times 4 \times 3 \times 2 \times 1} + \cdots$$

$$= 1 - \frac{x^2}{2 \times 1} + \frac{x^4}{4 \times 3 \times 2 \times 1}$$

$$- \frac{x^6}{6 \times 5 \times 4 \times 3 \times 2 \times 1} + \cdots = \cos(x).$$

$$\frac{\mathrm{d}}{\mathrm{d}x}(\cos(x)) = 0 - \frac{2x^1}{2 \times 1} + \frac{4x^3}{4 \times 3 \times 2 \times 1}$$

$$- \frac{6x^5}{6 \times 5 \times 4 \times 3 \times 2 \times 1} + \cdots$$

$$= -x + \frac{x^3}{3 \times 2 \times 1} - \frac{x^5}{5 \times 4 \times 3 \times 2 \times 1}$$

$$+ \cdots = -\sin(x).$$

## Answer 17

(a) $N(t)$ will decrease during the period $0 \leqslant t \leqslant 10$, because $\mathrm{d}N/\mathrm{d}t$ remains negative throughout.

(b) When $t = +1$ day, $\mathrm{d}N/\mathrm{d}t = -1000$ bacteria per milliliter per day.

## Answer 18

(a) 1/2
(b) 1/2

## Answer 19

The Michaelis–Menten equation should be rewritten in the form:

$$v_0 = \frac{V_{\max}[S]}{K_M + [S]} = \frac{V_{\max}[S]}{[S](1 + K_M/[S])} = \frac{V_{\max}}{1 + K_M/[S]}.$$

When the substrate concentration is increased, the $K_M/[S]$ term tends toward zero and the initial rate tends toward its maximum value, $V_{max}$:

$$\lim_{S \to \infty}(v_0) = \lim_{S \to \infty}\left(\frac{V_{max}}{1 + K_M/[S]}\right) = V_{max}.$$

**Answer 20**

5802 liters

# Chapter 8

**Answer 1**

(a) $-\dfrac{3}{4-3x}$

(b) $-2\exp(3-2x)$

(c) $2x\exp(x^2+1)$

(d) $\ln(x)+1$

(e) $\left(1 + x + \dfrac{1}{x} - \dfrac{1}{x^2}\right)e^x$

(f) $-\dfrac{e^{-3x}(1+3x)}{3x^2}$

**Answer 2**

(a) $\dfrac{2\exp(6x)}{9}$

(b) $7e^x(e^x+1)^6$

(c) $-\dfrac{3}{1-3x} - \dfrac{1}{x+4} - \dfrac{2}{x-1}$

(d) $\dfrac{2x}{1+x^2}$

(e) $(\sin(x) + x\sin(x) + x\cos(x))e^x$

(f) $-\dfrac{\sin(\ln(x))}{\ln(x)}$

**Answer 3**

(a) $\dfrac{e^{+3x}}{3} - \dfrac{e^{-3x}}{3} + c$

(b) $-\dfrac{\ln(1-2x)}{2} + c$

(c) $-\dfrac{\exp(-x^2)}{2} + c$

**Answer 4**

(a) $x - 1 + \ln(x-1) + c$

(b) $\dfrac{3x^2}{2} + c$

(c) $\dfrac{x^2}{2} + c$

**Answer 5**

(a) 2

(b) $e - 0.1 = 2.62$

**Answer 6**

(a) Let $u = x+1$;

$$I = \left[\frac{u^3}{3} - \frac{3u^2}{2} + 3u - \ln(u)\right]_1^2$$
$$= \frac{7}{3} - \frac{9}{2} + 3 - \ln(2) = 0.14$$

(b) Let $u = e^x + e^{-x}$,

$$I = [\ln(e^x + e^{-x})]_{x=-1}^{x=+1}$$
$$= \ln(e^{+1} + e^{-1}) - \ln(e^{-1} + e^{+1}) = 0$$

**Answer 7**

$(0, 0)$ is a stationary point of inflexion; there is a maximum turning point at $(3, 270e^{-3})$.

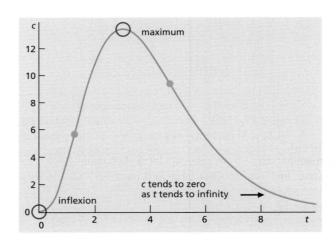

**Answer 8**

Velocity,

$$v = \frac{dx}{dt} = v_{max}(1 - e^{-kt})$$

and acceleration,

$$a = \frac{dv}{dx} = \frac{d^2x}{dt^2} = v_{max}\,k\,e^{-kt}.$$

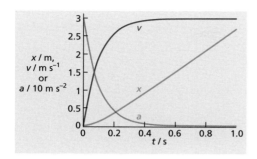

## Answer 9

$$\frac{d}{dT}\{\ln(p)\} = \frac{dp}{dT} \times \frac{d}{dp}\{\ln(p)\} = \frac{dp}{dT} \times \frac{1}{p} = \frac{\Delta H_{vap}}{RT^2},$$

therefore $\frac{dp}{dT} = \frac{p\Delta H_{vap}}{RT^2}$, so the two statements are equivalent.

## Answer 10

(a)

| $\delta x$ | 0.01 | 0.1 | 0.5 |
|---|---|---|---|
| Approximation | 0.01 | 0.1 | 0.5 |
| Correct result | 0.00995 | 0.0953 | 0.4055 |
| Error (%) | 0.5 | 4.9 | 23.3 |

(b)

| $\delta x$ | 0.01 | 0.1 | 0.5 |
|---|---|---|---|
| Approximation | 0.99 | 0.9 | 0.5 |
| Correct result | 0.99005 | 0.9048 | 0.6065 |
| Accuracy (%) | 0.005 | 0.53 | 17.6 |

## Answer 11

Intercepts, boundary points, and asymptotes: as $x$ approaches zero from above, $\ln(x)$ becomes large and negative, and $1/x$ becomes large and positive. This causes a vertical asymptote at $x = 0$ as $y \to \infty$.

When $y = 0$, $\ln(x)/x = 0$, which can only occur when $x = 1$.

When $x = 10$, $y = \ln(10)/10 = 0.23$.

The curve would reach $y = 0.5$ when $\ln(x)/x = 0.5$, that is, when $\ln(x^2) = x$

that is, when $x^2 = e^x$. This equation is never true, because the graph of $y = e^x$ always lies above that of $y = x^2$.

The curve reaches $y = -1$ when $\ln(x)/x = -1$, that is, when $\ln(x) = -x$,

that is, when $x = e^{-x}$. If $x$ is positive, then $0 < e^{-x} < 1$, which implies that $0 < x < 1$ when $y = -1$. (In fact, numerical analysis of the equation $y = \ln(x)/x$ shows that $x = 0.58$ when $y = -1$.)

Stationary points:

if $y = \frac{\ln(x)}{x}$, $\frac{dy}{dx} = \frac{1 - \ln(x)}{x^2}$ and $\frac{d^2y}{dx^2} = \frac{2\ln(x) - 3}{x^3}$.

Stationary points occur when the slope is zero, that is, when

$$\frac{dy}{dx} = \frac{1 - \ln(x)}{x^2} = 0.$$

This can only occur when $1 - \ln(x) = 0$, that is, when $\ln(x) = 1$, that is, when $x = e$.

At $x = e$,

$$\frac{d^2y}{dx^2} = \frac{2\ln(e) - 3}{e^3} = -e^{-3},$$

which is negative,

so there is a maximum turning point at $(e, e^{-1}) = (2.72, 0.37)$.

Points of inflexion: these occur when $d^2y/dx^2 = 0$, that is, when $2\ln(x) - 3 = 0$

that is, when $\ln(x) = 1.5$, that is, when $x = 4.48$ and $y = 0.33$.

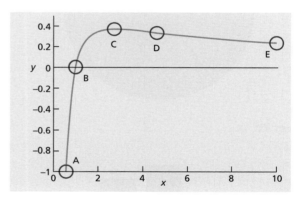

$A = (0.58, -1)$; $B = (1, 0)$; $C = (2.72, 0.37)$; $D = (4.48, 0.33)$; $E = (10, 0.23)$

## Answer 12
(a) $0.869 = 86.9\%$
(b) $0.754 = 75.4\%$

## Answer 13
6 h 29 min, 0.0015 %

## Answer 14
(a) 37.6 years
(b) 46 %

## Answer 15
The second derivative is $\frac{d^2x}{dt^2} = abc^2 e^{-be^{-ct}} e^{-ct}(be^{-ct} - 1)$

# Chapter 9

### Answer 1a
Sex: nominal; SVL and mass: ratio; temperature: interval; weather: ordinal (in terms of 'worse weather is higher'), or nominal.

### Answer 1b
Discrete: legs, litter size, lever press responses. Continuous: body mass, $O_2$ concentration.

### Answer 1c
Blood group and payroll: nominal; employment start date: interval; days sick leave: ratio.

### Answer 2a

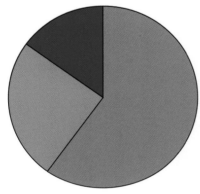

### Answer 2b

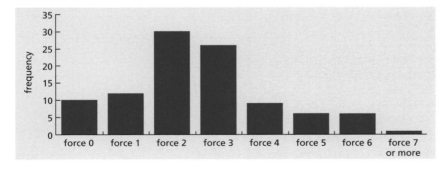

### Answer 2c

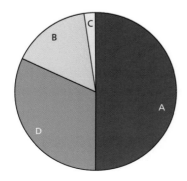

**Answer 3a**

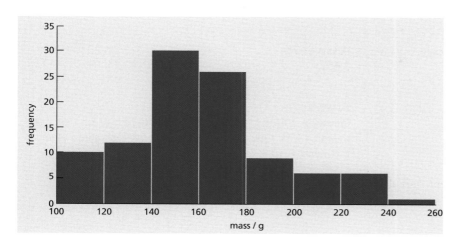

**Answer 3b**
(iii) It is unimodal (one peak) and positively skewed.

**Answer 3c**

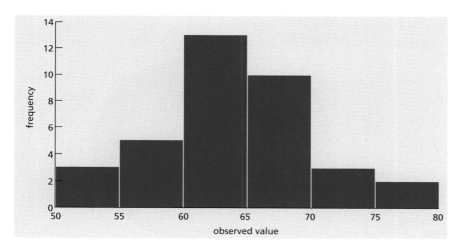

**Answer 4a**
14.5

**Answer 4b**
Sample variance = 6.5, standard deviation = 2.55.

**Answer 4c**
Sample mean = 34, sample variance = 763.8, sample standard deviation = 27.6.

**Answer 5a**

14; 26

**Answer 5b**

4(a): range = 10–18;  IQR = 13–16;  4(c): range = 9–83;
IQR = 12–44.

**Answer 5c**

Minimum = 20;   LQB = 24;   median = 39;   UQB = 92,
maximum = 114.

**Answer 6a**

Minimum = 11, median = 52, maximum = 95.

**Answer 6b**

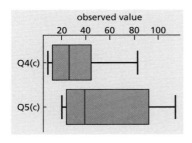

**Answer 6c**

| Q5(c) | | Q4(c) |
|---|---|---|
| | 00 | 9 |
| | 10 | 262 |
| 4420 | 20 | 48 |
| 48 | 30 | 0 |
| 0 | 40 | 4 |
| 4 | 50 | |
| | 60 | |
| | 70 | |
| | 80 | 23 |
| 22 | 90 | |
| 5 | 100 | |
| 4 | 110 | |

**Answer 7a**

6.325

**Answer 7b**

3.651

**Answer 7c**

Question 4(c): mean = 34, standard error of the mean = 8.74; Question 5(c): mean = 54.9, standard error of the mean = 10.3.

**Answer 8a**

$r = -0.823$

**Answer 8b**

Sample covariance = 316.25 mm$^2$;  $r = 0.791$; the literal covariance, cov$(x, y) = 281.11$ mm$^2$.

**Answer 8c**

0.25

**Answer 9**

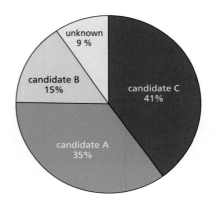

## Answer 10

Histogram is unimodal, and roughly symmetrical (some positive skew).

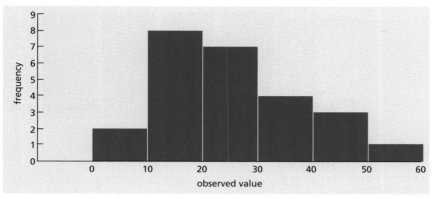

## Answer 11

Minimum = 9;  LQB = 17;  median = 21;  UQB = 33; maximum = 50.

## Answer 12

Sample mean = 25.12; sample variance = 142.69; standard deviation = 11.95; standard error of the mean = 2.39.

## Answer 13

All have sensible means apart from payroll and blood group.

## Answer 14

Skewness is close to zero; sample median is in the interval 30–40, IQR is from somewhere in the 20–30 interval to somewhere in the 40–60 range.

## Answer 15

No. The visual impression is that the consumption of apples is falling, whereas that of bananas is increasing. However, this is due to the use of a percentage increase measure. The most apples are consumed in the final year, and the absolute increase in apple consumption in 2004 is greater than the increase in banana consumption.

## Answer 16

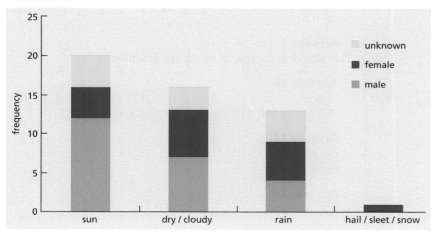

## Answer 17

$r = -0.094$

## Answer 18

No. Correlation does not indicate the direction of any causal relationship. The plant may be being eaten *as a result of sickness*, e.g. for analgesia.

## Answer 19

(a) Mean different height = 15 cm; sample variance in difference in height = 450 cm$^2$.

(b) Mean average height = 167.5 cm; sample variance = 112.5 cm$^2$.

## Answer 20

(a) Mean different height = 15 cm; sample variance in difference in height = 390 cm$^2$.

(b) Mean average height = 167.5 cm; sample variance = 127.5 cm$^2$.

## Answer 21

Correlation = 0 in Meanland; correlation = 0.133 in Modenia.

# Chapter 10

**Answer 1**

4/9

**Answer 2**

4/5

**Answer 3**

44/52

**Answer 4**

(a) No

(b) No

(c) Yes

**Answer 5**

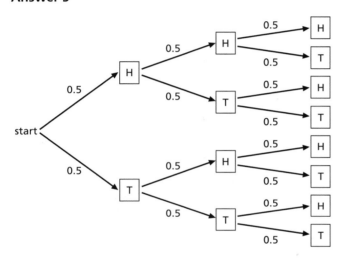

**Answer 6**

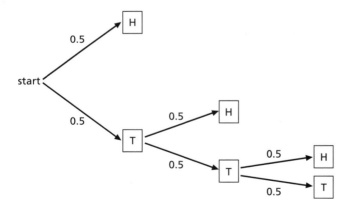

**Answer 7**

28/52

**Answer 8**

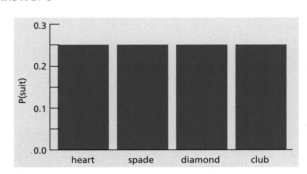

**Answer 9**

$f(1) = f(2) = f(3) = f(4) = f(5) = f(6) = 1/6.$  $f(x) = 0$  for all other values.

**Answer 10**

E|roll| = 3.5, Var(roll) = 2.917

**Answer 11**

(a) 0.252

(b) 0.103

**Answer 12**

50 %

**Answer 13**

71.8 %

**Answer 14**

(a) 36.9 %

(b) 24.7 %

(c) 47.3 %

**Answer 15**

(a) 50 %

(b) 20 %

**Answer 16**

(a) 14 %

(b) 67 %

(c) 79.2 %

(d) 0.81

**Answer 17**

0.125 for each trial, 50 % chance of failure.

**Answer 18**

$P(x) = 7^{x-1}/8^x$:    P(1st) = 0.125,    P(2nd) = 0.109, P(3rd) = 0.096, P(4th) = 0.084, P(failure) = 58.6 %.

**Answer 19**

(a) 2.75 trials

(b) 2.90 trials

**Answer 20**

A = 3

## Answer 21

(a) $f_{pdf}(x) = 2x$

(b) The pdf is highest when $x = 1$

(c) Mean $= 2/3$, variance $= 1/18$

(d) $P(X > \mu) = 1 - F_{cdf}(2/3) = 5/9$

(e) $P(|X-\mu| > \sigma) = P(X > \mu + \sigma) + P(X < \mu - \sigma) = 0.371$

## Answer 22

(a) $F_{cdf}(t) = t/900$

(b) 450 s

(c) $F_{cdf}(t) = t/900$

(d) At each time point, $P(A \wedge B) = P(A) \times P(B)$, thus $F_{cdf}(t) = (t/900)^2$

(e) 600 s

(f) At each time point, $P(A \vee B) = P(A) + P(B) - P(A) \times P(B)$, thus $F_{cdf}(t) = 2t/900 - (t/900)^2$

(g) 300 s

## Answer 23

(a) $x_{max} = 4$

(b) $f(x) = 0.25x^{-0.5}$

(c) $1\frac{1}{3}$ m

## Answer 24

(a) Transformation is $x = \ln(r/cm)$.

The mean of the distribution is the transformed value of $(r = 4$ cm$) \Rightarrow \mu_x = \ln(4) = 1.3863$.

When $r = 6$ cm, $\ln(6) = x = \mu_x + s_x = 1.7917$, $\Rightarrow \sigma_x = 0.4055 \Rightarrow \sigma_x^2 = 0.1644$.

The transformed radii have mean $\mu_x$ 1.39, and variance $\sigma_x^2 = 0.164$ (three significant figures).

(b) Let $X$ be the distribution of log(radii), $Y$ be the distribution of log(area).

$X$ follows $N$ with mean 1.386, s.d. $= 0.1644$.

$Y = \ln(\pi r^2) = \ln(\pi) + 2\ln(r) = \ln(\pi) + 2X$ (using $X = \ln(r)$).

$Y$ is a linear transformation of $X$, so follows the same shape (normal).

Mean of $Y = \ln(\pi) + 2\mu_x = 3.92$.

Var $(Y) = $ Var$(\ln(\pi)) + $ Var$(2X) = 4$Var$(X)$.

$\sigma_y^2 = 4\sigma_x^2 = 0.658$.

Distribution is normal with mean $= 3.92$, variance $= 0.658$.

(c) $y_{250} = \ln(25) = 3.912$. Distribution has mean $= 3.92$, variance $= 0.658$.

Expressing as a $z$ score, $z = -0.861$. The proportion of $Z$ that is below $z = +0.861 = 0.8054$ (from table). Normal curve is symmetrical, thus the proportion of leaves with area above 25 cm$^2$ is approximately 81 % (two significant figures).

*Alternatively*: let $r_1$ be the radius of a leaf with area 25 cm$^2 \Rightarrow 25 = \pi r_1^2$, i.e. $r_1 = \sqrt{25 \text{ cm}^2/\pi} = 2.821$ cm. The proportion of leaves with area over 25 cm$^2$ is the proportion with transformed radii that are above

$x_1 = \ln(2.821) = 1.037$. Expressing $x_1$ as a $z$ score, $z = (x_1 - \mu_x)/\sigma_x = (1.037 - 1.386)/0.4055 = -0.861$ (as above).

(d) Normal, mean $= 0$, variance $= 2 \times 0.658 = 1.32$ (three significant figures).

(e) The probability of the first leaf being twice the size of the second is the probability of sampling a difference in log area of above $\ln(2)$. The difference in log area between first and second leaf is normal, with mean $= 0$, standard deviation $= \sqrt{1.32}$.

Expressing $\ln(2)$ as a $z$ value: $z = \ln(2)/\sqrt{1.32} = 0.604$. Probability of $(Z > +0.604) = 1 - 0.727 = 0.273$.

There is an equal probability that the second leaf is more than twice the size of the first, so P(one more than twice the other) $= 2 \times (0.273) = 0.546$.

# Chapter 11

## Answer 1

(a) 1.04 to 4.96

(b) $\mu = 4 \pm 1.96$

(c) $\mu = 4 \pm 4.9$

(d) $\mu = 5 \pm 5.88$

(e) $\mu = 17 \pm 3.92$

## Answer 2

(a) 0.112

(b) 0.281 to 0.719

(c) $\pi = 0.4 \pm 0.215$

(d) $\pi = 0.4 \pm 0.175$

(e) $\pi = 0.75 \pm 0.085$

## Answer 3

(a) $z = 2$, and is significant at $\alpha = 0.05$

(b) $z = 0.4$, not significant at $\alpha = 0.05$

(c) $z = -0.667$, not significant at $\alpha = 0.05$

(d) $z = -5.0$, significant at $\alpha = 0.05$

(e) The test is significant when the null hypothesis value is outside the confidence interval.

## Answer 4

(a) $t_{15} = 0.2$, not significant at $\alpha = 0.05$

(b) $t_{31} = -0.94$, not significant at $\alpha = 0.05$

(c) 95 % confidence interval: $\mu = 12 \pm 3.50$; mean is significantly above 7 ($t_{39} = 2.89, p < 0.01$)

(d) 95 % confidence interval: $\mu = 10.14 \pm 3.30$

(e) 95 % confidence interval $= 83.21$ to 97.09; 99 % confidence interval $= 79.88$ to 100.42.

Significantly different from 100 at 5 %, not at 1 %.

## Answer 5

(a) $t_{29} = 2.74$; significantly above zero, $p < 0.05$ ($p = 0.010$)

(b) 95 % confidence interval: $\mu_d = 5 \pm 3.73$

(c) $t_6 = -1.62$; not significantly below zero, $p > 0.05$ ($p = 0.157$)

(d) 95 % confidence interval: $\mu_d = 0.657 \pm 0.994$

(e) 95 % confidence interval: $\mu_d = 9.00 \pm 8.20$: mean change is significantly above zero $p < 0.05$ ($t_8 = 2.53$; $p = 0.035$)

## Answer 6

(a) $t_{18} = -0.40$; difference is not significant, $p > 0.05$ ($p = 0.690$)

(b) 95 % confidence interval: $(\mu_A - \mu_B) = -2 \pm 10.37$

(c) 95 % confidence interval: $(\mu_1 - \mu_2) = -4.56 \pm 3.74$: difference between means is significantly larger than zero, $p > 0.05$ ($t_{16} = -2.58$; $p = 0.020$)

(d) 95 % confidence interval: $(\mu_1 - \mu_2) = -0.5 \pm 6.60$; difference between means is not significantly different from zero, $p > 0.05$ ($t_{14} = -0.16$; $p = 0.873$)

(e) 95 % confidence interval: $(\mu_A - \mu_B) = 4.5 \pm 3.94$; difference between means is significantly larger than zero, $p < 0.05$ ($t_{22} = 2.37$; $p = 0.027$)

## Answer 7

(a) $\chi_1^2 = 25$; pattern differs significantly from 50:50 ($p < 0.001$)

(b) $\chi_1^2 = 9$; pattern differs significantly from 10:90 ($p = 0.003$)

(c) $\chi_2^2 = 6.4$; pattern differs significantly from 1:1:1 ($p = 0.041$)

## Answer 8

(a) $\chi_1^2 = 5.13$; pattern differs significantly from independence ($p = 0.024$)

(b) $\chi_1^2 = 53.14$; pattern differs significantly from independence ($p < 0.001$)

(c) $\chi_2^2 = 7.62$; pattern differs significantly from independence ($p = 0.022$)

## Answer 9

Yes: $\chi^2 = 8.07$; $df = 1$; $p < 0.01$

## Answer 10

(a) $t_6 = 1.77$, non-significant

(b) $t_8 = 2.53$, $p < 0.05$

(c) $t_7 = 1.48$, non-significant

## Answer 11

$t_{16} = 2.64$, $p < 0.05$

## Answer 12

Yes: $\chi^2 = 15.09$; $df = 5$; $p < 0.01$

## Answer 13

4.51 to 5.52

## Answer 14

Yes: $\chi^2 = 7.14$; $df = 1$; $p < 0.01$

## Answer 15

(a) 591

(b) 95 % confidence interval = 506 ms to 675 ms; 99 % confidence interval = 471 ms to 710 ms

(c) Yes

## Answer 16

No significant difference: $t_{20} = 0.140$, $p > 0.05$

## Answer 17

(a) 0.736 to 0.864

(b)

|            | nest | no nest |
|------------|------|---------|
| native     | 72   | 48      |
| non-native | 15   | 15      |

(c) No, $\chi^2 = 0.99$, $df = 1$; $p > 0.05$

## Answer 18

95 % confidence interval is 29.4 % to 44.6 %. This is entirely within the estimated range, thus we should be *more than* 95 % confident in the prediction.

## Answer 19

(a) 11

(b) Yes, value is significant at $p < 0.05$, two-tailed.

(c) If the test was being conducted one-tailed.

## Answer 20

No, $\chi^2 = 7.64$, $df = 4$; $p > 0.05$

## Answer 21

Yes: $t_{11} = 2.218$; $p < 0.05$. (95 % confidence interval on increase = 0.03 to 7.30 more pronouns in 10 min interval.)

## Answer 22

No significant difference, $t_{18} = 1.83$, $p > 0.05$.

95 % confidence interval on increase due to A over B = $-6.9$ g to 100.7 g.

## Answer 23

(a) 95 % confidence interval cannot be broader than 99 % confidence interval.

(b) This value of $t$, with 7 $df$, is non-significant (has a two-tailed probability $> 0.05$).

(c) The reported $\chi^2$ is not valid. The categories are neither mutually exclusive, nor exhaustive (thus we do not have each person recorded once and only once!).

## Answer 24

(a) No: $t_{28} = 0.492$, $p > 0.05$

(b) 95 % confidence interval (for difference college 1 − college 2) = $-17.90$ to $+10.97$

(c) Non-significant: $\chi^2 = 0.136$, $df = 1$, so conclude nothing.

## Answer 25

(a) Positive

(b) Mean egg number = 17.84; 95 % confidence interval = $17.84 \pm 8.37 = 9.47$ to 26.21

(c) Mean $\log_{10}$(egg number) $= 1.064$; 95 % confidence interval $= 0.895$ to $1.233$

(d) Antilog of mean $\log_{10}$(egg number) $= 11.56$; 95 % confidence interval $= 7.85$ to $17.09$. Confidence interval is narrower when analyzing log data, thus more informative.

## Answer 26

(a) Male: mean $= 63.9$. 95 % confidence interval $= 57.1$ to $70.6$.

Female: mean $= 67.3$. 95 % confidence interval $61.1$ to $73.5$.

(b) No evidence for this, $t_{28} = 0.81$, $p > 0.05$.

(c) Yes, $\chi^2 = 5.40$, $df = 1$; significant evidence that females obtain more firsts.

(d) No, $\chi^2 = 2.08$, $df = 2$; no evidence that class marks differ from guidelines.

# Chapter 12

## Answer 1
$M = 3t + 4$

## Answer 2
$N = 60 \times 2^t = 60 e^{0.69t}$

## Answer 3
$$N(t) = \frac{1900}{1 + 18 \times 2^{-t}}$$

## Answer 4
$95.349$ units$^2$

## Answer 5
(a) First order, linear, and inhomogeneous.
(b) Second order, linear, and homogeneous.
(c) First order, nonlinear, and homogeneous.
(d) Second order, nonlinear, and inhomogeneous.
(e) First order, linear, and inhomogeneous.

## Answer 6
(a) $y = -3x^{-1/3} + c$

(b) $y = \dfrac{x^3}{3} - 5x^2 + c_1 x + c_2$

(c) $y = \dfrac{e^{2x}}{4} + c_1 x + c_2$

## Answer 7
$$k = \frac{\ln 2}{15 \, \text{min}} = 4 \ln(2) \text{h}^{-1}$$

## Answer 8
$$\frac{dy}{dx} = -4e^{-4t} = -4y$$

## Answer 9
(a) $P(t) = 2(4 + 0.2t)^{5/2} + c$
(b) $P(t) = 2(4 + 0.2t)^{5/2} + 36$
(c) $295\,284$

## Answer 10
(a) $y = \ln(x^2 + 4) + c$
(b) $y = e^{x^4 + c}$
(c) $y = (\ln(x + 2) + c)^{-1}$

## Answer 11
(a) Amount remaining

$$= 0.2 \text{ mg } 2^{-t/1.5\,\text{h}} \text{ or } 0.2 \text{ mg } e^{-0.69t/1.5\,\text{h}}$$

(b) 6 h 29 min
(c) 0.0015 %

## Answer 12
37.6 years ($K = \ln(1.0186) = 0.0184$)

## Answer 13
$y = x$

## Answer 14

(a) $\dfrac{d^2 x}{dt^2} = \dfrac{d}{dt}(y) = k^2 x$ and $\dfrac{d^2 y}{dt^2} = \dfrac{d}{dt}(k^2 x) = k^2 \dfrac{dx}{dt} = k^2 y$

(b) $\dfrac{dx}{dt} = \dfrac{1}{k} \times \dfrac{d}{dt}(e^{kt} - e^{-kt}) = \dfrac{1}{k}(ke^{kt} - (-k)e^{-kt}) = y$

$\dfrac{dy}{dt} = \dfrac{d}{dt}(e^{kt} + e^{-kt}) = k(e^{kt} - e^{-kt}) = k^2 x$

## Answer 15
$Y = 1.50X + 5.00$

## Answer 16
(a) 55.7 units$^2$
(b) 43.5 units$^2$
(c) 43.6 units$^2$, so (b) gives best fit
(d) $k = 1.024$; $SSe = 42.13$ units$^2$

## Answer 17
(a) Decreases as $dN/dt$ is always negative for positive or zero $t$.
(b) $-1500$ bacteria per milliliter per day at $t = 1$
(c) $N_t = N_0 - 1500 \times \ln(1 + t^2)$
(d) 3077 bacteria per milliliter
(e) After 23.7 days

## Answer 18
(a) $E = \dfrac{3t}{2} + 10e^{(0.5t-5)} + c$

(b) $E = 0.15t^2 + t + c$

(c) $E = \dfrac{3t}{2} + 10e^{(0.5t-5)} - 0.067$ and $E = 0.15t^2 + t$

(d) Approximation $SSe = 11.6$ units$^2$

(e) Linear approximation gives a poorer fit $SSe = 38.6$ units$^2$.

## Answer 19

(a) Rate of change in total *amount* of new blood is the rate of addition, less the rate of removal: $(Q - Q \times c) = Q(1 - c)$.

Thus rate of change in the proportion of new blood, $dc/dt = Q(1 - c)/V$.

The model states that the proportion of new blood present increases faster:

(i) if more new blood is introduced: $dc/dt \propto Q$;

(ii) if more old blood is present in the blood removed: $dc/dt \propto (1 - c)$;

(ii) if the total volume of blood in the baby is small: $dc/dt \propto 1/V$.

(b) The model assumes that the new and old blood is mixed instantly.

The model also states that blood is pumped into and out of the baby at a constant rate $Q$ and that the total volume of blood in the baby, $V$, is constant (which implies the assumptions that pumping is not influenced by the baby's heart rate, and that there is no other blood loss).

(c) 2303 s (38 minutes, 23 seconds)

(d) 2.30 liters

## Answer 20

(a) $K_a = \dfrac{N_a}{(N_T - N_a) \times [a]}$

$$(N_T - N_a) \times K_a \times [a] = N_a$$

$$N_T \times K_a \times [a] = N_a + N_a \times K_a \times [a]$$

$$N_a = \dfrac{N_T \times K_a \times [a]}{1 + K_a \times [a]}$$

(b) $SSe = 2426.6$ (pmol)$^2$

(c) $K_a = 0.0187\ \mu M^{-1}$; $N_T = 203.0$ pmol; $SSe = 120.4$ (pmol)$^2$

(d)

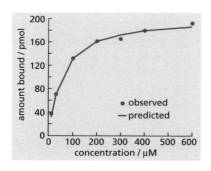

# Appendix 1 Summary of the Basic Operations of Arithmetic and Algebra

When you first encounter this summary, think of $a$, $b$, and $c$ as integers until you meet division, where you will need to let $c$ be a fraction. All of these rules still apply when $a$, $b$, and $c$ are any real numbers or any expressions that can be evaluated to produce real numbers.

The arithmetic you can actually do with pen and paper or a calculator is the arithmetic of rational numbers: that is, numbers that can be expressed as fractions that are the ratio of two integers. Thus the rules up to EQA1.28 with $a$, $b$, and $c$ as integers or fractions cover what you can actually do. Many roots of numbers (real or rational) are not rational numbers and only a rational approximation can be written down or displayed on a calculator. To discuss the actual values of roots we need to use real numbers.

An identity is an equation that is true not just for some particular values of the variables, but for each and every value. To emphasize this, in an identity the equals sign, $=$, can be replaced by $\equiv$, which is read 'is identical to'. For the simple statements here, using formal algebraic identities is a little like taking a sledge hammer to crack a nut but, because you already know what they must be saying, these identities do provide useful practice in 'reading' algebraic equations.

## Addition and subtraction

(i) When the number zero is added to any number, the number or quantity is unchanged,

$$a + 0 \equiv a. \qquad \text{(EQA1.1)}$$

(ii) We can add in any order,

$$a + b \equiv b + a \qquad \text{(EQA1.2)}$$

and $a + b + c \equiv (a + b) + c \equiv a + (b + c)$. (EQA1.3)

(iii) Subtraction and addition reverse each other, they are 'inverse operations',

$$a + b - b \equiv a - b + b \equiv a. \qquad \text{(EQA1.4)}$$

Adding a negative number and subtracting a positive number produce the same result,

$$a + (-b) \equiv a - b. \qquad \text{(EQA1.5)}$$

(iv) Subtracting a negative number produces the same result as adding a positive number,

$$a - (-b) \equiv a + b. \qquad \text{(EQA1.6)}$$

## Multiplication

Let $a$, $b$, and $c$ be any numbers (you add the descriptions, e.g. when the number one multiplies any number (or quantity), the number is unchanged):

$$a \times 1 \equiv a, \qquad \text{(EQA1.7)}$$
$$ab \equiv ba, \qquad \text{(EQA1.8)}$$
$$abc \equiv (ab)c \equiv a(bc), \qquad \text{(EQA1.9)}$$
$$a \times 0 \equiv 0, \qquad \text{(EQA1.10)}$$
$$a(b + c) \equiv ab + ac, \qquad \text{(EQA1.11)}$$

and $(-a) \times b \equiv -ab$ (EQA1.12)

which implies that $(-a) \times (-b) \equiv ab$. (EQA1.13)

Expressions like $ab$ or $abc$ in these equations are *not* ambiguous. The symbols $a$, $b$, and $c$ have been defined, but $ab$ and $abc$ have not. Thus $ab$ must mean multiplication of $a$ and $b$: that is, $a \times b$, and $abc = a \times b \times c$. The brackets in these equations mean 'do whatever is inside first'.

## Reciprocals and fractions

Let $a$ and $b \neq 0$ be any integers. The reciprocal of $b$ is defined so that

$$\frac{1}{b} \times b \equiv b \times \frac{1}{b} \equiv 1. \qquad \text{(EQA1.14)}$$

Taking a reciprocal can be thought of as the result of 'dividing' or slicing 1 into $b$ equal pieces. The fraction $a/b$ is defined as

$$\frac{a}{b} \equiv a \times \frac{1}{b} \equiv \frac{1}{b} \times a \qquad \text{(EQA1.15)}$$

where we can multiply the integer and the reciprocal in either order. You can think of a fraction, $a/b$, as first slicing a whole into $b$ pieces, then gathering up $a$ of these. Alternatively you could put together $a$ wholes and then slice this into $b$ pieces. You end up with the same amount.

## Reciprocals, fractions, and division

Let $a$ and $b \neq 0$ be integers. We want to show that the result of dividing $a$ by $b$, which we will call $c$, is the same as the fraction $a/b$. To do this note that by the definition of division if the result of dividing $a$ by $b$ is multiplied by $b$ this must take us back to $a$, i.e.

if

$$a \div b = c \qquad \text{(EQA1.16)}$$

then

$$b \times c = a. \qquad \text{(EQA1.17)}$$

Multiply both sides of the last equation by $1/b$, make use of the fact that we can multiply in any order on the left-hand side and, finally, use the definition of the fraction,

$$c \times b \times \frac{1}{b} = a \times \frac{1}{b}$$
$$c = a \times \frac{1}{b} = \frac{a}{b}. \qquad \text{(EQA1.18)}$$

This is the result we wanted. Dividing by any integer is the same as multiplying by its reciprocal; the answer is a fraction.

## Arithmetic with fractions

Addition

$$\frac{a}{b} + \frac{c}{d} \equiv \frac{ad}{bd} + \frac{cb}{db} \equiv \frac{ad + cb}{bd}. \qquad \text{(EQA1.19)}$$

Subtraction

$$\frac{a}{b} - \frac{c}{d} \equiv \frac{a}{b} + \frac{-c}{d} \equiv \frac{ad - cb}{bd}. \qquad \text{(EQA1.20)}$$

Multiplication

$$\frac{a}{b} \times \frac{c}{d} \equiv \frac{a \times c}{b \times d} \equiv \frac{ac}{bd}. \qquad \text{(EQA1.21)}$$

Reciprocal

$$\frac{1}{\left(\dfrac{a}{b}\right)} \equiv \frac{b}{a} \qquad \text{(EQA1.22)}$$

$$\frac{a}{b} \div \frac{c}{d} \equiv \frac{a}{b} \times \frac{d}{c} \equiv \frac{ad}{bc}. \qquad \text{(EQA1.23)}$$

## Powers

$$a^{-n} \equiv \frac{1}{a^n}, \quad \text{provided that } a \neq 0. \qquad \text{(EQA1.24)}$$

$$a^x \times a^y \equiv a^{x+y}. \qquad \text{(EQA1.25)}$$

$$\frac{a^x}{a^y} \equiv a^{x-y}. \qquad \text{(EQA1.26)}$$

$$a^x \times b^x \equiv (ab)^x. \qquad \text{(EQA1.27)}$$

$$(a^x)^y \equiv (a^y)^x \equiv a^{x \times y}. \qquad \text{(EQA1.28)}$$

## Powers and roots

In the following, $a$ and $b$ can be any real numbers, whereas $n$ and $m$ are integers.

If $a^n = b$, then $a$ is an $n$th root of $b$. (EQA1.29)

Whenever we need to specify only one $n$th root, we require that the root be positive and call it the principal $n$th root, $\sqrt[n]{b}$. With $a$ restricted to being real, it is not always possible to find an $n$th root of $b$ but provided that $b$ is positive, there is always at least a principal $n$th root.

$$a^{1/n} \equiv \sqrt[n]{a} \qquad \text{(EQA1.30)}$$

$$a^{m/n} \equiv (a^{1/n})^m \qquad \text{(EQA1.31)}$$

For the general case of a real number raised to a real power, $a^b$, provided we can calculate the values of $a^{m/n}$, we can get as close to the answer as we like by choosing values of $m/n$ that are sufficiently close to $b$.

## Order of precedence
The agreed order of precedence is:

1. **B**rackets (which include the brackets surrounding the arguments of functions).
2. **E**xponentiation, that is raising to a power or taking a root.
3. **D**ivision or **M**ultiplication.
4. **A**ddition or **S**ubtraction.
5. If there is still any ambiguity, proceed from left to right.

Some people find it helps to remember the acronym **BEDMAS** to get the operations in points 1–4 in the right order.

# Appendix 2 Four-figure Log Table

A four-figure log table allows you to find common logarithms of numbers expressed to four significant figures. To look up log(2.37) you first read down the left column until you find 2.3 then go across that line until you are under the 0.07 and read off the answer 0.3747. You estimate intermediate values by linear interpolation; for instance log(2.373), between log(2.37) = 0.3747 and log(2.38) = 0.3766, is

$$0.3747 + (3/10)(0.3766 - 0.3747) = 0.3753.$$

You can also reverse the procedure. What is the antilog of 0.7752, which is the same as asking what is $10^{0.7752}$?

Look up 0.7752 in the body of the table. In the log table it is in the row for 5.9 and the column for 0.06 so the answer is 5.96. Interpolation is used to find the antilog for values between those listed. For instance 0.5619 is between the values 0.5611 and 0.5623 in the columns for 0.04 and 0.05 in the row for 3.6, so the antilog will be between 3.64 and 3.65. The estimate by linear interpolation is

$$3.64 + 0.01 \times (0.5619 - 0.5611)/(0.5623 - 0.5611)$$
$$= 3.64 + 0.007 = 3.647.$$

|     | 0 | 0.01 | 0.02 | 0.03 | 0.04 | 0.05 | 0.06 | 0.07 | 0.08 | 0.09 |
|-----|--------|--------|--------|--------|--------|--------|--------|--------|--------|--------|
| 1   | 0.0000 | 0.0043 | 0.0086 | 0.0128 | 0.0170 | 0.0212 | 0.0253 | 0.0294 | 0.0334 | 0.0374 |
| 1.1 | 0.0414 | 0.0453 | 0.0492 | 0.0531 | 0.0569 | 0.0607 | 0.0645 | 0.0682 | 0.0719 | 0.0755 |
| 1.2 | 0.0792 | 0.0828 | 0.0864 | 0.0899 | 0.0934 | 0.0969 | 0.1004 | 0.1038 | 0.1072 | 0.1106 |
| 1.3 | 0.1139 | 0.1173 | 0.1206 | 0.1239 | 0.1271 | 0.1303 | 0.1335 | 0.1367 | 0.1399 | 0.1430 |
| 1.4 | 0.1461 | 0.1492 | 0.1523 | 0.1553 | 0.1584 | 0.1614 | 0.1644 | 0.1673 | 0.1703 | 0.1732 |
| 1.5 | 0.1761 | 0.1790 | 0.1818 | 0.1847 | 0.1875 | 0.1903 | 0.1931 | 0.1959 | 0.1987 | 0.2014 |
| 1.6 | 0.2041 | 0.2068 | 0.2095 | 0.2122 | 0.2148 | 0.2175 | 0.2201 | 0.2227 | 0.2253 | 0.2279 |
| 1.7 | 0.2304 | 0.2330 | 0.2355 | 0.2380 | 0.2405 | 0.2430 | 0.2455 | 0.2480 | 0.2504 | 0.2529 |
| 1.8 | 0.2553 | 0.2577 | 0.2601 | 0.2625 | 0.2648 | 0.2672 | 0.2695 | 0.2718 | 0.2742 | 0.2765 |
| 1.9 | 0.2788 | 0.2810 | 0.2833 | 0.2856 | 0.2878 | 0.2900 | 0.2923 | 0.2945 | 0.2967 | 0.2989 |
| 2   | 0.3010 | 0.3032 | 0.3054 | 0.3075 | 0.3096 | 0.3118 | 0.3139 | 0.3160 | 0.3181 | 0.3201 |
| 2.1 | 0.3222 | 0.3243 | 0.3263 | 0.3284 | 0.3304 | 0.3324 | 0.3345 | 0.3365 | 0.3385 | 0.3404 |
| 2.2 | 0.3424 | 0.3444 | 0.3464 | 0.3483 | 0.3502 | 0.3522 | 0.3541 | 0.3560 | 0.3579 | 0.3598 |
| 2.3 | 0.3617 | 0.3636 | 0.3655 | 0.3674 | 0.3692 | 0.3711 | 0.3729 | 0.3747 | 0.3766 | 0.3784 |
| 2.4 | 0.3802 | 0.3820 | 0.3838 | 0.3856 | 0.3874 | 0.3892 | 0.3909 | 0.3927 | 0.3945 | 0.3962 |
| 2.5 | 0.3979 | 0.3997 | 0.4014 | 0.4031 | 0.4048 | 0.4065 | 0.4082 | 0.4099 | 0.4116 | 0.4133 |
| 2.6 | 0.4150 | 0.4166 | 0.4183 | 0.4200 | 0.4216 | 0.4232 | 0.4249 | 0.4265 | 0.4281 | 0.4298 |
| 2.7 | 0.4314 | 0.4330 | 0.4346 | 0.4362 | 0.4378 | 0.4393 | 0.4409 | 0.4425 | 0.4440 | 0.4456 |
| 2.8 | 0.4472 | 0.4487 | 0.4502 | 0.4518 | 0.4533 | 0.4548 | 0.4564 | 0.4579 | 0.4594 | 0.4609 |
| 2.9 | 0.4624 | 0.4639 | 0.4654 | 0.4669 | 0.4683 | 0.4698 | 0.4713 | 0.4728 | 0.4742 | 0.4757 |
| 3   | 0.4771 | 0.4786 | 0.4800 | 0.4814 | 0.4829 | 0.4843 | 0.4857 | 0.4871 | 0.4886 | 0.4900 |
| 3.1 | 0.4914 | 0.4928 | 0.4942 | 0.4955 | 0.4969 | 0.4983 | 0.4997 | 0.5011 | 0.5024 | 0.5038 |
| 3.2 | 0.5051 | 0.5065 | 0.5079 | 0.5092 | 0.5105 | 0.5119 | 0.5132 | 0.5145 | 0.5159 | 0.5172 |
| 3.3 | 0.5185 | 0.5198 | 0.5211 | 0.5224 | 0.5237 | 0.5250 | 0.5263 | 0.5276 | 0.5289 | 0.5302 |
| 3.4 | 0.5315 | 0.5328 | 0.5340 | 0.5353 | 0.5366 | 0.5378 | 0.5391 | 0.5403 | 0.5416 | 0.5428 |
| 3.5 | 0.5441 | 0.5453 | 0.5465 | 0.5478 | 0.5490 | 0.5502 | 0.5514 | 0.5527 | 0.5539 | 0.5551 |
| 3.6 | 0.5563 | 0.5575 | 0.5587 | 0.5599 | 0.5611 | 0.5623 | 0.5635 | 0.5647 | 0.5658 | 0.5670 |
| 3.7 | 0.5682 | 0.5694 | 0.5705 | 0.5717 | 0.5729 | 0.5740 | 0.5752 | 0.5763 | 0.5775 | 0.5786 |
| 3.8 | 0.5798 | 0.5809 | 0.5821 | 0.5832 | 0.5843 | 0.5855 | 0.5866 | 0.5877 | 0.5888 | 0.5899 |
| 3.9 | 0.5911 | 0.5922 | 0.5933 | 0.5944 | 0.5955 | 0.5966 | 0.5977 | 0.5988 | 0.5999 | 0.6010 |
| 4   | 0.6021 | 0.6031 | 0.6042 | 0.6053 | 0.6064 | 0.6075 | 0.6085 | 0.6096 | 0.6107 | 0.6117 |
| 4.1 | 0.6128 | 0.6138 | 0.6149 | 0.6160 | 0.6170 | 0.6180 | 0.6191 | 0.6201 | 0.6212 | 0.6222 |
| 4.2 | 0.6232 | 0.6243 | 0.6253 | 0.6263 | 0.6274 | 0.6284 | 0.6294 | 0.6304 | 0.6314 | 0.6325 |
| 4.3 | 0.6335 | 0.6345 | 0.6355 | 0.6365 | 0.6375 | 0.6385 | 0.6395 | 0.6405 | 0.6415 | 0.6425 |
| 4.4 | 0.6435 | 0.6444 | 0.6454 | 0.6464 | 0.6474 | 0.6484 | 0.6493 | 0.6503 | 0.6513 | 0.6522 |
| 4.5 | 0.6532 | 0.6542 | 0.6551 | 0.6561 | 0.6571 | 0.6580 | 0.6590 | 0.6599 | 0.6609 | 0.6618 |
| 4.6 | 0.6628 | 0.6637 | 0.6646 | 0.6656 | 0.6665 | 0.6675 | 0.6684 | 0.6693 | 0.6702 | 0.6712 |
| 4.7 | 0.6721 | 0.6730 | 0.6739 | 0.6749 | 0.6758 | 0.6767 | 0.6776 | 0.6785 | 0.6794 | 0.6803 |
| 4.8 | 0.6812 | 0.6821 | 0.6830 | 0.6839 | 0.6848 | 0.6857 | 0.6866 | 0.6875 | 0.6884 | 0.6893 |
| 4.9 | 0.6902 | 0.6911 | 0.6920 | 0.6928 | 0.6937 | 0.6946 | 0.6955 | 0.6964 | 0.6972 | 0.6981 |

| | 0 | 0.01 | 0.02 | 0.03 | 0.04 | 0.05 | 0.06 | 0.07 | 0.08 | 0.09 |
|---|---|---|---|---|---|---|---|---|---|---|
| **5** | 0.6990 | 0.6998 | 0.7007 | 0.7016 | 0.7024 | 0.7033 | 0.7042 | 0.7050 | 0.7059 | 0.7067 |
| **5.1** | 0.7076 | 0.7084 | 0.7093 | 0.7101 | 0.7110 | 0.7118 | 0.7126 | 0.7135 | 0.7143 | 0.7152 |
| **5.2** | 0.7160 | 0.7168 | 0.7177 | 0.7185 | 0.7193 | 0.7202 | 0.7210 | 0.7218 | 0.7226 | 0.7235 |
| **5.3** | 0.7243 | 0.7251 | 0.7259 | 0.7267 | 0.7275 | 0.7284 | 0.7292 | 0.7300 | 0.7308 | 0.7316 |
| **5.4** | 0.7324 | 0.7332 | 0.7340 | 0.7348 | 0.7356 | 0.7364 | 0.7372 | 0.7380 | 0.7388 | 0.7396 |
| **5.5** | 0.7404 | 0.7412 | 0.7419 | 0.7427 | 0.7435 | 0.7443 | 0.7451 | 0.7459 | 0.7466 | 0.7474 |
| **5.6** | 0.7482 | 0.7490 | 0.7497 | 0.7505 | 0.7513 | 0.7520 | 0.7528 | 0.7536 | 0.7543 | 0.7551 |
| **5.7** | 0.7559 | 0.7566 | 0.7574 | 0.7582 | 0.7589 | 0.7597 | 0.7604 | 0.7612 | 0.7619 | 0.7627 |
| **5.8** | 0.7634 | 0.7642 | 0.7649 | 0.7657 | 0.7664 | 0.7672 | 0.7679 | 0.7686 | 0.7694 | 0.7701 |
| **5.9** | 0.7709 | 0.7716 | 0.7723 | 0.7731 | 0.7738 | 0.7745 | 0.7752 | 0.7760 | 0.7767 | 0.7774 |
| **6** | 0.7782 | 0.7789 | 0.7796 | 0.7803 | 0.7810 | 0.7818 | 0.7825 | 0.7832 | 0.7839 | 0.7846 |
| **6.1** | 0.7853 | 0.7860 | 0.7868 | 0.7875 | 0.7882 | 0.7889 | 0.7896 | 0.7903 | 0.7910 | 0.7917 |
| **6.2** | 0.7924 | 0.7931 | 0.7938 | 0.7945 | 0.7952 | 0.7959 | 0.7966 | 0.7973 | 0.7980 | 0.7987 |
| **6.3** | 0.7993 | 0.8000 | 0.8007 | 0.8014 | 0.8021 | 0.8028 | 0.8035 | 0.8041 | 0.8048 | 0.8055 |
| **6.4** | 0.8062 | 0.8069 | 0.8075 | 0.8082 | 0.8089 | 0.8096 | 0.8102 | 0.8109 | 0.8116 | 0.8122 |
| **6.5** | 0.8129 | 0.8136 | 0.8142 | 0.8149 | 0.8156 | 0.8162 | 0.8169 | 0.8176 | 0.8182 | 0.8189 |
| **6.6** | 0.8195 | 0.8202 | 0.8209 | 0.8215 | 0.8222 | 0.8228 | 0.8235 | 0.8241 | 0.8248 | 0.8254 |
| **6.7** | 0.8261 | 0.8267 | 0.8274 | 0.8280 | 0.8287 | 0.8293 | 0.8299 | 0.8306 | 0.8312 | 0.8319 |
| **6.8** | 0.8325 | 0.8331 | 0.8338 | 0.8344 | 0.8351 | 0.8357 | 0.8363 | 0.8370 | 0.8376 | 0.8382 |
| **6.9** | 0.8388 | 0.8395 | 0.8401 | 0.8407 | 0.8414 | 0.8420 | 0.8426 | 0.8432 | 0.8439 | 0.8445 |
| **7** | 0.8451 | 0.8457 | 0.8463 | 0.8470 | 0.8476 | 0.8482 | 0.8488 | 0.8494 | 0.8500 | 0.8506 |
| **7.1** | 0.8513 | 0.8519 | 0.8525 | 0.8531 | 0.8537 | 0.8543 | 0.8549 | 0.8555 | 0.8561 | 0.8567 |
| **7.2** | 0.8573 | 0.8579 | 0.8585 | 0.8591 | 0.8597 | 0.8603 | 0.8609 | 0.8615 | 0.8621 | 0.8627 |
| **7.3** | 0.8633 | 0.8639 | 0.8645 | 0.8651 | 0.8657 | 0.8663 | 0.8669 | 0.8675 | 0.8681 | 0.8686 |
| **7.4** | 0.8692 | 0.8698 | 0.8704 | 0.8710 | 0.8716 | 0.8722 | 0.8727 | 0.8733 | 0.8739 | 0.8745 |
| **7.5** | 0.8751 | 0.8756 | 0.8762 | 0.8768 | 0.8774 | 0.8779 | 0.8785 | 0.8791 | 0.8797 | 0.8802 |
| **7.6** | 0.8808 | 0.8814 | 0.8820 | 0.8825 | 0.8831 | 0.8837 | 0.8842 | 0.8848 | 0.8854 | 0.8859 |
| **7.7** | 0.8865 | 0.8871 | 0.8876 | 0.8882 | 0.8887 | 0.8893 | 0.8899 | 0.8904 | 0.8910 | 0.8915 |
| **7.8** | 0.8921 | 0.8927 | 0.8932 | 0.8938 | 0.8943 | 0.8949 | 0.8954 | 0.8960 | 0.8965 | 0.8971 |
| **7.9** | 0.8976 | 0.8982 | 0.8987 | 0.8993 | 0.8998 | 0.9004 | 0.9009 | 0.9015 | 0.9020 | 0.9025 |
| **8** | 0.9031 | 0.9036 | 0.9042 | 0.9047 | 0.9053 | 0.9058 | 0.9063 | 0.9069 | 0.9074 | 0.9079 |
| **8.1** | 0.9085 | 0.9090 | 0.9096 | 0.9101 | 0.9106 | 0.9112 | 0.9117 | 0.9122 | 0.9128 | 0.9133 |
| **8.2** | 0.9138 | 0.9143 | 0.9149 | 0.9154 | 0.9159 | 0.9165 | 0.9170 | 0.9175 | 0.9180 | 0.9186 |
| **8.3** | 0.9191 | 0.9196 | 0.9201 | 0.9206 | 0.9212 | 0.9217 | 0.9222 | 0.9227 | 0.9232 | 0.9238 |
| **8.4** | 0.9243 | 0.9248 | 0.9253 | 0.9258 | 0.9263 | 0.9269 | 0.9274 | 0.9279 | 0.9284 | 0.9289 |
| **8.5** | 0.9294 | 0.9299 | 0.9304 | 0.9309 | 0.9315 | 0.9320 | 0.9325 | 0.9330 | 0.9335 | 0.9340 |
| **8.6** | 0.9345 | 0.9350 | 0.9355 | 0.9360 | 0.9365 | 0.9370 | 0.9375 | 0.9380 | 0.9385 | 0.9390 |
| **8.7** | 0.9395 | 0.9400 | 0.9405 | 0.9410 | 0.9415 | 0.9420 | 0.9425 | 0.9430 | 0.9435 | 0.9440 |
| **8.8** | 0.9445 | 0.9450 | 0.9455 | 0.9460 | 0.9465 | 0.9469 | 0.9474 | 0.9479 | 0.9484 | 0.9489 |
| **8.9** | 0.9494 | 0.9499 | 0.9504 | 0.9509 | 0.9513 | 0.9518 | 0.9523 | 0.9528 | 0.9533 | 0.9538 |
| **9** | 0.9542 | 0.9547 | 0.9552 | 0.9557 | 0.9562 | 0.9566 | 0.9571 | 0.9576 | 0.9581 | 0.9586 |
| **9.1** | 0.9590 | 0.9595 | 0.9600 | 0.9605 | 0.9609 | 0.9614 | 0.9619 | 0.9624 | 0.9628 | 0.9633 |
| **9.2** | 0.9638 | 0.9643 | 0.9647 | 0.9652 | 0.9657 | 0.9661 | 0.9666 | 0.9671 | 0.9675 | 0.9680 |
| **9.3** | 0.9685 | 0.9689 | 0.9694 | 0.9699 | 0.9703 | 0.9708 | 0.9713 | 0.9717 | 0.9722 | 0.9727 |
| **9.4** | 0.9731 | 0.9736 | 0.9741 | 0.9745 | 0.9750 | 0.9754 | 0.9759 | 0.9763 | 0.9768 | 0.9773 |
| **9.5** | 0.9777 | 0.9782 | 0.9786 | 0.9791 | 0.9795 | 0.9800 | 0.9805 | 0.9809 | 0.9814 | 0.9818 |
| **9.6** | 0.9823 | 0.9827 | 0.9832 | 0.9836 | 0.9841 | 0.9845 | 0.9850 | 0.9854 | 0.9859 | 0.9863 |
| **9.7** | 0.9868 | 0.9872 | 0.9877 | 0.9881 | 0.9886 | 0.9890 | 0.9894 | 0.9899 | 0.9903 | 0.9908 |
| **9.8** | 0.9912 | 0.9917 | 0.9921 | 0.9926 | 0.9930 | 0.9934 | 0.9939 | 0.9943 | 0.9948 | 0.9952 |
| **9.9** | 0.9956 | 0.9961 | 0.9965 | 0.9969 | 0.9974 | 0.9978 | 0.9983 | 0.9987 | 0.9991 | 0.9996 |
| **10** | 1.0000 | 1.0004 | 1.0009 | 1.0013 | 1.0017 | 1.0022 | 1.0026 | 1.0030 | 1.0035 | 1.0039 |

# Appendix 3 Basic Trigonometric Functions

The table lists the values of the basic functions for angles between 0 and 90°. For the sine and cosine functions we can make use of the following identities, repeatedly if necessary, to relate the sine or cosine of any angle to the sine or cosine of an angle between 0 and $90° = \pi/2$.

$$\sin(-\theta) \equiv -\sin(\theta)$$

$$\cos(-\theta) \equiv \cos(\theta)$$

$$\sin(\theta) \equiv \cos(\theta - \pi/2)$$

$$\cos(\theta) \equiv -\sin(\theta - \pi/2)$$

The tangent can be calculated as $\tan(\theta) = \sin(\theta)/\cos(\theta)$ except when $\cos(\theta) = 0$.

| degrees | radians | sin($\theta$) | cos($\theta$) | tan($\theta$) |
|---|---|---|---|---|
| 0 | 0.0000 | 0.0000 | 1.0000 | 0.0000 |
| 1 | 0.0175 | 0.0175 | 0.9998 | 0.0175 |
| 2 | 0.0349 | 0.0349 | 0.9994 | 0.0349 |
| 3 | 0.0524 | 0.0523 | 0.9986 | 0.0524 |
| 4 | 0.0698 | 0.0698 | 0.9976 | 0.0699 |
| 5 | 0.0873 | 0.0872 | 0.9962 | 0.0875 |
| 6 | 0.1047 | 0.1045 | 0.9945 | 0.1051 |
| 7 | 0.1222 | 0.1219 | 0.9925 | 0.1228 |
| 8 | 0.1396 | 0.1392 | 0.9903 | 0.1405 |
| 9 | 0.1571 | 0.1564 | 0.9877 | 0.1584 |
| 10 | 0.1745 | 0.1736 | 0.9848 | 0.1763 |
| 11 | 0.1920 | 0.1908 | 0.9816 | 0.1944 |
| 12 | 0.2094 | 0.2079 | 0.9781 | 0.2126 |
| 13 | 0.2269 | 0.2250 | 0.9744 | 0.2309 |
| 14 | 0.2443 | 0.2419 | 0.9703 | 0.2493 |
| 15 | 0.2618 | 0.2588 | 0.9659 | 0.2679 |
| 16 | 0.2793 | 0.2756 | 0.9613 | 0.2867 |
| 17 | 0.2967 | 0.2924 | 0.9563 | 0.3057 |
| 18 | 0.3142 | 0.3090 | 0.9511 | 0.3249 |
| 19 | 0.3316 | 0.3256 | 0.9455 | 0.3443 |
| 20 | 0.3491 | 0.3420 | 0.9397 | 0.3640 |
| 21 | 0.3665 | 0.3584 | 0.9336 | 0.3839 |
| 22 | 0.3840 | 0.3746 | 0.9272 | 0.4040 |
| 23 | 0.4014 | 0.3907 | 0.9205 | 0.4245 |
| 24 | 0.4189 | 0.4067 | 0.9135 | 0.4452 |
| 25 | 0.4363 | 0.4226 | 0.9063 | 0.4663 |
| 26 | 0.4538 | 0.4384 | 0.8988 | 0.4877 |
| 27 | 0.4712 | 0.4540 | 0.8910 | 0.5095 |
| 28 | 0.4887 | 0.4695 | 0.8829 | 0.5317 |
| 29 | 0.5061 | 0.4848 | 0.8746 | 0.5543 |
| 30 | 0.5236 | 0.5000 | 0.8660 | 0.5774 |
| 31 | 0.5411 | 0.5150 | 0.8572 | 0.6009 |
| 32 | 0.5585 | 0.5299 | 0.8480 | 0.6249 |
| 33 | 0.5760 | 0.5446 | 0.8387 | 0.6494 |
| 34 | 0.5934 | 0.5592 | 0.8290 | 0.6745 |

| degrees | radians | sin($\theta$) | cos($\theta$) | tan($\theta$) |
|---|---|---|---|---|
| 35 | 0.6109 | 0.5736 | 0.8192 | 0.7002 |
| 36 | 0.6283 | 0.5878 | 0.8090 | 0.7265 |
| 37 | 0.6458 | 0.6018 | 0.7986 | 0.7536 |
| 38 | 0.6632 | 0.6157 | 0.7880 | 0.7813 |
| 39 | 0.6807 | 0.6293 | 0.7771 | 0.8098 |
| 40 | 0.6981 | 0.6428 | 0.7660 | 0.8391 |
| 41 | 0.7156 | 0.6561 | 0.7547 | 0.8693 |
| 42 | 0.7330 | 0.6691 | 0.7431 | 0.9004 |
| 43 | 0.7505 | 0.6820 | 0.7314 | 0.9325 |
| 44 | 0.7679 | 0.6947 | 0.7193 | 0.9657 |
| 45 | 0.7854 | 0.7071 | 0.7071 | 1.0000 |
| 46 | 0.8029 | 0.7193 | 0.6947 | 1.0355 |
| 47 | 0.8203 | 0.7314 | 0.6820 | 1.0724 |
| 48 | 0.8378 | 0.7431 | 0.6691 | 1.1106 |
| 49 | 0.8552 | 0.7547 | 0.6561 | 1.1504 |
| 50 | 0.8727 | 0.7660 | 0.6428 | 1.1918 |
| 51 | 0.8901 | 0.7771 | 0.6293 | 1.2349 |
| 52 | 0.9076 | 0.7880 | 0.6157 | 1.2799 |
| 53 | 0.9250 | 0.7986 | 0.6018 | 1.3270 |
| 54 | 0.9425 | 0.8090 | 0.5878 | 1.3764 |
| 55 | 0.9599 | 0.8192 | 0.5736 | 1.4281 |
| 56 | 0.9774 | 0.8290 | 0.5592 | 1.4826 |
| 57 | 0.9948 | 0.8387 | 0.5446 | 1.5399 |
| 58 | 1.0123 | 0.8480 | 0.5299 | 1.6003 |
| 59 | 1.0297 | 0.8572 | 0.5150 | 1.6643 |
| 60 | 1.0472 | 0.8660 | 0.5000 | 1.7321 |
| 61 | 1.0647 | 0.8746 | 0.4848 | 1.8040 |
| 62 | 1.0821 | 0.8829 | 0.4695 | 1.8807 |
| 63 | 1.0996 | 0.8910 | 0.4540 | 1.9626 |
| 64 | 1.1170 | 0.8988 | 0.4384 | 2.0503 |
| 65 | 1.1345 | 0.9063 | 0.4226 | 2.1445 |
| 66 | 1.1519 | 0.9135 | 0.4067 | 2.2460 |
| 67 | 1.1694 | 0.9205 | 0.3907 | 2.3559 |
| 68 | 1.1868 | 0.9272 | 0.3746 | 2.4751 |
| 69 | 1.2043 | 0.9336 | 0.3584 | 2.6051 |
| 70 | 1.2217 | 0.9397 | 0.3420 | 2.7475 |
| 71 | 1.2392 | 0.9455 | 0.3256 | 2.9042 |
| 72 | 1.2566 | 0.9511 | 0.3090 | 3.0777 |
| 73 | 1.2741 | 0.9563 | 0.2924 | 3.2709 |
| 74 | 1.2915 | 0.9613 | 0.2756 | 3.4874 |
| 75 | 1.3090 | 0.9659 | 0.2588 | 3.7321 |
| 76 | 1.3265 | 0.9703 | 0.2419 | 4.0108 |
| 77 | 1.3439 | 0.9744 | 0.2250 | 4.3315 |
| 78 | 1.3614 | 0.9781 | 0.2079 | 4.7046 |
| 79 | 1.3788 | 0.9816 | 0.1908 | 5.1446 |
| 80 | 1.3963 | 0.9848 | 0.1736 | 5.6713 |
| 81 | 1.4137 | 0.9877 | 0.1564 | 6.3138 |
| 82 | 1.4312 | 0.9903 | 0.1392 | 7.1154 |
| 83 | 1.4486 | 0.9925 | 0.1219 | 8.1443 |
| 84 | 1.4661 | 0.9945 | 0.1045 | 9.5144 |
| 85 | 1.4835 | 0.9962 | 0.08721 | 1.4301 |
| 86 | 1.5010 | 0.9976 | 0.0698 | 14.3007 |
| 87 | 1.5184 | 0.9986 | 0.0523 | 19.0811 |
| 88 | 1.5359 | 0.9994 | 0.0349 | 28.6363 |
| 89 | 1.5533 | 0.9998 | 0.0175 | 57.2900 |
| 90 | 1.5708 | 1.0000 | 0.0000 | infinite |

# Appendix 4 The Standard Normal Distribution

Cumulative distribution function $\Phi(z)$ is the area under the probability density function to the left of $z$ (see Figure).

This table thus gives the answer to the question:

'What is the probability that a number $\leqslant z$ will be sampled from a standard normal distribution?'

These values can be obtained in Microsoft Excel® by the worksheet function '=NORMSDIST($z$)'.

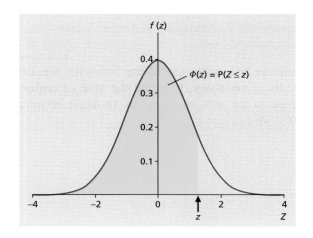

| Z | 0 | 0.01 | 0.02 | 0.03 | 0.04 | 0.05 | 0.06 | 0.07 | 0.08 | 0.09 |
|---|---|------|------|------|------|------|------|------|------|------|
| 0 | 0.5000 | 0.5040 | 0.5080 | 0.5120 | 0.5160 | 0.5199 | 0.5239 | 0.5279 | 0.5319 | 0.5359 |
| 0.1 | 0.5398 | 0.5438 | 0.5478 | 0.5517 | 0.5557 | 0.5596 | 0.5636 | 0.5675 | 0.5714 | 0.5753 |
| 0.2 | 0.5793 | 0.5832 | 0.5871 | 0.5910 | 0.5948 | 0.5987 | 0.6026 | 0.6064 | 0.6103 | 0.6141 |
| 0.3 | 0.6179 | 0.6217 | 0.6255 | 0.6293 | 0.6331 | 0.6368 | 0.6406 | 0.6443 | 0.6480 | 0.6517 |
| 0.4 | 0.6554 | 0.6591 | 0.6628 | 0.6664 | 0.6700 | 0.6736 | 0.6772 | 0.6808 | 0.6844 | 0.6879 |
| 0.5 | 0.6915 | 0.6950 | 0.6985 | 0.7019 | 0.7054 | 0.7088 | 0.7123 | 0.7157 | 0.7190 | 0.7224 |
| 0.6 | 0.7257 | 0.7291 | 0.7324 | 0.7357 | 0.7389 | 0.7422 | 0.7454 | 0.7486 | 0.7517 | 0.7549 |
| 0.7 | 0.7580 | 0.7611 | 0.7642 | 0.7673 | 0.7704 | 0.7734 | 0.7764 | 0.7794 | 0.7823 | 0.7852 |
| 0.8 | 0.7881 | 0.7910 | 0.7939 | 0.7967 | 0.7995 | 0.8023 | 0.8051 | 0.8078 | 0.8106 | 0.8133 |
| 0.9 | 0.8159 | 0.8186 | 0.8212 | 0.8238 | 0.8264 | 0.8289 | 0.8315 | 0.8340 | 0.8365 | 0.8389 |
| 1 | 0.8413 | 0.8438 | 0.8461 | 0.8485 | 0.8508 | 0.8531 | 0.8554 | 0.8577 | 0.8599 | 0.8621 |
| 1.1 | 0.8643 | 0.8665 | 0.8686 | 0.8708 | 0.8729 | 0.8749 | 0.8770 | 0.8790 | 0.8810 | 0.8830 |
| 1.2 | 0.8849 | 0.8869 | 0.8888 | 0.8907 | 0.8925 | 0.8944 | 0.8962 | 0.8980 | 0.8997 | 0.9015 |
| 1.3 | 0.9032 | 0.9049 | 0.9066 | 0.9082 | 0.9099 | 0.9115 | 0.9131 | 0.9147 | 0.9162 | 0.9177 |
| 1.4 | 0.9192 | 0.9207 | 0.9222 | 0.9236 | 0.9251 | 0.9265 | 0.9279 | 0.9292 | 0.9306 | 0.9319 |
| 1.5 | 0.9332 | 0.9345 | 0.9357 | 0.9370 | 0.9382 | 0.9394 | 0.9406 | 0.9418 | 0.9429 | 0.9441 |
| 1.6 | 0.9452 | 0.9463 | 0.9474 | 0.9484 | 0.9495 | 0.9505 | 0.9515 | 0.9525 | 0.9535 | 0.9545 |
| 1.7 | 0.9554 | 0.9564 | 0.9573 | 0.9582 | 0.9591 | 0.9599 | 0.9608 | 0.9616 | 0.9625 | 0.9633 |
| 1.8 | 0.9641 | 0.9649 | 0.9656 | 0.9664 | 0.9671 | 0.9678 | 0.9686 | 0.9693 | 0.9699 | 0.9706 |
| 1.9 | 0.9713 | 0.9719 | 0.9726 | 0.9732 | 0.9738 | 0.9744 | **0.9750** | 0.9756 | 0.9761 | 0.9767 |
| 2 | 0.9772 | 0.9778 | 0.9783 | 0.9788 | 0.9793 | 0.9798 | 0.9803 | 0.9808 | 0.9812 | 0.9817 |
| 2.1 | 0.9821 | 0.9826 | 0.9830 | 0.9834 | 0.9838 | 0.9842 | 0.9846 | 0.9850 | 0.9854 | 0.9857 |
| 2.2 | 0.9861 | 0.9864 | 0.9868 | 0.9871 | 0.9875 | 0.9878 | 0.9881 | 0.9884 | 0.9887 | 0.9890 |
| 2.3 | 0.9893 | 0.9896 | 0.9898 | 0.9901 | 0.9904 | 0.9906 | 0.9909 | 0.9911 | 0.9913 | 0.9916 |
| 2.4 | 0.9918 | 0.9920 | 0.9922 | 0.9925 | 0.9927 | 0.9929 | 0.9931 | 0.9932 | 0.9934 | 0.9936 |
| 2.5 | 0.9938 | 0.9940 | 0.9941 | 0.9943 | 0.9945 | 0.9946 | 0.9948 | 0.9949 | **0.9951** | 0.9952 |
| 2.6 | 0.9953 | 0.9955 | 0.9956 | 0.9957 | 0.9959 | 0.9960 | 0.9961 | 0.9962 | 0.9963 | 0.9964 |
| 2.7 | 0.9965 | 0.9966 | 0.9967 | 0.9968 | 0.9969 | 0.9970 | 0.9971 | 0.9972 | 0.9973 | 0.9974 |
| 2.8 | 0.9974 | 0.9975 | 0.9976 | 0.9977 | 0.9977 | 0.9978 | 0.9979 | 0.9979 | 0.9980 | 0.9981 |
| 2.9 | 0.9981 | 0.9982 | 0.9982 | 0.9983 | 0.9984 | 0.9984 | 0.9985 | 0.9985 | 0.9986 | 0.9986 |
| 3 | 0.9987 | 0.9987 | 0.9987 | 0.9988 | 0.9988 | 0.9989 | 0.9989 | 0.9989 | 0.9990 | 0.9990 |
| 3.1 | 0.9990 | 0.9991 | 0.9991 | 0.9991 | 0.9992 | 0.9992 | 0.9992 | 0.9992 | 0.9993 | 0.9993 |
| 3.2 | 0.9993 | 0.9993 | 0.9994 | 0.9994 | 0.9994 | 0.9994 | 0.9994 | 0.9995 | **0.9995** | 0.9995 |
| 3.3 | 0.9995 | 0.9995 | 0.9995 | 0.9996 | 0.9996 | 0.9996 | 0.9996 | 0.9996 | 0.9996 | 0.9997 |
| 3.4 | 0.9997 | 0.9997 | 0.9997 | 0.9997 | 0.9997 | 0.9997 | 0.9997 | 0.9997 | 0.9997 | 0.9998 |
| 3.5 | 0.9998 | 0.9998 | 0.9998 | 0.9998 | 0.9998 | 0.9998 | 0.9998 | 0.9998 | 0.9998 | 0.9998 |
| 3.6 | 0.9998 | 0.9998 | 0.9999 | 0.9999 | 0.9999 | 0.9999 | 0.9999 | 0.9999 | 0.9999 | 0.9999 |
| 3.7 | 0.9999 | 0.9999 | 0.9999 | 0.9999 | 0.9999 | 0.9999 | 0.9999 | 0.9999 | 0.9999 | 0.9999 |
| 3.8 | 0.9999 | 0.9999 | 0.9999 | 0.9999 | 0.9999 | 0.9999 | 0.9999 | 0.9999 | 0.9999 | 0.9999 |
| 3.9 | 1.0000 | 1.0000 | 1.0000 | 1.0000 | 1.0000 | 1.0000 | 1.0000 | 1.0000 | 1.0000 | 1.0000 |

If you have a $z$ value of 1.14, read down the left-hand side until you get to the row labeled '1.1', then read across until you get to the column labeled '0.04'. The number you reach is $\Phi(1.14)$.

The tail probability can be obtained from $1 - \Phi(z)$. This is the probability of sampling a value $\geqslant z$ (and is equal to the probability of sampling a number *below* $-z$). This is often easiest to see by drawing a quick sketch of the curve.

Values shown in bold represent the two-tailed critical values for $\alpha = 0.05$, 0.01, and 0.001 (corresponding to tail probabilities of 0.025, 0.005, and 0.0005, respectively).

# Appendix 5 The *t* Distribution

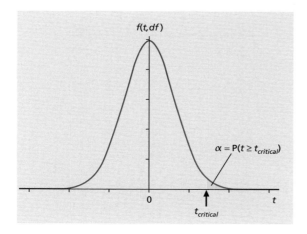

There would not be space here to give a *p* value for every possible combination of a *t* score and a certain number of degrees of freedom. Precise *p* values can be obtained from Microsoft Excel® using the worksheet formula '=TDIST(t,*df,tails*)' where *tails* = 1 for a one-tailed probability, 2 for a two-tailed probability.

These are the **critical values** of *t* for different values of degrees of freedom and $\alpha$. The critical values for two-tailed $\alpha = 0.05$ are used for generating 95 % confidence intervals.

In a hypothesis test, your value of *t* must be **bigger than or equal to** the critical value for you to take your data as evidence against the null hypothesis at the given significance level. If you have a **negative** value of *t*, you should ignore the minus sign (the *t* distribution is symmetrical about *t* = 0).

| One-tailed $\alpha$: | 0.05 | (0.025) | 0.01 | (0.005) |
|---|---|---|---|---|
| Two-tailed $\alpha$: | 0.1 | 0.05 | (0.02) | 0.01 |
| *df* | | | | |
| 1 | 6.314 | 12.706 | 31.821 | 63.656 |
| 2 | 2.920 | 4.303 | 6.965 | 9.925 |
| 3 | 2.353 | 3.182 | 4.541 | 5.841 |
| 4 | 2.132 | 2.776 | 3.747 | 4.604 |
| 5 | 2.015 | 2.571 | 3.365 | 4.032 |
| 6 | 1.943 | 2.447 | 3.143 | 3.707 |
| 7 | 1.895 | 2.365 | 2.998 | 3.499 |
| 8 | 1.860 | 2.306 | 2.896 | 3.355 |
| 9 | 1.833 | 2.262 | 2.821 | 3.250 |
| 10 | 1.812 | 2.228 | 2.764 | 3.169 |
| 11 | 1.796 | 2.201 | 2.718 | 3.106 |
| 12 | 1.782 | 2.179 | 2.681 | 3.055 |
| 13 | 1.771 | 2.160 | 2.650 | 3.012 |
| 14 | 1.761 | 2.145 | 2.624 | 2.977 |
| 15 | 1.753 | 2.131 | 2.602 | 2.947 |
| 16 | 1.746 | 2.120 | 2.583 | 2.921 |
| 17 | 1.740 | 2.110 | 2.567 | 2.898 |
| 18 | 1.734 | 2.101 | 2.552 | 2.878 |
| 19 | 1.729 | 2.093 | 2.539 | 2.861 |
| 20 | 1.725 | 2.086 | 2.528 | 2.845 |
| 21 | 1.721 | 2.080 | 2.518 | 2.831 |
| 22 | 1.717 | 2.074 | 2.508 | 2.819 |
| 23 | 1.714 | 2.069 | 2.500 | 2.807 |
| 24 | 1.711 | 2.064 | 2.492 | 2.797 |
| 25 | 1.708 | 2.060 | 2.485 | 2.787 |
| 26 | 1.706 | 2.056 | 2.479 | 2.779 |
| 27 | 1.703 | 2.052 | 2.473 | 2.771 |
| 28 | 1.701 | 2.048 | 2.467 | 2.763 |
| 29 | 1.699 | 2.045 | 2.462 | 2.756 |
| 30 | 1.697 | 2.042 | 2.457 | 2.750 |
| ⋮ | ⋮ | ⋮ | ⋮ | ⋮ |
| ∞ | 1.645 | 1.960 | 2.326 | 2.576 |

# Appendix 6 The $\chi^2$ Distribution

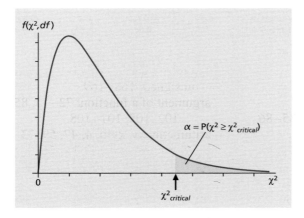

There would not be space here to give a $p$ value for every possible combination of a $\chi^2$ score and a certain number of degrees of freedom.

Precise $p$ values can be obtained in Microsoft Excel® using the worksheet formula '=CHIDIST($x,df$)'.

These are the **critical values** of $\chi^2$ for different values of degrees of freedom ($df$) and $\alpha$. If the value of $\chi^2$ is **bigger than** the critical value, the data can be taken as evidence against the null hypothesis at the relevant significance level.

| df | $\alpha$ 0.05 | 0.01 | 0.001 |
|----|------|------|-------|
| 1 | 3.84 | 6.63 | 10.83 |
| 2 | 5.99 | 9.21 | 13.82 |
| 3 | 7.81 | 11.34 | 16.27 |
| 4 | 9.49 | 13.28 | 18.47 |
| 5 | 11.07 | 15.09 | 20.51 |
| 6 | 12.59 | 16.81 | 22.46 |
| 7 | 14.07 | 18.48 | 24.32 |
| 8 | 15.51 | 20.09 | 26.12 |
| 9 | 16.92 | 21.67 | 27.88 |
| 10 | 18.31 | 23.21 | 29.59 |
| 11 | 19.68 | 24.73 | 31.26 |
| 12 | 21.03 | 26.22 | 32.91 |
| 13 | 22.36 | 27.69 | 34.53 |
| 14 | 23.68 | 29.14 | 36.12 |
| 15 | 25.00 | 30.58 | 37.70 |
| 16 | 26.30 | 32.00 | 39.25 |
| 17 | 27.59 | 33.41 | 40.79 |
| 18 | 28.87 | 34.81 | 42.31 |
| 19 | 30.14 | 36.19 | 43.82 |
| 20 | 31.41 | 37.57 | 45.31 |
| 21 | 32.67 | 38.93 | 46.80 |
| 22 | 33.92 | 40.29 | 48.27 |
| 23 | 35.17 | 41.64 | 49.73 |
| 24 | 36.42 | 42.98 | 51.18 |
| 25 | 37.65 | 44.31 | 52.62 |
| 26 | 38.89 | 45.64 | 54.05 |
| 27 | 40.11 | 46.96 | 55.48 |
| 28 | 41.34 | 48.28 | 56.89 |
| 29 | 42.56 | 49.59 | 58.30 |
| 30 | 43.77 | 50.89 | 59.70 |
| ⋮ | ⋮ | ⋮ | ⋮ |
| 40 | 55.76 | 63.69 | 73.40 |
| 50 | 67.50 | 76.15 | 86.66 |
| 60 | 79.08 | 88.38 | 99.61 |
| 70 | 90.53 | 100.43 | 112.32 |
| 80 | 101.88 | 112.33 | 124.84 |

# Index